J. C. BURGE
AF545236

WELDMENTS: PHYSICAL METALLURGY AND FAILURE PHENOMENA

Proceedings of the Fifth Bolton Landing Conference

August 1978

Edited by: R.J. Christoffel, E.F. Nippes, H.D. Solomon

Published and Distributed by

GENERAL ELECTRIC COMPANY
Technology Marketing Operation
120 Erie Boulevard
Schenectady, New York 12305

Copyright © 1979 by General Electric Co.
Technology Marketing Operation
120 Erie Boulevard, Schenectady, NY 12305

All rights reserved. No part of this book may be reproduced in any form without permission from the publisher, except by a reviewer who may quote brief passages in a review to be printed in a magazine or newspaper.

Distributed worldwide by General Electric Co.
Manufactured in the United States of America

Library of Congress Cataloging in Publication Data

Christoffel, R.J.
Bolton Landing Conference, 5th, 1978
Weldments.

1. Steel - Welding - Congresses
2. Welded Joints - Fatique - Congresses
I. Christoffel, R.J.
II. Nippes, Ernest Frederick
III. Solomon, Harvey D.
IV. Title
TS227.A1B64 1978 672'.5'2 79-14269
ISBN 0-931690-07-2

5th Bolton Landing Conference

SPONSORED BY:

HUDSON-MOHAWK SECTION AIME

R. Wright	Chairman
A. Petrovich	Vice-Chairman
R. Gangloff	Past-Chairman
R. VanStone	Secretary
C. Lyman	Treasurer

EASTERN NEW YORK CHAPTER ASM

A. Urguhart	Chairman
F. Buschmann	Vice Chairman
R. Diefendorf	Past Chairman
R. Horowitz	Secretary
J. McCarthy	Treasurer

NORTHERN NEW YORK SECTION AWS

J. Van Ullen	Chairman
G. Young	Vice-Chairman
K. Flood	Past-Chairman
S. Hayes	Secretary-Treasurer

CONFERENCE COMMITTEE

E. F. Nippes	Chairman
H. D. Solomon	Vice Chairman
R. J. Christoffel	Program Chairman
J. J. McCarthy	Treasurer

COMMITTEES

A. M. Beltran
T. F. Chase
P. J. Dembowski
R. P. Gangloff
J. J. Pepe
W. F. Savage

Preface

The concept of the Bolton Landing Conferences, first established in 1960, was to bring together in a relaxing atmosphere scientists and engineers for the purpose of discussing a topic of metallurgical significance. Its intent was to permit the presentation of scientific advances to the engineering community and at the same time make the scientific community aware of the engineering needs.

The topic selected for this Fifth Bolton Landing Conference, "Weldments - Physical Metallurgy and Failure Phenomena," was in keeping with this concept. It provided ample opportunity for interaction between material scientists concerned with metallurgical phenomena occurring in weldments and engineers who must utilize these weldments to provide a useful function.

With a topic as diverse as Weldments, no attempt was made to focus merely on a few critical questions, rather, the sessions were devoted to the various aspects of welding. The Conference opened with the keynote address and session dealing with the phenomena associated with weld solidification. The ultimate usefulness of any weld is frequently established at the time of solidification and an understanding and control of the operating mechanism is vital to the successful application of welding.

The thermal gradients encountered with the welding process produce a complex residual stress state and the measurement of these stresses and their effects was the subject of another session. These weld thermal conditions may also produce embrittling effects and these were discussed in detail in relation to the heat affected zone in austenitic stainless steel.

As with any material, its properties are controlled by its microstructure and weldments are no exception. These relationships, particularly as they are influenced by the welding thermal cycle, were explored for high strength low-alloy steel in still another of the Conference sessions.

Finally considered were the behavior of weldments at elevated temperatures and the unique problems associated with the welding of dissimilar materials.

For three days, in the gracious surroundings of the Sagamore Hotel and amid the natural splendors of Lake George, 130 Conferees from 5 countries and 22 States listened to and discussed these subjects. These deliberations are recorded in this volume to provide a permanent record of the many contributions which were made.

The Steering Committee wishes to thank not only those listed in the front of this book as members of the Committee but also Doreen Ball, Rose Marie Wadsworth and Thomas Natale whose enthusiastic efforts contributed significantly to the smooth operation of the Conference. Finally, the support of the local sections of the American Welding Society, the American Society for Metals, and the American Institute of Mining and Metallurgical Engineers is gratefully acknowledged.

Schenectady, NY
March, 1979

E. F. Nippes
H. D. Solomon
R. J. Christoffel

Table of Contents

SENSITIZATION OF AUSTENITIC STAINLESS STEELS

STRUCTURE-PROPERTY RELATIONSHIP IN HIGH STRENGTH LOW-ALLOY STEELS

Continued on next page.

"Solidification, Segregation and Weld Defects"

Warren F. Savage

Rensselaer Polytechnic Institute
Troy, New York

ABSTRACT

A fusion weld is a microcosm which involves melting, solidification, and virtually every microstructural change that can be produced by thermal treatments. If filler metal is added the fusion zone consists of a composite zone within which the composition is modified by dilution with base metal. This zone is surrounded by an unmixed zone, comprized exclusively of base metal melted and solidified in situ without mixing with the added filler metal.

Beyond the unmixed zone, partial melting occurs and moving grain boundaries become decorated with liquid films which have effective solidus temperatures significantly below that of the nominal composition. These low-melting grain boundaries not only are responsible for hot-cracking and microfissuring, but also serve as "pipelines" for diffusion of hydrogen and other dissolved gases derived from the fusion zone into the surrounding base metal.

This paper summarizes the origin and nature of microsegregation in the various regions of a fusion weld and correlates the microsegregation present in each region with various types of welding defects.

INTRODUCTION

Consider for a moment the dynamics of producing an autogenous gas-tunsten-arc weld (GTAW). Shortly after the arc is established, a pool of molten metal is created which is bounded by the locus of the liquidus for the alloy being welded. If the welding parameters remain constant, a dynamic equilibrium is soon established and the pool assumes a characteristic size and shape as it proceeds along the joint beneath the moving heat source.

On the average, the solid at the trailing boundary of the weld pool tends to grow into the moving pool at a rate, R, defined by the following relationship:

$$R = V \cos \theta \qquad \text{Eq. 1}$$

where:

R = growth rate (cm/sec)

V = the welding velocity (cm/sec)
θ = angle between a normal to the solid-liquid interface at the point in question and the welding direction.

Thus, the growth rate for a circular or elliptical pool varies from a maximum value equal to V at the weld centerline, where $\theta = 0$, to 0 at the side of the weld, where $\theta = 90^o$.

WELD METAL SOLIDIFICATION MECHANICS

Requisites for Solidification of an Alloy To solidify an alloy, it is necessary to:

1. redistribute the solutes present between the phases in contact with one another at the moving solid-liquid interface.
2. dissipate the latent heat of fusion which is liberated as the more energetic atoms in the liquid join the more "regimented" atoms in the crystal structure of the growing solid.

Boundary Conditions for Weld Metal Solidification The following boundary conditions best approximate the conditions under which weld metal solidification occurs:

1. no diffusion in the solid - (The cooling times are so short that this is a close approximation.)
2. no mechanical mixing in the liquid - (this is also a close approximation since the extent of the diffusion zone in the liquid is small in comparison to the thickness of the stagnant layer in the liquid at the advancing solid-liquid interface)
3. concentration changes in the liquid occur by diffusion alone - (this follows from (2) above)

Growth Mechanism Before considering the implications of the above boundary conditions, let us consider the mechanism by which growth of the solid occurs during welding. Recall that the atoms in the liquid are highly mobile, and since they exhibit no periodicity in their spacial arrangement; their potential energy is greater than that of atoms in the solid state. None-the-less, atoms in the liquid tend to have approximately the same number of near neighbors as do those in the solid phase. Therefore, to extend the solid phase, atoms in the liquid need only assume positions which correspond to an extension of the crystal lattice of the solid grain on the opposite side of the solid-liquid interface and surrender their excess energy (the latent heat) to the surroundings.

The above mechanism of growth corresponds to perfect heterogeneous nucleation where, in effect, the critical nucleus size shrinks to that of a single atom. This growth mechanism involving attachment of atoms from the liquid on the ideal substrate is known as epitaxial growth (From the Greek "taxi"-to arrange and "epi"-upon.)

Figure 1 (at the top of the next page) is a photomacrograph of a GTAW on a 1/8-in. thick slab from a nickel ingot. The dashed lines, AA and BB delineate the boundaries of the weld fusion zone. Note that the grain boundaries which intersect the fusion boundaries are continuous across the fusion line and that no grain boundary is created across the grains at the fusion line. This metallographic evidence of epitaxial growth has been supported conclusively by microbeam x-ray diffraction studies[1] which showed

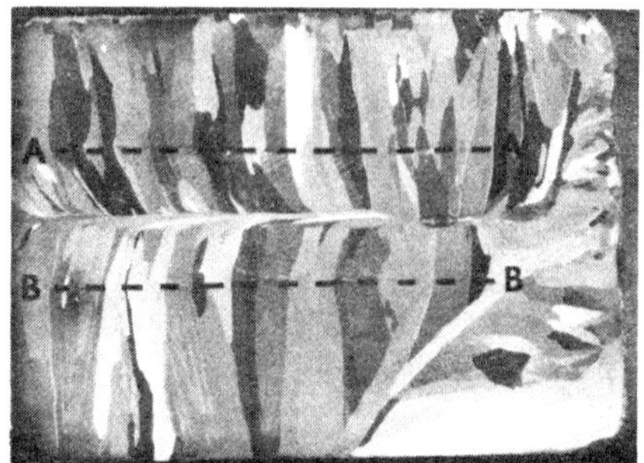

Fig. 1. GTAW on a slab from a Nickel ingot.

the crystallographic orientation of each grain to be continuous across the weld fusion boundaries.

The Shape of the Weld Pool The dissipation of the latent heat of fusion liberated at the moving solid-liquid interface exerts a major influence on the shape of the weld pool. Note that with a circular or elliptical weld pool the maximum growth velocity corresponds to the welding velocity and occurs at the centerline of the weld where $\theta = 0^o$. Since the instantaneous rate of liberation of latent heat is directly proportional to the rate of growth of the solid, the maximum rate of liberation of latent heat occurs at the centerline of the weld. Unfortunately, the temperature gradient, and therefore the rate of heat transfer is minimum at this location.

Therefore, above some critical welding velocity, the angle θ assumes a minimum value greater than 0°, and the weld pool assumes a tear drop shape. As a result, the maximum growth rate is reduced from $R = V\cos 0 = V$ to $R = V\cos\theta$ where $\cos\theta$ is inversely proportional to θ. Increasing the welding velocity above this critical velocity causes a corresponding increase in the minimum value of θ, and the apex angle at the center of the trailing edge of the pool becomes increasingly acute.

Competitive Growth Each crystalline structure exhibits a characteristic preferred growth direction. In both body-centered cubic (bcc) and face-centered cubic metals, the preferred growth directions are of the <100> type which are parallel to the edges of the unit cell. In hexagonal close packed metals, the preferred growth directions are of the $<10\bar{1}0>$ type which are parallel to the close packed directions in the basal plane.

In general, growth of any given grain at the trailing edge of the weld pool occurs parallel to the preferred growth direction with the greatest resolved component of the temperature gradient in the liquid. The arrows perpendicular to the trailing edge of the weld pools shown in Fig. 2 represent the orientation of the maximum temperature gradient at various points along the advancing solid-liquid interface. Note that the orientation relative to the welding direction is constant over a significant portion of the trailing edge of the tear-drop shaped pool in Fig. 2a. Therefore, grains which chance to have a preferred growth direction parallel or nearly parallel to the maximum temperature gradient expand laterally as they grow inward and crowd out less favorably oriented neighboring grains as indicated in Fig. 2a.

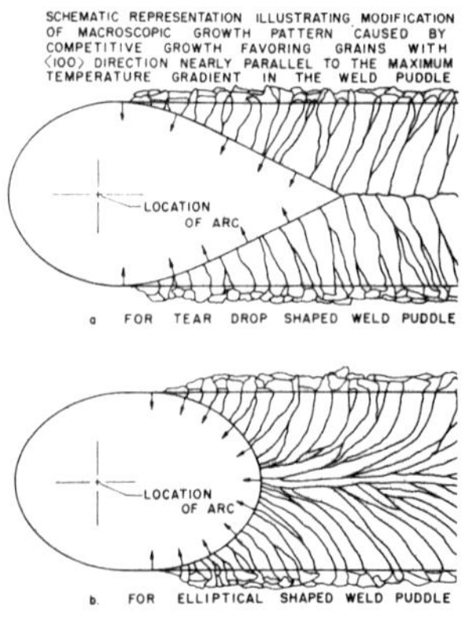

Fig. 2

On the other hand with an elliptical weld pool the orientation of the maximum temperature gradient changes continually along the trailing edge of the weld pool as shown in Fig. 2b. Thus, there is continuous competition amoung the grains as first one and then another has a preferred growth direction in alignment with the maximum temperature gradient as growth proceeds.

Note that as a result of the competitive growth process, the favorably oriented grains converge along the weld centerline with a tear drop shaped pool as indicated in Fig. 2a. Thus, there is a tendency for entrappment of low-melting solute-enriched liquid between the converging S-L interfaces to cause centerline hot-cracking problems.

However, with an elliptical weld pool the grains curve in the direction of welding as indicated in Fig. 2b and because of the more competitive nature of the growth process tend to have a smaller average grain size. Furthermore, no centerline segregation occurs and thus the problem of centerline hot-cracking is eliminated.

Microscopic Equilibrium Solidification of metals from their melts is closely approximated by the assumption that equilibrium prevails at the S-L interface during growth. Thus, although large concentration gradients may be present in both the solid and liquid phases near an advancing S-L interface, the equilibrium concentration for the solid and liquid phases will be present over narrow regions in contact with one another at the interface.

Figure 3, an idealized portion of an equilibrium constitutional diagram for a binary solid solution alloy, will be used to illustrate the application of this principle, known as microscopic equilibrium.

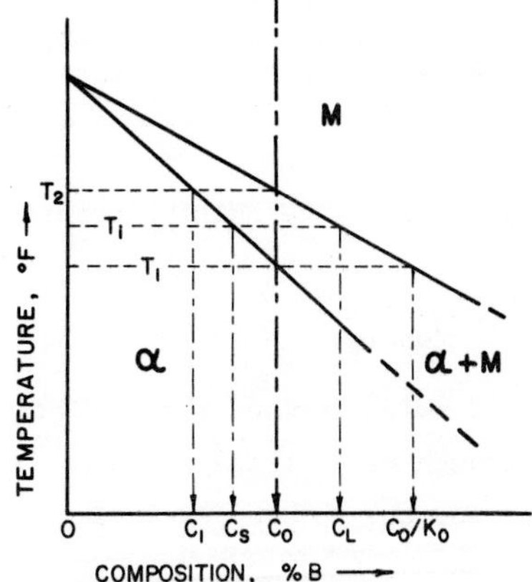

IDEALIZED PORTION OF A TYPICAL CONSTITUTIONAL DIAGRAM FOR A BINARY SOLID SOLUTION ALLOY

Fig. 3

The Equilibrium Distribution Coefficient In describing solidification, it is convenient to refer to the equilibrium distribution coefficient, k_o which is defined with reference to Fig. 3 as follows:

$$k_o = \frac{C_S}{C_L}\bigg|_{T_i} \qquad \text{Eq. 2}$$

where C_S= the equilibrium concentration of the solid phase at T_i
C_L= the equilibrium concentration of the liquid phase at T_i
T_i= the instantaneous temperature of the interface and can range from the liquidus to the solidus

With reference to Fig. 3, when a liquid of composition C_o is cooled to T_2, the liquidus temperature, a minute quantity of solid of composition C_1 forms. This composition can be related to that of the liquid by rearranging Eq. 2 as follows: $C_1 = k_oC_o$ (at T_2) Eq. 3

The Initial Transient Period The formation of a minute volume of solid of composition C_1 requires that the excess solute be rejected to the adjacent liquid. Thus a concentration gradient is established between the solute enriched liquid at the S-L interface and the bulk liquid of composition C_o.

This solute enrichment decreases the effective liquidus and causes an increase in the solute content of the next layer of solid to form.

Thus, at any intermediate temperature, T_i, the solid will exhibit a concentration gradient from k_oC_o in the first to form to k_oC_L at the S-L interface. By the same token, the liquid will exhibit a concentration gradient from C_L

at the interface to C_o in the bulk liquid. Figure 4a shows the distribution of solute at an early stage during unidirectional solidification of an alloy with $k_o < 1.0$. Note the concentration gradient "frozen" into the solid and the diffusion gradient in the solute enriched liquid at the solid-liquid interface. Figure 4b shows the solute distribution at a similar stage in the solidification of an alloy with $k_o > 1.0$. Note that the formation of solute enriched solid causes depletion of the solute in the diffusion zone produced in the adjacent liquid. Since this type of solid redistribution is characteristic of solutes whose presence lowers the melting temperature of the solvent species, depletion of the solute in the liquid lowers the effective liquidus of the material in the diffusion zone.

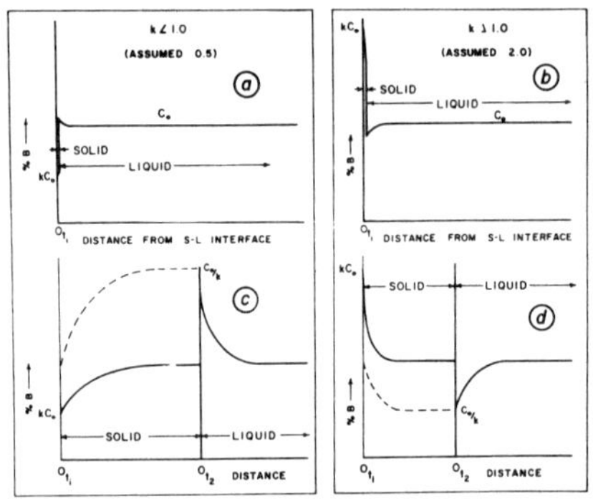

Fig. 4

Ultimately, the interface temperature will fall to the solidus temperature, T_1, and a dynamic equilibrium will be extablished whereby solid of composition C_o forms continuously from liquid of composition C_o/k_o. The interval required to establish this dynamic equilibrium is called the initial transient period.

The distance, d_i, that the S-L interface moves during the initial transient period is approximated by the relationship:

$$d_i \cong \frac{5D}{k_o R} \qquad \text{where:} \qquad \text{Eq. 4}$$

D= the diffusion coefficient in the liquid $\cong 5 \times 10^{-5}$ cm^2/sec for most solutes

R= the velocity of the S-L interface = the growth rate of the solid (cm/sec)

k_o= the equilibrium distribution coefficient for the solute involved

The Steady State Period   Once the concentration of the liquid at the S-L interface reaches C_o/k_o, the solidification process enters what is called the steady state period. During this period, the concentration gradient in the liquid diffuses ahead of the moving S-L interface unchanged in shape. Thus, a dynamic equilibrium is established whereby solid of nominal composition, C_o, forms continuously from the liquid of composition C_o/k_o on the opposite side of the moving S-L interface.

Figures 4c and 4d show the solute distribution during a portion of the steady state period for unidirectionally solidified alloys with $k < 1.0$ and $k_o > 1.0$, respectively. The dashed curves show the locus of the solute content of the liquid at the S-L interface as a function of the growth distance. Note that the liquid composition is C_o/k_o at the S-L interface in both cases, but the concentration gradient has a negative slope for $k_o < 1.0$.

The breadth of the diffusion zone, d_d, can be approximated by:

$$d_d = \frac{5D}{R} \qquad \text{Eq. 5}$$

where D and R are as defined earlier. If typical values for R are inserted in the above equation, the resulting values of d_d range from about 5.8×10^{-4}

in. (1.5 x 10^{-3}cm) for a growth rate of 4 ipm (0.17 cm/sec) to about 2.32 x 10^{-5}in. (5.9 x 10^{-5}cm) for a growth rate of 100 ipm (4.23 cm/sec).

These estimates of the breadth of the diffusion zone are presented to justify the earlier assumption that no mechanical mixing occurs in the liquid during weld metal solidification. It is quite probable that even with the convection currents and turbulence present in weld pool, a stagnant layer of liquid with a thickness greater than the breadth of the diffusion zone exists at the S-L interface. Therefore, the redistribution of the solute within such a stagnant layer would have to occur by diffusion.

The Terminal Transient Period When a pair of S-L interfaces approach one another from opposite directions (as would the case near the center of a weld made with the tear-drop shaped pool shown in Fig. 2a) their diffusion zones ultimately overlap. When this occurs, the solute content of the liquid in advance of each solid-liquid interface exceeds C_o/k_o for $k_o < 1.0$ and is depleted below C_o/k_o for $k_o > 1.0$. In both cases this causes the effective liquidus to fall below the nominal solidus, and the solidification process enters the terminal transient period. Thus, as the S-L interfaces converge, the last solid to form will unvariably be of such solute content as to exhibit an effective solidus considerably below that predicted from equilibrium for the nominal composition of the alloy.

If the melting temperature of the solvent is T_E, the slope of the liquidus for the particular solute is m_L, and the distribution coefficient is k_o, the equilibrium solidus for an alloy of composition, C_o, can be calculated as follows:

$$T_s = T_E + C_L m_L = T_E + \frac{C_o}{K_o} m_L \qquad \text{Eq. 6}$$

However under the conditions prevailing during weld metal solidification, the last solid to form during the terminal transient period will be of composition, C_M, the maximum solid solubility of the solute in the given solvent metal.

Take as an example the Fe-P system: T_E for Fe = 2795°F (1538°C) $m_L \cong -86.4°$ F/%, and C_M = 2.55w%, $k_o \cong 0.25$; if C_o = 0.020%, T_{S_E}= 2795 + (0.020/0.25) (-86.4) $\cong$ 2788°F (1531°C)

However, the effective solidus, T_F, of the last solid to form during the terminal transient period would be: $T_{S_F} = 2795 + \frac{2.55}{0.25}(-86.4) \cong 1920°F$ (1049°C)

From the above it should be apparent that solutes which have distribution coefficients that differ significantly from 1.0 and have a marked influence on the slope of the liquidus are most likely to cause hot-cracking and microfissuring problems by depressing the effective solidus.

The Origin of Constitutional Supercooling Figure 5 summarizes, for an alloy with $k_o < 1.0$, the influence of the solute gradient within the diffusion zone on the effective liquidus during steady state solidification. At the upper left, the solute concentration is shown as a function of the distance from the instantaneous location of the S-L interface. The composition scale in this portion of Fig. 5 is identical to that of the abscissa of the binary consitiutional diagram at lower right. Thus the effective liquidus for each point on the concentration gradient in the diffusion zone can be established by cross-referencing of the data shown in these two plots.

The figure at lower left shares, as its ordinate, the temperature scale of the binary diagram at the right, and, as its abscissa, the distance scale of the concentration gradient diagram above it. Thus, by cross referencing,

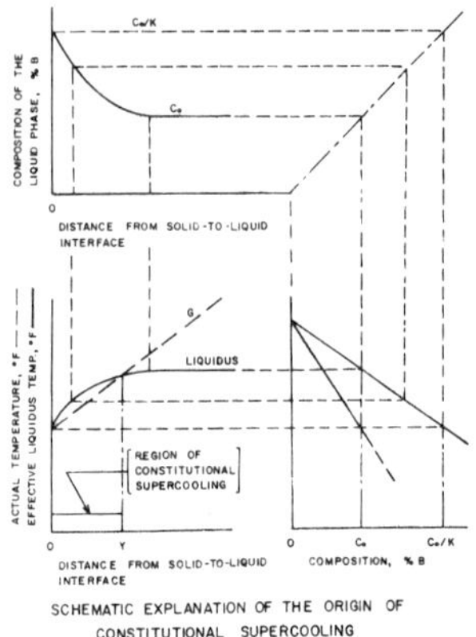

SCHEMATIC EXPLANATION OF THE ORIGIN OF CONSTITUTIONAL SUPERCOOLING

Fig. 5

as indicated by the dashed lines, the plot of the effective liquidus vs. distance from the moving S-L interface has been constructed (solid curve).

The sloping dashed line, labelled G, in the figure at lower right in Fig. 5 is a schematic representation of the actual temperature in the liquid for a hypothetical set of welding conditions. Note that the actual temperature in the liquid (dashed line) is below the effective liquidus from the S-L interface to the intersection of the dashed line with the effective liquidus curve (solid line) at a distance x. This indicates that the liquid is <u>effectively</u> supercooled in this region by virtue of the influence of its constitution on the effective liquidus. Thus, the phenomenon is called <u>constitutional supercooling</u>.

EFFECT OF WELDING VARIABLES ON CONSTITUTIONAL SUPERCOOLING

<u>Effect of Welding Velocity</u> It was noted (Eq. 1) that the growth rate, R, at the trailing edge of a weld pool is given by R = V cosθ. Thus, in general, increasing the welding velocity will increase the growth rate, R. Since the width of the diffusion zone is inversely proportional to R, (Refer to Eq. 4) it follows that the steepness of the curve of the effective liquidus vs. distance from the S-L interface will be directly proportional to R (and therefore to V), as shown schematically in Fig. 6 by the solid curves. The slope of the temperature gradient is inversely proportional to the size of the weld pool. Thus increasing the energy input, which tends to increase the size of the weld pool, tends to decrease the temperature gradient. Unfortunately, although both increasing the travel speed and increasing the energy input are cost-effective from a production point of view, both increase the extent of constitutional supercooling, and therefore have a detrimental effect on the microstructure.

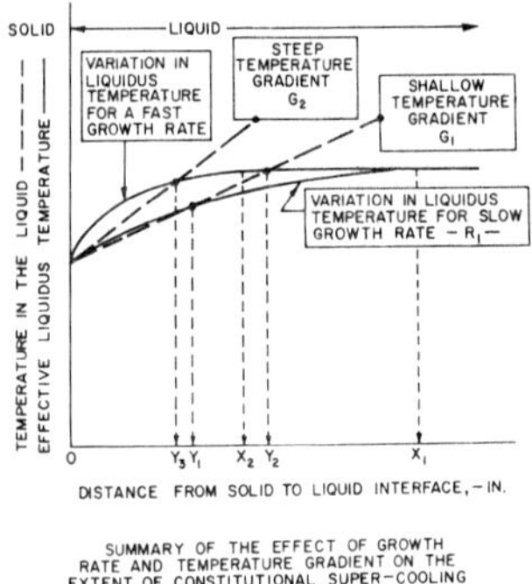

SUMMARY OF THE EFFECT OF GROWTH RATE AND TEMPERATURE GRADIENT ON THE EXTENT OF CONSTITUTIONAL SUPER-COOLING

Fig. 6

Increasing the initial plate temperature (and/or interpass temperature) also tends to decrease the steepness of the temperature gradient and therefore to increase the extent of constitutional supercooling. Thus, high preheat temperatures, which may be necessary to combat problems such as hydrogen-induced cracking in steels may have an adverse effect on the solidification process.

<u>Measurement of Distribution Coefficients</u> Figure 7 is a schematic diagram of a hot-cracking test developed at Rensselaer. This test, known as the Varestraint test,[2] subjects a specimen to a known augmented strain as it is being welded. This is accomplished by applying a force, F, to the outer end of the specimen in order to force it to conform to the radius of curvature, R, of the die-block, B, just before the arc passes the point of tangency at C. The augmented strain, ε_1, in the top surface is given approximately by:

$$\varepsilon_1 \cong \frac{t}{2R} \quad \text{where} \qquad \text{Eq. 7}$$

t = the specimen thickness

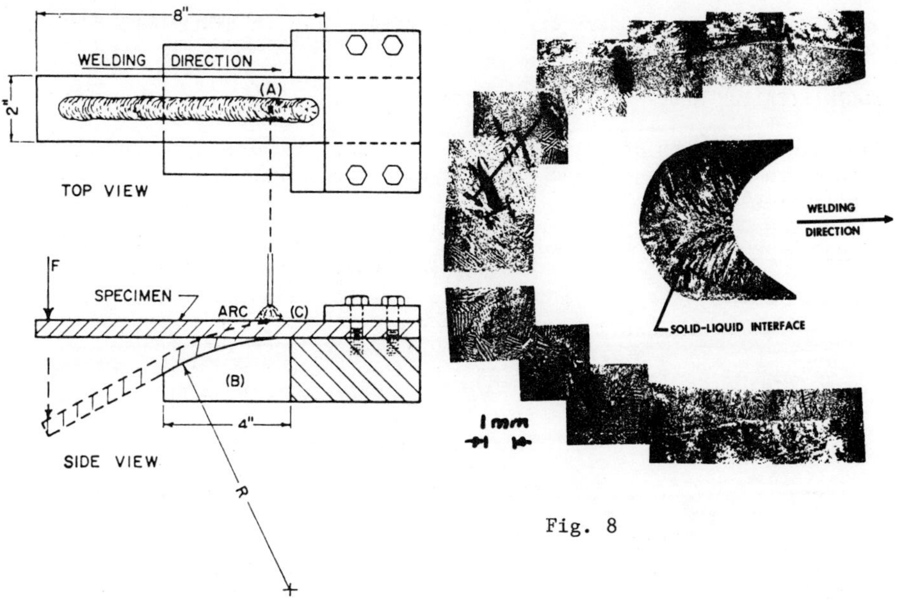

Fig. 8

Fig. 7

R = the radius of curvature of the die blocks (R >> t)

The sudden application of the augmented strain causes hot-cracking to occur along liquated grain boundaries adjacent to the trailing edge of the weld pool as shown in Fig. 8. Often these cracks are backfilled with molten metal from the diffusion zone at the S-L interface. This provides a captive sample of sufficient size to be analyzed successfully with the electron beam microprobe.

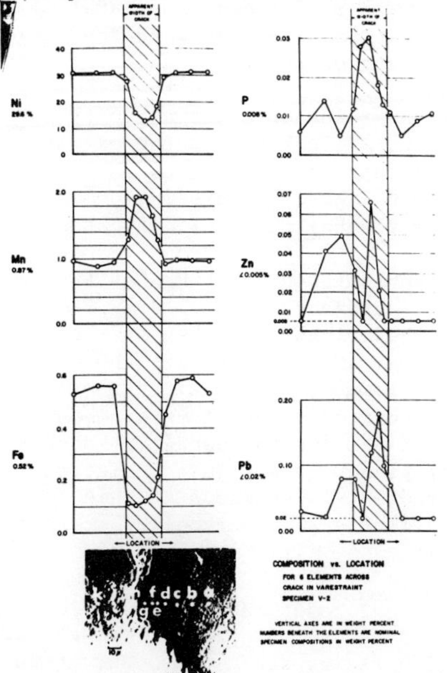

Figure 9 summarizes the results of such an analysis. The backfilled crack is shown in the lower portion of this figure with the location of the excited spots shown as white dots. The results of the point counts at the twelve locations are plotted above for the 6 elements analyzed in the 70 Cu-30 Ni alloy.

Note that the Ni falls from 30% in the base metal to about 12.5% in the crack, indicating an effective distribution coefficient of approximately 2.4 for Ni in Cu. The other data indicate k for Mn to be about 0.52, that for Fe to be about 5.5, that for P to be about 0.27, that for Zn to be about 0.08, and that for Pb, about 0.11. This technique has since been used in other systems and has proved to be a valuable

Fig. 9: Microprobe results on Back-filled Hot Crack.

way of obtaining estimates of the effective values of distribution coefficients in complex alloys.

SOLIDIFICATION MODE

Five distinct modes of solidification have been proposed from purely theoretical considerations,[3] and all five have been observed in welds made in our laboratory.

Planar Growth In the lower part of Fig. 10, a schematic plot indicates a condition whereby the actual temperature lies above the effective liquidus at all distances from the S-L interface. Under such a condition, if a chance protuberance were to form on the interface, it would enter a region of the liquid phase whose effective liquidus (solid line) was below the actual temperature (dashed line) and would thus remelt. Therefore, the conditions shown would cause planar growth of the S-L interface as shown in the upper portion of Fig. 10.

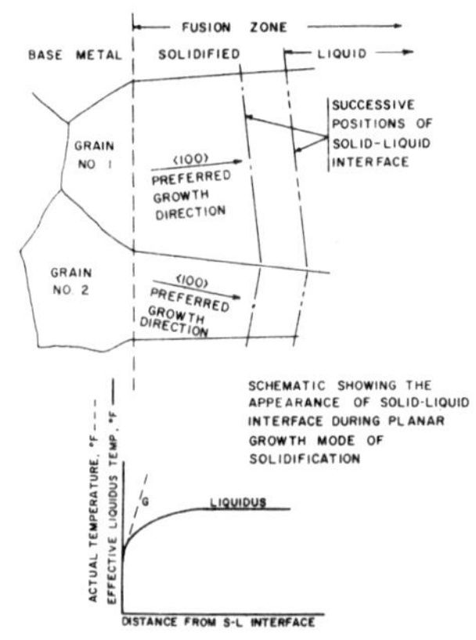

Fig. 10

Cellular Growth In Figure 11 the steepness of the temperature gradient has been decreased so that a limited amount of constitutional supercooling is present. A chance pretuberance on the S-L interface now enters a region of constitutionally supercooled liquid and is thus not only able to survive but can also dissipate its latent heat to the surrounding supercooled liquid. If the extent of the constitutionally supercooled liquid, x, is small compared with the average grain diameter, solidification will now occur by the simultaneous growth of a number of elongated subgrains of roughly hexagonal cross section with their long axes parallel to the preferred growth direction with the largest resolved component of the temperature gradient. Thus, each grain will grow into the liquid as a cluster of subgrains or cells whose transverse dimensions are related to the extent of constitutional supercooling, x, as shown in the upper portion of Fig. 11. This is commonly referred to as the cellular solidification mode.

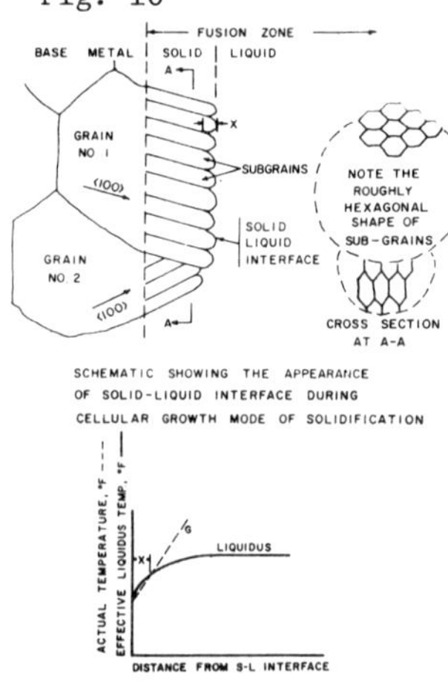

Fig. 11

Fig. 12 is a photomicrograph taken of the fusion boundary of an autogenous GTA weld in Fe-49Ni sheet. Note the epitaxial growth of the grains in the base metal at the left into the weld metal without change in orientation as evidenced by the continuity of the grain boundaries across the vertical fusion boundary. Note also the region of planar growth just inside the fusion boundary. Since the temperature gradient is steepest at this location and the growth rate is nearly zero, no constitutional supercooling was present and planar growth occurred as a result.

After a brief period of planar growth, however, constitutional supercooling occurred and the solidification mode changed to one of cellular growth parallel to the preferred growth direction with the greatest component of the

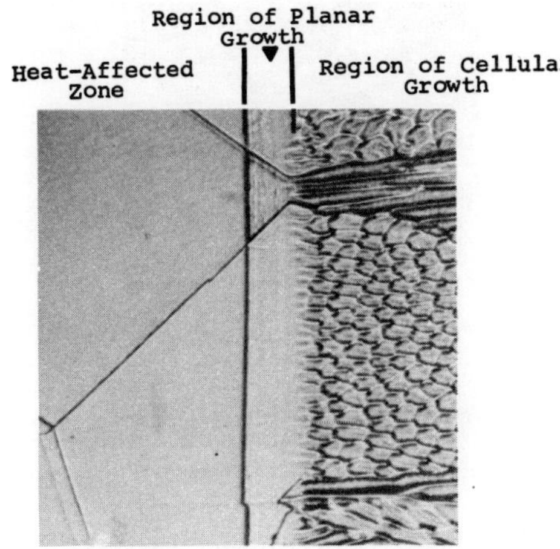

Fig. 12

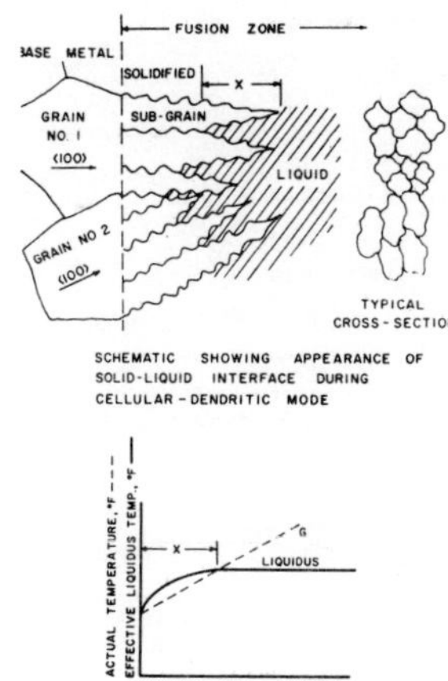

Fig. 13

temperature gradient. Note that the cells within the grain at the center of the photomicrograph are inclined to the surface whereas those in the grain directly above have their long axes nearly parallel to the surface.

<u>Cellular Dendritic Growth</u>   Figure 13 shows, in schematic form, a more extensive amount of constitutional supercooling. In this case the extent of supercooling is sufficient to support branched growth along preferred growth directions normal to the cell axes. This mode, shown schematically at the top of Fig. 13, is referred to as the <u>cellular</u> <u>dendritic</u> solidifica-mode.

Figure 14 shows the microstructure produced by this mode of solidification in an 18Ni Maraging steel, as revealed by a solute sensitive etching reagent. Note the dark etching walls and crossbranches of the cellular dendrites. Also visible in this photomicrograph are microfissures at the grain boundaries where the terminal transient caused excessive depression of the effective liquidus and solidus.

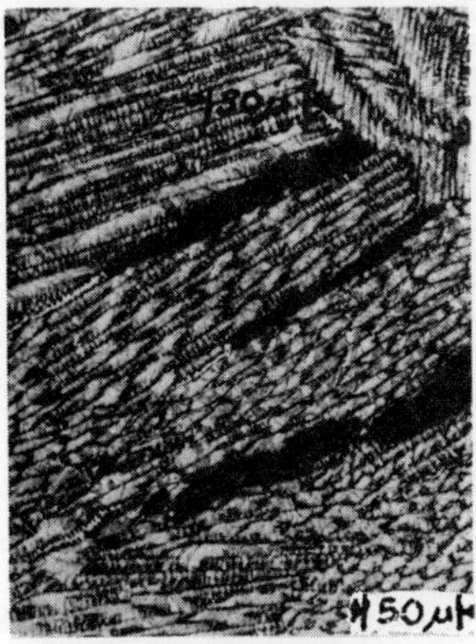

<u>Columnar Dendritic</u>   Although most texts refer to weld metal as having a columnar dendritic structure, this is not rigorously correct. In fact, although most welds have a columnar grain structure, the solidification substructure within the grains is usually either cellular or cellular dendritic.

A true columnar dendritic solidification mode requires that the extent of the constitutional supercooling be of the same order of magnitude as the grain diameter as shown in Fig. 15. The resultant growth occurs by branching, and often crossbranching from a single central stalk as shown schematically in the upper portion of Fig. 15. Figure 16 shows an actual example of this rarely encountered growth mode in a Ti-6Mn alloy. Note that the entire grain in the center of this photomicrograph appears to have grown by branching and cross-branching from a single central stalk.

50μ

Fig. 14: Cellular Dendrites

<u>Equiaxed Dendritic</u>   This mode of solidification requires the most

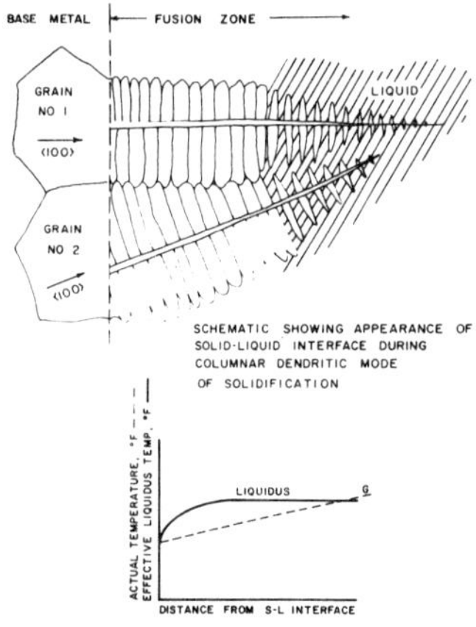

Fig. 15

Fig. 16: Columnar Dendrite

extensive degree of constitutional supercooling. It is believed to result when the fragile tips of growing dendrites are broken away from the main dendrite and carried into the supercooled region by convection currents. They are then able to grow in all three preferred growth directions to form equiaxed grains as shown schematically in Fig. 17.

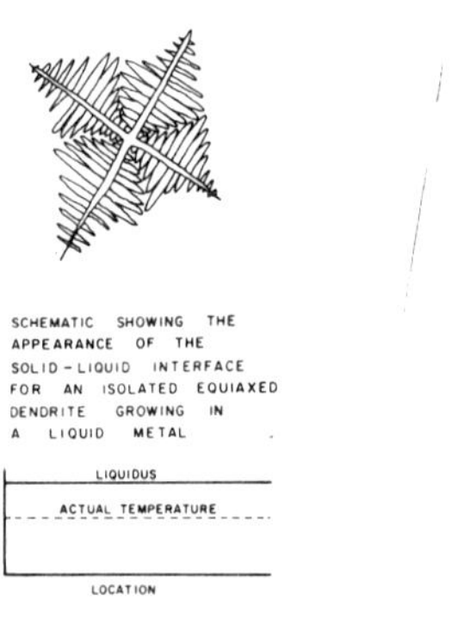

Fig. 17

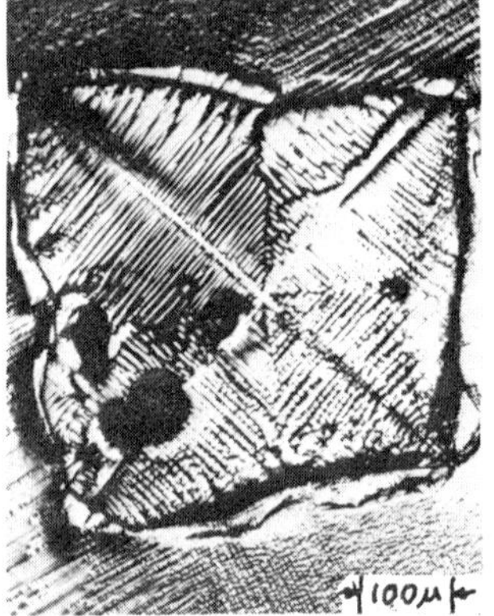

Fig. 18: Equiaxed Dendrite

Figure 18 shows a typical equiaxed dendrite grain in the terminal crater of a weld in silicon-iron sheet. This mode of solidification is usually found only in the terminal crater and promotes severe microsegregation problems which often lead to the formation of so called star or crater cracks.

It was discovered empirically that gradually decreasing the arc current and "filling" the crater would usually eliminate this form of cracking. It is now apparent that the gradual decrease in current maintains a positive temperature gradient until the solidification is complete and thus eliminates the extensive constitutional supercooling required to cause this undesirable solidification mode in the terminal crater.

Summary Plot   Figure 19, adapted from Tiller & Rutter[3], summarizes, in schematic form, the factors which influence the solidification mode. The ordinate represents the concentration of the solute (or solutes) which are being redistributed during solidification. Since, in general, the level of residuals and other elements which contribute to welding problems can only be controlled by the producer, this might well be identified as the "producer's" axis.

It is important to note that, within a given specification, a wide variation in the effective location of individual alloys on the producer's axis is possible. In general, controlling the concentration of elements which depress the effective liquidus will minimize hot-cracking and microfissuring. However, because of heat-to-heat variations possible within a given specification, it is dangerous to categorize a given type of material as either "readily weldable" or "difficult to weld" on the basis of a limited sample. It is quite possible that variations in residuals well within the established specification limits could either raise or lower the effective value of C_o in Fig. 19 and so drastically change the weldability.

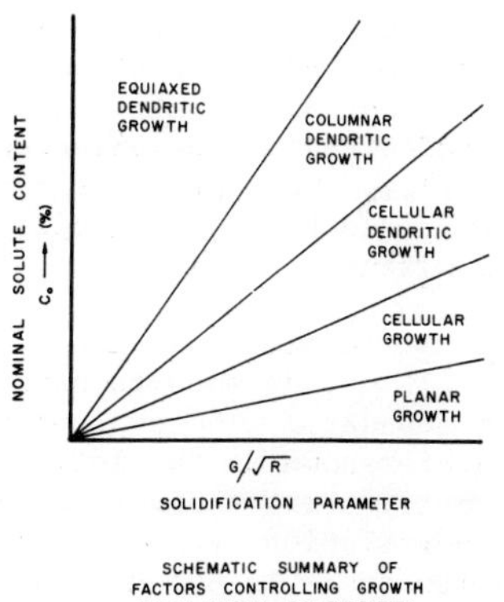

Fig. 19

The abscissa of Fig. 19 represents the value of the parameter $G/\sqrt{R}$. Since both G, the temperature gradient, and R, the growth rate, were shown earlier to depend upon the energy input, preheat temperature and welding velocity, this axis might well be labelled the "fabricator's" axis. In general, high energy inputs and fast growth rates tend to shift the value of $G/\sqrt{R}$ to the left in Fig. 19.

Note that the field in Fig. 19 is subdivided into 5 regions, each identified by a specific solidification mode. Thus, a combination of C_o and $G/\sqrt{R}$ locates a particular welding operation within a specific field. In general, microsegregation related problems become more severe as one moves upward and to the left toward less desirable solidification substructures. Conversely, the solidification substructure tends to become finer and the pattern of microsegregation less injurious as one moves downward and to the right in Fig. 19.

This implies that excessive energy inputs, excessive travel speeds, high deposition rate processes tend to cause less desirable microstructures. Unfortunately, all of the above improve cost effectiveness and so fabricators are often tempted to utilize processes and procedures which increase the liklihood of encountering welding problems.

THE ZONES REDEFINED

The Composite Zone When filler metal is added to a weld, the composition of the majority of the weld bead is modified by dilution with material melted from the underlying solid material and intimately mixed with the filler metal. Therefore, this region of the fusion zone has a composition intermediate between that of the base metal (or underlying composite zone in a multipass weld) and that of the added filler metal. As a result, its response to solute sensitive etches is normally quite different from that of its surroundings and the boundaries of this composite zone are readily identified metallographically.

The Unmixed Zone Szekeres,[4] while studying the solidification substructure of welds in HY80 with the aid of solute sensitive etches, observed the microstructures at the fusion boundary shown in Fig. 20. In this photomicrograph, the alternate dark and light vertical bands differ in solute content as a consequence of the segregation produced during solidification of the original ingot. Although "homogenized" by a high temperature soaking treatment, and subjected to extensive hot rolling, the variation in solute content was still sufficient to be revealed by solute-sensitive etching procedures. The dark-etching area at the top of Fig. 20 is part of the composite region of the weld as defined above.

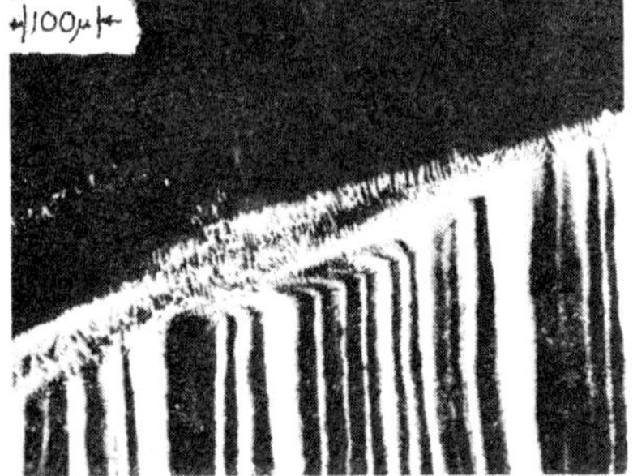

Fig. 20: The Fusion Boundary

Attention is now drawn to the diagonal band of material across the central portion of the Fig. 20. Within this region, the etchant reveals a solidification substructure, indicating it to have once been completely molten. Note that the overall contrast of this region as revealed by the etchant is essentially the same as that of the adjacent base metal. In addition, note that while molten, the banded regions experienced only lamellar flow. This provides metallographic evidence that the liquid at the fusion boundary doesn't experience mechanical mixing.

Electron Beam Microanalysis of the Fusion Boundary An electron beam traverse, identified as C-C' at the right of Fig. 21, may be seen to have extended across the microstructures at the fusion boundary. The concentration profiles for Mn and Cr obtained by this scan are shown at the left. The concentration scale (microprobe analyser intensity corrected for background intensity) is relative and an approximate correlation between both Mn and Cr composition and the relative concentration scale is shown. This correlation was obtained by equating the Mn and Cr composition of the base metal to the average output intensity of the microprobe analyser during the period the beam was impinging on the base metal.

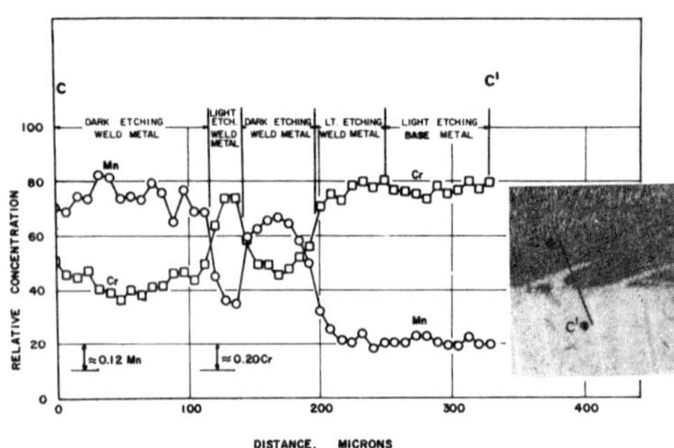

Fig. 21: Microprobe Analysis of the Fusion Boundary Region.

Across the top of the graph in Fig. 21, the extents of the various regions are identified as they correspond to the path of the beam (C-C') shown at right in Fig. 21. It is apparent from the concentration profiles that a significant portion of the light etching weld metal, including the fold, has the same concentration of both Mn and Cr as does the base metal.

It is thus obvious that the fold was a portion of the melted but unmixed base metal that was washed away from the fusion boundary by convection currents in the liquid. Such folds appeared frequently in all of the sections of weld beads examined and some of the folds extended a considerable distance into the composite zone.

The existence of the unmixed zone is of considerable practical significance. Careful reexamination of many so called heat-affected zone cracks has revealed many of them to be intimately associated with the unmixed zone. Thus, the microsegregation which occurs during solidification of the base metal within the unmixed zone appears to contribute to the formation of both hot and cold cracks at the fusion boundary.

THE PARTIALLY MELTED ZONE

In the presence of point-to-point variations in solute content in the base metal, the effective melting temperature would vary from place to place. A schematic representation of the variation in melting temperature caused by residual segregation in the base metal is shown at the top of Fig. 22. Two typical temperature gradients have been superimposed on the schematic showing the point-to-point variation in effective melting temperature. Note that the localities where the effective melting temperature is below the instantaneous temperature will experience localized melting. These melted areas are shown schematically in the center of Fig. 22 for the shallow temperature gradient.(line AB) It is assumed that no movement of the grain boundaries has yet occurred.

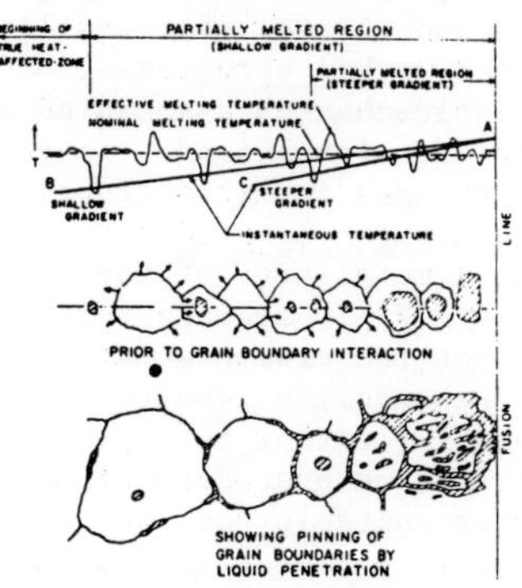

Fig. 22: Proposed Mechanism of Partial Melting.

At the high temperature experienced near the fusion boundary, grain boundary migration occurs as a result of normal grain growth. However, the melted regions act as a second phase to retard grain boundary migration. In fact, the liquid phase, being similar in composition, wets the grain boundaries and effectively pins all those that intersect a melted region.

The extent of localized melting present and the separation between localized molten regions determine the distance the boundaries migrate before becoming pinned. Thus, as indicated at the bottom of Fig. 22, the interaction between the molten regions and the grain boundaries controls the extent of grain coarsening. Since the distance between molten patches decreases as the distance from the fusion boundary decreases, the grains near the fusion line are smaller that those in the nearby region of the true heat-affected zone where no localized melting was present.

Figure 23 summarizes the microprobe analyses performed on a weld in a 70Cu-30Ni alloy to confirm the proposed mechanism. The photograph at the top reveals evidence of localized melting and grain boundary penetration. Note particularly the locations T' and S' which are separated by only 10 microns. T', in the matrix of a grain, contained 37.2%Ni while S' in a boundary which appears to have been penetrated by a molten film contained only 14.5 %Ni. This difference in composition would cause the effective liquidus to be depressed by about 180°F (100°C). Attention is drawn to the continuous network of broadened grain boundaries visible in Fig. 23. The composition of the material in these broadened boundaries is always lower in melting temperature than that of the matrix of the grains. Therefore, the evidence summarized in Fig. 23 proves that liquated grain boundary films are created in the _partially melted zone_ adjacent to the fusion boundary where peak temperatures range from the effective liquidus to the effective solidus for the alloy.

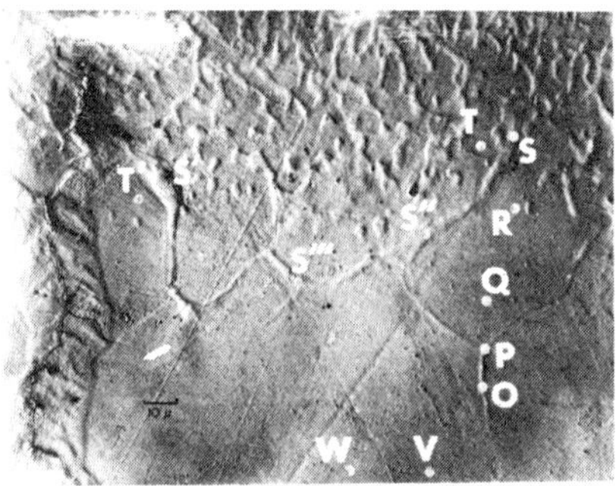

COMPOSITIONS AT VARIOUS LOCATIONS IN PARTIALLY MELTED REGION IN VARESTRAINT SAMPLE V-2

	Ni	Mn	Fe	P	Zn	Pb
O GRAIN BOUNDARY	28.4	1.02	0.49	0.010	<0.005	<0.02
P GRAIN BOUNDARY	27.6	0.99	0.43	0.023	0.043	<0.02
Q MATRIX	31.2	1.15	0.54	0.020	0.020	<0.02
R MATRIX	33.3	0.82	0.63	0.009	<0.005	<0.02
S GRAIN BOUNDARY	32.0	0.87	0.54	0.007	<0.005	0.06
S' GRAIN BOUNDARY	14.5	0.87	0.16	0.020	<0.005	<0.02
S" GRAIN BOUNDARY	22.2	1.00	0.41	0.013	<0.005	<0.02
S‴ GRAIN BOUNDARY	26.1	0.96	0.46	0.005	0.012	0.02
T MATRIX	36.4	0.91	0.76	0.004	0.049	0.02
T' MATRIX	37.2	0.89	0.73	0.001	0.022	0.04
V MATRIX	30.9	0.97	0.53	0.009	<0.005	0.04
W MATRIX	30.9	0.96	0.55	0.007	0.010	0.03

3µ EXCITED SPOTS ARE DRAWN TO SCALE IN PHOTO

Fig. 23: The Partially Melted Zone.

THE TRUE HEAT-AFFECTED ZONE

Beyond the fusion boundary, material within which microstructural changes occur entirely within the solid state constitutes the true heat affected zone. Depending upon the alloy, a wide variety of thermally activated changes in microstructure and properties is possible. Included are normal grain growth, solution and precipitation of second phases, recrystallization of cold worked structures, and metallographic changes associated with allotropic transformations. In general, the changes in microstructure and properties can be predicted from the physical metallurgy of the alloy in question. If it appears that the properties of the true heat-affected zone will be severely impaired, it is often possible, either by appropriate choice of welding procedure or by application of suitable post-weld heat-treatments, to minimize the extent and severity of the impairment.

THE ORIGIN OF WELD DEFECTS

Hot Cracking and Microfissuring Hot cracks and microfissures are believed to initiate at liquated grain and subgrain boundaries in the fusion zone and at the molten grain boundaries within the partially melted zone at temperatures above the effective solidus. Thus, the mechanisms of microsegregation described in the preceeding sections play an important role in that they always tend to occur in such a direction as to depress the effective liquidus of the material at the grain and subgrain boundaries.

In multipass welds, the microsegregation present in underlying weld beads tends to extend the region of partial melting produced adjacent to subsequent passes. Thus, microfissuring, such as was shown in Fig. 14, is often observed in the portions of underlying passes which are adjacent to the fusion boundary of a subsequent pass.

The subject of centerline hot-cracking has been discussed earlier and can not only be explained in terms of solidification mechanics, but can be prevented by proper selection of welding procedures.

Cold Cracking Cold cracking, more properly called hydrogen-induced cracking, requires the simultaneous presence of three requisites:

1. A stress of approximately yield strength magnitude.
2. A crack sensitive microstructure.
3. A critical level of diffusible hydrogen.

The combined action of residual and service stresses nearly always provides the first requisite, and thus the control of hydrogen-induced cracking requires elimination of one or both of the remaining requisites.

Unfortunately, the microsegregation present in the unmixed zone and the partially melted zone produces grain boundaries which are enriched in both carbon and other solutes which cause a local increase in the hardenability of the steel. Thus, the microstructure present in the grain boundaries within both regions is made susceptible to hydrogen-induced cracking.

In addition, a network of low melting liquid on the grain boundaries extends from the composite zone, through the unmixed zone into the partially melted zone and has a high solubility for hydrogen. Thus, such liquated grain boundaries serve both to transport the hydrogen from the composite zone and to concentrate the diffusible hydrogen in the segregated, crack sensitive microstructure produced when these liquated films solidify.

It is therefore not surprising that recent studies of hydrogen-induced cracking, utilizing solute sensitive etching techniques, almost invariably indicate that these cracks initiate at grain boundaries in either the unmixed zone or the partially melted zone. Furthermore, until a critical stress intensity is reached, such cracks usually propagate preferentially along these boundaries.

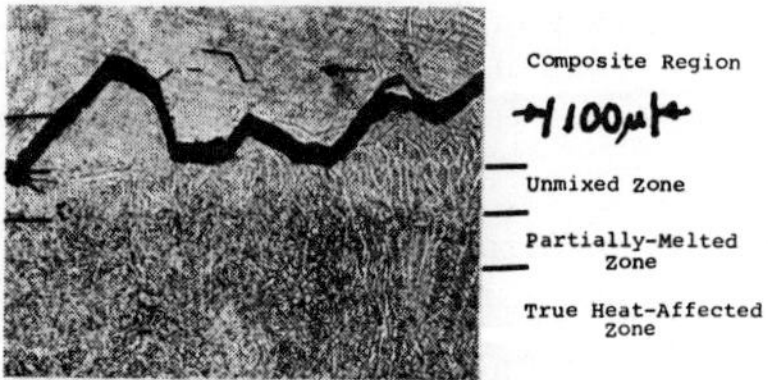

Fig. 24: Typical Hydrogen-Induced Crack in HY-80 Steel.

Figure 24 shows a typical hydrogen-induced crack in an HY-80 weldment after etching with a solute sensitive etchant. Attention is drawn to the fact that the cracking is exclusively confined to grain boundaries in the unmixed and the partially melted zones. With normal etching practices, using structure-sensitive etches such as nital or picral, the dark etching composite zone in Fig. 24 would have appeared to be the fusion boundary and the cracking would have been erroniously classified as heat-affected zone cracking.

Figure 25 summarizes in schematic form the zones in a weldment as redefined in the preceeding sections. It is important to note the location of two important new zones at the fusion boundary.

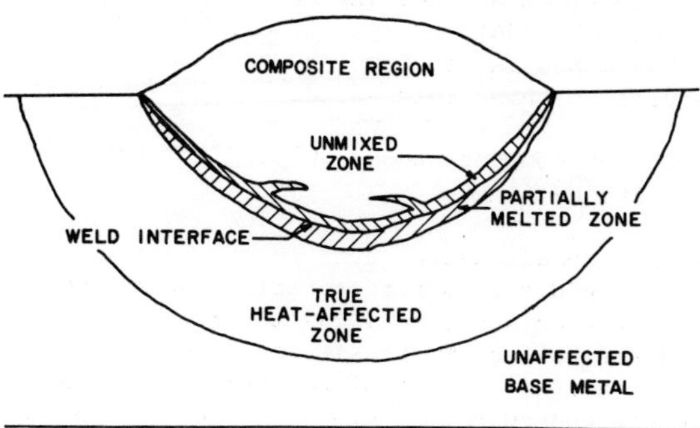

Fig. 25

Note that the microsegregation produced both within the fusion zone during solidification of the weld metal and the penetration of grain boundaries in the partially melted zone by low melting liquid films can play a major role in the formation weld defects. Hopefully, through improved understanding of the influence of welding procedures on the mechanics of solidification and through control of the level of residuals, such as phosphorus and sulfur, which contribute to the localized melting in the partially melted zone, progress will be made in minimizing the incidence of both hot and cold cracks.

REFERENCES

1. Savage, W. F. and Aronson, A. H.,"Preferred Orientation in the Weld Fusion Zone", Welding Journal, Res. Suppl. 45:No. 2, February, 1966.

2. Savage, W. F. and Lundin, C. D., "The Varestraint Test" Ibid. 44:No. 10, October, 1965.

3. Tiller, W. A. and Rutter, J. W., "The Effect of Growth Conditions Upon Solidification of a Binary Alloy", Canadian Journal of Physics 34, 96, 1956.

4. Savage, W. F., Nippes, E. F., and Szekeres, E. F., "A Study of Weld Interface Phenomena in a Low Alloy Steel", Welding Journal, Res. Suppl. 55: 9, September, 1976.

DISCUSSION

W. Borges (Supship, Boston): Observation of first pass weld beads in Cr Ni and Ni alloys often indicate a hairline longitudinal dark crack in the center of a groove or fillet. To the inexperienced layman this is a crack, when in fact it is an inherent mark.

Would you comment on this phenomena?

Author: I am afraid I would have to see this phenomenon before I would care to comment, I am not sure I understand exactly what type of artifact you are describing.

Z. P. Saperstein (Modine Manufacturing Co.): What is the significance of surface ripples on weld?

Author: Surface ripples can arise from two sources:
1. Variations in arc pressure on the surface of the weld pool as a result of ripple on the power supply current.
2. Surface free energy effects such as those described by Ishizaki in his paper entitled "Surface Tension Theory on Arc Welding Phenomena", Transaction of Japan Institute of Welding, Vol. 30, 1961, p. 703 (Report 1) and Vol. 31, 1962 (Report 2) p. 184. This phenomenon is too involved to be discussed here, but I recommend these reports to those interested.

R. A. Kelsey (Alcoa Labs): Could you comment on how solidification mechanics relates to pore formation during welding?

Author: Porosity nucleates within the diffusion layer in the liquid at the moving solid liquid interface as a result of supersaturation caused by the rejection of gaseous solutes from the adjacent solid as it grows. If examined metallographically, the distribution of porosity outlines the location of the solid-liquid interface at the instant of formation. If the solute gas is present in sufficient concentration growth is continuous over apprecialbe distances to form "wormhole" type porosity rather than isolated spherical porosity.

A. C. Lingenfelter (Huntington Alloys): I noted that the grain boundaries in your columnar dendritic example did not follow the dendrite boundaries. I have noted the same situation in cellular structures. Have you seen hot cracking occur anywhere buy along these grain boundaries?

Author: The hot cracking always occurs along the boundaries at which the terminal transient phase of solidification occurred. This is not necessarily the location of the grain boundaries after the weld cools to room temperature. Structure sensitive etches will reveal where the crystallographic grain boundaries are located but solute sensitive etches will reveal the segregation caused during solidification as what are sometimes referred to as ghost grain boundaries.

By acceptance of this article, the publisher or recipient acknowledges the U.S. Government's right to retain a nonexclusive, royalty-free license in and to any copyright covering the article.

SOLIDIFICATION BEHAVIOR OF TYPE 308 STAINLESS STEEL FILLER METAL*

S. A. David and G. M. Goodwin

Oak Ridge National Laboratory
Metals and Ceramics Division
Oak Ridge, Tennessee 37830

ABSTRACT

A series of experiments that combined thermal analysis and interrupted solidification was carried out to understand the solidification behavior of type 308 stainless steel weld metal. Results indicate the following sequence of phase separations for the alloy investigated:

$$L \rightarrow L + \delta \rightarrow L + \delta + \gamma \rightarrow \gamma + \delta \ .$$

Ferrite observed at room temperature was identified as the untransformed primary δ-ferrite that formed during the initial stages of solidification. The mode of freezing and the extent of solid-state diffusion both control the amount and distribution of ferrite. Microstructural and microprobe analyses indicated both extensive solid-state transformation and solute redistribution during solidification and cooling from the nonequilibrium solidus to room temperature.

INTRODUCTION

During the past three decades the subject of ferrite† in austenitic stainless steels has received and continues to receive a great deal of attention.[1–3] Austenitic stainless steel weld metal normally has a duplex

*Research sponsored by the Division of Materials Sciences, Department of Energy under contract W-7405-eng-26 with the Union Carbide Corporation.

†The usage of the terms ferrite and δ-ferrite will be synonymous in this article and used interchangeably.

structure that contains varying amounts of ferrite. We recognize that if sufficient ferrite is present in the weld, the ferrite effectively prevents hot cracking.[4–6] Several theories have been advanced to explain this phenomenon.[4–10] Yet, to date, our understanding of the hot cracking phenomenon is still incomplete. Knowledge of the sequence of phase separation leading to the microstructure and solute distribution observed during and after solidification may significantly improve our understanding of the role of ferrite in preventing hot cracking. The alloy compositions in Fig. 1 show that solidification of austenitic stainless steel weld metal can take place by primary precipitation of austenite or ferrite. In certain cases, subsequent to the formation of ferrite, austenite envelopment could occur and thus interrupt the formation of ferrite.

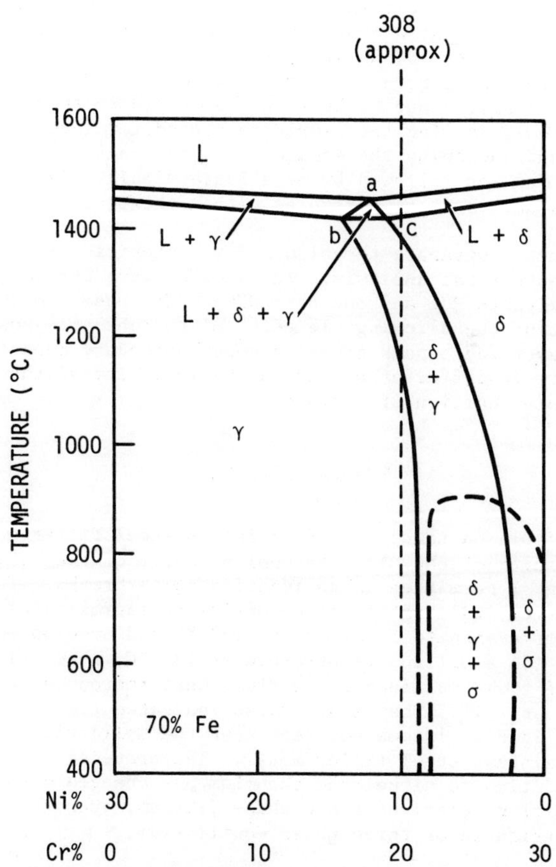

Fig. 1. Vertical Section (at 70 wt % Fe) of Ternary Equilibrium System Fe-Cr-Ni. Source: I. Masumoto, K. Tamaki, and M. Katsuna, *Jap. Weld. Soc. J.* 41(11): 1306–14 (1972).

The purpose of this investigation was to examine in detail the solidification behavior of type 308 stainless steel filler metal. Thermal analysis and interrupted solidification experiments were designed to correlate the observed microstructural features and solute distribution with solidification behavior.

EXPERIMENTAL PROCEDURE

Differential Thermal Analysis (DTA) and interrupted solidification experiments were carried out by suspending 25-g samples contained in an alumina crucible in a vertical Pt—40% Rh resistance heated tube furnace. The samples contained by weight percent: 0.065 C, 1.52 Mn, 0.44 Si, 0.02 P, 0.008 S, 20.51 Cr, and 10.53 Ni, the balance being iron. The furnace tube was sealed at the bottom and top and a constant rate of argon flow was maintained in the furnace. A pure platinum sample was used as the standard during DTA. After the sample was heated to about 50°C above the presumed liquidus, it was cooled at a constant rate of about 20°C/min. Thermocouples made of Pt vs Pt—10% Rh were used to record the temperature differential and the cooling curves on an x-y recorder. Interrupted solidification experiments were carried out by heating the sample to above its liquidus temperature, slowly cooling the sample to a predetermined temperature below the liquidus, and quenching the sample in an ice bath. The progress of the solidification was followed by metallographically analyzing the samples quenched from a series of temperatures within the freezing range of the alloy.

Standard metallographic techniques for austenitic stainless steel were used for microstructural analysis. The samples were etched with a solution containing five parts HCl and one part HNO_3. To determine the segregation characteristics of the alloying elements, microprobe analyses were done on the samples traversing a path crossing a few secondary dendrite arms. Ratios of the intensity at different points to the intensity of the pure element were plotted as a function of distance.

THERMAL ANALYSES

Results of DTA on the type 308 stainless steel filler metal are shown in Fig. 2. Results of the DTA, the cooling curve of the sample, and the guidance of Fig. 1 permitted us to identify three breaks. These indicated: (1) the primary crystallization of δ-ferrite at approximately 1435°C; (2) the start of austenite formation at 1370°C and envelopment of δ-ferrite; and (3) a solidus temperature at 1339°C, respectively. The temperatures of these multiphase reactions were reproduced within ±2°C on other samples, as well. The results also indicated that the alloy line position shown in Fig. 1 does not represent the solidification sequences of type 308 stainless steel filler metal. The presence of other alloying elements in addition to nickel and chromium has changed considerably the alloy line position relative to the phase diagram, such that it partly traverses the triangle of three phase equilibrium.

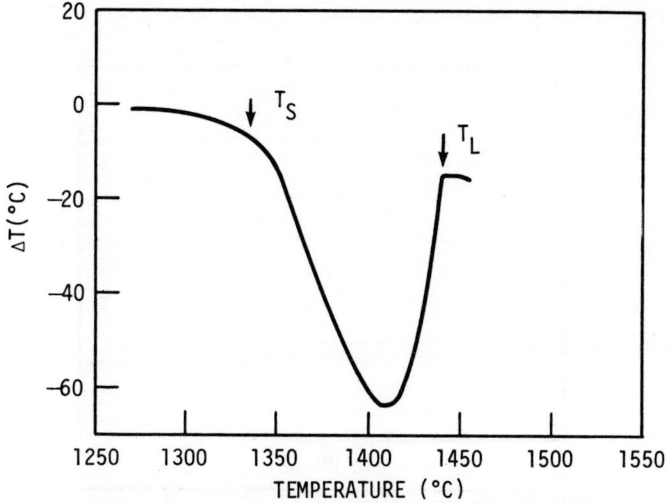

Fig. 2. Results of Differential Thermal Analysis of Type 308 Stainless Steel Filler Metal. Plot of temperature differential between standard platinum and sample vs sample temperature.

MICROSTRUCTURE

Figure 3 shows low-magnification photomicrographs of the samples quenched from various temperatures below the liquidus temperature. Precipitation of primary δ-ferrite occurs on cooling the sample to a temperature just below the liquidus temperature [Fig. 3(a)]. Here the dark-etching blocky phase is the primary δ-ferrite. The fine structure surrounding the primary phase indicates that the rest of the area was liquid at the moment of quenching. The dark appearance of ferrite is mainly due to the transformation of what was primary δ-ferrite to Widemanstätten austenite. Fredriksson made a similar observation in 18-8 stainless steel ingots.[12] The volume fraction of the primary δ-ferrite continues to increase as the temperature decreases below the liquidus. However, at 1370°C, as shown in Fig. 3(b), an envelopment of austenite around the primary δ-ferrite occurs. A high-magnification photomicrograph of the transformed ferrite enveloped by austenite is shown in Fig. 4. Magnetic etching identified the dark phase in the photomicrograph as δ-ferrite. In a way the transformation of δ-ferrite to Widmanstätten austenite clearly marks the boundary between primary ferrite and austenite at the time of quenching. From the point of complete envelopment further transformation — $L \rightarrow \gamma$ and $\delta \rightarrow \gamma$ — proceeds at the γ-L and γ-δ interfaces. As the sample cools to a temperature slightly below that of the solidus (1339°C), the transformation at the γ-L interface goes to completion, leaving behind a skeletal network of untransformed delta ferrite along the core of the primary and secondary dendrite arms [Fig. 3(c)]. This residual ferrite appears to be very stable on quenching. Interdendritic ferrite due to solute segregation was absent in the samples examined. Additionally, the amount of residual δ-ferrite decreased further on slowly cooling the sample from solidus to room temperature.

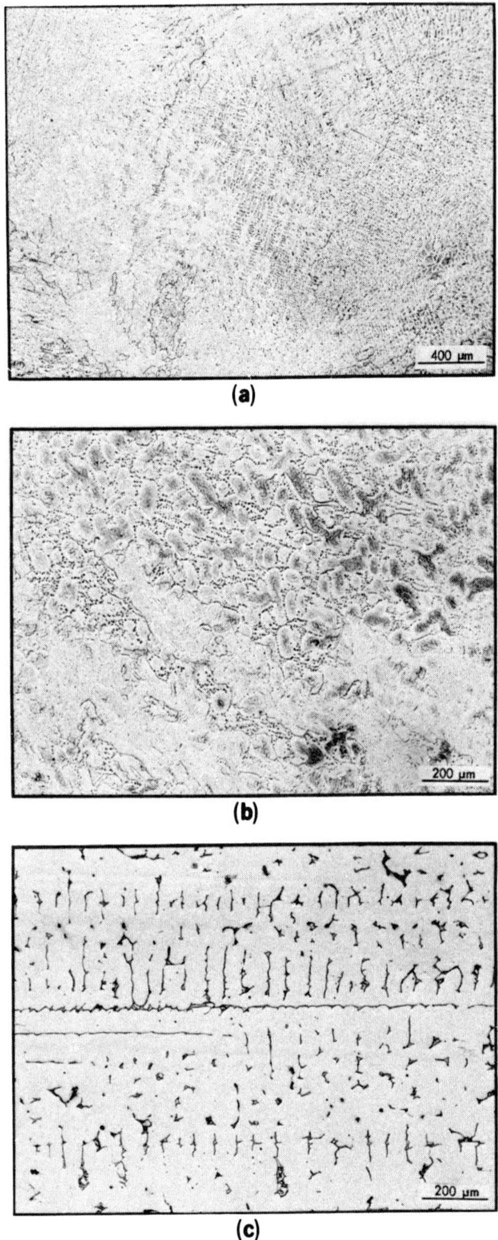

Fig. 3. Photomicrographs of Interrupted Solidification on Samples Quenched from (a) 1430°C, (b) 1365°C, and (c) 1335°C.

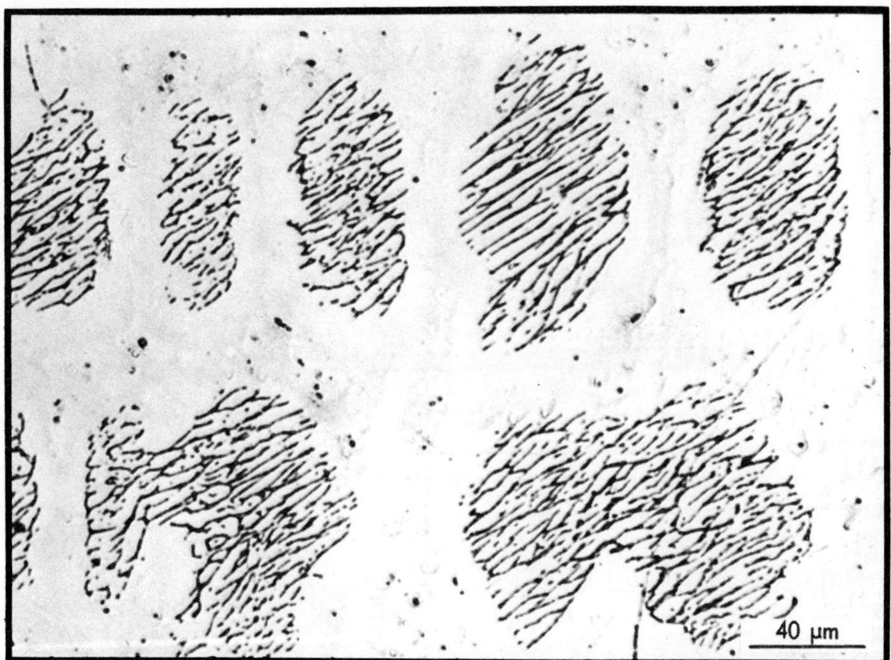

Fig. 4. High-Magnification Photomicrograph of Interrupted Solidification Sample Quenched from 1365°C.

SOLUTE DISTRIBUTION

Figure 5 shows the distribution of chromium, nickel, silicon, and manganese across secondary dendrite arms in the same area of the sample shown in Fig. 3(a). From the intensity ratios with the sample as an internal standard, the primary δ-ferrite regions have been found to contain 2 wt % more chromium and 2 wt % less nickel than the austenite. The distribution of manganese and silicon was fairly uniform. Figure 6 shows the distribution of chromium, nickel, silicon, and manganese across a ferrite arm in the same area of the sample as shown in Fig. 3(c). The ferrite regions contained approximately 5 wt % more chromium and 4 wt % less nickel than the austenite. Apparently, the level of chromium increased significantly in the ferrite, while that of nickel decreased during the $\delta \rightarrow \gamma$ transformation. The high level of chromium in the residual ferrite may have been due to the transport of partitioned chromium into δ-ferrite, with further redistribution of chromium occurring within the ferrite by volume diffusion during $\delta \rightarrow \gamma$ solid-state transformation.[13] This high level of chromium seemed to stabilize the ferrite and thus prevented it from transforming to Widmanstätten austenite during the quench, as discussed earlier.

SUMMARY

In the light of these results we may conjecture that the type 308 stainless steel filler metal studied generally solidifies by $L \rightarrow L + \delta \rightarrow L + \delta + \gamma \rightarrow \gamma + \delta$ transformations. Delta-ferrite in the duplex structure observed at room temperature has been identified as the residual ferrite

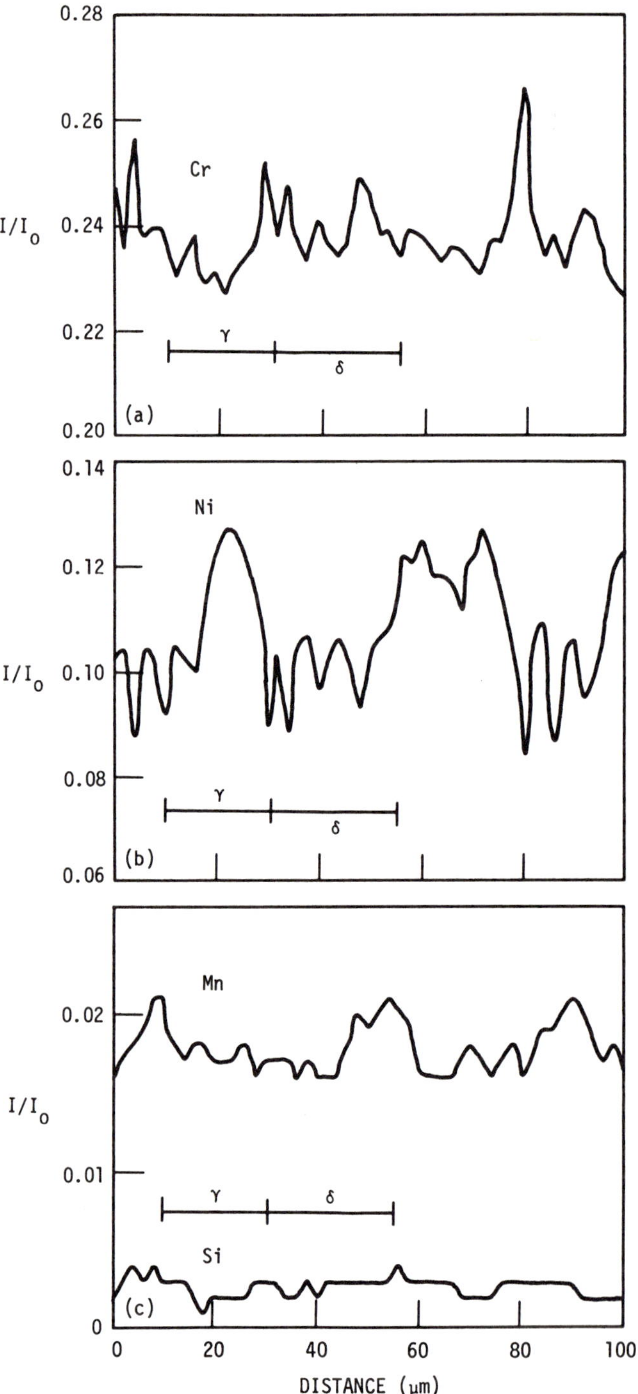

Fig. 5. The Distribution of (a) Chromium, (b) Nickel, and (c) Silicon and Manganese Across Secondary Dendrite Arms in the Same Area of the Sample as Shown in Fig. 3(a).

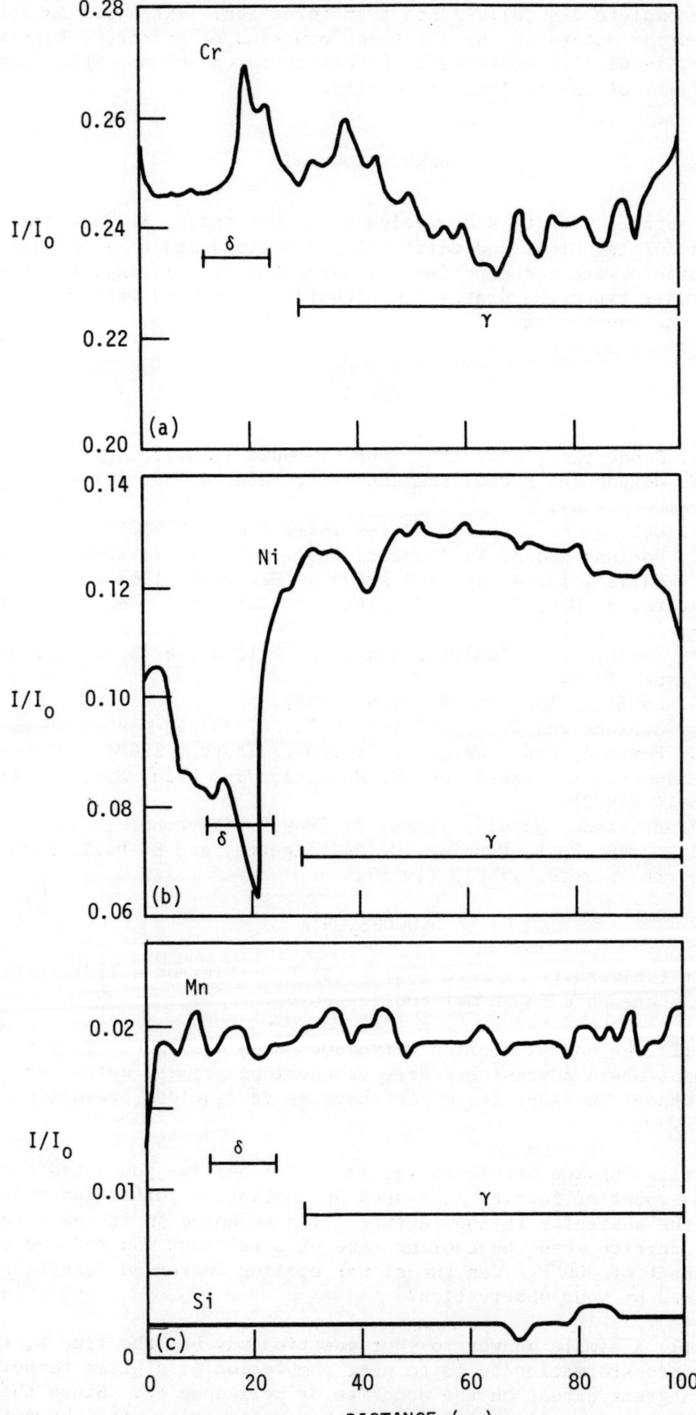

Fig. 6. The Distribution of (a) Chromium, (b) Nickel, and (c) Silicon and Manganese Across Secondary Dendrite Arms in the Same Area of the Sample as Shown in Fig. 3(c).

due to incomplete δ-γ solid-state transformation. Extensive solute redistribution occurs during the transformation of primary δ-ferrite to austenite, leading to enrichment of ferrite by chromium. This increases the stability of the residual δ-ferrite.

ACKNOWLEDGMENTS

The authors wish to acknowledge C. T. Liu and R. A. Vandermeer for their helpful technical and editorial suggestions and to J. D. Hudson for his assistance with the experimental setup during this investigation. Kathy Gardner typed the draft; Nan Richards edited and Denise Jackson prepared the manuscript.

REFERENCES

1. A. L. Schaeffler, *Met. Prog.* 56: 680—680B (1949).
2. W. T. DeLong and E.O.G. Szumachowski, *Weld. J.* 35: 521-s—528-s (November 1956).
3. W. T. DeLong, *Weld. J.* 53: 273-s—286-s (July 1974).
4. J. C. Borland and R. N. Younger, *Brit. Weld. J.* 7: 22—60 (January 1960).
5. F. C. Hull, *Weld. J.* 46: 399-s—409-s (September 1967).
6. Y. Arata, F. Matsuda, and S. Katayama, *Jap. Weld. Res. Inst. Trans.* 5(2): 35 (1976).
7. P. P. Puzak, W. R. Apblett, and W. S. Pellini, *Weld. J.* 35: 9-s—17-s (December 1976).
8. W. S. Pellini, *Foundry,* 80: 125 (1952).
9. W. R. Apblett and W. S. Pellini, *Weld. J.* 33: 83-s—90-s (December 1954).
10. J. C. Borland, *Brit. Weld. J.* 7: 508—12 (August 1960).
11. I. Masumoto, K. Tamaki, and M. Katsuna, *Jap. Weld. Soc. J.* 41(11): 1306—14 (1972).
12. H. Fredriksson, *Metall. Trans.* 3: 2989—97 (November 1972).
13. C. E. Lyman, P. E. Mauning, D. J. Daugette, and E. Hall, *Scanning Electron Microsc.* 1: 213 (1978).

DISCUSSION

H. W. Kerr (University of Waterloo): (1) Did start of solidification of austenite phase show on normal cooling curve?

Author: (1) The normal cooling curve does show a slight change in the slope at the point where austenite starts to envelope primary delta-ferrite. However the change in slope is not as sharp as at liquidus or solidus temperatures.

L. M. Petrick (Ebasco Services, Inc.): (1) I believe you cited a condition where the amount of ferrite increased on cooling, by diffusion of the chromium from the austenite to the ferrite. I have noted an increase in the amount of ferrite when the cooling rate of a weldment was reduced by applying a preheat of 300°F. Can the slower cooling increased ferrite content be explained by your observation?

Author: (1) A simple answer to your question may be, the time t, required for $\delta \rightarrow \gamma$ transformation to go to near completion at a given temperature depends to a great extent on the dendrite or cell spacing. Since the dendrite or cell spacing is closely related to the cooling rate, the above time t is small for a fine dendrite spacing (fast cooling rate) and large for a coarse dendrite spacing (slow cooling rate). Therefore the amount of residual ferrite left behind at a certain period during cooling may be higher in a structure solidified at slow cooling rates than that solidified at fast cooling rates. However the actual situation is more complicated than this.

CORRELATION OF SOLIDIFICATION PHENOMENA AND ALLOY CHEMISTRY ON THE PROPERTIES OF DIRECT AGED PRECIPITATION HARDENED STAINLESS STEEL WELDS*

T. J. Bosworth

Boeing Marine Systems

Seattle, Washington

Precipitation hardened stainless steels are a family of corrosion resistant alloys characterized by the combination of a high strength to weight ratio and good corrosion resistance, plus good fabricability and moderate cost; a combination that is attractive for many applications. The struts and foils of the Boeing built hydrofoil craft represent a weight critical application where 15-5PH and 17-4PH stainless steels were selected because of the favorable combination of the above factors. The design criteria for these weldments include high strength, sea water fatigue and flaw growth characteristics, fracture toughness, erosion resistance, and stringent dimensional control for hydrodynamic purposes.

It is standard industry practice when fabricating heat treatable steel structures to fully heat treat following welding so as to achieve the optimum combination of properties in the welds and base metal. However, in the case of the hydrofoil struts and foils, which are large complex weldments where precise dimensional and contour control are mandatory; the post weld heat treatment became an undesirable, and costly necessity. Complex post weld heat treatments are also undesirable following any field welding application including both erection and repair.

An effort to develop a more cost effective method of manufacture and to eliminate post weld heat treatment of repair welds indicated the feasibility of minimizing post weld heat treatment to a direct aging step only. This investigation demonstrated that direct aging of welds would produce high tensile properties and good corrosion resistance, however impact properties and microstructure varied widely.

Subsequently, an investigation was conducted on the effects of the individual alloying elements and their synergistic effect on the microstructure and properties of the weld deposit. The objective was to develop an alloy chemistry that would yield the appropriate combination of strength, thoughness and corrosion resistance following a simple post weld aging cycle.

*Manuscript not received at time of printing so only abstract is included.

The properties of direct aged welds are influenced by each of the following factors that are controlled by alloy chemistry.

1. Nucleation and dendritic growth.
2. Development of martensitic structure without stringent restrictions on cooling rates.
3. Distribution of the precipitating phases.
4. Formation of appropriate metal carbides.
5. Presence of embrittling phases or compounds.

Solidification of the weld puddle was found to have a significant influence on the impact characteristics of the aged weld metal. Because of the directional cooling characteristics of a weld, there is a strong tendency for columnar grain formation to occur. When strong dendrite formers are present in the alloy chemistry, the dendrites from each preceeding weld passes can act as nuclei for the rapidly freezing puddle, thus producing continuous columnar grains. These dendrites are little affected by the aging process and heavy precipitation often occurs on their surface. Invariably, the predominant failure mode in such welds is cleavage. By minimizing the presence of the dendrite and ferrite formers, the solidification characteristics are altered such that solidification occurs on a more random basis and the stability of the dendrites is reduced. The predominant failure mode becomes quasicleavage interspersed with broad islands exhibiting ductile failure.

Stringent control of the dendrite and ferrite formers also contributes to the development of a fully martensitic structure even after multiple weld passes. To achieve a fully martensitic structure after welding also requires a balance between the ferrite and austenite forming elements. An unsatisfactory level of ferrite formers can produce high strength but with little ductility and little resistance to impact. An excess of austenite stabilizers will produce good impact resistance and ductility, however, the structure will be partially martensitic and the resulting tensile properties low.

The last three factors are interrelated involving elements that contribute to dendrite and ferrite formation and form the precipitating phases. The roll of columbium, and titanium as carbide stabilizers is discussed. It is shown that control of these elements is critical to the development of good impact strength and fracture toughness in the direct aged welds.

Finally, the influence of the weld process variables including heat input and shielding gases on the properties of direct aged welds are discussed.

Oxygen and Nitrogen Contamination

During Arc Welding

T. W. Eagar

Department of Materials Science and Engineering
Massachusetts Institute of Technology
Cambridge, MA 02139

Abstract

The sources, mechanisms, and expected levels of oxygen and nitrogen contamination during gas tungsten arc, gas metal arc, shielded metal arc, self-shielded metal arc, and submerged arc welding are reviewed. Calculations indicating the importance of decomposition of SiO_2 into silicon monoxide and oxygen are presented, indicating that silicon transfer between the slag and metal occurs by a gas-metal rather than a slag-metal reaction mechanism. A model suggesting that arc stabilizing additions to fluxes should provide volatile subspecies upon heating is also discussed.

Introduction

Oxygen and nitrogen contamination of weld metal is of concern in that relatively small quantities of these elements may affect the cleanliness, toughness and/or porosity of the resulting weldment. Generally speaking, these elements should be maintained at as low a level as possible; except perhaps in the case of austenitic stainless steels, where the percentage of delta ferrite may be manipulated by changing the nitrogen content of the weld metal. In any case, the sources and mechanisms which control the oxygen and nitrogen contents during welding must be understood in order to provide control over the process and the resultant properties of the weldment.

Providing that high quality starting materials are used, there are basically only two sources of oxygen and nitrogen contamination of the weld metal; viz., absorption from the surrounding atmosphere and reaction with any flux which may be present. Absorption from the atmosphere depends largely on plasma jet phenomena;flow rate, composition,and ionization state of the shielding gas; and the absorption mechanisms at the molten metal-gas interface. Fluxes may contribute oxygen by decomposition of less stable oxides, while nitrogen contamination from the flux, which is of potential concern,[1] is not generally important. The purpose of the present paper is to discuss these sources and their relative importance in the various arc welding processes.

Arc Phenomena

Rein has described the oxygen and nitrogen levels to be expected when using several different arc welding processes.[2] His summary, reproduced in Figure 1, illustrates the wide variations in nitrogen and oxygen levels which may be found among arc welding processes generally, and within the individual processes specifically. In order to understand these differences, it is necessary to distinguish the basic differences between the processes. First, however, several fundamental aspects of arc welding should be reviewed.

Welding arcs consist of ionized gases conducting large electrical currents. The arc maintains a delicate balance between the heat generated by the electrical resistance of the plasma and the heat lost to the surrounding atmosphere, workpiece and/or electrodes by conduction, convection, and radiation. The size of the arc is determined by balancing the pressure of thermal expansion of the plasma and the pressure of electromagnetic constriction on the conducting ions. If either of these balances is upset, the arc becomes unstable, leading to wide variations in welding current and possible extinction of the arc.*

A simple calculation, using estimates of electron mobility, plasma density, and arc geometry,shows that it is only necessary for approximately 3 to 30% of the arc plasma to become ionized in order to conduct the welding current, hence relatively small additions of easily ionizable elements can lower the arc temperature and change the arc profile near the molten metal surface, where the contamination reaction is occurring. If these easily ionizable additions are not continuous or uniform, arc instabilities may occur. It is interesting to note that oxygen and nitrogen have lower ionization potentials than either Ar or He. Carbon dioxide has an ionization potential that lies between oxygen and nitrogen. Since atomic or ionized oxygen and nitrogen are chemically more active than the respective diatomic molecules, the choice of shielding gas can have a profound effect on the resultant contamination levels of these elements. An equimolar mixture of nitrogen-CO_2 should produce strongly ionized CO and O and weakly ionized N_2 near the molten metal surface, creating a plasma of lower temperature and lower nitrogen activity than an equimolar mixture of nitrogen-Ar, in which the N_2 near the surface must be strongly ionized. On this basis alone, the N_2-CO_2 mixture might be expected to produce lower nitrogen levels in the weld metal. Such an effect has, in fact, been noted.[3]

Another arc phenomenon that is of general importance is the formation of plasma jets. This phenomenon was described originally by Maecker[4] as originating from a difference in magnetic pressure caused by the divergence of electrical current traveling from the electrode to the workpiece. These jets attain velocities of up to 600 m/sec, causing the entrainment of large quantities of gases into the arc plasma. If insufficient inert shielding is available, atmospheric gases will be pulled into the plasma by these jets. Jet formation requires less than 50 microseconds, hence any arc fluctuations below approximately ten kilohertz will alter the jet behavior leading to instabilities and possible atmospheric contamination. Such fluctuations are to be expected in all consumable electrode welding processes.

---

* It should be noted that the present discussion is concerned primarily with DC arcs. AC phenomena, while more complicated, follow the same basic principles.

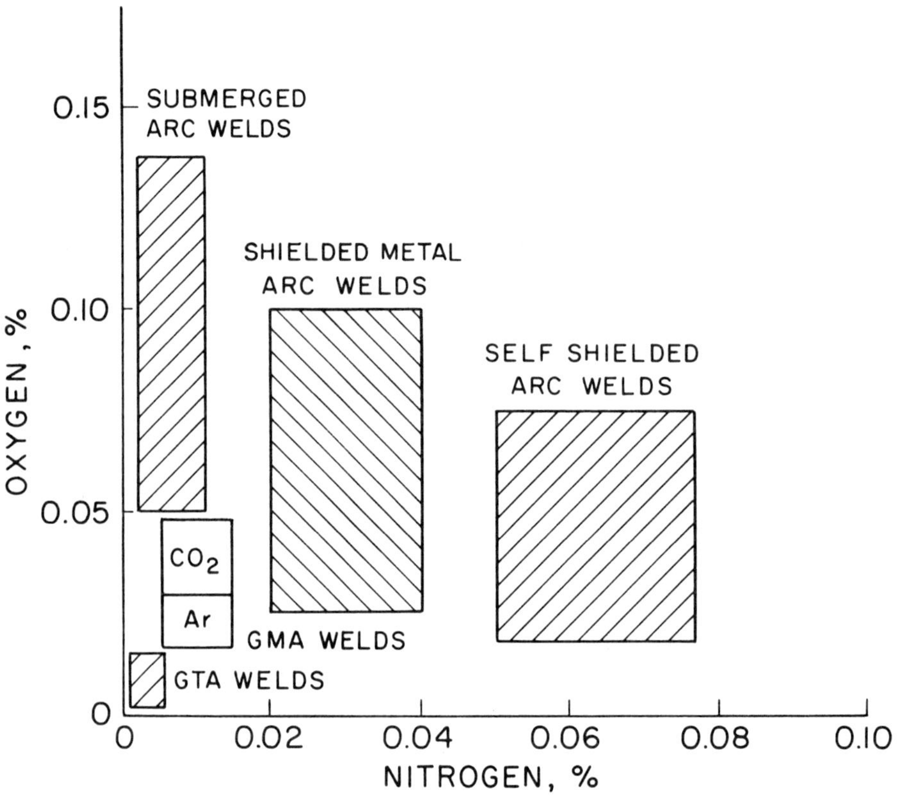

Figure 1. Oxygen and nitrogen levels expected from several arc welding processes. (After Rein.[2])

Welding Processes

Given the above brief discussion of shielding gas composition and plasma jet behavior, it is now worthwhile to review Figure 1 in more detail, describing the phenomena particular to each welding process.

1. Gas tungsten arc welding

The gas tungsten arc process produces the cleanest arc welds possible. The oxygen and nitrogen levels depicted in Figure 1 for GTA welding represent variations in initial base metal purity.

The cleanliness of GTA welding results from the absence of a flux, coupled with a short, very stable non-consumable arc. The stable arc provides smooth, uniform jet flow and the short arc length permits efficient shielding with gases that are required to be free of either oxygen or nitrogen.

2. Gas metal arc welding

The gas metal argon arc process differs from GTA welding only in the use of a consumable electrode. The shielding gas flow rate is increased to compensate for the longer arc length, however, oxygen and nitrogen contamination arises from plasma jet instabilities, which are inherent in any consumable electrode process. An oscilloscope trace of the current and voltage during GMA welding reveals arc interruptions lasting a few milliseconds. As noted previously, fluctuations on this time scale may cause the plasma to be extinguished, upsetting local gas flow behavior in the arc. When the arc is reignited only milliseconds later, large amounts of gas are aspirated into the arc cavity. Often atmospheric gases are included and contamination results. The higher oxygen levels observed when using CO_2 gas shielding are obviously caused by creation of CO and O in the welding arc. At these temperatures, carbon monoxide is a reducing gas; however, the oxygen combines with the weld metal causing loss of carbon in steels, formation of additional CO gas, and a net increase in the weld metal oxygen content.

Neither GTA nor GMA welding processes utilize fluxes for molten metal protection, and hence any oxygen or nitrogen contamination must result from either decomposition of the shielding gas or incomplete atmospheric shielding. As shown in Figure 1, these processes produce the least contaminated deposits. It will be shown subsequently that the greater oxygen contamination levels of the remaining three processes, SAW, SMAW, and SSMAW, are partially due to inadequate atmospheric shielding, but also depend significantly upon the breakdown of less stable oxides in the flux.

3. Shielded metal arc welding

Due to extensive empirical flux formulation studies and a moderate level of heat input, shielded metal arc welds are capable of unusually high quality weld deposits. As such, this process continues to find favor in many applications; however, continued progress is likely to be extremely slow.

Shielded metal arc welding is an open arc process in that no external shielding gas is applied. Exclusion of atmospheric contaminants is dependent upon a slag cover over the solidifying weld pool as well as gas evolution from decomposing flux. Since this is a consumable electrode process, arc fluctuations are inherent in the process. Atmospheric contamination cannot be avoided; it can only be minimized by choosing carbonate or cellulosic coatings which evolve copious amounts of carbon dioxide or hydrogen upon decomposition. As will be discussed later, part of the oxygen contamination in this process results from breakdown of less stable oxides in the flux.

4. Self-shielded arc welding

Self-shielded arc welding is very similar to shielded metal arc welding except for the addition of strong nitride forming elements (e.g., Al, Ti, Zr) to the welding wire. These nitride forming elements are also strong deoxidizers, which results in slightly lower oxygen levels than are found in SMA welding; however, this improved deoxidation leads to an increased rate of nitrogen absorption, as seen in Figure 1.

The rate of nitrogen absorption on iron surfaces is strongly dependent on the oxygen content of the iron, decreasing by a factor of six as the weld metal oxygen content increases from 100 ppm to 750 ppm.[5] Hence, any alloy additions which tend to lower the oxygen content of the weld metal will also increase the rate at which nitrogen is absorbed into the weld metal. Even modest bulk concentrations of oxygen cause saturation of the iron surface by oxygen atoms, presenting a barrier to nitrogen transfer from the gas to the metal phase.[6]

The competition of nitrogen and oxygen for surface active sites on iron provides an explanation of the observation made by Sis,[3] that carbonate self-shielded electrodes result in lower weld metal nitrogen levels than do basic (CaF_2) self-shielded electrodes. The carbonate electrode not only releases CO_2 shielding gas, but increases the oxygen content of the weld metal, significantly reducing the weld nitrogen content.

5. Submerged arc welding

Although commonly believed to provide complete exclusion of atmospheric contamination, the submerged arc welding process can be shown to achieve weld metal oxygen levels which are inexplicable in terms of flux decomposition reactions.[7] Evidence of this can be seen most readily in an analysis of weld metals produced with basic fluxes. These weld metals commonly are reduced in all deoxidizing elements: C, Mn, Si, Al, etc., compared with the base metal and wire analyses; yet the weld metal oxygen level is much higher than the starting materials. A mass balance on such a system requires a source of oxygen separate from the flux. Atmospheric entrainment due to arc fluctuations and instabilities in the slag cover is the most likely explanation.[7] Nonetheless, the very high oxygen levels associated with acidic flux submerged arc welding cannot be caused solely by atmospheric contamination. Such a large quantity of atmospheric oxygen would be expected to produce significant increases in the nitrogen content as well, producing an oxygen-nitrogen distribution similar to that observed in SMAW. The bulk of the oxygen contamination occurring during submerged arc welding must be related to decomposition of the oxide fluxes.

Flux Decomposition Reactions

The empirical fact that the oxygen content of commercial submerged arc welds can be related to a basicity index is well known,[8] if not appreciated. Several experts, failing to see any logical connection between flux basicity and expected oxygen content of the weld metal, have proposed instead that fluxes be classified according to their "oxygen potential." One of the difficulties with this proposal is that the concept of oxygen potential is ill-defined. In steelmaking slags, the oxygen potential is controlled by the FeO activity in the slag, although this is known to be a complex function of the basicity.[7] In arc welding, however, much higher temperatures are attained and at these temperatures several other flux components may achieve higher oxygen potentials than FeO. Consider, for instance, the decomposition of silica liquid into silicon monoxide gas plus oxygen, as given in the reaction

$$SiO_2\ (\ell) \rightarrow SiO\ (g) + 1/2\ O_2\ (g). \qquad (1)$$

$$K = \frac{P_{SiO}\ P_{O_2}^{\ 1/2}}{a_{SiO_2}}$$

At normal steelmaking temperatures this reaction is of little importance, although in the presence of carbon its importance in the blast furnace reaction has been suggested.[9] In arc welding systems, Heile and Hill have demonstrated that this reaction may be responsible for the generation of silica fumes.[10] As the temperature of interest increases above that required for steelmaking, many of the oxides which are normally considered to be most stable are actually among the least stable. North, et al., have claimed that CaO is chemically active in arc welding fluxes.[11]

An important consideration in reviewing reaction (1), is the fact that one oxygen atom is released for every silicon monoxide molecule formed, hence, the O_2 partial pressure for this reaction must be one-half the SiO partial pressure. This constraint permits the equilibrium constant for reaction (1) to be rewritten as

$$K = \frac{(0.5)^{1/2}\ (P_{SiO})^{3/2}}{a_{SiO_2}} = \frac{2P_{O_2}^{\ 3/2}}{a_{SiO_2}}\ .$$

Choosing 2000°C as the temperature of interest for the silica reaction during arc welding,[7] and calculating an equilibrium oxygen potential for reaction (1) above, as well as the FeO reaction equilibrium in CaO-SiO_2-FeO slags with 1% and 5% FeO, results in the graph presented in Figure 2. This figure presents the oxygen pressure (or oxygen potential) as a function of basicity. Dashed reference lines are given for the oxygen potential of air and pure FeO at this temperature. In addition, the observed oxygen potential of the weld metal is shown by a dotted line. The fact that this dotted line does not correspond directly to any other line shown illustrates that the weld metal and slag cannot be considered to be in equilibrium at 2000°C.[11] It will be noted that, at 2000°C, the oxygen potentials of both FeO and SiO_2 are generally greater than the weld metal oxygen potential, implying that either compound is potentially oxidizing to the weld metal at this temperature. It will also be noted that the oxygen potential of the SiO_2 decomposition reaction is significantly greater than the FeO oxygen potential, hence it cannot be assumed that FeO controls the oxygen potential of complex oxide fluxes at the elevated temperatures occurring during arc welding. It is probable that the bulk of the oxygen found in submerged arc weld metal made with acidic fluxes results from the decomposition of SiO_2 rather than FeO. This is evidenced by the large increase in weld metal silicon content (up to one percent) when acid fluxes are used.[8] Furthermore, the extremely rapid kinetics observed for the SiO_2 decomposition reaction imply that it is a gas-metal, rather than slag-metal, reaction; which is again evidence that reaction (1), i.e., formation of silicon monoxide gas from the slag, is the first step in oxygen and silicon transfer to the weld metal when employing high silica

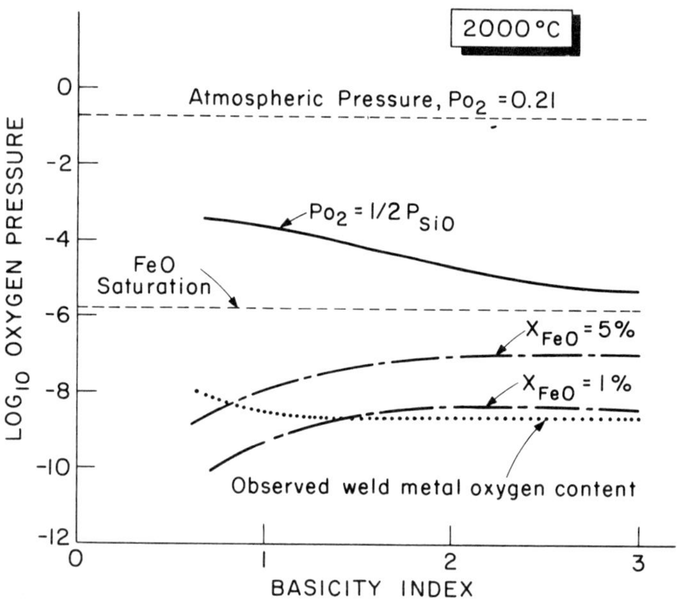

Figure 2. Oxygen potential as a function of basicity for SiO_2-CaO (solid line) and SiO_2-FeO (dashed lines) based slags at 2000°C.

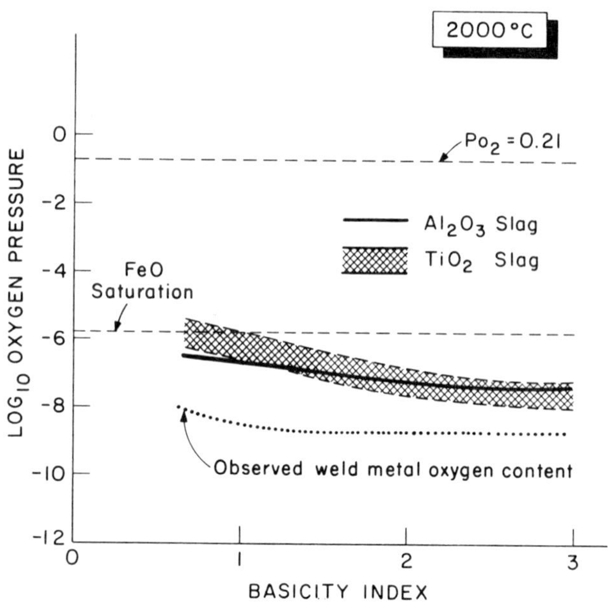

Figure 3. Oxygen potential as a function of basicity for TiO_2-CaO and Al_2O_3-CaO based slags at 2000°C.

activity (acid) fluxes.

It is possible to formulate acid fluxes which are low in silica, primarily by using TiO_2 or Al_2O_3 as major constituents. An analysis was performed for each of these components in a manner similar to that described above for SiO_2. The results of this calculation are shown in Figure 3.* It is readily apparent in comparing Figures 2 and 3 that Al_2O_3 and TiO_2 provide a significant reduction in the oxygen potential when compared with SiO_2; hence, either of these components would be expected to lower the oxygen content of the weld metal when used as a replacement for silica. This has been observed in rutile based slags by North, et al.[11] Furthermore, the gaseous suboxides TiO and AlO form much less readily than SiO, which implies that the kinetics of the TiO_2 and Al_2O_3 decomposition reactions will be much slower than the SiO_2 reaction.

The thermodynamic data for the preceding reactions have all been calculated at 2000°C for reasons discussed elsewhere[7]; however, the same general trends and conclusions would be applicable at any temperature above approximately 1800°C. The analysis presented in no way implies that 2000°C is the only temperature of interest. Extreme thermal gradients exist during arc welding, and the chosen temperature merely represents a convenient reference from which to evaluate the relative thermodynamic stability of the various oxides.

While the above discussion on flux decomposition during arc welding applies most directly to the submerged arc process, the conclusion that SiO is an important constituent in the arc atmosphere of any silica based flux welding process is a general one.[10] In fact, the realization that SiO_2 is capable of evolving large quantities of a gaseous suboxide aids one in rationalizing the relative arc stabilizing effect observed for various compounds by Hazlett, as reproduced in Table 1.[14]

Table 1. Relative Arc Stability Produced by Various Compounds (from Hazlett[14])

Arc Stability Index	Compound
1	Commercial Flux
2	K_2CO_3
	$CaCO_3$
3	Fe_2O_3
	SiO_2
	Na_2CO_3
4	MnO_2
	$MgCO_3$
	MgO
	Al_2O_3
	CaF_2
5	Na_2SiO_3

* The thermodynamic data for Al_2O_3-CaO slags were taken from Rein and Chipman.[12] Activities for the TiO_2-CaO system are not available, although the data of Martin, et al.[13] suggest that the activity of TiO_2 in CaO is similar to that of SiO_2 in CaO. Hence, the calculation for TiO_2-CaO melts was done for unit activity and given the same slope as shown in Figure 2 for the SiO_2-CaO system.

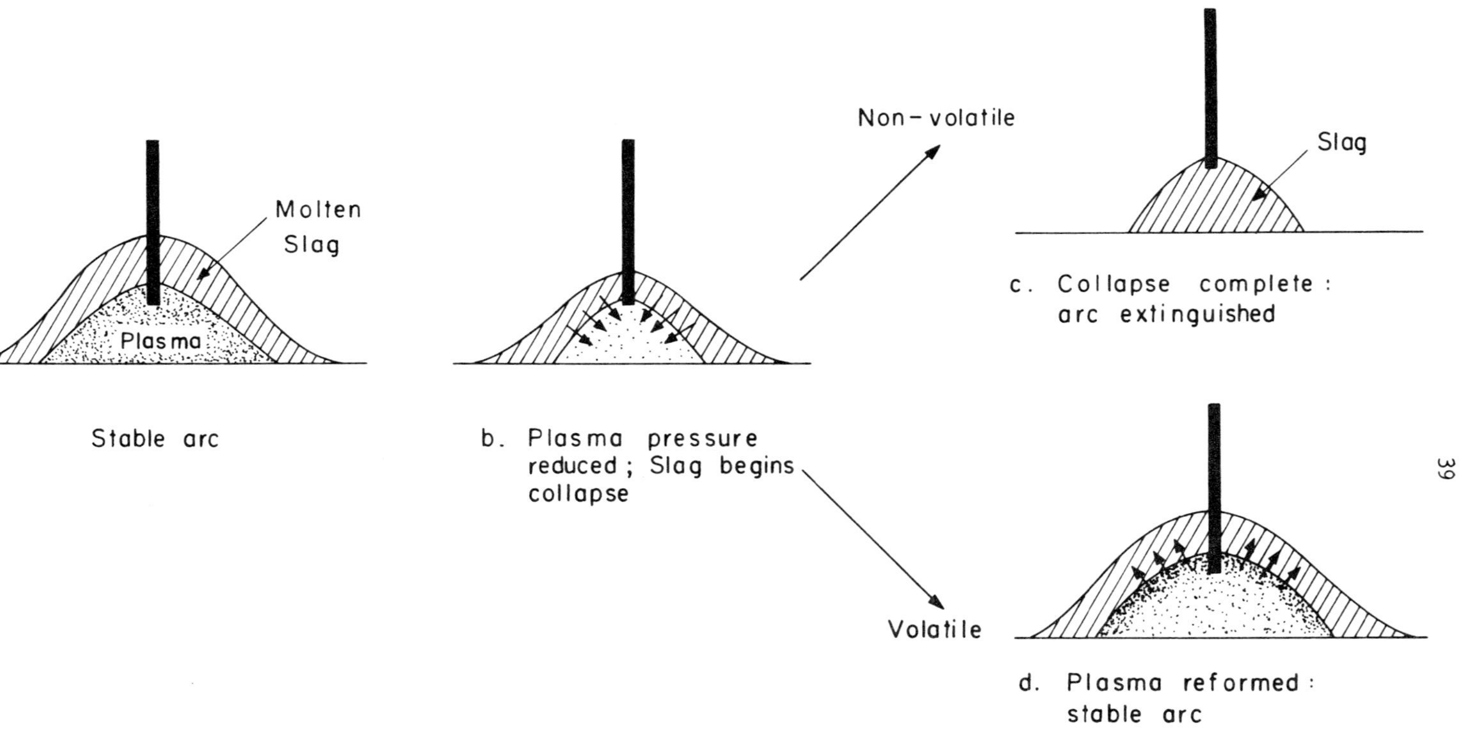

Figure 4. Schematic of the effect of volatile flux components on the stability of the submerged arc welding process. a. Stable arc. b. Plasma begins to be consumed by reaction with the weld metal; slag begins to collapse. c. Flux without volatile components; plasma is consumed; slag collapses and arc is extinguished. d. Flux with volatile components; heated slag produces plasma ions; pressure increases, slag is pushed back to stable arc position.

Although arc stability is dependent upon numerous factors, the most stable submerged arc fluxes shown in Table 1 tend to produce a gaseous reaction product upon heating. It is thought that this gas evolution process provides a stabilizing effect on the possible extinction of the arc plasma by the molten slag cover. As depicted schematically in Figure 4, a flux which does not evolve gas upon heating would tend to swamp out the arc if any decrease in plasma pressure were encountered. A gas evolving flux, on the other hand, would rapidly heat up if more closely exposed to the arc, thereby evolving gas, increasing the local plasma pressure, and pushing back the molten slag. This latter process would provide a stable arc plasma free of extinction by collapse of the molten slag. The most stable compounds in Table 1 are generally carbonates which evolve CO_2 upon heating; but Fe_2O_3 and SiO_2, which are both greater stabilizers than several of the carbonates, may also evolve gases; the former by changes in stoichiometry and the latter by formation of a volatile suboxide. It is currently believed that a volatile flux component, essentially a source of plasma ions, is a requirement of stable submerged arc welding fluxes. Russian investigators have found it necessary to add chloride salts for improved arc stability when welding highly reactive titanium alloys under fluoride based fluxes.[15] Under these conditions, the titanium weld metal will consume any oxygen ions in the plasma. If a new source of ions is not provided, the plasma will be extinguished for lack of a carrier gas. Chlorides have much lower vapor pressures than fluorides, and hence, would provide a gaseous plasma which would not be consumed by the titanium.

Conclusions

The extent of oxygen and nitrogen contamination during arc welding is dependent upon the process employed. These differences can be rationalized in terms of plasma jet phenomena, shielding gas composition, surface absorption phenomena, alloy additions, arc stability, and flux chemistry. It has been shown that silicon monoxide formation from the flux may provide an explanation of the rapid transfor of silicon between the slag and the metal. Substitutions of TiO_2 and Al_2O_3, being more stable than SiO_2, have the potential of reducing oxygen contamination. One possible requirement of arc stabilizing elements in submerged arc welding is that they provide a source of gaseous ions upon heating.

Acknowledgements

The author would like to acknowledge support from the Office of Naval Research, Grant N0014-77-C-0569, under which portions of this study have been performed. He is also indebted to R. Gage for initially suggesting the possible importance of sub-oxide formation in welding fluxes.

References

1. K. Schwerdtfeger and H. G. Schubert, "Solubility of Nitrogen and Carbon in ESR Type Slag," Met. Trans. 8B, 1977, p. 689.

2. R. H. Rein in Proceedings of a Workshop on Welding Research Opportunities, B. A. MacDonald, ed., Office of Naval Research, 1974, p. 92. AD-A028395.

3. L. B. Sis, "Welding Flux Cored Electrodes in N_2-CO_2 and N_2-Ar Atmospheres," Welding J, July 1977, p. 211-s.

4. W. Finkelnburg and H. Maecker, "Electric Arcs and Thermal Plasma," in Handbuch der Physik (translation available ARL62-302), January 1962.

5. R. D. Pehlke and J. F. Elliott, "Solubility of Nitrogen in Liquid Iron Alloys, II. Kinetics" Trans. AIME, 227, 1963, p. 844.

6. E. T. Turkdogen and P. Grieveson, "Kinetics of Nitrogen Transfer through an Iron Surface," J.Electrochem. Soc., January 1967, p. 59.

7. T. W. Eagar,"Sources of Weld Metal Oxygen Contamination During Submerged Arc Welding," Welding J.,March 1978, p. 76-s.

8. S. S. Tuliani, T. Boniszewski, and N. F. Eaton, "Notch Toughness of Commercial Submerged Arc Weld Metal," Weld. and Metal. Fab., August 1969, p. 327.

9. N. Tsuchiya, M. Tokuda, and M. Ohtani, "The Transfer of Silicon from the Gas Phase to Molten Iron in the Blast Furnace," Met. Trans. 7B, 1976, p. 315.

10. R. F. Heile and D. C. Hill, "Particulate Fume Generation in Arc Welding Processes," Welding J., July 1975, p. 201-s.

11. T. H. North, H. B. Bell, A. Nowicki, and I. Craig, "Slag-Metal Interaction, Oxygen and Toughness in Submerged Arc Welding," Welding J., March 1978, p. 63-s.

12. R. H. Rein and J. Chipman, "Activities in the Liquid Solution SiO_2-CaO-MgO-Al_2O_3 at 1600°C," Trans. AIME, 233, 1965, p. 415.

13. E. Martin, O. I. H. Abdelkarim, I. D. Somerville, and H. B. Bell, "The Thermodynamics of Melts Containing MnO, FeO, CaO, TiO_2 and SiO_2" in Metal-Slag-Gas Reactions and Processes, Z. A. Foroulis and W. W. Smeltzer, eds., The Electrochemical Society, Princeton, N. J., 1975, p. 1.

14. T. H. Hazlett, "Coating Ingredients' Influence on Surface Tension, Arc Stability and Bead Shape," Welding J., January 1957, p. 18-s.

15. L. K. Bosak and S. M. Gurevich, "A Non-Hygroscopic Flux for Welding Titanium and Its Alloys," Avt. Svarka., 1966, No. 5, p. 70.

DISCUSSION

B. W. Schaaf (E. I. duPont de Nemours): Your bar graph on O_2 content of welds as a function of process shows the SMA process having a much higher O_2 content than the GMA process. How do you rationalize the better resistance of the SMA process to porosity as a result of cross drafts when its deposit was a higher O_2 content under normal conditions?

Author: Porosity in steel weldments is due primarily to nitrogen contamination which comes primarily from the atmosphere. SMA electrodes have a relatively short unshielded length, essentially equal to the arc length - compared with GMA torches whose unshielded length in a cross draft is the arc length plus the electrode stickout. Hence, during cross drafts the shielding around the GMA electrode is exposed over a greater distance, resulting in more contamination and more porosity. The higher oxygen content of the SMA weld as shown in Fig. 1 is due to flux decomposition reactions and not atmospheric contamination.

It should also be noted that the gaseous contamination levels depicted in Fig. 1 for the GTA and GMA processes assume reasonable precautions have been taken against external contamination sources such as dirt or cross drafts.

W. E. Lawrence (Rockwell Science Center): (1) Does the data presented for GMAW reflect only bare wire electrode material? (2) Where would you classify gas-shielded flux core electrode materials and wouldn't they be better than bare wire if correctly formulated?

Author: (1) Yes. (2) Gas shielded flux core electrodes are not on the graph mainly because they could fit anywhere depending on the flux formulation. I certainly agree that a properly formulated gas-shielded flux core electrode could be the equal of a bare electrode GMA deposit. In some cases they might even be better, but generally I do not believe the improvement would be great.

J. T. McGrath (Dept. of Energy Mines & Resources): Would you comment on the role of CaF_2 in subarc fluxes on the reduction of oxygen content in weld metal?

Author: I would prefer not to comment on this at this time since I really do not have any data to back me up; however, in my opinion, CaF_2 itself does not reduce the oxygen content in submerged arc weld metal. It does lower the viscosity and melting temperature of the flux, leading to greater flux consumption and perhaps more effective refining on the part of other flux constituents. The Japanese data which I have seen claiming beneficial effects of CaF_2 on reduction of weld metal oxygen content has not been extensive, nor can it be stated conclusively that the relatively small reductions in oxygen content which were found, were not within the experimental scatter. I personally do not believe that any effect of CaF_2 in SAW fluxes is due to the chemical reactivity of CaF_2 with the weld metal. There may be secondary influences of CaF_2 as noted above, but I believe these to be small.

W. F. Savage (Rensselaer Polytechnic Inst.): Would you care to comment on the potential of treated wires for straight polarity MIG in order to eliminate O_2 pickup from Argon and Oxygen shielding gas?

Author: I am not really familiar with these treatments. We have found in our laboratory that treatments such as anodizing titanium electrodes reduces the nitrogen levels of the weld deposit. I suppose that a similar type of treatment (obviously not anodizing) might reduce the oxygen content of steel weld deposits, but not knowing exactly which treatments are being experimented upon, I cannot really comment.

RESIDUAL STRESS CONSIDERATIONS IN WELDMENTS FOR THE NUCLEAR INDUSTRY

E. Landerman, Advisory Engineer
Westinghouse Nuclear Energy Systems
Pittsburgh, PA 15230

and

G. Grotke, Fellow Engineer
Westinghouse Research and Development Laboratories
Pittsburgh, PA 15235

Reference to consideration of residual stresses associated with weldments in nuclear power plant components is stated in the Inservice Inspection Code of Section XI of the ASME Boiler Code. Four examples of these considerations are reviewed in this paper: (1) Type 304 pipe sections failed due to transgranular stress corrosion cracks starting from the pipe inside diameter due to residual stresses induced by closure welds as well as those associated with attachment welds on the outside diameter of the pipe. (2) The assessment of residual stresses associated with the location of underclad micro-cracks in SA508 Class 2 was made. Residual stress measurements using hole drilling techniques were made at locations where micro-cracking was determined metallographically. These micro-crack locations were at the overlapped areas of the coarse grain section of the weld heat affected zone where the second pass reheated the coarse grained area of the first pass to just below the A1 of the clad base material. These micro-cracks were only observed after postweld heat treatment of the clad part. Residual stress measurements were made at the surface and in the base metal underneath the weld overlay cladding. The stresses below the cladding became compressive and, therefore, these underclad micro-cracks were self limiting. Different weld overlay cladding techniques have been evaluated. (3) An 11-inch thick nozzle cutout containing a full thickness submerged arc weldment from a production vessel was sectioned to measure the weld residual stresses after postweld heat treatment. The techniques were (a) measurement of test strain gauges inserted into 1/4 inch diameter holes (up to 11 inches deep in the weldment) and (b) use of hole drilling technique. The maximum residual tensile stress measured was approximately seven ksi. (4) ASME Section XI repair weld technique after components after service permits the use of non-postweld heat treated welds. An evaluation of the level of stresses in test weldments was made with and without peening. The weldments were made using the technique detailed in Section XI of the ASME Boiler Code. These weldments were made using the 1/2 bead technique.

The 2" x 2" x 10" long repair welds were made in A533 Grade B Class 1 blocks which were 10" x 17" x 9" thick. The peening was performed manually. The level of stresses determined by the hole drilling technique in the peened repair weld was measured in the unpeened weld.

INTRODUCTION

The nuclear industry has a history of innovation added to the sound engineering technology developed through the years in pressure vessel technology. The Boiler Code of the American Society of Mechanical Engineers (ASME) and in particular Section III covering Rules for Construction of Nuclear Power Plant Components and Section XI covering Rules for Inservice Inspection of Nuclear Plant Components standardize these innovations in these construction Codes on an ongoing basis. The nuclear sections of the Code have included the need for consideration of residual stresses especially in the area of repair welds using the one-half bead "temper pass" technique where postweld heat treatment (PWHT) is not required.

Residual stresses play an insignificant role in vessel integrity consideration for PWHT vessels. The residual stresses after PWHT in the heavy wall vessels are quite low, i.e. approximately 7 ksi (48 MPa). It is well known that residual stresses, being self-equilibrating, do not influence ductile failure load;[1] the Heavy Section Steel Technology (HSST) program on intermediate vessels[2,3] has verified these conclusions. For nonductile failure, it is generally agreed that the effect of residual stresses is small associated with the large critical flaw sizes that would be necessary for failure. In general, residual stress effects are covered in the conservatisms already present in the nonductile failure analyses performed under Section III requirements. Further information on these effects will be available from the results of testing of the HSST intermediate vessel V-8.

This paper reports the residual stress measurement efforts performed by Westinghouse in order to assess the magnitude and location and their engineering assessment in welding applications in the nuclear industry. The prime thrust of the nuclear industry efforts is directed toward non-PWHT repair welds. Current programs by NRC[4] and EPRI[5] will provide specific data in the area.

Residual Stress Considerations - Four Examples

Four examples of residual stress considerations in weldments for the nuclear industry which will be covered are:

1. Service failure of Type 304 stainless steel associated with weld residual stresses,

2. Residual stress measurements associated with underclad cracking,

3. Measurement of residual stresses of PWHT 11-inch thick vessel weld, and

4. Reduction of residual stresses by peening in an ASME Section XI non-PWHT repair weld.

1. Type 304 Stainless Steel Corrosion Cracking

Sections of 8-inch Schedule 10 Type 304 stainless steel pipe containing a 2-inch long through wall crack were evaluated for causes of failure. The pipe was part of a stagnant line containing boric acid solution attaining temperatures of 175°F and less. A routine metallurgical analysis was performed of the cracked area which was adjacent to the weld. The cracking observed was transgranular. Sections of the weld and adjacent base materials were evaluated by the Strauss Test, boiling copper sulfate--sulphuric acid per ASTM A262 Reference E. The weld zone was not sensitized as determined by A262.

The cracks were primarily located about 1/4 inch from the circumferential welds. One minor crack was close to the fusion line of the weld and one branch extended about 20 mils in the weld. The primary cracking observed was tight, multiple transgranular branch cracks initiated from the inner diameter (ID) and were not located in the heat affected zone. Other ID transgranular cracks (Figure 1) were located in the area of the pipe where the welders identification tags were fillet welded to the pipe, Figure 2. It was concluded that the cracks were caused by stress corrosion cracking, probably chloride contamination on the ID from an unknown source. The cracking was associated with locations where residual stresses were induced by welding.

2. Residual Stress Measurement Associated with Underclad Cracking.

In 1970, underclad heat-affected zone discontinuities (microcracks) were reported in pressure vessel components. The pressure vessel industry formed a Task Group on Underclad Cracking under the Subcommittee on Thermal and Mechanical Effects of the Fabrication Division of the Pressure Vessel Research (PVRC) to review this observed metallurgical condition[6] The underclad cracks were restricted to narrow bands within the heat-affected region beneath the stainless steel overlay cladding where the clad overlapped, Figures 3, 4. Since the plane of the flaws was essentially perpendicular to the length of the weld deposits, the stress orientation of principal interest is tension in the "X" direction, Figure 5.

The microcracks occurred immediately in a zone below the clad interface in the region coarsened during the first bead deposition which zone was subsequently heat (tempered) to approximately 500 to 700°C (932° to 1292°F) when the second bead was deposited. The PVRC Task Group[4] concluded that the combination of high residual stresses at the overlap area coupled with tempering of this zone by the second bead provided the conditions for stress relief cracking of susceptible material, SA508 Class 2. These microcracks were observed primarily with the high heat input weld cladding methods using the one layer technique (e.g. 6 wire technique and strip cladding). The subclad cracks were observed to be less than 0.20 (5 mm) deep[6] and were shown "not critical to safe operation."[7,8,9]

The level of residual stresses at this overlap area was measured in the as welded overlay condition and in the post weld heat treated (PWHT) condition. The stresses were determined by measuring strains with small 3-gauge rosettes, Figure 6, with an effective diameter of 0.3 inches (8 mm). After the gauges were bonded to the specimens, a centrally located hole 0.06 inches (1.5 mm) in both depth and diameter, cut by an abrasive-arc-jet, Figure 7, provided local strain relaxation. This method has been described in detail by Bush.[10]

Residual stress measurements were made on strip, 6 wire, and 3 wire cladding (both commercial and experimentally clad) on SA508 Class steel (susceptible analysis) and SA533 Grade B Class 1 (non-susceptible analysis). The specimen identifications are shown in Table 1.

The strain guage locations are shown in Figure 8. The stresses in the X direction (σ_X) which would be most significant for crack initiation and propagation are shown in Figures 9, 10, 11, and 12. The distribution of stress prior to stress relief is shown in Figure 9. Stresses in the range 19 to 30 ksi (131 to 207 MPa) were measured at the 0.1 inch (3 mm) location. However, the gradient of underclad stresses is steep, becoming compressive about 0.25 inch (6 mm) below the clad interface. Figure 10 which presents data for two experimentally 6-wire clad specimens illustrates the effect of stress relief. Stress relaxation occurred only in the ferritic steel, resulting in retention of the steep gradient in the region immediately below the stainless clad. However, the extent of the tensile field is extremely limited. Inspection of Figure 11 which summarizes the results for four stress-relieved specimens, confirms this observation. Data points for the three specimens commercially clad with high heat-input techniques (6-wire and strip) fall within a reasonably tight scatterband with negligible stress at the 0.1 inch (3 mm) location. The results for specimen 3, experimentally 6-wire clad, are in substantial agreement.

Figure 12 compares the commercial specimen scatterband for high heat-input practice (from Figure 11) and the data points for commercially 3-wire clad specimen 8. There is no indication of a relationship between cladding practice and the residual stress distribution.

The scatterband for all commercially clad specimens (strip, 3-wire, 6-wire) is presented in Figure 13 together with the results obtained from an experimentally strip-clad specimen examined by Mitsubishi. The Japanese work is part of an extensive cooperative effort which employed a variety of strain gauge and X-ray diffraction techniques. This, and other tests, demonstrated that peak tensile stresses occur within the clad deposit, not at the clad surface. Within the ferritic steel the same precipitous drop in stress may be observed.

It should be noted that discussion of significant underclad tensile stress refers only to the regions where clad beads overlap. Japanese investigations[11] also established that the underclad stresses peak at the crack-susceptible regions and fall to a minima below the middle of beads, see Figure 14. The data, for stresses under 60 mm (2.4 inches) strip deposits, show localized peaks above 50 ksi (345 MPa) as deposited, and to 23 ksi (159 MPa) after stress relief.

Although it has been shown that microcracking condition does not effect the integrity of clad vessels,[7,8,9] the acceptable cladding techniques used in the nuclear industry for pressure vessel components are currently qualified by metallographic examination (for detection of underclad cracking).

3. Measurement of Residual Stresses of PWHT 11-inch Thick Vessel Weld.

Residual stress measurements were on an 11-inch thick weldment obtained from a nozzle-cut removed from the nozzle shell course of a commercially fabricated pressure vessel.[12] The nozzle cut out was 53 inches (1320 mm) in diameter and slightly over 11 inches (280 mm) thick, Figure 15. The vessel diameter was about 11 feet (3400 mm) outside diameter (OD). The SA533 Grade B Class 1 steel vessel shell had been clad with stainless steel and had been given an immediate PWHT of one hour at 1100°F ± 50°F(590 ± 30°C).

The weld groove was essentially straight walled with 1 3/8 groove width, double U-groove with the welding sequence being 3 inches from the inside and 8 inches from the outside. The weld was made by submerged arc welding (both single and tandem arc); the root or back weld was made by shielded metal arc. Residual stresses were measured (a) after the initial one hour intermediate stress relief for surface stresses only and (b) after an additional 5 1/2 hours at 1125° ± 25°F (610 ± 14°C).

The hole drilling technique described under "Consideration 2" was used to measure surface stresses. The stresses through the weldment were determined using the "tube method." Strain gauges are bonded to the inside of a 0.250 inch (0.35 mm) diameter, .006 inch (0.152 mm) wall thickness stainless steel tube through a small cut out, Figure 16. These tubes are then secured by epoxy to holes drilled in the weldment specimen. The tube holes and the "hole drilling" gauges are shown in Figure 17 and 18. A total of forty-one tubes containing 168 gauges were used in this test. Parallelpipes were sawed from the weldment, Figure 19.

The coordinate stresses σ_X (transverse to the weld length) and σ_Z (parallel to the weld length) versus distance from the weld centerline are plotted in Figure 20. The stresses parallel to the weld length (usually greater than the transverse stresses) did not follow the expected pattern. It is believed that this is associated with unbalanced weld-groove geometry with greater shrinkage near the surface, 1 3/8 (35.5 mm), initial joint gap versus measured 1.2 inches (30.5 mm).

The measured residual stresses are shown in Figure 21, 22, 23, 24, 25, and 26. The maximum stress measured was 7.4 ksi (51 MPa) in the full relieved condition, 6 1/2 hours (considering the latest ASME Section III requirements of 1 hour per inch of thickness to 5 inches and 15 min. for each inch over 5 inches). These results are similar to those obtained in an earlier study[13] in a 5 inch (127 mm) thick welded specimen of comparable materials and weld procedure. The maximum stress in the earlier study was 3.7 ksi (26 MPa).

The low level of residual stresses measured in PWHT A533 weldments is similarly confirmed from a stress relaxation test of the nozzle cut out material at 1150°F, Figure 27; the strength after 6 1/2 hours dropped to 7.3 ksi (50 MPa).

4. Reduction of Residual Stresses by Peening of a Section XI non-PWHT Repair Weld.

A test program was initiated to determine the effect of peening on an ASME Section XI repair weld. Section IWB-4421, "Procedure Number 4 Welding of Low Alloy Steels," permit a "half bead technique" (temper pass technique) for repair of components where PWHT is not required. The essential features are:

1. Pre-repair procedure qualification using a mockup that simulates the cavity to be filled and any obstructions that limit access.

2. Use of small-diameter electrodes, 5/32 inch (3.97 mm) diameter maximum, to insure good weld-metal toughness.

3. Use of the "half-bead" buttering technique (3/32 inch and 1/8 inch [2.38 and 3.18 mm] electrodes) to enhance the toughness of base-metal heat-affected zones.

4. Maintaining of preheat, and post-weld heat treatment between 450 and 550°F (230 and 290°C) to facilitate diffusion of hydrogen.

5. Mechanical tests of material from the qualification mockup to demonstrate acceptable tensile properties, notch toughness and soundness.

6. Controlled peening, to minimize distortion and residual stress, is permitted.

The sequence for the half bead repair is shown schematically in Figure 28. The effectiveness of this half bead technique has been demonstrated in a large scale test[14,15,16] and in a vessel test program under the Heavy Section Steel Technology Program.

The details of the peening procedure are not detailed in Section XI; guidelines in Section XI include the following:

1. No peening of initial or final layers.

2. Cleaned weld beads shall be given a light peen with a blunt-nose tool, about 1/2 inch (12 mm) minimum radius, with a medium size air-operated tool capable of delivering a 9 ft-lb (12 Joule) blow at 90 psi (0.62 MPa).

3. The temperature of the weld bead during peening shall be approximately the same as the interpass temperature, 350 - 500°F (175 - 260°C).

Reference to "controlled" or "light" peening recognizes that cold work can reduce notch toughness and that extensive cold work can initiate cracks. The plastic deformation by peening must occur in a depth greater than the depth remelted by subsequent weld beads. Calamare et. al.[17] and Pellini and Eachbacher[18] did provide an engineering base for the art of peening and was considered in this program. The peening hammer used in this program was an Atlas-Capco Model RRC12 (Serial 250134) operated at 80 psi (0.55 MPa) with a 1/2 inch (12.7 mm) radius peening tip.

The time required to obtain significant elongation of weld metal was established by uniform peening of weld beads deposited on one face of thin hot-rolled steel specimens. The specimens, Figure 29, were 1/8 inch thick x 1 inch wide x 7 inches long (0.33 x 2.5 x 17 cm) with a central 6 1/4 inch (16 cm) long weld. The welds were peened, for times ranging from one-half to ten minutes, on a massive steel backup. Bow measurements in the as-deposited and the peened conditions are summarized in Table 2. Figure 30 shows the percent reduction in a bow as a function of peening rate. A bow reduction of 40% or more was obtained by peening from 30 to 60 seconds per inch of length (12-20 seconds/cm). Therefore, a rate of 40 seconds/inch (16 sec./cm) was selected for the peened weldment. All surfaces were examined for flaws at 30X. Small cracks were found in the specimen peened about 100 seconds per inch (39 sec./cm).

The half bead repair specimens were 10 inches wide by 1.8 inches long x 8 1/4 inches thick (25 x 46 x 21.6 cm) plates made from SA533 Grade B Class 1 material. The grooved plate and its dimensions are shown in Figure 31 and 32. The specimen after the initial butter layer is shown in Figure 33, the removal of 1/2 of the first layer in Figure 34, and the completed weldment in Figure 35.

The strain gauge may be noted in Figure 33, 34, and 35; the actual locations are shown in Figure 36. The hole drilling technique described under consideration 2 was used to determine residual stresses. The comparison of the residual stresses with and without peening are shown in Figure 37. The comparison of the mean values show that the stress of primary interest, σ_x at the sides and ends of the repair is about 35% lower in the peened weldment. Lower stresses may be possible since the final layer of weld metal was not peened. These stresses could be reduced further by depositing an additional layer, therefore, leaving the surface as the next to last weld layer or by peening the final layer and removing it, thus meeting the intent of the Code.

CONCLUSIONS AND RECOMMENDATIONS

These four considerations of residual stresses in weldments indicate an ongoing interest in this important engineering factor in welding. The prime current interest is in consideration of non-post weld heat treated welds. The PVRC has an active Task Group on the Effect of PWHT of Welds under the Subcommittee on Thermal and Mechanical Effects. Several important programs under NRC, ORNL (HSST program), and EPRI as well as programs I am sure are underway throughout the world will provide engineering data with which residual stresses will be included in the evaluation of welds. These programs whould be encouraged and supported.

REFERENCES

1. Neal, B.G., "Plastic Methods of Structural Analysis," 2nd Edition 1965, Wiley and Sons.

2. Bryan, R.H., et. al., "Test of 6-inch Thick Pressure Vessels Series 3: Intermediate Test Vessel V-7A Under Sustained Loading," ORNL/NUREG-9 dated February 1978.

3. Merkle, J.G., et. al., "Test of 6-inch Thick Pressure Vessels Series 3: Intermediate Test Vessel V-7," ORNL/NUREG-1, August 1976.

4. Rybicki, E.F., "Residual Stresses at Weld Repairs in Pressure Vessels," NUREG/CR40037, dated March 1978.

5. EPRI Program RP-1974-L, "Effects of Weld Parameters on Residual Stresses in BWR Piping System--Contractor Battelle Columbus Laboratories," (R. Smith) EPRI Journal April 1978, page 37.

6. Vinckier, A.G. and Pense, A.W., "A Review of Underclad Cracking in Pressure Vessel Components," WRC Bulletin 197, August 1974.

7. Mager, T.R., Landerman, E., and Kubit, C.J., "Reactor Vessels Weld Cladding--Base Metal Intersections," Westinghouse Electric Corporation Nuclear Energy Systems Report, WCAP 7733, July 1971.

8. Agnes, D.J. and Siddall, W.F., "Finite Element Analysis of Structure Integrity of a Reactor Vessel During Emergency Core Cooling," ASME Petroleum Mechanical Engineering Pressure Vessel and Piping Conference, Denver, CO, September 13 - 17, 1970.

9. Balladome, G., and Dorner, H., "Safety Evaluation of Undercladding Cracks," Heavy Section Steel Technology Program, 6th Annual Information Routing, April 25 - 26, 1972 (Paper 1109).

10. Bush, A.J. and Kramer, F.J., "Simplification of the Hole-Drilling Method of Residual Stress Measurement," ISA Transactions 12, 1973.

11. Kume, R., Okabayashi, H., and Amano, M., "Mechanism of Underclad Cracking--Combined Effects of Residual Strains and Heat Affected Zone Ductility," Transactions of the ASME, October 1976.

12. Grotke, G.E., Bush, A.J., and Kramer, F.J., "Residual Welding Stresses in an 11-Inch Thick Pressure Vessel," Westinghouse Research Laboratories 73-1D4-WELST-R2, July 1973.

13. Ferrill, D.A., Juhl, P.B., and Miller, D.R., "Measurement of Residual Stresses in a Heavy Weldment," Welding Journal Res. Supp., November 1966.

14. Wismer, S.W. and Holz, P.H., "Half Bead (Temper) Repair Welding for Heavy Section Steel Technology Program," NUREG/CR0113, ORNL/NUREG/TM-177, July 1978.

15. Goins, W.D. and Butler, W.L., "Weld Repair of Heavy Section Steel Technology Program Vessel V-7," EPRI NP-179 Project 604-1, Final Report, August 1976.

16. Holz, P.P., "Half-Bead Weld Repairs for In-Service Applications," ASME 78 PVP-10.

17. Calamari, P.L., Crum, F.J., and Place, G.W., "An Investigation of Peening," Welding Journal, Research Suppl., 387S-402S, August 1973.

18. Pellini, W.S. and Eachbacher, E.W., "Effects of Peening Last Pass of Welds," Welding Journal, Research Suppl., 71S-76S, February 1954.

TABLE I

CLAD SPECIMEN IDENTIFICATION

Spec.	Steel	Cladding	Process	Condition
1	A508	Experimental	6-Wire	AW
2	A533	Experimental	6-Wire	AW
3	A508	Experimental	6-Wire	SR
4	A508	Commercial	6-Wire	SR
5	A508	Experimental	Strip	AW
6	A508	Commercial	Strip	SR
7	A508	Commercial	Strip	SR
8	A533	Commercial	3-Wire	SR

NOTES: SA-508 Class 2 and SA-533 Grade B, Class 1 Steels

Stress relieved at 1150°F (620°C)

Strip cladding 60 mm (2.4 in.) wide

TABLE II

PEENING EVALUATION

Specimen	Peening Time, min.	Bow, in. Original	Bow, in. Final	Bow Reduction, %	Peening Rate Sec./in.
1	0.50	0.69	0.56	19	5
2	0.92	0.50	0.44	12	9
3	2.0	0.63	0.44	30	19
4	3.0	0.63	0.38	40	29
7	4.0	0.56	0.25	55	38
11	4.0	0.50	0.25	50	38
5	5.0	0.50	0.19	62	48
10	5.0	0.75	0.44	41	48
8	6.0	0.38	0.06	84	58
9	6.0	0.63	0.38	40	58
6	10.0	0.69	0.25	64	96

NOTES:

1. 1/8" x 1" x 7" hot-rolled steel specimen with a centrally located 6 1/4" long weld bead (5/32" dia. electrode).
2. Hammer radius 1/2".
3. 1 in. = 25.4 mm.

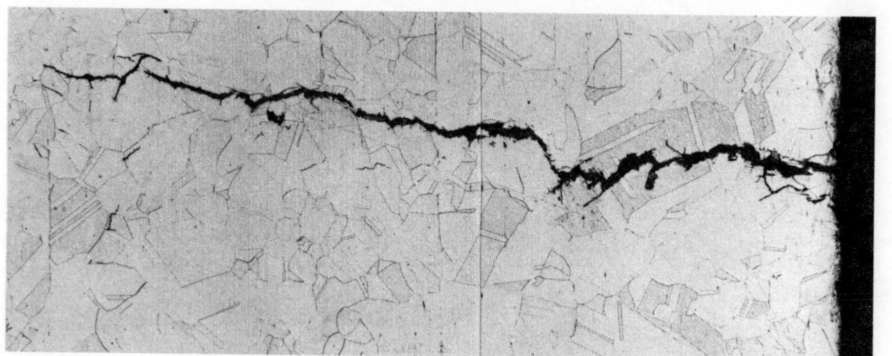

Figure 1 - Transgranular Cracking in Type 304 Stainless Steel Piping (75X).

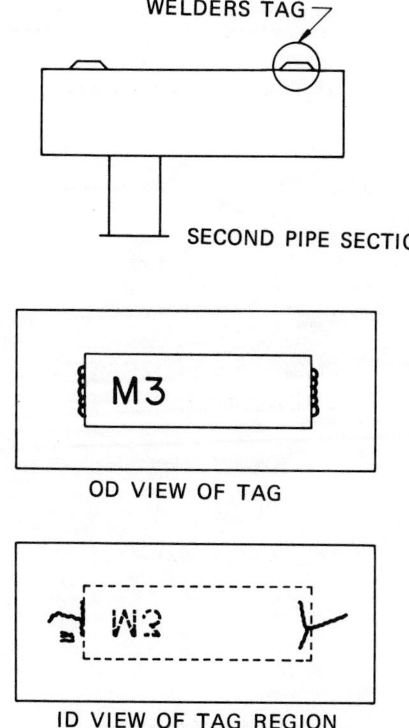

Figure 2 - Location of Cracks Associated with Fillet Welding of Welders' Tags.

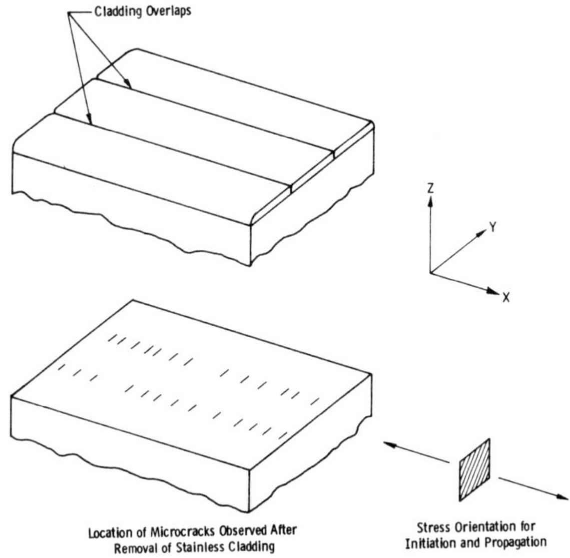

Figure 3 - Location of Microcracks Observed After Removal of Stainless Cladding.

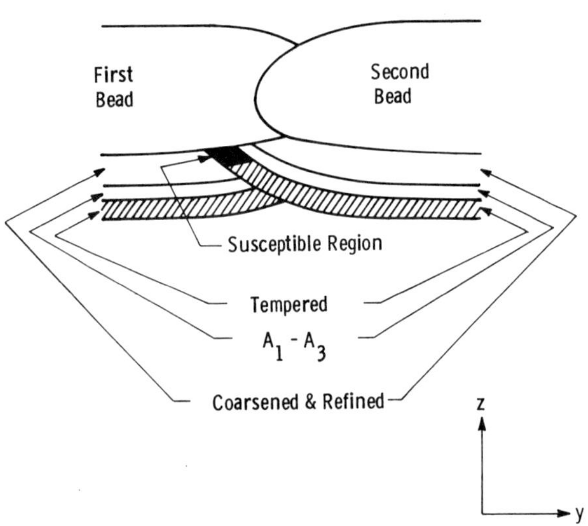

Figure 4 - Location of the Crack Susceptible Microstructure.

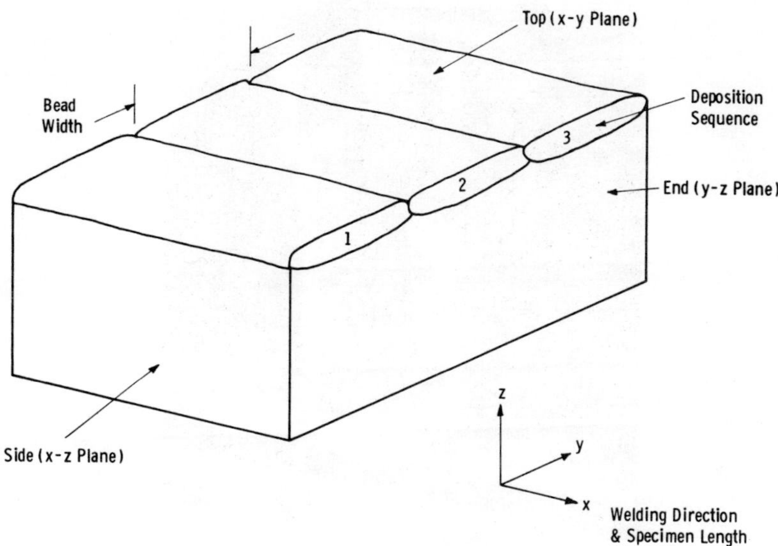

Figure 5 - Identification of Planes and Stress Direction.

Figure 6 - Three-Gage Rosette Used to Determine Residual Stress.

Figure 7 - Equipment for Abrasive Jet Machining of Hole (Center of Three-Gage Rosette).

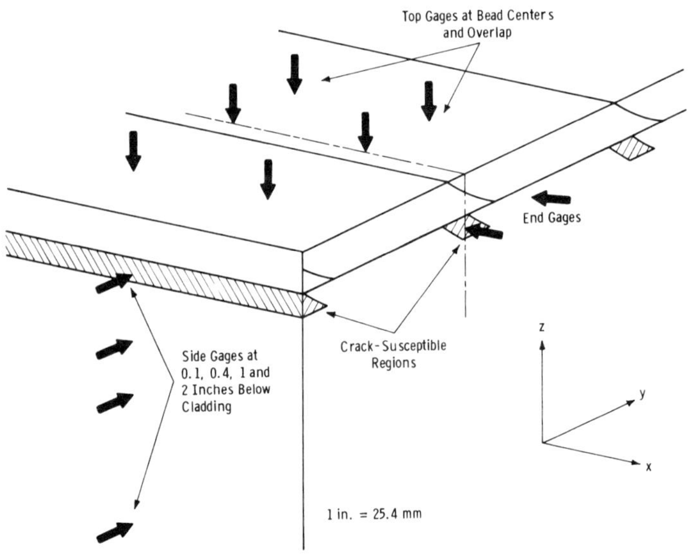

Figure 8 - Strain Gage Locations.

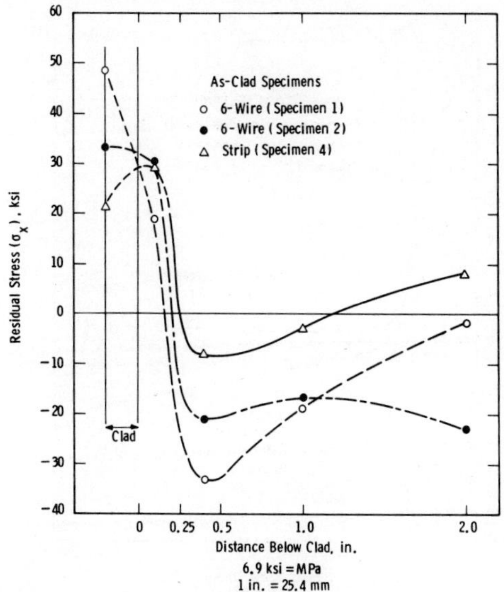

Figure 9 - Distribution of X Direction Underclad Stress in the X-Z Plane -- As Clad Specimens.

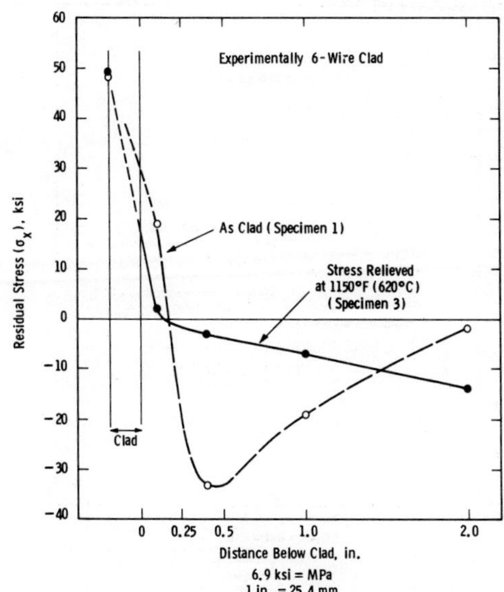

Figure 10 - Distribution of X Direction Underclad Stress in the X-Z Plane -- Effect of Stress Relief.

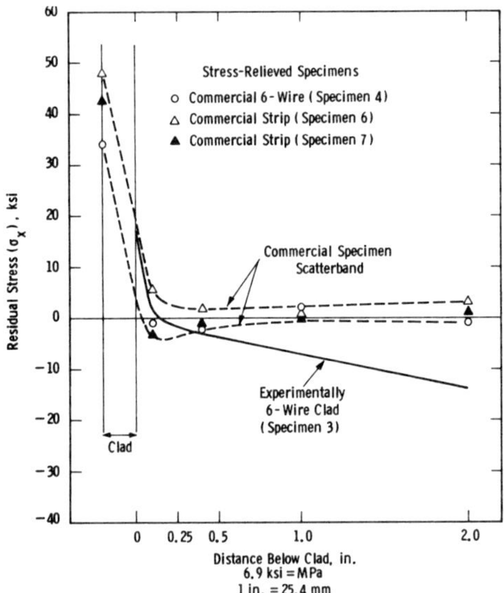

Figure 11 - Distribution of X Direction Underclad Stress in the X-Z Plane -- Stress-Relieved Specimens.

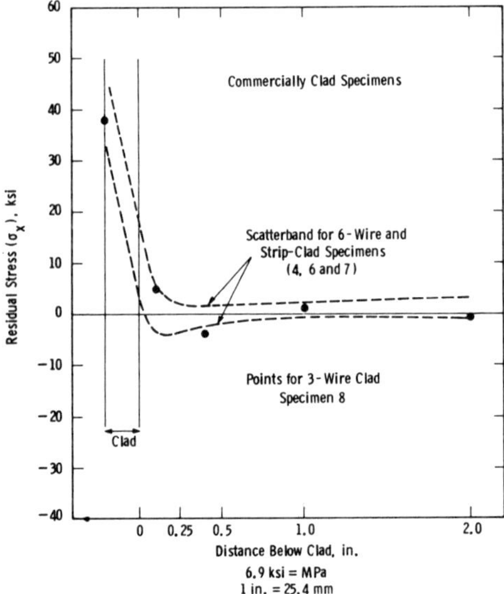

Figure 12 - Distribution of X Direction Underclad Stress in the X-Z Plane -- Commercially Clad Specimens.

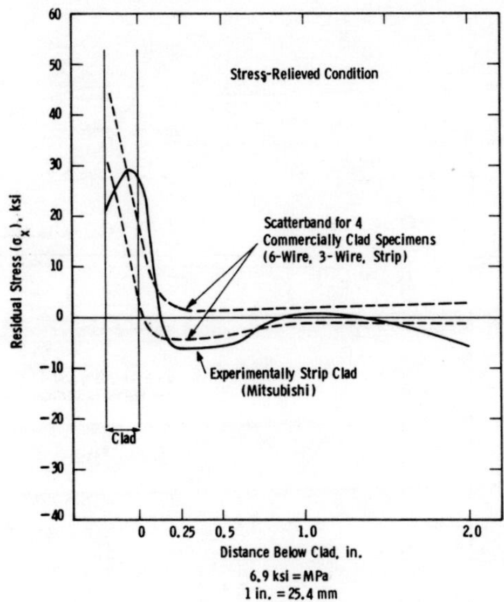

Figure 13 - Distribution of X Direction Underclad Stress in the X-Z Plane -- Comparison of Westinghouse and Mitsubishi Data.

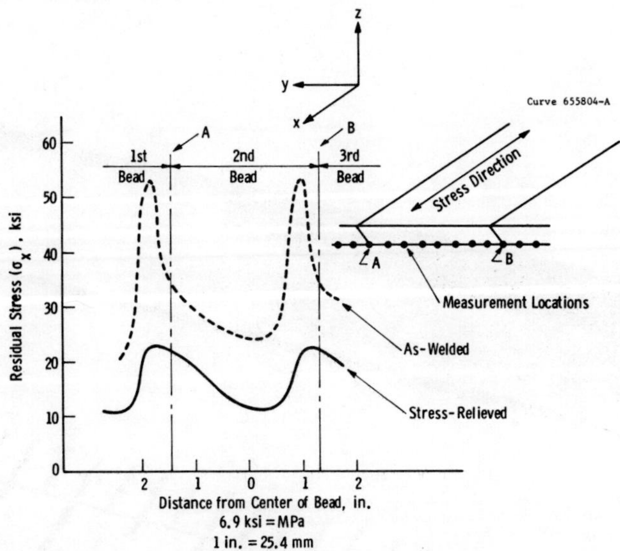

Figure 14 - Distribution of X Direction Underclad Stress in the Y-Z Plane.

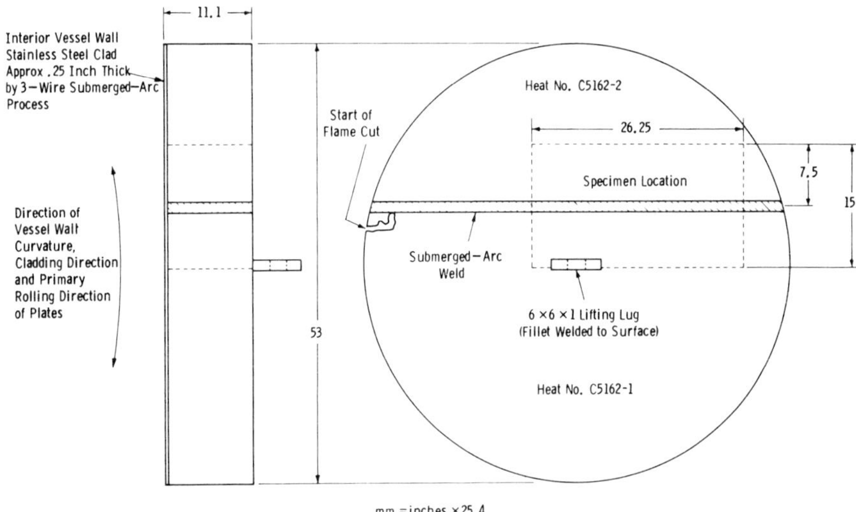

Figure 15 - Schematic of the Nozzle Cut-Out Showing Location of the Residual-Stress Specimen.

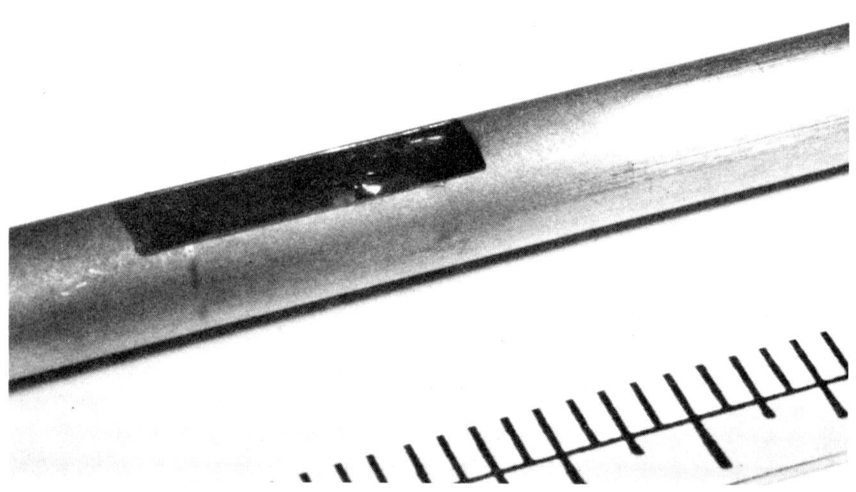

Figure 16 - Strain-Gage Installation on the Inner Wall of a 0.250-inch (6.35 mm)-Diameter Tube.

Figure 17 - End View of the Weldment After Drilling and Mounting of Terminal Strips (REF is Reference Corner).

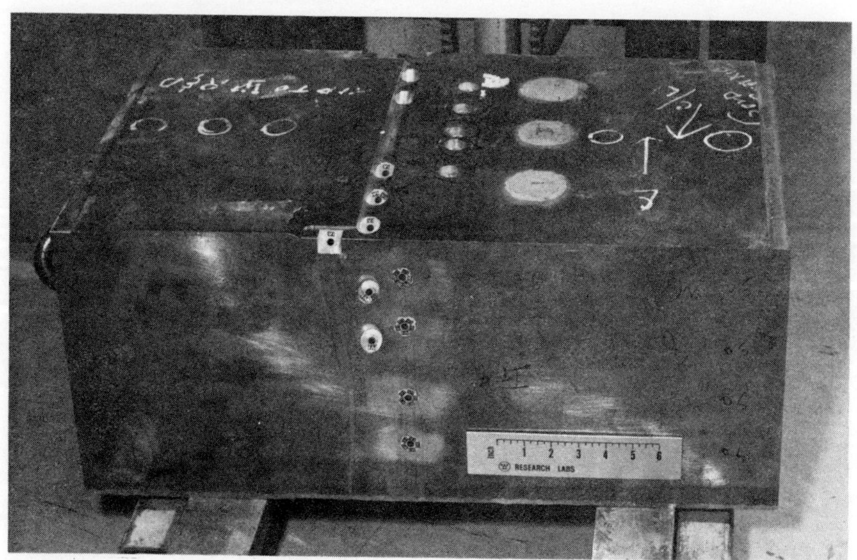

Figure 18 - Side View of Weldment.

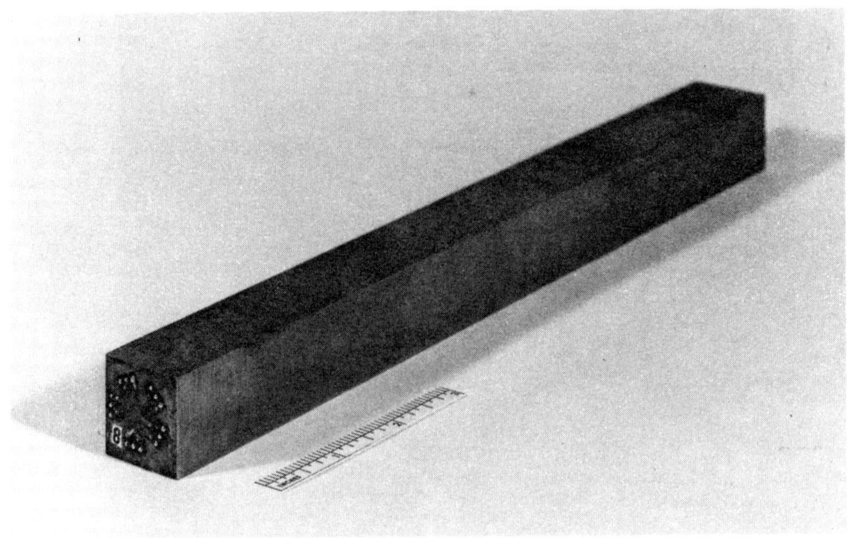

Figure 19 - Typical Parallelepiped Sawed From the Weldment.

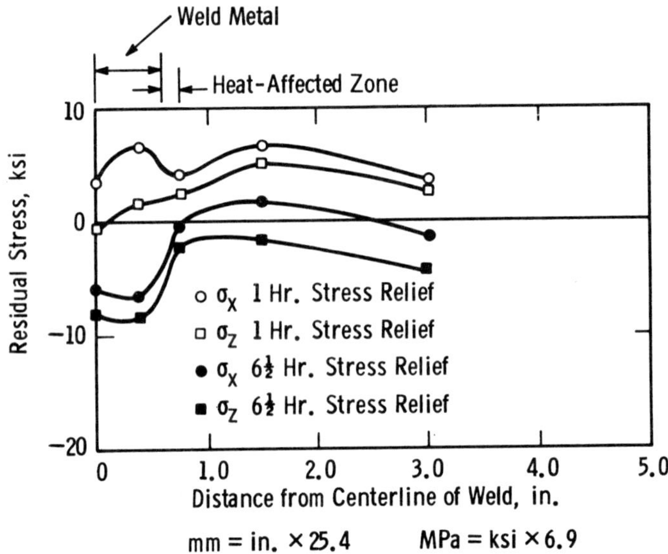

Figure 20 - Coordinate Residual Stresses on the Nozzle Cut-Out Surface Prior To and After Final Stress Relief.

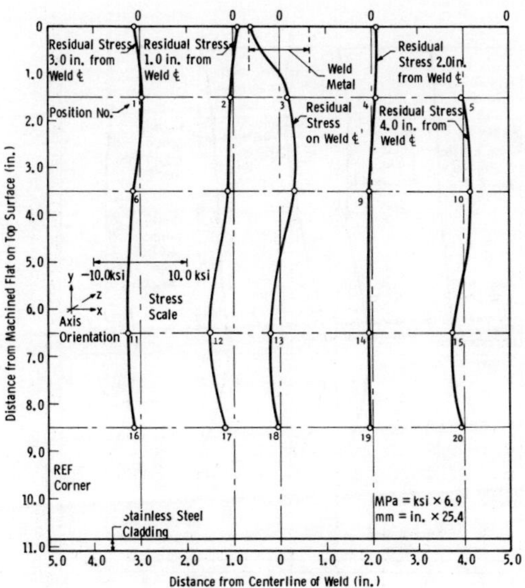

Figure 21 - σ_x Residual Stresses (Transverse to the Weld Length) Plotted with Reference to the y Axis.

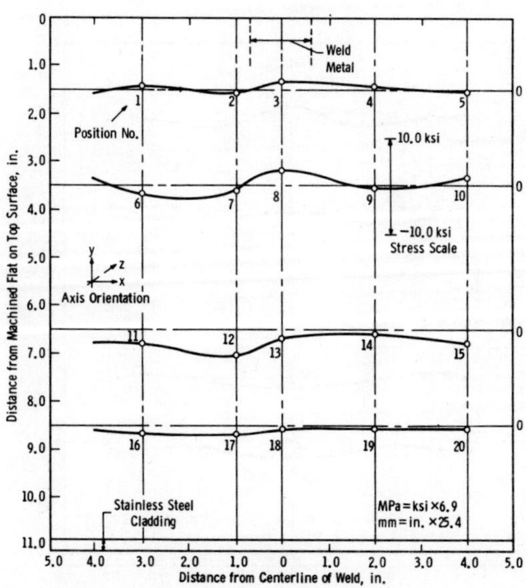

Figure 22 - σ_x Residual Stresses (Transverse to the Weld Length) Plotted with Reference to the X Axis.

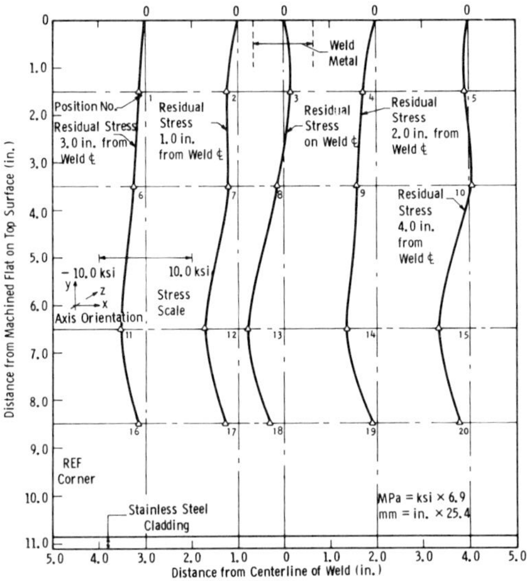

Figure 23 - σ_y Residual Stresses (Through Thickness) Plotted With Reference to the y Axis.

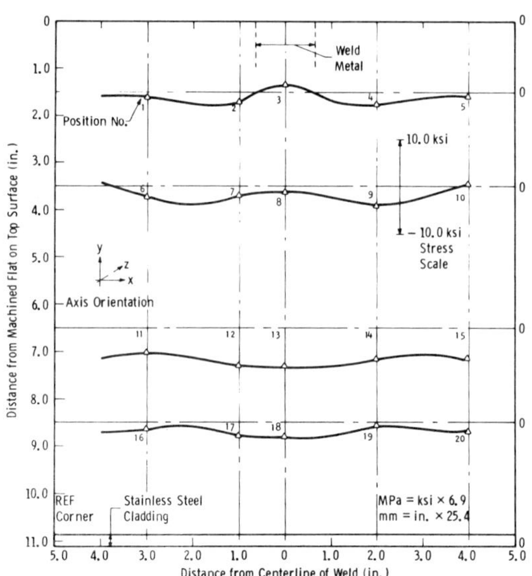

Figure 24 - σ_y Residual Stresses (Through Thickness) Plotted With Reference to the X Axis.

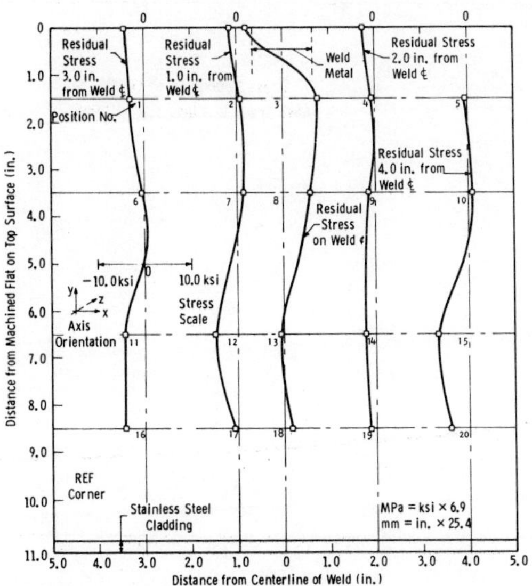

Figure 25 - σ_z Residual Stresses (Parallel to the Weld Length) Plotted With Reference to the y Axis.

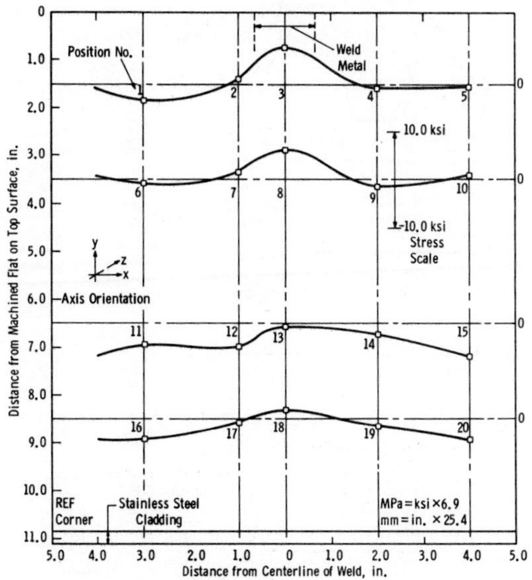

Figure 26 - σ_z Residual Stresses (Parallel to the Weld Length) Plotted With Reference to the X Axis.

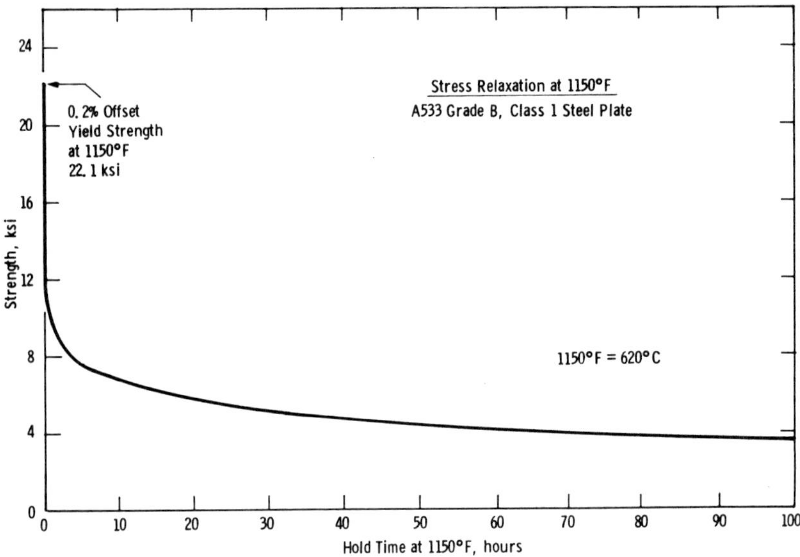

Figure 27 - Stress-Relation Behavior of SA 533 Class 1 Steel Tested at 1150°F (620°C).

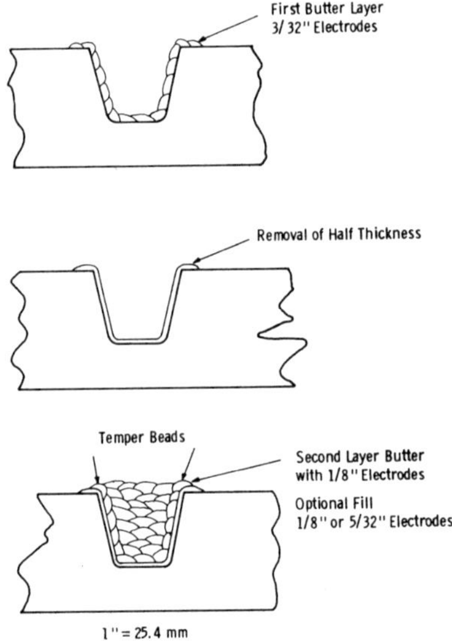

Figure 28 - Sequence for Half Bead Repairs (Schematic).

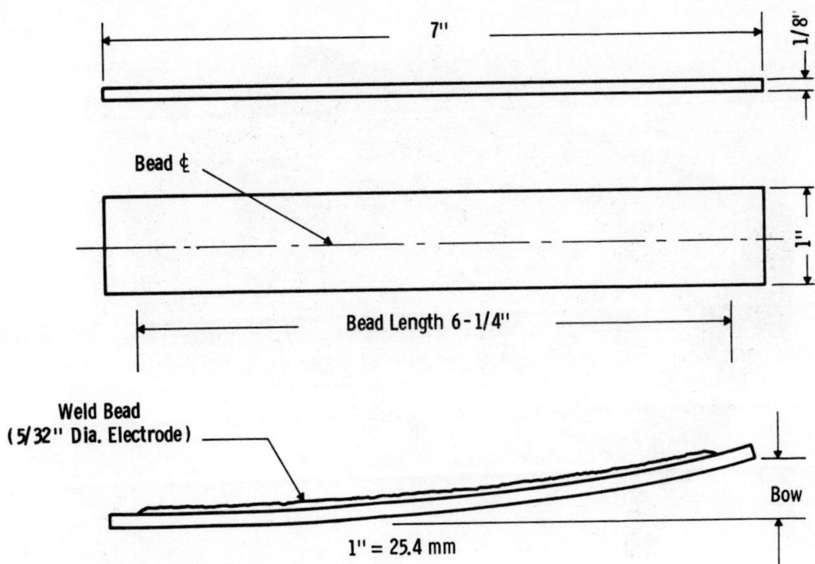

Figure 29 - Specimens Used for Peening Evaluation.

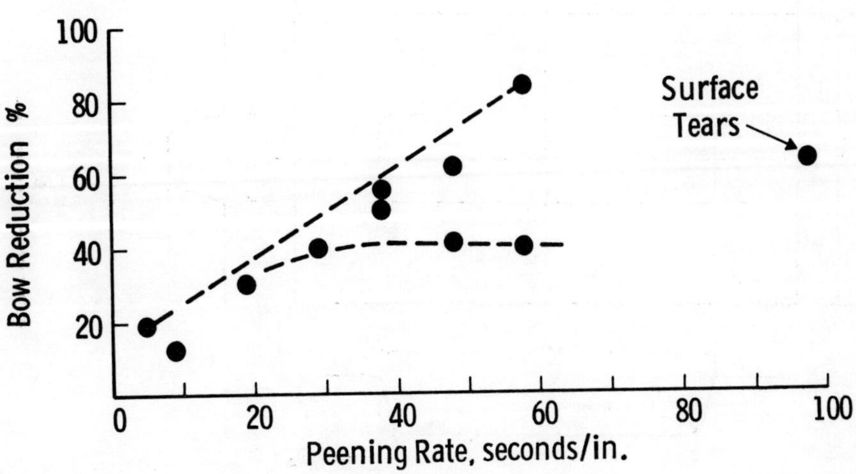

Figure 30 - Effect of Peening on Bow Reductions.

Figure 31 - Grooved Specimen for Simulated Repair.

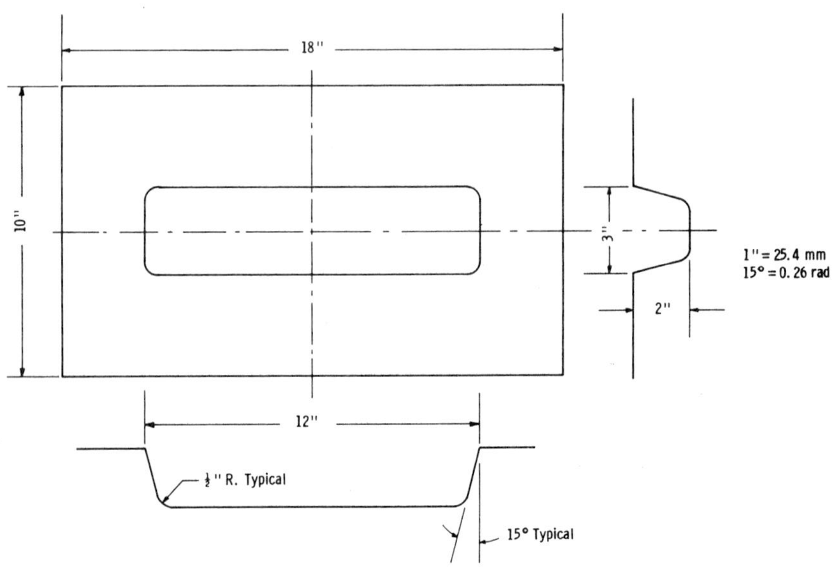

Figure 32 - Groove Dimensions.

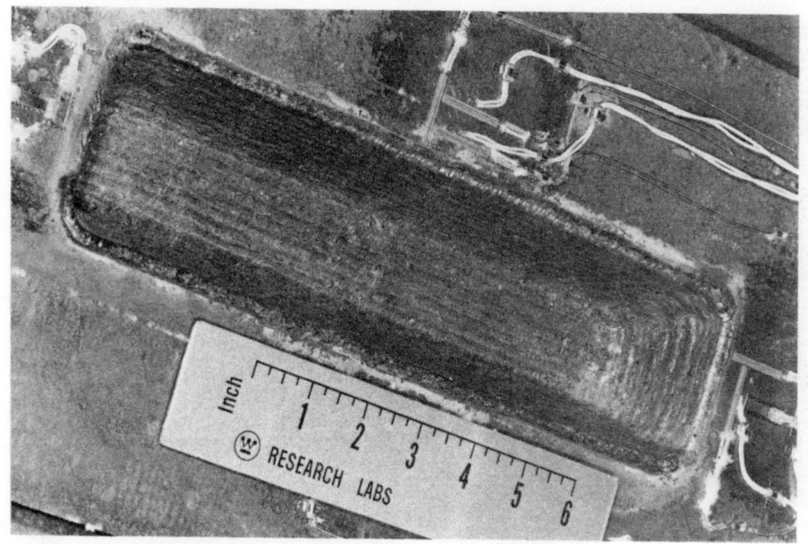

Figure 33 - Appearance of the Initial Buttering Layer.

Figure 34 - Initial Layer After "Half-Bead" Grinding.

Figure 35 - A Completed Repair

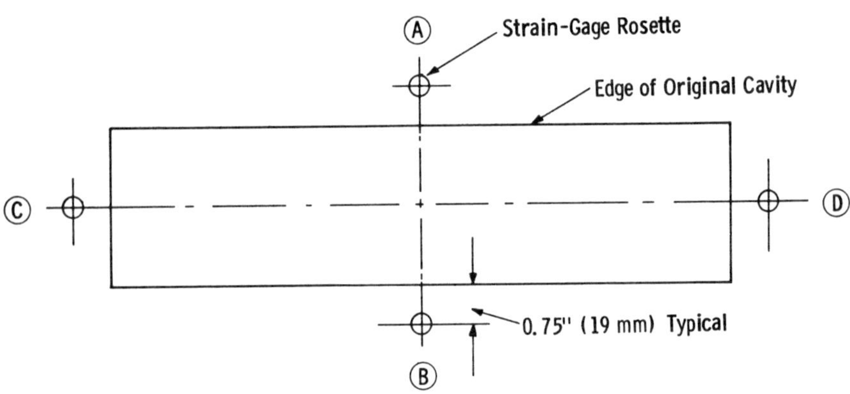

Figure 36 - Strain-Gage Locations for Residual Stress Measurement.

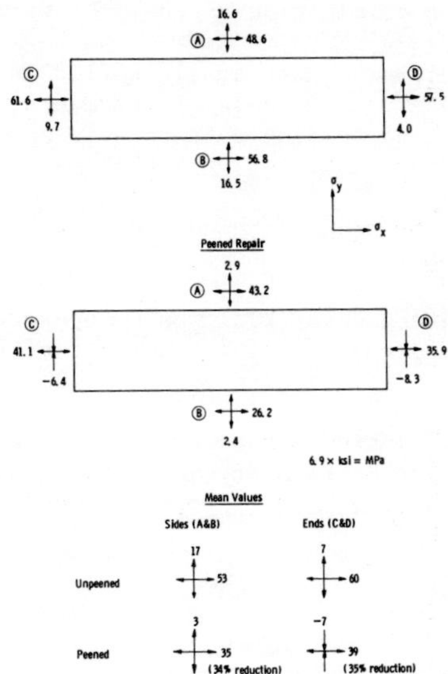

Figure 37 - Comparison of Residual Stresses With and Without Peening.

DISCUSSION

M. C. Coleman (CEGB): In your work on the ASME Section XI half bead repair technique, have you ever experienced cracking problems either in the as-welded repairs or the peened repairs?

Author: The data available on the limited number of actual repairs and weld repair mockups did not reveal cracking. We do not know of any peened ASME Section XI repair welds made at the present. We did mention the possibility of cracking in "over-peened" welds based on our tests.

L. M. Petrick (Ebasco Services, Inc.): Has there been any attempt to control the peening process energy, by controlled air pressure, as one method?

(Not asked, but pertinent to above)
Would it be possible to achieve the desired energy control in peening by employing a special air actuated peening tool? - many different peening units now available perhaps causing much of the confusions.

Author: We do not know of any controlled air pressure techniques. There have been a few attempts in the past on controlling variables for peening (Welding Journals, 1953 and 1954). It would seem that automation would lend itself to the controls you are suggesting. Alternatively, one is left with the weld mockup required by ASME Section XI.

Kenneth MacKay (Dominion Engineering Works): With respect to your problem on stainless steel overlay where you encountered reheat cracking, can you tell me what practical steps you took to overcome the problem?

Author: The reheat cracking was restricted to A508 Class 2 material with high heat input cladding processes. Reduction in heat input avoided the problem. Two-cladding layer techniques (where the base material heat affected zone was refined) also avoided the problem.

Questioner; Were there welding defects associated with the use of the ½ bead technique and the elimination of the post weld heat treatment (PWHT)?

Author: A special weld procedure qualification is required for weld repair where PWHT is not used. In the several weldments that have been evaluated using this technique, no welding defects were observed in the heat affected zone related to this approach.

Questioner: (1) What peening techniques did you use? (2) Have you made the peening process automatic?

Author: (Grotke): The peening tool was air operated using a blunt-nose tool with a ½ inch (1.2 cm) minimum radius. Details of the technique are included in the paper.

Author: (Landerman): There is no single set of peening parameters described in the ASME Code Section XI; a special qualification of the peening procedure would be required. This procedure would then be part of the repair weld qualification.

RESIDUAL STRESS MEASUREMENTS

IN WELDED ALUMINIUM ALLOYS

Peter Furrer and Milan Koutny

Swiss Aluminium Ltd.
Research and Development
CH-8212 Neuhausen am Rheinfall
Switzerland

The growing interest for the application of welded aluminium structures has put forward an increasing demand for non-destructive residual stress measurements in the heat affected zone of the welds. The present paper describes the procedures for measuring residual stresses by X-ray diffraction methods and shows experimental results obtained with welded sheets and extruded sections.

The residual stress distribution within the heat affected zone depends on the alloy composition and the microstructure of the material before welding as well as on the welding procedure (welding method and parameters) and the resulting microstructural variations due to the heat absorption. The height of the residual stress level may be reduced considerably by an additional heat treatment of the component after welding. The experimental results obtained by X-ray methods compare fairly well with mechanical stress measurements on the same samples.

Introduction

The formation of residual stresses may not be excluded during the fabrication of metallic engineering components under normal circumstances. They result from inhomogeneous plastic deformations and may often not be determined theoretically due to their complicated way of formation. However, the service behavior of a component is strongly influenced by the residual stresses within the material. Therefore it is of major interest to perform stress measurements in the final state of fabrication. Normally two methods are used for determining residual stresses, dissection (based on the relaxation of stress via hole-drilling or material-removing techniques) and X-ray diffraction. However, only the X-ray diffraction method allows local, nondestructive measurements.

Residual stress measurements by X-ray diffraction methods are strictly valid only for elastic, homogeneous and isotropic materials. Generally, polycrystalline metals can be considered to satisfy these requirements. It must also be assumed that the stress normal to the sample surface is zero for the whole of the sample which is examined by X-rays (penetration depth of the radiation normally used is of the order of 0.1 mm in aluminium alloys).

Stress measurements by X-ray diffraction methods are commonly established for steel alloys. Similar measurements for aluminium alloys have been made relatively seldom (1-4) and there are practically no systematic studies available. The present paper describes the experimental methods which have been used for the determination of the residual stresses in welded structures of aluminium sheets and extruded sections and concentrates on the problems imposed by the microstructure in the measured area.

Experimental Methods

If a metal is deformed elastically the interplanar spacings of the crystal lattice change from their stress-free values to new values which depend on the magnitude and direction of the stress. These changes in spacings may be measured by X-ray diffraction techniques and it is possible to calculate the stress in a given direction from the change in diffraction angle using either the bulk elastic constants of the material or experimentally determined elastic constant obtained from X-ray diffraction measurements of the same material under known stress conditions. Extensive descriptions of the basic principles may be found in the literature (5).

Strain $\varepsilon_{\phi,\psi}$ in the direction defined by the angles ϕ and ψ is given by:

$$\varepsilon_{\phi,\psi} = \frac{S_2}{2} \sigma_\phi \sin^2 \psi + S_1 (\sigma_1+\sigma_2) = \frac{d_{\phi,\psi} - d_o}{d_o} \quad (1)$$

where $d_{\phi,\psi}$ is the lattice spacing in the direction defined by ϕ and ψ, d_o is the lattice spacing in the unstressed state, σ_ϕ is the surface stress in the direction defined by ϕ, and σ_1 and σ_2 are principal stresses parallel to the specimen surface. The angle ϕ is the angle between the direction of the principal stress σ_1 and the projection of the direction of $\varepsilon_{\phi,\psi}$ onto the specimen surface, the angle ψ is the angle between the surface normal and the direction of $\varepsilon_{\phi,\psi}$. The elastic constants S_1 and $\frac{S_2}{2}$ are given by $S_1 = -\frac{\nu}{E}$ and $\frac{S_2}{2} = \frac{1+\nu}{E}$, where E = Young's modulus and ν = Poisson's ratio.

The lattice spacings are determined by the Debye-Scherrer method. The specimen is irradiated with a collimated monochromatic X-ray beam and the diffraction angle is measured either by film techniques or with a diffractometer using the conventional step-scan technique or a position-sensitive detector. In practice it is only possible to measure within the range $o \leq \psi \lesssim 50^o$ as for $\psi > 50^o$ the penetration depth of the incident beam normal to the specimen surface gets too small due to the absorption of the X-rays in the specimen. Stress measurements by film techniques are usually made with a flat back-reflection camera. Various methods are generally known: a single exposure with $\psi = 45^o$ 1, two exposures with $\psi = o^o$ and $\psi = 45^o$ or two (or more) exposures with $\psi \neq 0^o$ (6). For the common "two exposure method" used in this work, where the lattice spacings are measured at $\psi = 0^o$ ($d_\perp$) and $\psi = 45^o$ (d_ϕ), the relation:

$$\sigma_\phi = \frac{1}{S_2} \frac{d_\phi - d_\perp}{d} \quad (2)$$

can be derived from Eq. (1). Fig. 1a shows the SIEMENS stress camera. The measurements on aluminium alloys have been made with $Cu_{K\alpha}$ radiation, the X-ray tube operating at 40 kV and 20 mA. The distance between the film and the specimen was about 50 mm, mostly the reflection (511)/(333) of the aluminium lattice was measured using silver crystal powder deposited onto the specimen surface for calibration. The exposure time is about 10 min. for $\psi = 0^o$, about 15 min. for $\psi = 45^o$.

Eq. (1) shows that the strain, $\varepsilon_{\phi,\psi}$, should be a linear function of $\sin^2\psi$ and that any surface stress component may be determined from:

$$\sigma_\phi = \frac{m^*}{\frac{S_2}{2}} \quad , \quad m^* = \frac{1}{d_o} \frac{\delta d_{\phi,\psi}}{\delta \sin^2\psi} \qquad (3)$$

This is the basis for the diffractometer method where the diffraction angles are determined under several tilt angles ψ and the residual stress is obtained from the slope of the resulting line in a plot d_ψ vs $\sin^2\psi$. The great advantage of the diffractometer method is the fact that it can easily be automated. Furthermore, the intensity correction factors for X-ray diffraction measurements such as the Lorentz factor, the polarization factor, the atomic scattering factor, the temperature, and an absorption factor (7) may be included in the calculation. Fig. 1b shows a SIEMENS diffractometer for stress measurements in large ingineering components. The completely computer controlled determination of a single stress component in an aluminum part takes about 20 - 40 min. depending on the number of tilt angles ψ examined. The irradiated area on the specimen has been limited to about 1 mm^2 for both the camera and the diffractometer method to allow the determination of detailed stress distribution curves near the weld (Fig. 2).

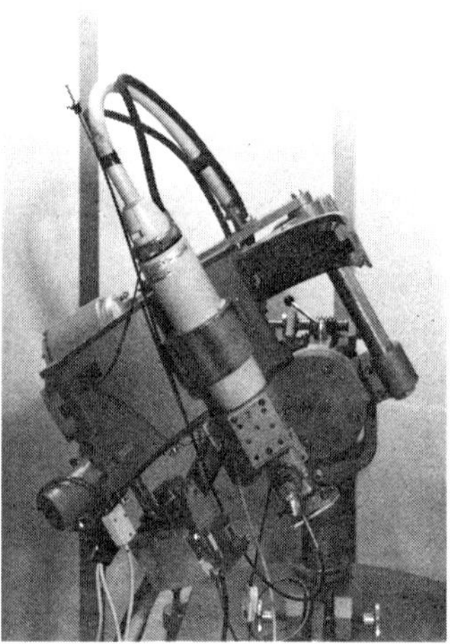

Fig. 1: Equipment for X-ray residual stress measurements
a) Camera
b) Diffractometer

Fig. 2: X-ray residual stress measurements in the heat-affected zone.

Aluminium is elastically a nearly isotropic material ($E_{[111]}/E_{[100]}$ 1.17). Preliminary experiments have shown that it is possible to use within the envisaged error limits mean values of the mechanically determined elastic constants:

$$E \approx 70{,}000\ N/mm^2$$

$$\nu \approx 0.34$$

for X-ray stress measurements in most commercial aluminium alloys. The elastic constants for X-ray measurements are normally determined from uniaxially loaded specimens parallel to the loading direction ($\phi = 0^o$), where the load is well within the elastic range. Such experiments have been made with tensile specimens for commercial AlZnMg and AlMgSi alloys. Table 1 shows the result for an AlZnMg alloy (σ = 100 N/mm^2), where in addition stress measurements have been made also in other directions to the tensile axis.

Table 1: Stress values measured by X-ray methods in a AlZnMg tensile specimen.

ϕ	Film Technique [N/mm^2]	Diffractometer Technique [N/mm^2]	Theoretical Values [N/mm^2]
0°	103	95	100
20°	86	-	94
45°	53	70	71
70°	22	-	34
90°	5	-4	0

Numerous experiments during the last five years have shown that it is possible to determine residual stresses in aluminium alloys with an error limit of $\pm$ 10-20 N/mm^2. These tests have been made either on specimens subjected to known stress conditions or by comparison with the results of subsequent measurements with destructive methods (stress relaxation by hole-drilling). Errors in X-ray diffraction stress measurements arise from a number of sources: elastic constants, intensity correction factors (7), geometry of the measuring arrangement (8), random counting statistics (9), etc. However, the major reason for the relatively large uncertainty in the determination of residual stresses in aluminium alloys is the microstructure involved.

An exact determination of the lattice parameter from the back reflection pattern is hindered, when the ring pattern is not complete, but consists only of several points or when the rings are broad and diffuse so that the lines corresponding to the $K_{\alpha 1}$ and $K_{\alpha 2}$ radiation are not resolvable. Incomplete ring patterns are the results of a large grain size, e.g. when the measured surface volume ($\approx 10^{-1}mm^3$) contains only a few grains, or of an inhomogeneous distribution of the grain orientations (texture effect). Broad, diffuse rings occur when the grain size is extremely small or when the specimen is strongly deformed. In certain cases, the influence of the large grain size or of a strong texture may be overcome by increasing the measured area or by a regular movement of either the X-ray tube or the specimen. However, the resulting loss in spatial resolution is not always acceptable.

The same difficulties arise also when the position of the X-ray reflection is determined with the diffractometer. Large grain sizes and texture effects may result in a very low intensity of the diffracted X-ray beam under certain ψ angles so that the X-ray reflection cannot be differentiated from the background. Broad and asymmetric peaks often do not allow the determination of the exact position of the maxima. This has led to the development of various correction methods either by theoretical calculations or by instrumental means (10).

Furthermore, the expected linear dependence of the interplanar spacings, d, on $\sin^2 \psi$ is not always observed. Large deviations occur especially in strongly deformed, textured materials (11), (12). These oscillations are caused by microstresses and require the application of special procedures to measure stresses with X-rays (11). In the case of aluminium alloys additional consideration must be given to the second phases present in most commercial alloys (3), (12). The following examples show that the microstructure in welded aluminium engineering components is mostly unfavourable for X-ray residual stress measurements.

Examples

a) Welding of extruded sections (alloy 6062)

Extruded flat sections with a thickness of 4 mm have been welded together along the extrusion direction (Fig. 3). The extruded alloy (Al 1 Mg 0.6 Si) has been age-hardened (180°C/6 h) before joining, using a inert-gas-shielded arc welding technique with Al5Si as filler metal. The resulting residual stresses in the heat affected zone have been measured on the top surface in two directions normal and parallel to the welding direction.

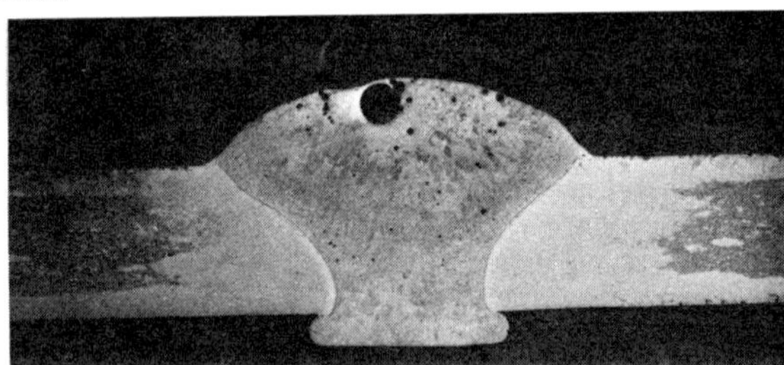

Fig. 3: Cross section through the non-machined weld joining two extrude section (enlargement 4.5 x)

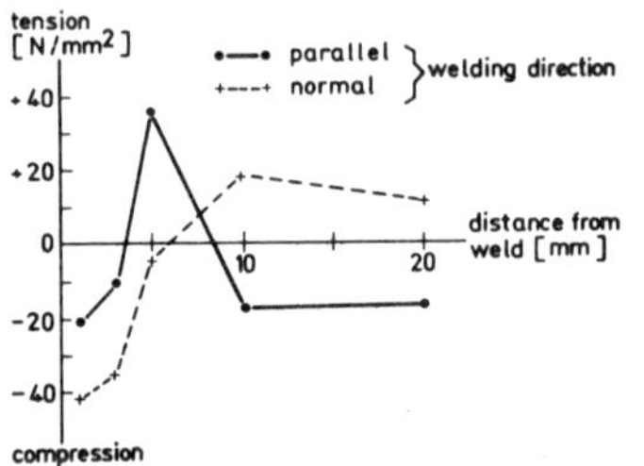

Fig. 4: Variation of the residual stresses measured in the heat-affected zone of two welded extruded sections (alloy 6062)

Residual stress measurements have been made on three similar samples, but only minor differences have been obtained in the measured stress distribution. Fig. 4 shows that the residual stresses are relatively small. Within the transition zone between the weld metal and the original microstructure of the extruded section, which shows a very coarse grain structure, negative stress values have been measured (compression). The highest stress level (in tension) is found in those regions where the heat input was just not sufficient to induce the recrystallization of the extruded structure.

Fig. 5: Grain structure of an extruded section alloy (6062) at the surface (cross section, enlargement 40 x)

A relatively coarse grain size exists not only in the heat affected zone near the weld, but also in the surface region of the original extruded section of this material (Fig. 5). Therefore, residual stress measurements have been found to be extremely difficult and time consuming, because automated measurements with the diffractometer were impossible. Usually the weld is finished by grinding the cover side in the welded area. This treatment generally reduces the residual stresses considerably and may even lead to relatively large negative values (compression) in most parts of the heat affected zone.

b) Welding of rolled plates (alloy 7020)

In this experiment, two cold rolled plates of an Al-Zn-Mg alloy (thickness 6 mm) have been welded together either parallel or normal to the rolling direction. An inert-gas-shielded arc welding technique has been used with Al5Mg as added metal. Light microscopic observations reveal only a thin, fine-grained recrystallized transition zone adjacent to the weld.

Preliminary measurements in the heat affected zone have shown that the highest stress values are found on the top side for the directions normal to the rolling direction. The stress values measured in the direction parallel to the rolling direction reach only about 50% of these values. It is interesting to note that the maximum residual stress level in the heat affected zone is considerably higher when the sheets are welded parallel to the rolling direction (Fig. 6a) instead of normal to the rolling direction (Fig. 6b). The influence of a subsequent two-step heat treatment is also shown in Fig. 6. The decrease of the residual stress values is much stronger for the welding direction parallel to

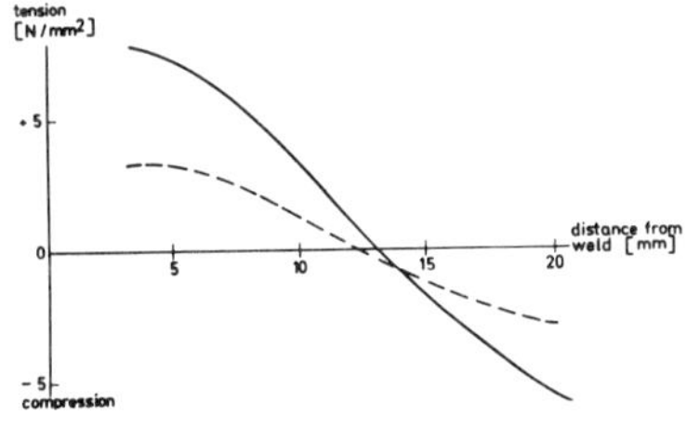

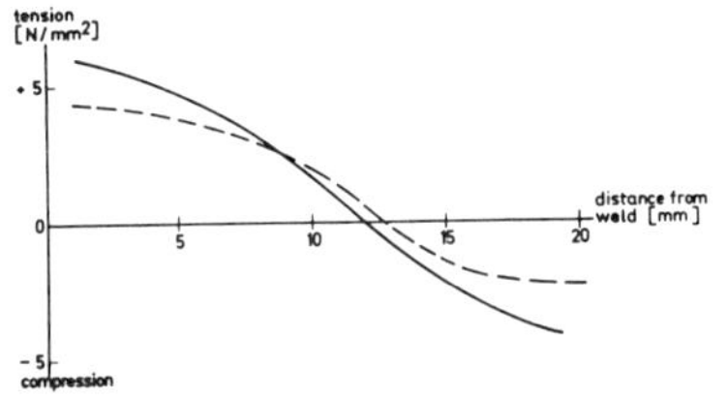

Fig. 6: Variation of the residual stress in the heat affected zone of welded Al-Zn-Mg sheets after welding (-) and after subsequent heat treatment 8 h/95°C + 16 h/145°C (---).

a) welding direction parallel to rolling direction (shown is stress component normal to the weld.)

b) welding direction normal to rolling direction (shown is stress component parallel to the weld).

the rolling direction leading to approximately the same retained residual stresses after heat treating in both cases.

Electron microscopic observations have shown that the strongly deformed microstructure of the cold rolled plate (Fig. 7a) recovers considerably in the areas of the heat affected zone where the highest stress levels exist (Fig. 7b). However, no recrystallization processes take place. The dense, irregular network of dislocations in the cold rolled state has rearranged to a subgrain structure whereas the distribution of the small

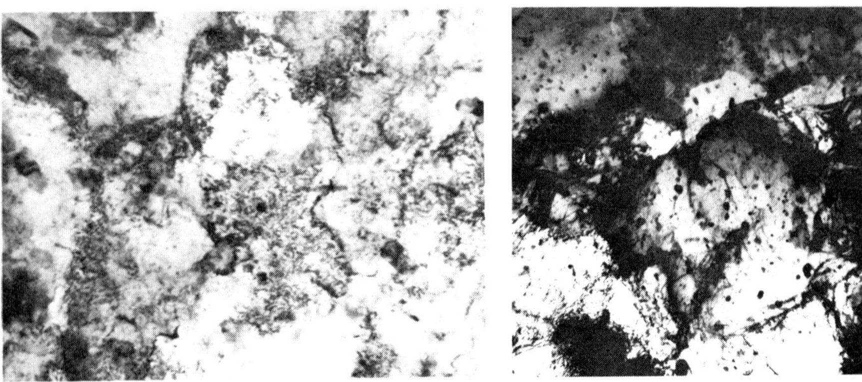

Fig. 7: Electron micrographs of the microstructure in the cold rolled plate (a) and in the heat affected zone near the weld (b) (enlargement 900 x).

precipitates has not changes appreciably. Exact residual stress measurements have been hindered in this example by the strongly distorted microstructure (Fig. 7) as well as by the pronounced texture of the material.

In addition, a large amount of residual stress measurements have been made either on welded test specimens with simple geometric dimensions or on actual engineering components. It is not yet possible to present a systematic description of the residual stresses present in welded aluminium constructions, as all the factors which determine the actual stress distribution (material, welding process and parameter, geometrical configuration, etc.) may vary in a wide range. However, a careful, but unfortunately often time consuming examination with X-ray diffraction methods allows the determination of the stress distribution with a relatively good reliability in most cases.

References

1. D. A. Bolstad and W. E. Quist, Adv. X-Ray Analysis 8 (1965) 26.
2. H. E. Franz and H. Kühnle, Metall 29 (1975) 771.
3. V. Hauk, A. Rinken and H. Sesemann, Z. Metallkde 67 (1976) 92.
4. M. Koutny and M. Pirner, Schweissen und Schneiden 27 (1975) 122.
5. H. P. Klug and L. E. Alexander, X-Ray Diffraction Procedures, John Wiley and Sons, Inc., New York (1954).
6. G. Faninger and W. Wolfstieg, Härterei-Techn. Mitt. 31 (1976) 14.
7. M. A. Short and C. J. Kelly, Adv. X-Ray Analysis 16 (1972) 379.
8. H. Zantopulos and C. F. Jatczak, Adv. X-Ray Analysis 14 (1970) 360.
9. C. J. Kelly and M. A. Short, Adv. X-Ray Analysis 14 (1972) 377.
10. U. Wolfstieg, Härterei-Techn. Mitt. 31 (1976) 23.
11. R. H. Marion and J. B. Cohen, Adv. X-Ray Anal. 18 (1974) 466.
12. V. Hauk and H. Sesemann, Z. Metallkde 67 (1976) 646.

DISCUSSION

J. S. Ahearn (Martin Marietta Labs): (1) To increase the signal to noise ratio in your measurements of residual stress, can you rotate either the sample or the X-ray source? (2) Can you use the X-ray peak broadening to determine mico-residual stresses?

Author: (1) The signal to noise ratio can be increased by the appropriate movement of either the sample or the X-ray source. However, a movement of the specimen is often not possible (large structures) and the possible movements of the X-ray tube are limited by the construction of the measuring equipment. (2) Microstresses (stresses which are homogeneous only over one grain or parts of a grain) are one of the different factors influencing the width of the X-ray peaks. But it is generally not possible to separate the influence of other factors (grain size, dislocation density, segregation effects, etc.) in commercial materials.

G. A. Knorovsky (Sandia Labs): Rather than attempting to compensate for large-grain size in DeBye-Scherrer methods, is it possible to treat individual grains via single-crystal methods to arrive at residual stress measurements?

Author: In principle, it would be possible to measure residual stresses in individual large grains by single-crystal methods. However, the application of this method to practical cases where the mean stress values over a number of grains are to be determined would be extremely time consuming.

R. W. Hertzberg (Lehigh University): How large a difference do you expect to find in elastic properties between various aluminum alloys?

Author: The elastic constants of various Al alloys ranging from pure Al to high-strength AlZnMgCu alloys vary not more than about 10%.

COMPARISON OF STRAIN GAUGE AND X-RAY DIFFRACTION MEASUREMENTS OF RESIDUAL STRESSES IN A WELDMENT OF HY130 STEEL

K.S. Milliken and C.M. Mitchell

Energy, Mines and Resources, Canada.

ABSTRACT

The residual stress distribution over the HAZ and the weld in a circular patch test plate of HY130 steel has been determined by a strain gauge method, using a rim mounted strain gauge and trepanning from the plate centre; and by X-ray diffraction using a dual diffractometer. The connection between the residual stresses and stress corrosion cracking in such plates is demonstrated.

INTRODUCTION

Residual stress measurements have been carried out on circular patch weld restraint specimens of HY130 steel by the strain gauge trepanning method and by X-ray diffraction, to compare the results of the two procedures. The trepanning method, in which rim mounted strain gauges record the change in strain as the radius of a central hole is increased, measures the strain distribution over the bulk of the section. In the X-ray diffraction method, the surface stress is determined from measurements of lattice strain in two directions in the material, and the diffraction occurs in a thin surface layer. The results of the two methods should be equivalent if the stress distribution is uniform throughout the section.

Stress corrosion cracking tests on similar circular patch weld restraint specimens have been carried out by J. Gilmour of CANMET Laboratories[1]. These tests show substantially different corrosion cracking behaviour on opposite faces of the plates, which could be related to non-uniform stress distribution.

The Trepanning Method

The stress measurements were carried out on a circular test weld in a circular plate, a "circular patch restraint specimen", which is shown schematically in Fig. 1. Strain gauges are mounted on the plate rim at opposite ends of a diameter. The centre of the plate is bored out progressively and the strain gauge readings are recorded at each successive radius.

The residual stress components were determined by the method of Adams and Corrigan[2]. This method assumes that a state of plane stress prevails, and that the stress has rotational symmetry. Radial and tangential residual stresses are calculated from strain measurements made at a point on the disc periphery.

The radial component, S_r, of the residual stress at a distance r from the disc centre is given by the equation

$$S_r = \frac{E\, e_{t_o}}{2} [r_o^2/r^2 - 1] \tag{1}$$

and the corresponding tangential component, S_t, of the residual stress by the equation

$$S_t = \frac{d}{dr}(r\, S_r) \tag{2}$$

where

r = the radius of the bored hole,

r_o = external radius of the circular plate,

e_{t_o} = the change in tangential strain, measured by the strain gauge at the plate periphery, and

E = Young's modulus.

The radial stress, S_r, is obtained directly by substituting the strain gauge readings, e_{t_o}, into equation 1. In order to obtain the tangential stress, S_t, from equation 2 it is necessary to differentiate the tabulated function S_r. This has been carried out by fitting a smoothed cubic spline to the e_{t_o} values.

A cubic spline is a set of N cubic polynomials each having their own range, which taken together form a continuous function, with continuous first and second derivatives, passing through the points. A smoothed cubic spline replaces the condition that the spline passes through the points, by two conditions, which are

$$\sum_i ((e^*_{t_o}(r_i) - e_{t_o}(r_i))/\sigma_e)^2 \leq T \tag{3}$$

and

$$\int_{r_1}^{r_N} \left(\frac{d^2 e^*_{t_o}(r)}{dr^2} \right)^2 dr = \text{minimum}$$

where

e_{t_o} = the observed change is tangential strain,

$e^*_{t_o}$ = the smoothed cubic spline,

r_i = the radii of the bored holes,

σ_e = estimated standard deviation of the strain gauge readings, and

N = the number of points to be fitted.

The natural values of T lie within the confidence interval corresponding to the left hand side of (3):

$$N-\sqrt{2N} < T < N + \sqrt{2N} .$$

Smoothing was carried out using a FORTRAN subroutine[3] based on the treatment of Reinsch [4] .

Trepanning Experiment and Results

The HY130 steel used in this work was electric furnace air melt vaccum degassed (AMVD). The specifications for the alloy are given in the U.S. Department of Defence Standard MIL-S-24371A (Ships) 21 Aug. 1975. The chemical analyses are given in Table 1 along with the analyses of the welding wire which contained 1.4 ppm hydrogen.

Table 1. Chemical analysis, wt%

Element	AMVD (12 mm)	Welding Wire (1.6 mm)
C	0.10	0.09
Ni	4.90	2.23
Cr	0.54	1.00
Mo	0.49	0.67
Mn	0.70	1.74
P	0.005	0.010
S	0.007	0.010
Cu	0.05	0.07
Si	0.23	0.39
V	0.08	0.010
Ti	0.01	0.010
Al	-	0.010
Zr	-	0.010

Analysis supplied by manufacturers.

"Circular patch" restraint specimens of 12 mm (0.47 in.) plate of HY130 steel, 299 mm (11.78 in.) in diameter were formed by machining circular grooves 152 mm (5.99 in.) diameter with the cross-section shown in Fig. 2. Specimens were welded in two passes by the automatic gas metal-arc process using a 1.6 mm (1/16 in.) electrode conforming to Type MIL 140S of specification MIL-E-24355. Specimens were mounted on a motor-driven turn table whose rotational speed could be accurately controlled. The weld was made in the grooves using an electric arc shielded by argon-2% O_2, and one pass was made on each face of the plate. The welding parameters were approximately 360 amp, at 24 V, and a travel speed of 394 mm/min. Each pass was deposited when the joint region was at 50 to 65°C.

The strain gauges used were Type EA-06-500BH - 120 and were attached with M bond 200 cement, as shown in Fig. 2. A milling machine was used rather than a lathe so that the specimen would be stationary, and the strain gauge leads could remain connected to the strain measuring instrument throughout the experiment.

The density of strain gauge readings was about one per mm. The estimated standard deviation of the strain readings was 3×10^{-6}.

The resulting radial and tangential stress distributions are shown in Fig. 3. The radial stress is tensile over the weld having a broad maximum near the outer weld edge of 152 MPa (22 ksi) compared with 83 MPa (12 ksi) near the inner weld edge. The tangential stress shows peaks near the inner and outer edges of the weld of 510 MPa (74 ksi) and 550 MPa (80 ksi) respectively, between which the stress falls to 200 MPa (29 ksi) at the weld centre. About 5 mm beyond the outer edge of the weld the stress is compressive at a level -345 MPa (-50 ksi).

The X-ray Diffraction Method

The residual surface stress distribution was measured using a stress diffractometer designed and constructed in the CANMET laboratories[5] for stress measurement on large specimens. The diffractometer uses dual fixed detectors and a scanning X-ray tube so that strain measurements can be made simlutaneously in two directions in the specimen. The principal stress in the equatorial plane of the instrument is determined from the difference in lattice strain in the two specimen directions.

The construction of the instrument is shown in Fig. 4 and schematically in Fig. 5. The diffractometer has three concentric shafts, the outer carrying the vertically mounted X-ray tube driven by a cone drive worm, a middle fixed shaft on which the detectors are mounted, and an inner shaft geared to the drive in a 1:2 ratio, on which the specimen is mounted. The X-ray tube and the detectors are mounted on columns to provide clearance for the specimens.

The angle subtended by the detectors at the specimen is adjusted to give coincident diffraction maxima when the X-ray beam is incident along the bisector.

Si(Li) semi-conductor detectors are used with single channel analysers which act as monochromators.

The back reflection diffraction angles ϕ_A and ϕ_B for the two detectors can be expressed in terms of the interdetector angle β and the scanning angle ω_o of the bisector of this angle.

$$\phi_A = \beta/2 - (\omega_A - \omega_o)$$

$$\phi_B = \beta/2 + (\omega_B - \omega_o)$$

In a strained specimen the strain difference $\delta\varepsilon$ is obtained from the diffraction maxima ϕ_A and ϕ_B using the differential form of the Bragg relation

$$\delta\varepsilon = (\varepsilon_A - \varepsilon_B) = 0.5\ (\phi_A - \phi_B)\ \tan \phi_H/2$$

where ϕ_H is the strain-free supplementary diffraction angle. The parameters ω_o and β are determined using a strain-free standard.

The measurement of the strain difference by simultaneous recording of the diffraction peaks eliminates the diffractometer zero error and, also, where the detector radii are equal, eliminates the eccentricity error.

The stress τ_z normal to the specimen surface is zero and the sample is oriented so that the principal stresses τ_x and τ_y are parallel and perpendicular to the equatorial (xz) plane.

In an isotropic material, using linear stress strain equations, the strain ε_i in a direction lying in the equatorial plane at an angle ψ_i to the surface normal is given by the relation

$$\varepsilon_i = \xi_x(\sin^2\psi_i - \nu\cos^2 \psi_i) - \nu\xi_y$$

where ν = Poisson's ratio,

$\xi_x = \tau_x/E$,

$\xi_y = \tau_y/E$,

and E = Young's modulus. ξ_x and ξ_y are referred to here as partial strains.

The coefficient of the partial strain ξ_x is

$$K_i = \sin^2\psi_i - \nu\cos^2\psi_i.$$

Measurements in the two directions in the equatorial plane, at angles ψ_A and ψ_B to the surface normal, give the principal stress

$$\tau_x = E(\varepsilon_A - \varepsilon_B)/(K_A - K_B).$$

X-ray Stress Measurements

Radial and tangential surface stress measurements were carried out on a circular patch restraint specimen of HY130 steel made by the same procedure as was used for the trepanning specimen. The plate was ground to a fine finish on both faces. The plate was electropolished on each face, along one radius, over an area 3 x 6 cm to remove grinding stresses.

The radial and tangential residual stress distributions were measured on the two faces using the stress diffractometer. Figure 8 shows the circular patch test plate mounted in the instrument. The irradiated specimen area was 1 mm in the equatorial plane by 2 mm normal to the plane. Measurements were made at approximately 1 mm intervals across the heat-affected zone and the weld, using iron Kα radiation.

The resulting distributions of residual stresses in the surfaces of the two faces of the plate are shown in Fig. 6 and 7. All stresses were calculated using linear stress-strain equations. The 0.2% yield stress of the weld material has been reported[6] as 1005 MPa (146 ksi). Three of the calculated stresses exceed this nominal yield stress. Because the present stress is biaxial, it is not certain that these stresses are above the actual yield stress, particularly as strain hardening can occur as a result of plastic flow.

Residual Stresses and Stress Corrosion Cracking

The surface stress distributions are markedly different on opposite faces of the plate for both the radial and tangential directions. There is a correlation between the surface tangential stress distribution and the cracking of the material due to stress corrosion.

Circular patch weld restraint specimens have been shown to have different stress-corrosion cracking behaviour on opposite faces of the plate. On one face, radial cracks occur extending across the weld metal. On the opposite face, radial cracks initiate in the heat-affected zone but do not penetrate the weld. This has been related to the welding sequence. Radial cracking through the weld occurs in the initial weld. This is illustrated in Fig. 9 for a plate in which face 1 was welded from A to B. The plate was reversed and a full weld made on face 2. Then the second half of face 1 was welded from C to D.

The corrosion cracking is consistent with the tangential surface stress distributions observed. On face 1, a very high tensile stress exists across the full width of the weld, which favours radial crack propagation through the weld. On face 2, high tensile stresses exist in the heat-affected zone and the fusion zone; however these are bounded by compressive maxima in the weld interior, which inhibit propagation of radial cracks initiating in the tensile regions at the weld edges.

Summary

The trepanning method shows that the residual stress distribution in the bulk of the material has a slowly varying radial stress distribution and a tensile peak of 152 MPa (22 ksi), and a tangential stress distribution having tensile peaks of 550 MPa (80 ksi) near the edges of the weld. The analysis used in the trepanning method was based on the assumption that the residual stress distribution is two-dimensional. X-ray measurements of surface stress indicate that this assumption does not apply to the present case.

The surface stress measured by X-ray diffraction shows different stress distributions on opposite faces. On one face the radial stress distribution is tensile of the order of 275 MPa (40 ksi) and the tangential distribution shows tensile peaks on 480 MPa (70 ksi) separated by compressive peaks of -410 MPa (-60 ksi) in the weld interior. On the opposite face the radial stress distribution is compressive, having a single maximum of -500 MPa (-72 ksi). The tangential stress shows a broad tensile distribution over the weld with a maximum approaching the ultimate tensile strength of the weld material.

The surface stress distributions observed by X-ray diffraction are in good general agreement with the observed stress corrosion cracking of the circular patch weld restraint specimens.

Acknowledgements

The authors wish to acknowledge the technical assistance of V.R. Chartrand, P. Fryzuk, and K.A. Rocque. The trepanning work was initiated by F.W. Marsh before his retirement. The welding of the plate was supervised by W.P. Campbell. The work was carried out in the Metal Physics Section of PMRL under the general direction of W.N. Roberts.

References

1. J.B. Gilmour, "Environmental Cracking of HY130 Steel". OTC 3192, Offshore Technology Conference, Houston, Texas, May 1978.

2. C.M. Adams Jr. and D.A. Corrigan, "Mechanical and Metallurgical Behaviour of Restrained Welds in Submarine Steels". Welding Lab., MIT. code 634B, May 1966.

3. Subroutine ICSSCU. International Mathematical and Statistical Libraries, Inc., Houston, Texas.

4. C.H. Reinsch, "Smoothing by Spline Functions". Numerische Mathematik, 10 (3) 1967, 177-183.

5. C.M. Mitchell, "A Dual Detector Diffractometer for Measurement of Residual Stress". Advances in X-ray Analysis, vol 20 (1977), Plenum Publishing Corp.

6. D.N. Shackleton, "Welding HY100 and HY130 Steels -- Literature Review". Report Series, Welding Institute, Sept. 1973, 73.

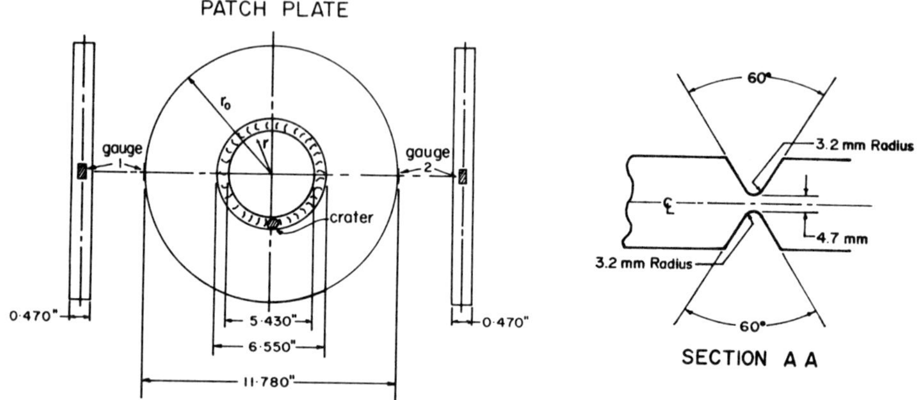

Fig. 1. Circular patch weld restraint specimen.

Fig. 2. Cross-section of specimen preparation for welding.

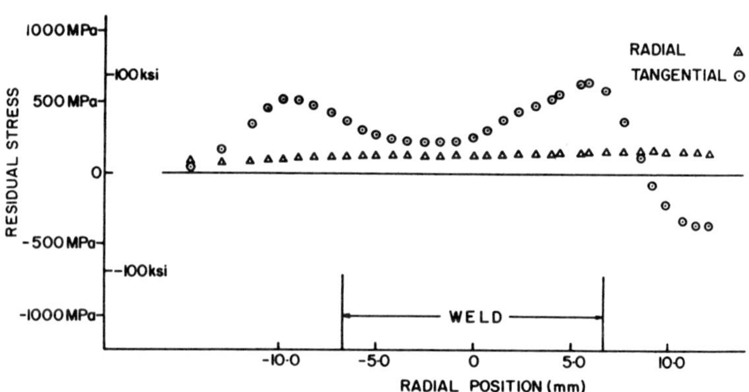

Fig. 3. Radial and tangential stress distributions obtained by trepanning.

Fig. 4. Stress diffractometer.

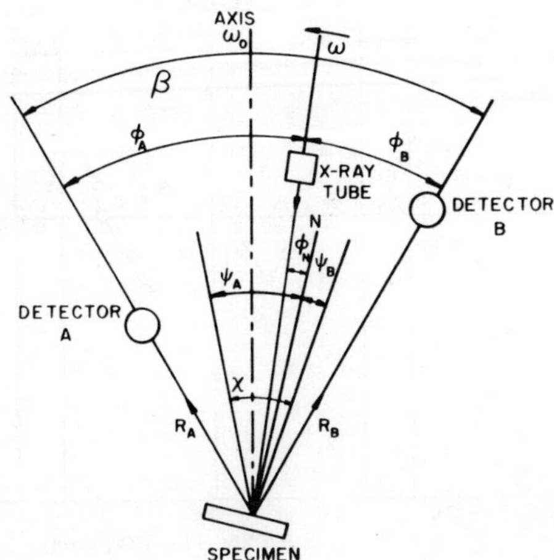

Fig. 5. Equatorial plane of the stress diffractometer.

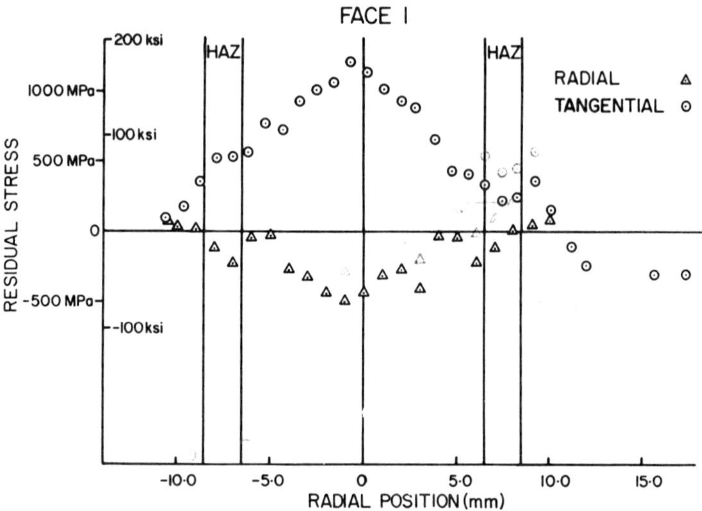

Fig. 6. Stress distributions obtained by diffraction from the surface of face 1.

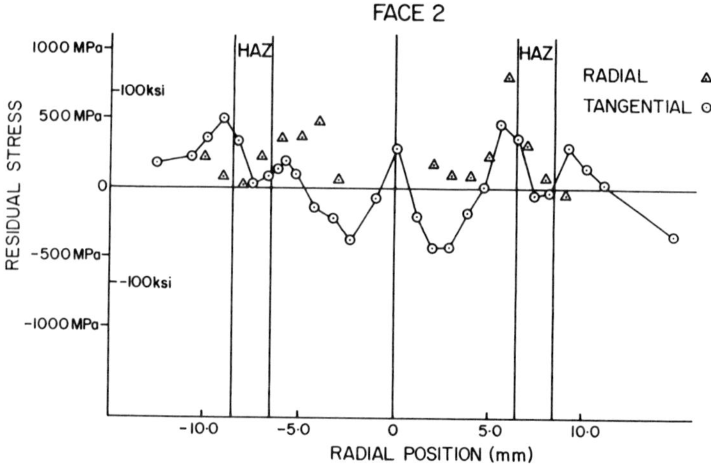

Fig. 7. Stress distribution obtained by diffraction from the surface of face 2.

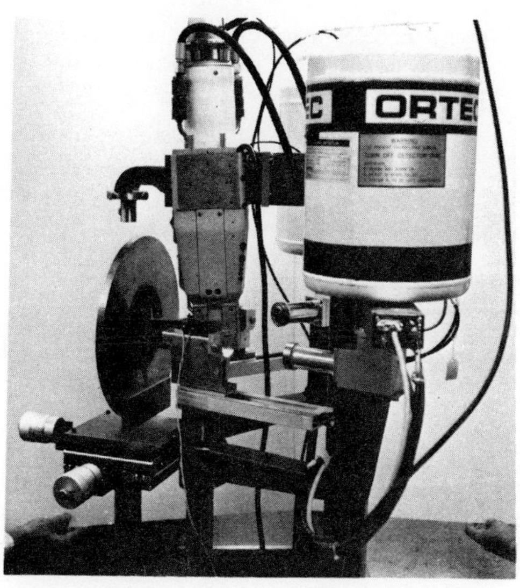

Fig. 8. Specimen mounted in stress diffractometer.

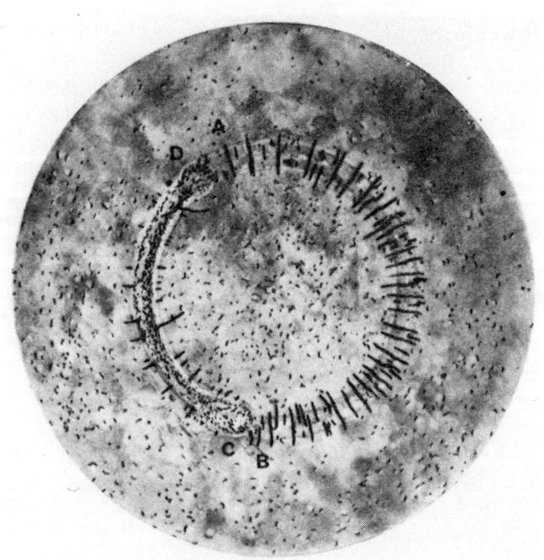

Fig. 9. A specimen after cracking while cathodically charged at 6 ma/cm^2 in 1N H_2SO_4. Corrosion has widened the cracks. (Courtesy of J.B. Gilmour).

DISCUSSION

M. C. Coleman (CEGB): You commented that in the X-ray work it was assumed that the principle stress directions were parallel and normal to the welding direction. Was any work carried out to confirm the validity of this assumption?

Author: No. The assumption of principal stresses parallel and normal to the radius was based on specimen symmetry, and is the assumption used by Adams and Corrigan[2] in their treatment.

A. B. Rothwell (Noranda Research Centre): Did you carry out any measurements by the X-ray technique which would have allowed you to detect variations in radial or tangential stresses depending on position along the weld bead, e.g., at or near start/stop positions as opposed to the mid-circumference position?

Author: No. Measurements were carried out on opposite faces along a single radius remote from the start/stop positions.

S. S. Glickstein (Westinghouse-Bettis Lab): Has there been any calculational effort to predict from just principles the experimental measured residual stresses?

Author: No. We have made no effort to derive a theoretical stress distribution.

R. M. Chrenko (G. E. Corporate Research & Development): In one of your last slides you showed stress corrosion cracking was different on the opposite sides of the weld. Could the SCC difference be due to a difference in metallurgical condition as well as residual stress?

Author: The existence of a difference in metallurgical condition on opposite sides of the plate cannot be ruled out, since this was not investigated. However, the observed cracking was confined to regions of high tensile stress and did not propagate into compressive stress regions.

A. G. Glover (Canadian Welding Inst.): In your X-ray analysis technique, what error is introduced by the effect of surface finish? Does the surface ripple of the weld introduce too large an error if it is not removed?

Author: Surface ripple produces distortion and displacement of the diffraction lines. The dual detector geometry compensates for line displacement due to specimen eccentricity so that a certain amount of ripple could probably be tolerated, particularly with a small irradiated area. However in the present experiment the plate surface was ground and electropolished to eliminate surface effects.

L. W. Sandor (Sun Shipbuilding and Dry Dock Co.): In your last slide, the photo-micrograph showed a great deal of "dark spots or phases" present in the weld in addition to the radial cracks; can you tell me what they are? Are they porosity?

Author: The "dark spots" are not porosity. These are locally corroded areas or corrosion pits which developed after the cracking when the power supply failed. In the eroded areas of the weld the anodic sites for corrosion were the cracks - thus widening the cracks. In the non-cracked regions of the weld corrosion occurred locally, resulting in the "dark spots" in the photograph.

ON THE STRESS STATE IN BONDED METAL JOINTS

A. S. Kao, A. S. Weinstein and C. L. Bauer

Center for the Joining of Materials
Carnegie-Mellon University
Pittsburgh, Pa. 15213

Abstract

The stress state in bonded metal joints has been investigated by a model based on continuum mechanics. This state is determined by first imposing a uniform uniaxial stress in bonded circular cylinders of dissimilar materials and then searching for the localized stresses which satisfy conditions of geometric compatibility. Results are presented for a hypothetical aluminum-copper joint and possible causes of ultimate failure suggested. In general, superposition of these localized stresses results in a complex stress distribution near the joint and a non-planar interfacial profile. Future efforts will combine derived stress states with postulated failure criteria in order to identify likely failure mechanisms.

Models of Bonded Metal Joints

In order to achieve basic understanding of the stress state in actual bonded metal joints, it is useful to consider extremes of joint behavior by analysis of two types of joint: The first type (I) is composed of cylinders of dissimilar materials (A and B) connected by an intermediate rigid layer, as illustrated schematically in Fig. 1, whereas the second type (II) is composed of the same two cylinders connected to each other without an intermediate layer, as illustrated schematically in Fig. 2. The cylinders are assumed to deform elastically, the rigid layer is assumed to be undeformable, and all connections are assumed to be ideal so that external load P is transmitted freely across the interface.

Procedurally, deformation of cylinders A and B under applied load P is considered separately (cf. Figs. 1b and 2b), and then the connecting face of each cylinder is manipulated by adjusting both radial and axial displacements in order to satisfy the imposed boundary conditions (cf. Figs. 1c and 2c). For case I, faces of each cylinder must be expanded radially to their initial dimensions, whereas, for case II, faces of each cylinder must be expanded radially by different amounts (in general), thus inducing a curvature of the joint interface. Moreover, compatible matching of the curvatures in cylinders A and B produces an additional localized axial stress.

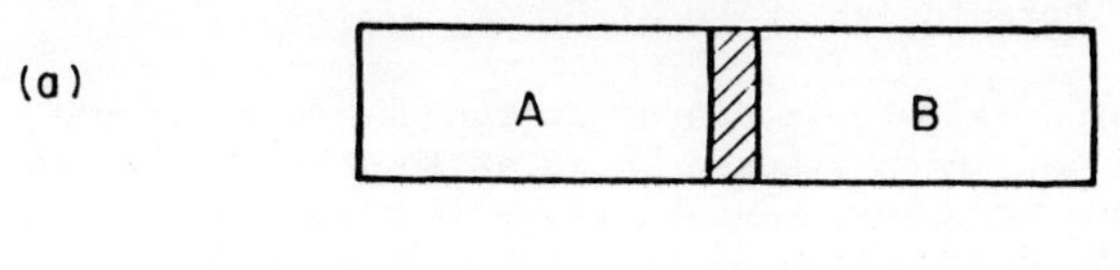

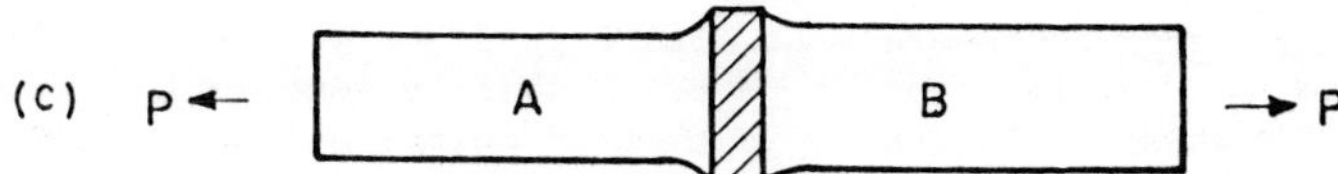

Fig. 1. Joint of the first type (I). (a) Circular cylinders of dissimilar materials (A and B) connected by an intermediate rigid layer (hatched region), (b) independent uniaxial (elastic) deformation of the cylinders under applied load P, and (c) final shape of the joint, which produces additional localized stresses.

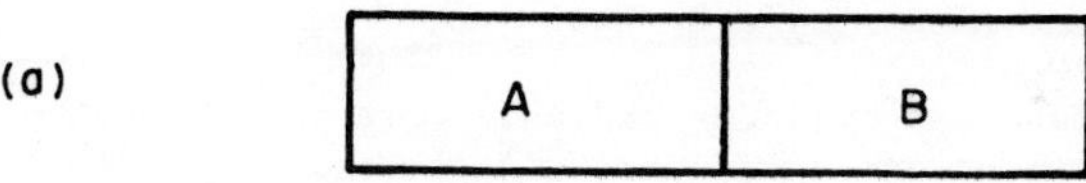

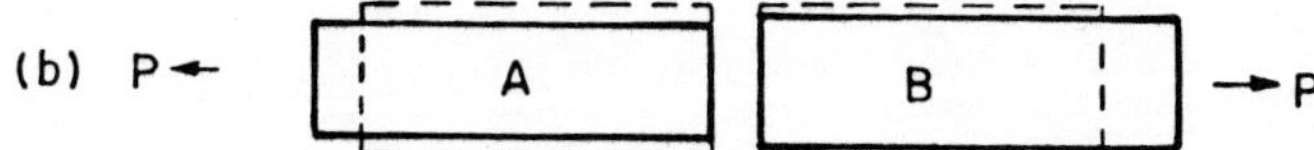

Fig. 2. Joint of the second type (II). (a) Circular cylinders of dissimilar materials (A and B) connected to each other without an intermediate layer, (b) independent uniaxial (elastic) deformation of the cylinders under applied load P, and (c) final shape of the joint, which produces additional localized stresses.

Formulation of Local Stress States

Formulation of Local stress states at the end of a circular cylinder is a problem that has long been studied in an attempt to develop an approximate analysis of adequate accuracy. In theory, a rigorous solution of the problem of axisymmetric deformation of circular cylinders is readily obtained;[1-3] however, extraction of practical engineering solutions remains elusive because of computational difficulties. Thus, a somewhat less rigorous method is developed, based on a technique suggested by Horvay and Mirabal,[4] wherein stress boundary conditions are satisfied at discrete points while constraints of the equilibrium equations are somewhat relaxed so that simple, rapidly converging and physically significant functions can be obtained.

It is assumed that the length-to-diameter ratio of the cylinders z/r is sufficiently large to insure that load P produces a uniform stress distribution over the end faces prior to compatibility formulation. When a set of self-equilibrating, rotationally symmetric normal and shear tractions are then applied to an end face of a cylinder, stresses are developed which decay as a function of distance from the loaded end as $\exp[-\alpha_k(a)]$, where α_k denotes the real part of the eigenvalue, which can be obtained from the eigenvalue equation for axisymmetric deformation of a cylindrical bar. The analysis is performed in a variational approximation and is based on Sadowsky-Sternberg stress functions $\Phi_k(z,r)$, which are modified to satisfy equilibrium conditions.[1,2]

For a cylinder free from tractions on its curved surface and subjected to self-equilibrating, radially symmetric shear and normal tractions, the normal stress in the axial direction $\sigma_z(r,z)$ must not produce a net moment at the end face (z = 0). This condition is stipulated by the expression

$$\int_0^1 \sigma_z(r,0)\ r\ dr = 0, \tag{1}$$

where r denotes normalized cylinder radius. Moreover, boundary conditions for stress-free surfaces and anti-symmetric shear distributions are given by the expressions

$$\tau(0,z) = \tau(1,z) = 0 \tag{2}$$

and

$$\sigma_r(1,z) = 0, \tag{3}$$

where $\tau(r,z)$ and $\sigma_r(r,z)$ denote, respectively, shear stress acting on the z plane in the r direction and normal stress in the r direction. The appropriate shear and normal stresses, τ, σ_z, σ_r and σ_θ, where σ_θ denotes the normal stress in the θ direction, can then be extracted by the aforementioned procedure and corresponding displacements can be determined from the compatibility relationships.

Results

While the rigorous solution for this problem is expressed by an infinite series of eigenfunctions, approximate solutions presented herein are based on only the first three eigenvalues, corresponding to the first three fundamental modes of deformation. The exact stress function is taken as a linear combination of individual eigenfunctions. Generally, case I is analyzed by restoring the radial displacement at z = 0 to its initial value and disallowing equivalent displacements in the rigid layer, whereas case II is analyzed by requiring axial displacements on each side of the interface to be equal in magnitude but opposite in sense and radial displacements to be compatible on the free surfaces. In either case, resulting stresses are adjusted to satisfy the imposed boundary conditions. Computer-generated radial distributions of localized stress for case I and II are presented in Figs. 3 and 4, respectively, for the case of an aluminum (ν = 0.30) - copper (ν = 0.33) joint (ν denotes Poisson's ratio) at various levels of applied stress. The corresponding profile of the interface is depicted in Fig. 5.

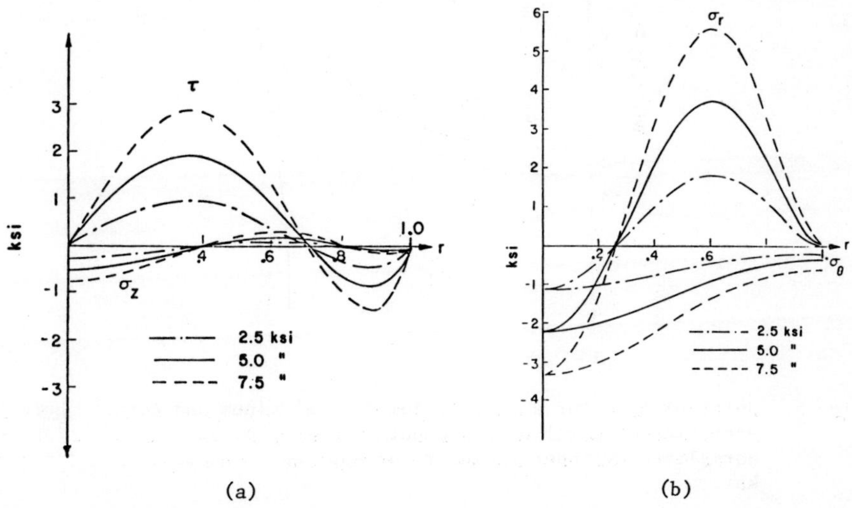

Fig. 3. Radial distribution of localized stresses for a type I joint of aluminum and copper. (a) Variation of axial σ_z and shear τ stresses and (b) radial σ_r and tangential σ_θ stresses as a function of normalized cylinder radius r for applied stresses of 2.5, 5.0, and 7.5 ksi.

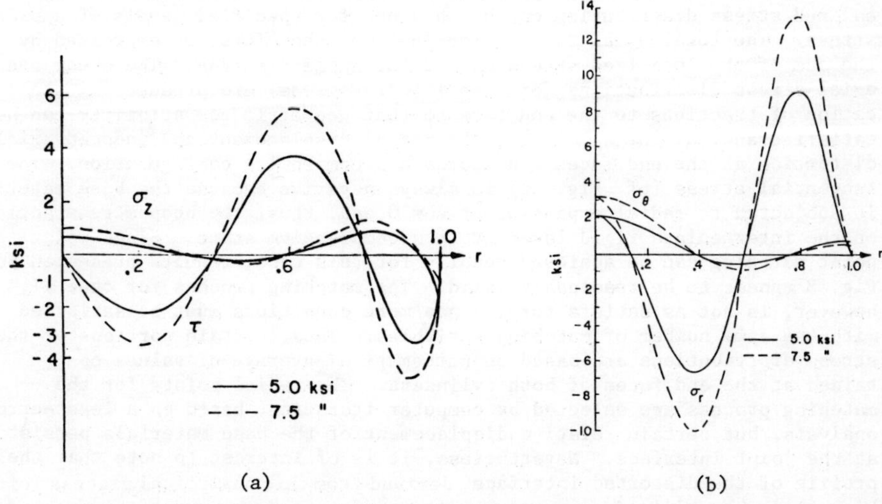

Fig. 4. Radial distribution of localized stresses for a type II joint of aluminum and copper. (a) Variation of axial σ_z and shear τ stresses and (b) radial σ_r and tangential σ_θ stresses as a function of normalized cylinder radius r for applied stresses of 5.0 and 7.5 ksi.

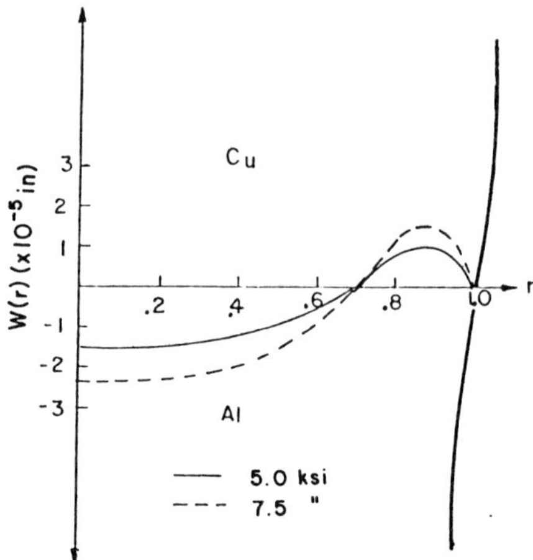

Fig. 5. Joint profile for a type II joint of aluminum and copper where variation of axial displacement w(r) is plotted as a function of normalized cylinder radius r for applied stresses of 5.0 and 7.5 ksi.

Discussion of Results

The principal purpose of this investigation is to apply techniques of continuum mechanics to determine the stress state in bonded metal joints, and thereby deduce possible failure mechanisms. By considering certain limiting cases (I and II), approximate analytical solutions for the localized stress distribution can be obtained for specified levels of applied stress. The total axial stress distribution, therefore, is expressed by the sum of the localized stress σ_z and the applied stress. The shear and axial stress distributions for case I (cf. Fig. 3a) are produced by application of tractions to the end face so that geometric compatibility can be satisfied and, at the same time, the radial displacement and induced axial distortion at the end faces can approach a compromise configuration. The tangential stress (cf. Fig. 3b) is always negative because the base material is subjected to radial expansion at z = 0 and, thus, the hoop stress acting on the intermediate rigid layer is in a compressive state. Since point-to-point matching can be achieved readily for this case, results presented in Fig. 3 appear to be reasonably valid. The matching process for case II, however, is not as satisfactory, since more conditions must be satisfied with the same number of matching variables. Thus, certain portions of the stress distributions are based on mathematical average of values obtained at the end faces of both cylinders. The radial points for the matching process are selected by computer iteration, based on a least-error analysis, but certain relative displacement of the base materials persist at the joint interface. Nevertheless, it is of interest to note that the profile of the distorted interface, derived from this analytical process (cf. Fig. 5), is in qualitative agreement with the distorted contour at the interface of a brazed steel joint.[5] Difference in material properties may result in different levels of deformation, but qualitative similarities provide confidence that the analytical procedure is basically valid.

Although it is difficult to identify possible failure mechanisms from an analysis based on continuum elasticity, several important features may be noted. First, localized normal stress distributions are comparable or sometimes exceed the magnitude of the applied uniaxial stress, implying that local yielding at the joint interface may occur before global yielding of even the weaker material. Second, the shear stress at the interface is also large in magnitude and generally oscillates in sense. Moreover, this sense is reversed for cases I and II. These shear stresses could cause nucleation and subsequent propagation of cracks along the joint interface, especially if an intermediate rigid layer is present (case I). Lastly, the interfacial profile may not remain planar (cf. Fig. 5). Although such distortions may only amount to fractions of a micrometer, the fact that interfacial profiles can be computed may lead to ultimate experimental verification. Future efforts will combine derived stress states with postulated failure criteria in order to identify likely failure mechanisms.

Summary

The stress state in bonded metal joints has been determined by first imposing a uniform uniaxial stress in bonded circular cylinders of dissimilar materials and then searching for the localized stresses which satisfy conditions of geometric compatibility. Results have been presented for a hypothetical aluminum-copper joint and possible causes of ultimate failure suggested. In general, superposition of these localized stresses results in a complex stress distribution near the joint and a non-planar interfacial profile.

Acknowledgement

Support from the Materials Research Laboratory Section, National Science Foundation under grant DMR76-81561 is gratefully acknowledged.

References

1. A. E. Love, "A Treatise on the Mathematical Theory of Elasticity," Cambridge University Press, 276-278 (1927).

2. R. W. Little and S. B. Childs, Quart. of Appl. Mathematics, 25, 261 (1967).

3. W. Flügge and V. S. Kelkar, Int. J. of Solid Struct., 4, 397 (1968).

4. G. Horvay and J. A. Mirabal, J. of Appl. Mech. 25, Trans. ASME 80, 561 (1958).

5. W. G. Moffatt and J. Wulff, Welding Journal, Research Supplement, 115S (1963).

WELD INDUCED CAMBER IN STRUCTURAL SECTIONS

M.J. Bibby

Associate Professor, Carleton University, Ottawa, Canada

J.A. Goldak

Professor, Carleton University, Ottawa, Canada

The "Okerblom method" of calculating structural camber and longitudinal strain as a result of welding is reviewed and compared to other conventional methods frequently used in practice. In particular the advantages of the basic "thermomechanics" approach used by Okerblom are described and contrasted to the fundamentally less rigorous semi-empirical approaches. In addition considerable experimental evidence in support of the validity of the Okerblom treatment is presented. The results are discussed in terms of the limits associated with heat flow and section considerations.

Introduction

One of the limits of welding as a fabrication technique is the distortion that inevitably occurs as a result of the thermal cycle. In general distortion is due to a net shrinkage as the weld cools. Shrinkage parallel to the weld (longitudinal shrinkage) gives rise to a relatively high tensile stress in the neighborhood of the weld which is balanced by compressive reactive stresses further out in the base material. This results in a longitudinal shrinkage and a structural deflection if the weld is not positioned along the neutral axis of the section. Such a deflection is often termed "camber"[1].

To describe the net effect of longitudinal shrinkage, the usual approach is to visualize a shrinkage force (P) acting in the neighborhood of the weld (Figure 1). In effect the shrinkage force can be replaced by

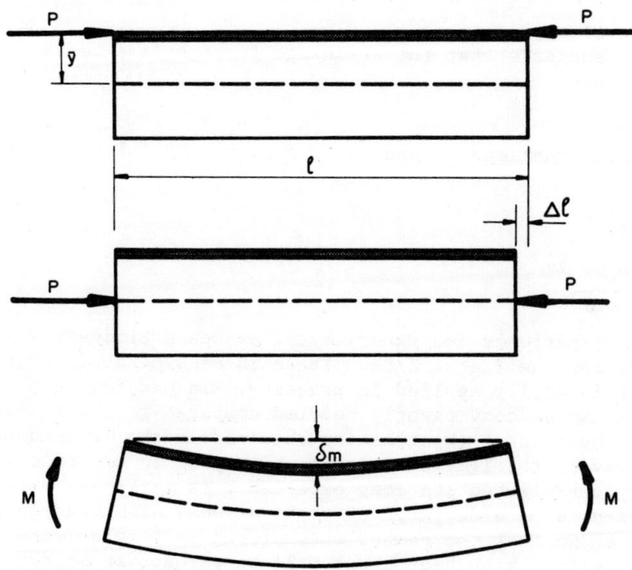

FIGURE 1 - Longitudinal shrinkage ($\Delta\ell$) and camber (δ_m) as a result of weld shrinkage.

an equivalent force acting at the neutral axis and a couple $M = P\,\bar{y}$ where $\bar{y}$ is the neutral axis - weld center of gravity distance. The force P causes a longitudinal shrinkage:

$$\Delta\ell = \frac{P\ell}{AE} \tag{1}$$

where ℓ is length, A is cross sectional area and E is Young's Modulus. Camber is given by the well known relationship:

$$\frac{1}{R} = \frac{M}{EI} \tag{2}$$

where R is the radius of curvature and I is the moment of inertia. For sections in pure bending it is easily shown that the maximum deflection δ_m is given by:

$$\delta_m = \frac{1}{R}\,\frac{\ell^2}{8} \tag{3}$$

Substituting Equation 2 into Equation 3:

$$\delta_m = \frac{P\bar{y}\ell^2}{8EI} \tag{4}$$

Equations 1 and 4 then, completely describe longitudinal distortion and the analytical problem reduces to one of determining the shrinkage force P. At first glance this might seem an impossible task. The shrinkage force develops as a result of thermal stresses caused by transient heat flow over a temperature range of 1500°C. Because of the complexity of the analysis many investigators have resorted to semi-empirical relationships based on experiment and in some cases long experience. Perhaps the best known and most recent semi-empirical work is that of Blodgett which will be described below[1]. On the other hand the "thermomechanics" approach to the problem has been developed by the Russian academician N.O. Okerblom[2]. While Okerblom's work does not seem to be very well known outside the Soviet Union, it is a classic and should occupy a prominent position in welding technology.

(i) The Blodgett Approach[1]

Blodgett suggests that the shrinkage force P is a linear function of weld area A_w i.e.

$$P = K\,A_w \tag{5}$$

In view of this, Equations 1 and 4 reduce to the following forms:

$$\Delta\ell = \frac{K\,A_w\,\ell}{AE} \tag{6}$$

$$\delta_m = \frac{K\,A_w\,\bar{y}\,\ell^2}{8EI} \tag{7}$$

Based on experience and experimental evidence Blodgett suggests a value of 0.005 for the factor K/8E. There is considerable merit in this approach as it is easily applied in practice. In particular A_w is a parameter that can be conveniently handled compared to the inherently more difficult heat input (Q) quantity required in the "thermomechanics" approach. However, the limits of applicability of an empirical approach are not easily established and long experience is necessary before it can reliably be used for a wide range of applications. For example it has recently been shown that the camber encountered in a relatively thin section welded with a high heat input will be perhaps an order of magnitude less than predicted[3].

(ii) The Okerblom "thermomechanics" Approach[2]

Essentially Okerblom combines the well known heat transfer theory of Rosenthal/Rykalin with conventional mechanics to establish the shrinkage force[4,5]. The treatment begins by considering the quasi-stationary heat flow situation for welding as shown in Figure 2. The temperature distribution is said to be stationary with respect to the heat source and can be calculated according to the following relationship:

$$T = T_o + \frac{Q}{\pi kt}\, e^{-\frac{vx}{2\alpha}}\, K_o\left(\frac{vr}{2\alpha}\right) \tag{8}$$

T is the temperature at any point r,x where x,y are the orthogonal distances from the heat source as shown (Figure 2), r is the radial distance from the heat source, Q is the heat input, T_o is the temperature at $r \rightarrow \infty$, k is the thermal conductivity, v is the welding speed, α is the thermal diffusivity and K_o is a modified Bessel function of the second kind. This assumes a heat source uniformly distributed along a line across the top of the bar, two dimensional heat flow, an infinite temperature at the heat source, an infinite bar width and $T = T_o$ at $r = \infty$. Isotherms represent the temperature distribution in the bar at any instant

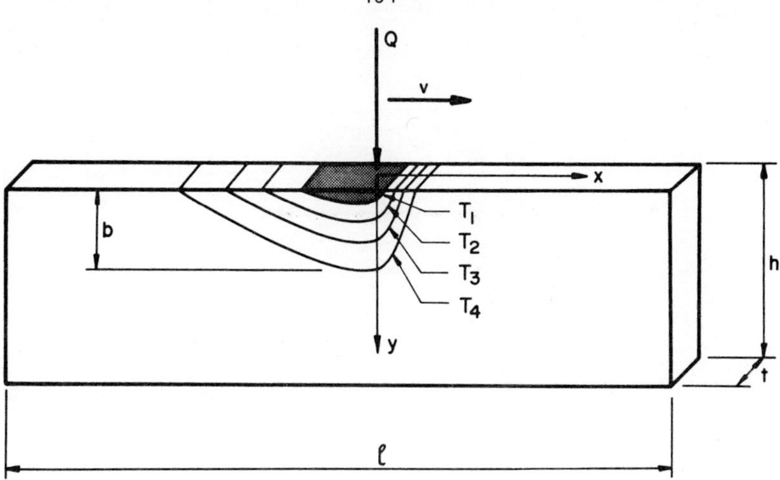

FIGURE 2 - Isotherm distribution as a result of welding along the edge of a bar section.

of time. Transient changes are apparent by virtue of the fact that the entire isotherm ensemble moves along the bar at the welding speed v.

Okerblom among others shows that the maximum width (w) of an isotherm is given by[6]:

$$w = \frac{1}{2} \left(\frac{Q}{vt}\right) \left(\frac{1}{c\rho}\right) \frac{1}{T} \tag{9}$$

where $c\rho$ is the volumetric heat capacity and T is the isotherm temperature. An inspection of Figure 2 shows that the isotherm temperature T is the maximum temperature reached by a point at a distance y = w from the weld centerline. It is therefore evident from Equation 9 that the maximum temperature distribution can be obtained from the relationship[6]:

$$T_{max}\rbrack_y = \frac{1}{2} \left(\frac{Q}{vt}\right) \left(\frac{1}{c\rho}\right) \frac{1}{y} \tag{10}$$

Initially the section is considered to be rigid, i.e. there is no strain at any point due to deflection. In addition Okerblom assumes an elastic-plastic material and a yield stress-temperature dependence such that there is no decrease in yield stress up to about 500°C (Figure 3). At points remote from the weld where the temperature rise is small, strain will develop according to $\varepsilon = \alpha\, T_{max}\rbrack_y$ where α is the coefficient of thermal expansion. In those fibers the compressive strain is small and reversible, and elastic compressive stresses will arise accordingly ($\sigma_c = E\varepsilon$). However, as the weld is approached, the strain eventually exceeds the yield strain which causes plastic deformation. The material inside this zone, is said to be "thermally upset". The distance y_2 (Figure 4) defines the outer limit of the upset area where the thermal strain is just equal to the yield strain (ε_y).

$$\varepsilon_y = \alpha\, T_{max}\rbrack_{y = y_2} \tag{11}$$

The distance y_2 is obtained by substituting Equation 11 into Equation 10:

$$y_2 = \frac{1}{2} \left(\frac{Q}{vt}\right) \left(\frac{\alpha}{c\rho}\right) \left(\frac{1}{\varepsilon_y}\right) \tag{12}$$

When the fibers inside y_2 cool they end up in tension. Consider, for example those fibers near the upset boundary zone. The strain $\alpha\, T_{max}\rbrack_y - \varepsilon_y$ appears as tensile strain and a tensile stress

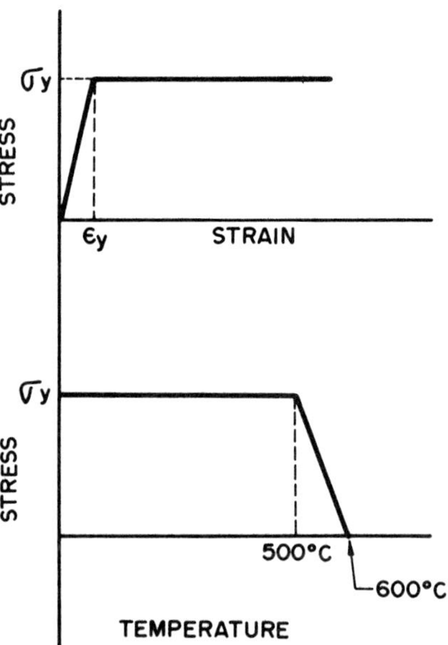

FIGURE 3 - Elastic-plastic model and yield stress - temperature relationship assumed by Okerblom.

can be calculated accordingly $\sigma_T = E(\alpha\ T_{max}]_y - \varepsilon_y)$. However, the value of $\alpha\ T_{max}]_y - \varepsilon_y$ increases as the weld is approached and eventually exceeds the yield strain in tension. The point y_1 in Figure 4 is defined by the condition $\alpha\ T_{max}]_y - \varepsilon_y = \varepsilon_y$

$$\alpha\ T_{max}]_{y = y_1} = 2\ \varepsilon_y \tag{13}$$

The distance y_1 is obtained by substituting Equation 13 into Equation 10:

$$y_1 = \frac{1}{2} \left(\frac{Q}{vt}\right) \left(\frac{\alpha}{c\rho}\right) \left(\frac{1}{2\ \varepsilon_y}\right) \tag{14}$$

Hence all fibers inside this zone undergo plastic deformation in tension and the residual stress in this region is simply σ_y by virtue of the fact that strain hardening is neglected. The residual stress distribution is shown in Figure 4 and the total force is obtained by summing the contributions from zones 1 and 2.

$$P = E\ \varepsilon_y \left[\frac{1}{2} \left(\frac{Q}{vt}\right) \left(\frac{\alpha}{c\rho}\right) \left(\frac{1}{2\ \varepsilon_y}\right)\right] t + t \int_{y_1}^{y_2} E\ (\alpha\ T_{max}]_y - \varepsilon_y)\ dy$$

from which:

$$P = \frac{E}{3}\ \left(\frac{Q}{v}\right)\ \left(\frac{\alpha}{c\rho}\right) \tag{15}$$

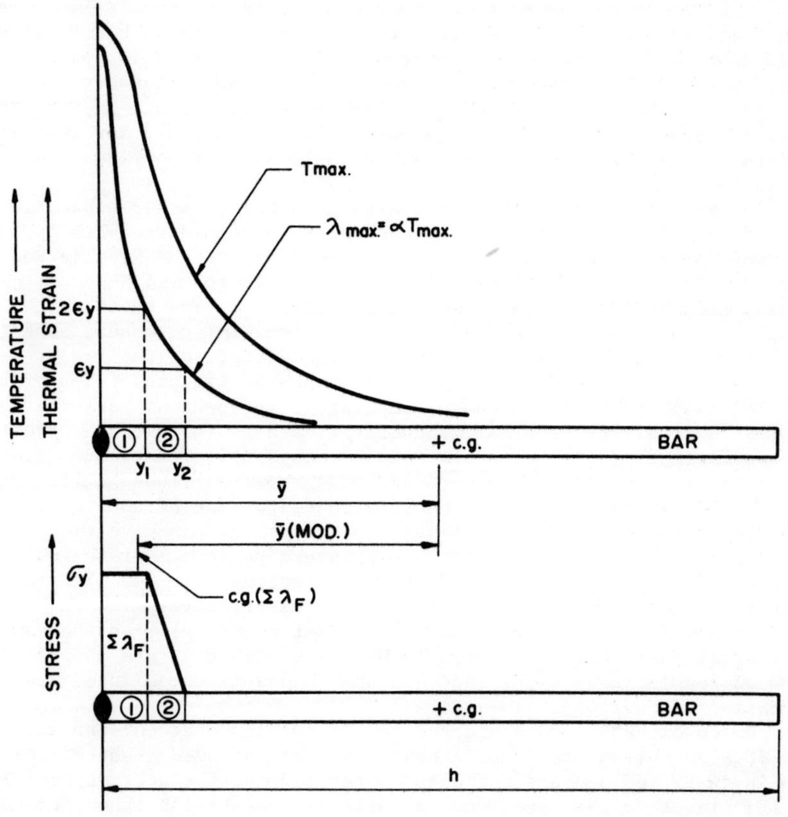

FIGURE 4 - Schematic diagram showing maximum temperature and thermal strain during welding together with the final longitudinal stress distribution.

Substituting this into Equations 1 and 4:

$$\Delta \ell = \frac{1}{3} \left(\frac{Q}{v}\right) \left(\frac{\alpha}{c\rho}\right) \left(\frac{\ell}{A}\right) \tag{16}$$

$$\dot{\delta}_m = \frac{1}{24} \left(\frac{Q}{v}\right) \left(\frac{\alpha}{c\rho}\right) \frac{\bar{y}\ell^2}{I} \tag{17}$$

Equations 16 and 17 are quite general and should apply to any material (mild steel, stainless, aluminum etc.) where the tensile characteristics approximate an elastic-plastic behaviour and where the yield strain does not change with temperature appreciably up to those temperatures corresponding to $T_{max}]_{y_1}$ and $T_{max}]_{y_2}$. While these assumptions must be tested wherever Equations 16 and 17 are to be applied, it is evident that they are reasonably well approximated for structural steels. Strain aging prevents a decrease in yield strength as the temperature increases. Therefore a consistent yield stress from room temperature to 500°C is a reasonably good approximation and this is well above $T_{max}]_{y_2}$. The total strain around the weld zone is small enough that strain hardening can be neglected and an elastic-plastic model is a reasonable approximation.

In addition to the above assumptions the Okerblom relationships (Equations 16 and 17) are arrived at by assuming a rigid bar section i.e. deflection strains are small compared to thermal strains. However, Okerblom shows that assuming a rigid section is not necessary and the same results are obtained if deflection strains are included. On the other hand, the arguments required to show this are relatively involved and readers are referred to the original translation for confirmation of this conclusion.[2]

It is apparent that there are also a number of other assumptions implied in this approach and they must also be considered when applying the Okerblom relationships. All the assumptions of the Rykalin/Rosenthal heat transfer theory are inherited since this is the basis of the thermal strain calculations. These assumptions include a line heat source, an infinite temperature at the heat source, a semi-infinite heat sink (i.e. the bar is large compared to the weld), the effects of the latent heats of fusion and transformation are small, and the thermal properties of the material can be characterized by temperature independent parameters (thermal diffusivity, thermal conductivity and in the Okerblom treatment the coefficient of thermal expansion is included). These assumptions certainly are a cause for concern and have repeatedly been called into question in the calculation of cooling rates for the purpose of predicting weld microstructure. While thermal approximations can give rise to considerable inaccuracy in higher temperature regions near the weld, they are less troublesome in lower temperature regions. In the Okerblom theory, thermal predictions are only necessary at the relatively low temperature remote locations (y_1 and y_2 - near the upset boundary). At the same time the thermal parameters suggested by Okerblom (Table 1) are apparently based on considerable experience and are thus well established mean values for distortion calculations.

Okerblom has also considered the assumption that the bar should provide a so-called semi-finite heat sink for the weld. Whether or not this is justified depends not only on the size of the section but also on the heat input to that section. A limit has been established for edge welded bars based on Russian experimental experience:

$$\left(\frac{Q}{vA}\right)\left(\frac{\alpha}{c\rho}\right) < 2 \times 10^{-3} \tag{18}$$

A modified technique is suggested for sections that do exceed this limit and readers are referred to the original translation for this analysis[2].

While this limit is probably accurate for edge welded bar sections, there is some question about its validity for other shapes[3]. In general Equations 16 and 17 will not hold whenever there is any significant heat buildup in the section. Further comment on this will be made after the results of this investigation have been examined.

It is evident that the calculations do not contain any biaxial or triaxial stress effects. In bar sections and other structural sections this is often justified. Holt has suggested that the results will perhaps be as much as some 30% low where biaxial or triaxial effects are apparent and some judgement must be applied[8].

Perhaps the most serious limitation of all is the effect of the $\alpha \rightleftarrows \gamma$ transformation on the extent and form of the residual stress distribution. While this is neglected in Okerblom's treatment, McParlan and Graville have recently shown that the magnitude and distribution of residual stress around a weld is very much influenced by the volume expansion that appears as the steel transforms during cooling[9]. In general the extent and form of residual stress patterns around welds are not well understood and the magnitude of this effect is not easily assessed.

In view of all the assumptions inherent in the Okerblom theory and the absence of supporting experimental data in available journals, the authors decided that experimental evaluation of the theoretical predictions was necessary. While Okerblom does make reference to some experimental data in the translation, few details are given. In addition, the references that are cited are largely either internal reports and totally unavailable or have not yet been translated. In any event further experimental verification is necessary if the method is to win general acceptance.

Experimental

While the Okerblom approach applies in principle to any structural shape, an edge weld on a simple rectangular bar provides the most convenient arrangement to conduct these kinds of experiments. A series of 12.7 mm (0.5 in.) SAE 1020 mild steel bars ranging from 305 mm (12 in.) to 2438 mm (96 in.) in length and from 102 mm (4 in.) to 254 mm (10 in.) in width were selected for study. Detailed dimensions of a representative group of these bars are contained in Table 1. The very short bars (305 mm - 12 in.) were selected to investigate the significance of transient

TABLE 1

EXPERIMENTAL AND CALCULATED CAMBER DEFLECTIONS

BAR[a]	h		ℓ		δ_m(EXP)[b]		δ_m(CAL.)[c]		$\bar{y}$(MOD)[d]		δ_m(MOD)[e]	
	MM	(IN)	MM	(IN)	MM	(IN)	MM	(IN)	MM	(IN)	MM	(IN)
1	102	(4)	305	(12)	0.16	(0.006)	0.20	(0.008)	45.2	(1.78)	0.18	(0.007)
2	102	(4)	305	(12)	0.17	(0.007)	0.20	0.007)	45.2	(1.78)	0.18	(0.007)
3	102	(4)	609	(24)	0.77	(0.030)	0.79	(0.031)	45.2	(1.78)	0.71	(0.028)
4	102	(4)	609	(24)	0.79	(0.031)	0.79	(0.031)	45.2	(1.78)	0.71	(0.028)
5	102	(4)	914	(36)	1.80	(0.070)	1.78	(0.070)	45.2	(1.78)	1.57	(0.062)
6	102	(4)	914	(36)	1.70	(0.065)	1.78	(0.070)	45.2	(1.78)	1.57	(0.062)
7	102	(4)	2438	(96)	14.1	(0.55)	12.6	(0.496)	45.2	(1.78)	11.2	(0.442)
8	102	(4)	2438	(96)	12.9	(0.51)	12.6	(0.496)	45.2	(1.78)	11.2	(0.442)
9	64	(2.5)	914	(36)	4.1	(0.16)	4.55	(0.179)	27.4	(1.08)	3.9	(0.154)
10	64	(2.5)	914	(36)	4.4	(0.17)	4.55	(0.179)	27.4	(1.08)	3.9	(0.154)
11	125	(5)	914	(36)	1.04	(0.041)	1.14	(0.045)	57.7	(2.27)	1.04	(0.041)
12	125	(5)	914	(36)	1.17	(0.046)	1.14	(0.045)	57.7	(2.27)	1.04	(0.041)
13	152	(6)	914	(36)	0.69	(0.027)	0.79	(0.031)	70	(2.75)	0.71	(0.028)
14	152	(6)	914	(36)	0.89	(0.035)	0.79	(0.031)	70	(2.75)	0.71	(0.028)
15	203	(8)	914	(36)	0.58	(0.023)	0.43	(0.017)	95	(3.73)	0.41	(0.016)
16	203	(8)	914	(36)	0.61	(0.024)	0.43	(0.017)	95	(3.73)	0.41	(0.016)
17	254	(10)	914	(36)	0.36	(0.014)	0.28	(0.011)	119.4	(4.71)	0.28	(0.011)
18	254	(10)	914	(36)	0.41	(0.016)	0.28	(0.011)	119.4	(4.71)	0.28	(0.011)

(a) Bar Thickness = 12.7 mm (0.5 IN.)

(b) Welding Conditions

Electrode	- E7016 (DCRP)
Voltage (V)	- 23 volts
Current (I)	- 110 amps
Welding Speed (v)	- 4.2 mm/sec (10 ipm)
Efficiency (η)	- 0.60
Heat Input (ηVI/v)	- 360 J/mm

(c) Calculated according to equation 17; weld - neutral axis distance $\bar{y}$ = h/2 Thermal parameters $(\alpha/c\rho)$-3.1 x $10^{12} m^3/J$

(d) Modified weld - neutral axis distance ($\bar{y}$) calculated according to Equations (19) and (20)

(e) Calculated according to equation 17 using modified $\bar{y}$

end effects on the predictive capability of the Okerblom relationships. In addition, bars of various widths were investigated to establish the effect of heat sink size on camber.

All bars were welded on edge as shown in Figure 2. The shielded metal arc process was used in all cases with E7016 electrodes (DCRP). Considerable care was taken to ensure a uniform welding speed and fixed welding conditions. The bars were mounted in asbestos jigs to minimize heat loss during cooling. In all cases the welding conditions were 110 amps, 23 volts, and a welding speed of 4.2 mm/sec. (10 ipm). While local variations in weld deposits are inevitable with this kind of welding technique, the deposit area obtained from cross sectioning various welds measured 0.13 cm^2 (0.020 $in.^2$) and was uniform throughout the investigation to within $\pm$ 5%. In addition thermocouples were "harpooned" into the molten pool and the thermal cycle for each bar measured. No siginificant difference in thermal cycles was detected throughout the investigation and this was once again taken as an indication of consistent welding. The 800°C - 500°C cooling time for these welds measured $\Delta t_{800-500} = 4.5$ seconds.

Okerblom also suggests that the results could be quite sensitive to any previous residual stress. Hence all bars were fully annealed at 900°C for one hour and furnace cooled over a period of 24 hours. The bars were then sand blasted to remove scale. Measurements were taken with dial gages along the entire bar length both before and after welding to establish the net deflection caused by the thermal cycle. Figure 5 shows the camber encountered in bars 7 & 8.

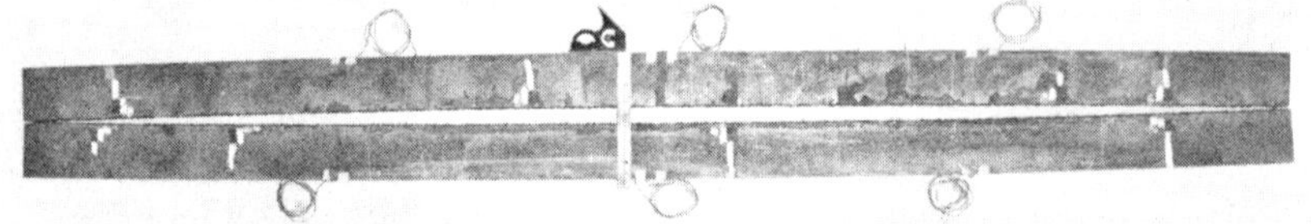

FIGURE 5 - Weld camber - see bars 7 and 8 in Table 1 for section dimensions and welding conditions.

Results and Discussion

The measured deflections for a 305 mm (12 in.) bar, a 609 mm (24 in.) bar and a 914 mm (36 in.) bar all 102 mm (4 in.) wide are given in Figure 6. The uniform smooth curves apparent for these runs indicates limited transient end effects and a uniform radius of curvature throughout although there is some scatter for the shorter lengths.

The measured maximum deflections for a number of bars are given in Table 1 along with the calculated values obtained from Equation 17. In making the calculations the value of $\alpha/c\rho$ was taken as 3.1×10^{-12} m^3/J - a value suggested by Okerblom. A value of $\bar{y} = h/2$ was used to obtain the first set of calculated values in column 5 of Table 1. This assumes that the shrinkage force acts at the centroid of the weld which is the usual way of handling the moment for these calculations. This is justified in many cases where there is base metal on both sides of the weld - for instance in a web to flange fillet weld. However, for an edge weld on a bar $\bar{y}$ should strictly speaking be viewed as the distance from the neutral axis to the centre of thermal deformations (Figure 4). Okerblom shows that the equivalent area (F) containing residual tensile stress at the yield strength is given by:

$$F = \frac{1}{\frac{1}{A} + \frac{\bar{y}^2}{I} + 3 \left(\frac{c\rho}{\alpha}\right) \left(\frac{\varepsilon_y v}{Q}\right)} \qquad (19)$$

If ε_y is taken as 1500 μ strain then the upset area can be calculated. It is interesting to note that the upset area ranges from a low of 1.1 cm^2 (0.17 $in.^2$) for bars 9 & 10 to 1.87 cm^2 (0.29 $in.^2$) for bars

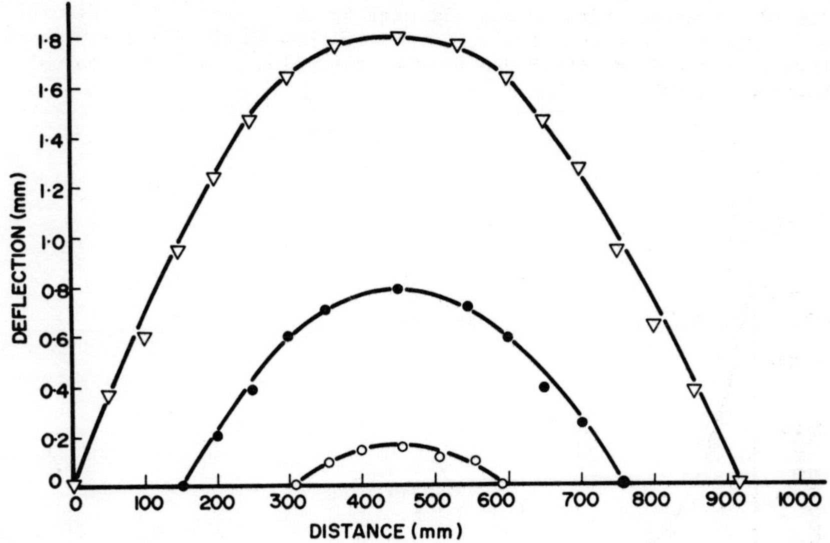

FIGURE 6 - Typical deflection curves for welded bar section (o - bar 2, ● - bar 4, ▽ - bar 5 in Table 1).

17 & 18 while the weld area is 0.13 cm^2 (0.02 $in.^2$) for these same bars. This indicates that the upset area around a weld is some 10 to 15 times the area of the weld itself.

A modified or corrected $\bar{y}$ can be calculated with the help of Equation 19 as shown in Figure 4. The following relationship is used for this purpose:

$$\bar{y} \text{ modified} = h/2 - F/2t \tag{20}$$

where h is bar width and t is bar thickness. The modified $\bar{y}$ values are contained in column 6 (Table 1) and the corrected deflection values contained in column 7.

While the evaluation of most of the other terms in Equation 17 is evident, arriving at a heat-input is less straightforward. In general the heat input is expressed as follows:

$$Q = \eta \, VI \tag{21}$$

The current I and voltage V are measured during the test but the efficiency η depends on a number of factors and is difficult to evaluate directly. Okerblom suggests an efficiency of 0.55 but fails to identify the welding conditions that give rise to this efficiency.

Christensen et. al. have conducted an extensive investigation of efficiencies and suggest a range from 0.55 to 0.75 depending on the conditions[10]. It is evident, however, that the arc efficiency is less for an edge weld than for either butt or fillet welds and an efficiency of 0.55 to 0.65 is expected[11].

In any event the procedure used in this investigation to arrive at an arc efficiency was to compare the measured centreline cooling curve for a large bar (bars 18 & 19) with that calculated using the Rosenthal/Rykalin approach. Equation 8 is used to generate the isotherms shown in Figure 2. At the same time the isotherms move along the bar with the heat source at the welding speed v. Hence for a fixed point the time it takes isotherms to reach this position is known and a cooling curve can be determined. Cooling curves calculated for arc efficiencies of 0.55 and 0.65 are shown in Figure 7 together with an experimental curve measured by harpooning a thermocouple into the weld pool. Evidently then an arc efficiency of 0.6 applies to these experiments and this is the value used in the calculated deflections in Table 1.

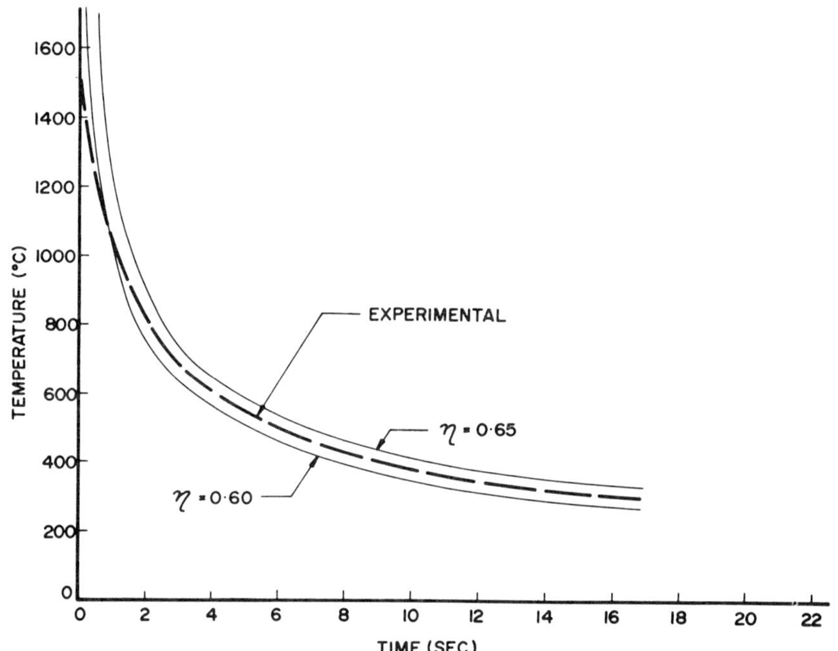

FIGURE 7 - Experimental and calculated cooling curves. Cooling curves calculated from the Rosenthal/Rykalin theory with heat transfer efficiencies η = 0.55 and η = 0.65.

Since non-uniform thermal conditions give rise to the cambering mechanism, it is evident that significant general heating would decrease the deflection. To ensure limited section heating the Okerblom limit (Equation 18) is applied. In the results reported in this investigation the largest value of $(Q/vA)\ (\alpha/c\rho) = 1.38 \times 10^{-3}$, significantly less than the 2×10^{-3} limit. Nevertheless considerable general heating of the section does occur. In bars 7 & 8 a saturation temperature level of $\sim$ 150°C is measured. The Okerblom limit, therefore, allows a surprisingly high general heat level in the section.

It is evident from Table 1 that the measured deflections are indeed surprisingly close to the calculated deflections. Figure 8 shows just how accurate the calculations really are for a $\bar{y} = h/2$. It is evident that there is less agreement when a corrected $\bar{y}$ is used (Figure 9). However, the difference in either case is not great and it is probably not worth while making the correction in practice.

While there is some scatter in the results for short bar lengths, the results do indicate that transient end conditions do not invalidate the Okerblom relationship. In addition the deflections of both narrow and wide bars have been accurately predicted and heat sink size, at least within the limits of this investigation, does not have an appreciable effect.

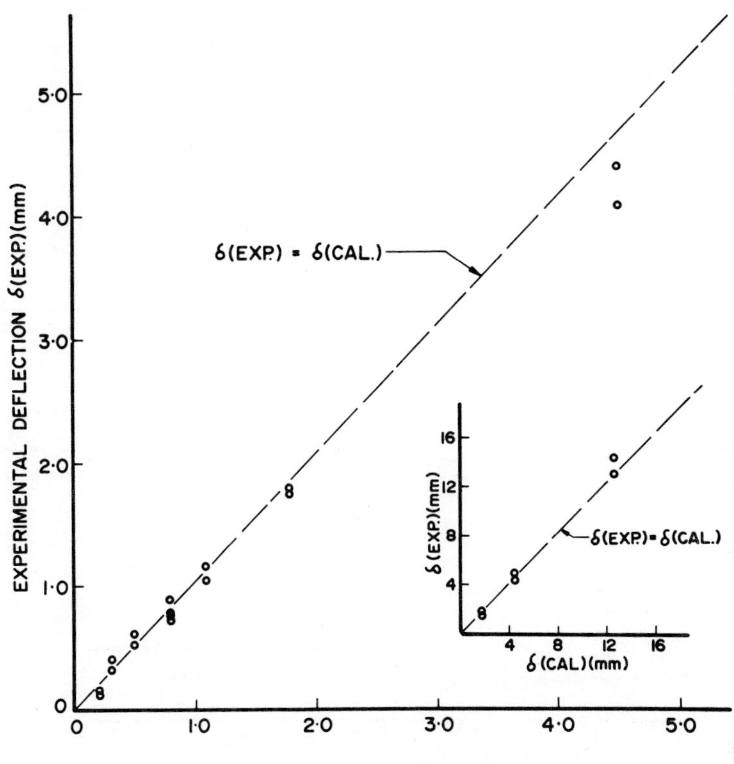

FIGURE 8 - Experimental and calculated camber deflections ($\bar{y}$ = h/2).

The agreement between measured and predicted values is truly surprising in view of all the assumptions needed to arrive at the Okerblom relationship. The assumptions involved in the Rykalin/Rosenthal heat flow theory are serious enough. Yet the temperature independence of yield strength, an elastic-plastic behaviour and the influence of the $\alpha \rightleftarrows \gamma$ transformation on the residual stress pattern are all added to this in the Okerblom approach.

While the agreement in this investigation is undeniable, preliminary experiments indicate that prior residual stress has a considerable effect on the magnitude of the camber. In addition, there is further evidence that while the Okerblom approach holds for a relatively wide range of annealed carbon steel bars it is limited for quenched and tempered bars. These effects are, in fact, predicted by Okerblom's theory and will be the subjects of further publications by the authors.

While only bar sections have been investigated in this presentation, the fundamental nature of the Okerblom approach provides reasonable confidence that it can be applied to a relatively wide variety of shapes and materials. Without doubt the limiting assumptions must be tested when applying this method in other cases. However these assumptions are relatively well understood and hence reasonable judgement is possible.

Conclusions

(1) Edge welded bar section experiments demonstrate that the Okerblom relationship:

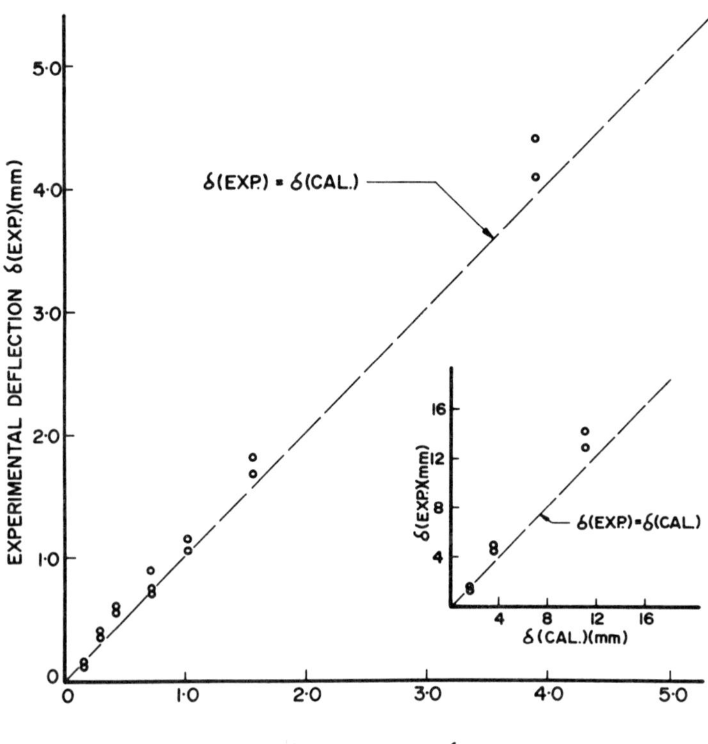

FIGURE 9 - Experimental and calculated camber deflections ($\bar{y}$ modified according to Equations 19 and 20).

$$\delta_m = \frac{1}{24} \left(\frac{Q}{v}\right) \left(\frac{\alpha}{c\rho}\right) \left(\frac{\bar{y}\ell^2}{I}\right)$$

correctly predicts weld camber deflections where:

$$\left(\frac{Q}{vA}\right) \left(\frac{\alpha}{c\rho}\right) < 2 \times 10^{-3}$$

(2) The above relationship appears to hold in spite of several rather questionable assumptions inherent in the theory. Among the most serious of these assumptions is the absence of anything in the theory to account for the large effect of the $\alpha \rightleftarrows \gamma$ transformation on the residual stress magnitude and distribution.

(3) While only bar sections have been examined in the investigation, the fundamental thermomechanical approach to the problem suggests that it should apply to other section shapes and other materials.

Acknowledgements

The financial support of the National Research Council NRCA4601 (M. Bibby) and NRCA2922 (J. Goldak) and of the Dominion Bridge Co. (Montreal) are gratefully acknowledged. Our thanks also to Mr. H. Haystead for his help with the experimental work and to Ms. P. McFarlane for typing the manuscript.

References

1. Blodgett, O., Distortion (Part III), Canadian Welder and Fabricator, March 1976, p. 15.
2. Okerblom, N.O., The Calculation of Deformations of Welded Metal Structures, Her Majesty's Stationery Office, London (1958).
3. Bibby, M.J. and Goldak, J.A., Camber in Electron Beam Welded Structural Sections, Accepted for publication in the AWS Welding Journal, 1977.
4. Rosenthal, D., Trans. ASME, Volume 68, 1946, p. 849.
5. Rykalin, N.N., Calculations of Heat Flow in Welding, Translated from Russian by Zvi Paley and C.M. Adams, 1963.
6. Gray, T., Spence, J., and North, T., Rational Welding Design, London Butterworth Publication, London, 1975, p. 45.
7. Bibby, M.J. and Goldak, J.A., Camber in Electron Beam Welded Stainless Steel Sections, CSME, Vol. 4, No. 2, 1976/77, p. 45.
8. Holt, R.E., Primary Concepts of Flame Bending, AWS Welding Journal, June 1971, p. 416.
9. McParlan, M. and Graville, B.A., Hydrogen Cracking in Weld Metals, Welding Research Supplement, April 1976, p. 955.
10. Christenson, N., Davies V. DeL., Gjermundsen, K., Distribution of Temperatures in Arc Welding, British Welding Journal, Feb. 1965, p. 64.
11. Tall, L., Residual Stresses in Welded Plates - A Theoretical Study, AWS Welding Journal, 1974, p. 105.

DISCUSSION

J. S. Haydon (General Electric Co., GTEMD): Was analysis limited to single pass welds?

Author: The analysis presented here is limited to single pass welds. However, it is not difficult to extend it to multipass situtations. Experiments are underway in our laboratory to test the validity of the multipass analysis.

B. Rothwell (Noranda Research Center): Is it possible to refine the model to take account of transformation strains? Clearly, these only apply to a small area, but can become considerable when welding conditions and materials combine to produce low transformation temperatures.

Author: There is no question that this is an important consideration. Experiments with the G-Bop test conducted by Graville[9] show that the transformation plays a considerable role in the development of longitudinal stress particularly at low transformation temperatures. It probably is possible to include this in the analysis. However, the Okerblom approach uses a maximum thermal strain criterion for the size of the upset zone and then deals only with final mechanical equilibrium. The transformation, on the other hand, depends on thermal history and hence transient changes would have to be traced to take it into account. This implies the use of numerical models and in the end this still may be the only way to provide sufficient detail for the analysis of residual stresses around a weld with high accuracy.

J. S. Ahearn (Martin Marietta Labs): Can you comment on why in the cold rolled material, the induced camber is smaller than in fully annealed material?

Author: Cold rolled material contains relatively high residual stresses from the bar forming operation. The forming residual stresses in the upset zone are eliminated and replaced by weld induced residual stresses. Distortion is, therefore, the result of the difference between these two residual stress distributions rather than due to weld induced stresses alone. In addition, Okerblom shows that the size of the upset zone is altered by the presence of stress. If the forming residual stresses in the upset zone are tensile the deflection is smaller than predicted by the Okerblom approach while compressive forming residual stresses leads to larger weld deflections than predicted.

EFFECT OF RESIDUAL STRESSES ON FATIGUE PROPERTIES OF ALUMINUM ALLOY BUTT WELDS

R. A. KELSEY & G. E. NORDMARK

ALCOA LABORATORIES

ALCOA CENTER, PA 15069

ABSTRACT

Residual stresses are produced in welds by local thermal expansion, plastic deformation, and subsequent shrinkage during cooling. The residual stress at the surface of the weld may reach a level close to the yield strength.

A review of the literature and the results of an investigation reported herein support the conclusion that residual welding stresses may, under certain conditions, have a significant effect on the number of cycles required to initiate a fatigue crack, as well as on fatigue crack growth rate. Under other conditions, the effect of residual welding stresses can be negligible.

Residual stresses are produced in weldments by local thermal expansion, plastic deformation, and subsequent shrinkage on cooling.[1-3*] In the region of the weld the resulting tensile residual stresses may reach a level close to the yield strength**. In general, the magnitude of residual welding stresses will depend on welding process, welding parameters, restraint on the joint, and the dimensions of the parts being joined.

In a summary published in 1964, Munse[6] noted that views expressed in the literature concerning the effect of residual welding stresses on fatigue differed markedly. There continues to be some disagreement on this subject although many investigators have reported that residual welding stresses do affect fatigue strength.[2,6-13] Little work appears to have been done, however, to evaluate quantitatively the effect of residual welding stresses on fatigue crack initiation, and even less on fatigue crack growth (FCG) rate. Furthermore, the conclusion that residual welding stresses do affect fatigue life is often based on a comparison with fatigue data obtained for welds in which residual stresses have been relieved or modified by thermal or mechanical means. The stress relief procedure may alter the microstructure and therefore affect the fatigue properties. The subject is further complicated because the magnitude and distribution of residual stresses are rarely reported and test specimens are often removed from the weldment in such a manner as to relieve all or most of the residual welding stresses.[10,14]

This paper reviews work which has been done to evaluate the effect of residual welding stresses on fatigue of aluminum butt welds and presents the results of an investigation at Alcoa Laboratories to obtain definitive data on the effect of residual welding stresses on FCG.

EFFECT OF RESIDUAL WELDING STRESSES ON FATIGUE CRACK INITIATION - A REVIEW

Welding produces residual stresses both normal to and in the direction of welding, as shown in Fig. 1. In multi-pass butt welds welded from both sides, transverse stresses (normal to the weld) will vary from tension in the outer weld passes to balancing compressive stresses across the root passes (Fig. 1a). The tensile stresses near the weld surface can approach the yield strength of the weld metal when the combination of number of passes and plate thickness are such as to cause large temperature gradients through the thickness. Welds made in thin plate or with relatively few passes in thick plate may have negligible residual stresses. For example, a weld made with four passes in 32 mm (1-1/4 in.) plate had essentially no transverse residual stresses whereas the residual stress was about 150 MPa (22 ksi) tension when the weld was made in 16 passes[15] (Fig. 2). In addition to the self-balancing residual stresses produced in multi-pass welds, residual stresses may also be produced transverse to the welding direction when the part being welded is restrained by adjacent members in a structure as indicated schematically in Fig. 1b.

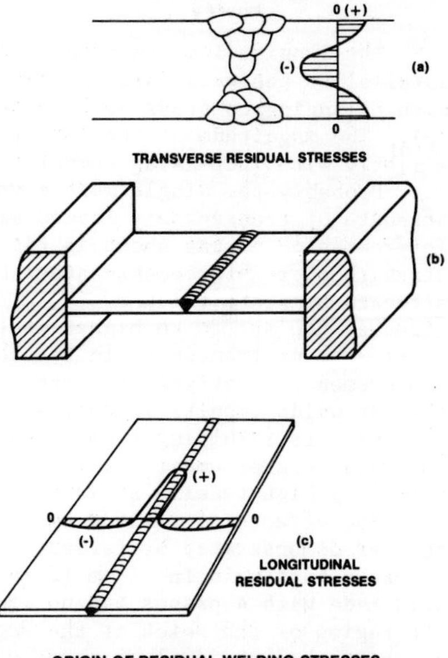

ORIGIN OF RESIDUAL WELDING STRESSES
FIGURE 1

*Numbers refer to references listed at end of text.

**Some specifications covering the fatigue strength of welded steel structures assume that all welds have residual tensile stresses equal to the yield strength.[4,5]

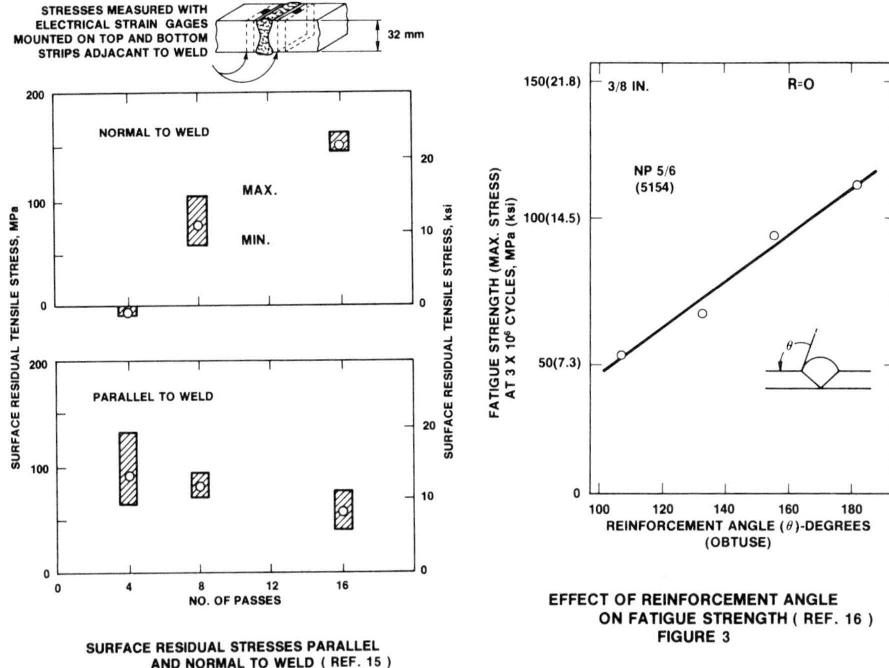

SURFACE RESIDUAL STRESSES PARALLEL AND NORMAL TO WELD (REF. 15)
FIGURE 2

EFFECT OF REINFORCEMENT ANGLE ON FATIGUE STRENGTH (REF. 16)
FIGURE 3

The longitudinal residual welding stresses, in the direction parallel to the weld, are tensile in the weld and adjacent base metal with balancing compressive stresses further away from the weld (Fig. 1c). The magnitude of the longitudinal residual stress decreases slightly with increasing number of passes (Fig. 2).

Probably the single most significant factor affecting fatigue strength of transversely loaded welds is the profile of the weld reinforcement[16-19], as shown in Fig. 3. For this reason, the fatigue strength of reinforcement-intact butt welds (Fig. 4) may be greatly affected by plate thickness[20], even though transverse residual welding stresses are apt to be higher in thicker plate. Removal of the weld reinforcement from welds in the thinner plate results in a significant improvement in fatigue strength, but removing the reinforcement from the thicker welds results in only a slight increase in fatigue strength. The benefits resulting from eliminating the stress concentration due to the weld reinforcement for the thicker plate are apparently partly offset by high tensile residual welding stresses at the surface.

The effects of reinforcement geometry and residual stresses are further demonstrated by fatigue tests of 6.2 mm (1/4 in.) slices taken across a butt weld in 51 mm (2 in.) 5456 plate, as shown in Fig. 5. The weld made with 4 passes has no measurable transverse residual stress in the region of the notch at the edge of the weld while the weld made with 12 passes has residual stresses of 90 MPa (13 ksi) tension at the edge of the weld. However, the higher weld crown accompanying the fewer number of passes results in a more severe notch at the edge of the weld. About five times as many cycles are required to initiate a fatigue crack in the specimen having the high residual stress. Weld profile appears to be the overriding factor.

Because of the overriding effect of weld geometry, thermal stress relief probably would not appreciably improve the transverse fatigue

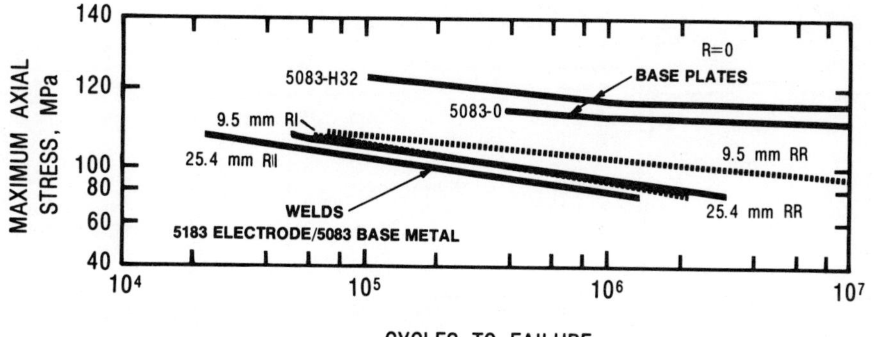

EFFECT OF REINFORCEMENT ON TRANSVERSE FATIGUE PROPERTIES OF BUTT WELDS IN ALLOY 5083 PLATE (REF. 20)

FIGURE 4

strength of reinforcement-intact welds. However, peening has been found to be beneficial,[27] probably because it reduces the severity of the notch at the edge of the weld, as well as introducing residual compressive stresses. Thermal stress relief or peening should improve the transverse fatigue properties of welds having the reinforcement removed.

Fig. 6 illustrates the effect of thermal stress relief and peening on residual stress and fatigue life of longitudinal butt welds in 5456 alloy plate welded with 5556 filler.[8] Surface residual welding stresses adjacent to the weld were 34 MPa (5 ksi) tension for the as-welded condition, zero for the specimen given a thermal stress relief, and 172 MPa (25 ksi) compression for the peened specimen. In the weld itself, the maximum longitudinal residual stress was about 117 MPa (17 ksi) tension. Using the fatigue strength of thermally stress relieved welds as a baseline, these data show that the surface residual tensile stresses reduce fatigue strength while surface compressive stresses result in a significant increase in fatigue life. The peened specimens all failed by fretting in a

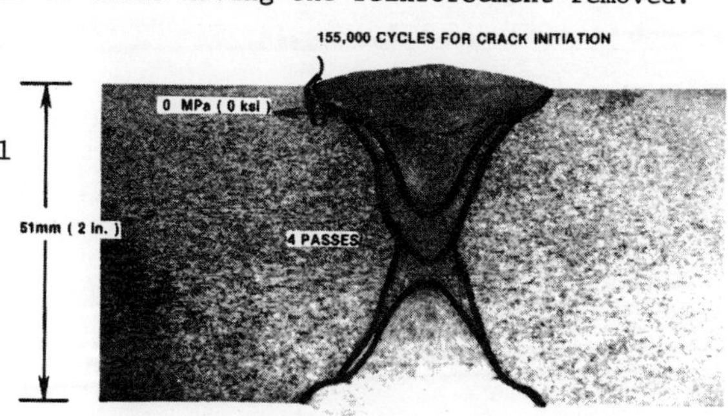

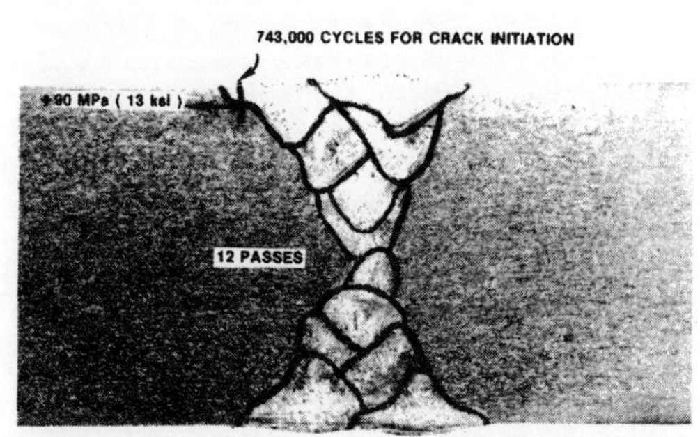

STRESS RANGE: 0 TO 103 MPa (15 ksi)
5456-H117 PLATE
5556 ELECTRODE

FIGURE 5 EFFECT OF WELD REINFORCEMENT CONTOUR AND RESIDUAL STRESS ON FATIGUE LIFE OF ALUMINUM ALLOY BUTT WELD

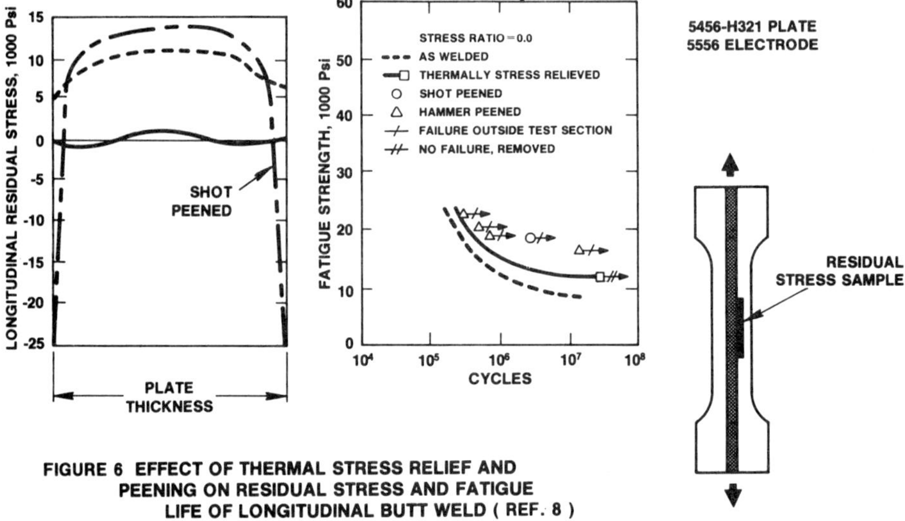

FIGURE 6 EFFECT OF THERMAL STRESS RELIEF AND PEENING ON RESIDUAL STRESS AND FATIGUE LIFE OF LONGITUDINAL BUTT WELD (REF. 8)

grip end, while the as-welded and stress-relieved specimens failed in the weld. Referring to Fig. 1c, one can visualize that removal of material from the edge of a longitudinally welded specimen will partially relieve the residual stresses. Thus, in preparing longitudinally welded specimens for fatigue test, it is important to minimize the amount of material machined from the specimen edges after welding.

EFFECT OF RESIDUAL WELDING STRESSES ON FATIGUE CRACK GROWTH

Although considerable work has been done in the last few years to develop FCG data for aluminum welds,[21-24] there does not appear to be any data of a quantitative nature relating to residual stresses. Dawes[25] and Nelson and Kaufman[26] reported difficulties in obtaining uniform crack fronts during fatigue precracking of fracture toughness specimens from multi-pass butt welds. Dawes did not have this problem when precracking specimens from single-pass welds or when thinner specimens were machined from multi-pass welds. In the first case, residual stresses transverse to the weld would be negligible while in the latter case residual stresses would be relaxed when removing the specimen from the weld. Nelson and Kaufman attempted, with only limited success, to overcome the problem of nonuniform crack front curvature by modifications of the starter notch, (Fig. 7).

Kelsey[27] demonstrated that the compressive residual stress at the mid-depth of a multi-pass weld causes the crack growth there to be slower than that in the surface regions containing residual tensile stresses (Fig. 8). These results are difficult to interpret quantitatively because of uncertainty concerning the stress intensity along the crack front. However, FCG rate based on the average crack length was found to be similar to that measured in specimens removed from a weld so as to be essentially free of residual stresses. Thus, this type of residual welding stress distribution (i.e., self-balancing) can be concluded as having no effect on the overall growth rate of a through crack.

The "H-plate" specimen. shown in Fig. 9. was used in this investigation to study the effects of residual stresses produced by external restraint on FCG. This specimen had originally been developed for evaluating the effect of residual welding stresses on stress-corrosion cracking.[28] The restraint in the H-plate specimen is such that the weld creates fairly uniform residual tensile stresses across the center

FIGURE 7 MODIFICATIONS OF MACHINED CRACK STARTERS MADE IN AN ATTEMPT TO OVERCOME HIGH COMPRESSION RESIDUAL STRESS IN CENTER OF THICKNESS OF NOTCH-BEND FRACTURE-TOUGHNESS SPECIMENS FROM GROOVE WELDED PANELS **(REF. 26)**

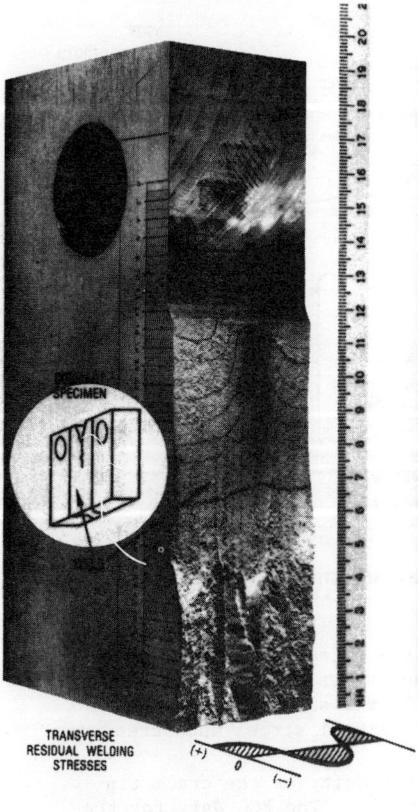

TRANSVERSE RESIDUAL WELDING STRESSES

FIGURE 8
EFFECT OF RESIDUAL WELDING STRESSES ON FATIGUE CRACK GROWTH THROUGH BUTT WELD (REF. 27)

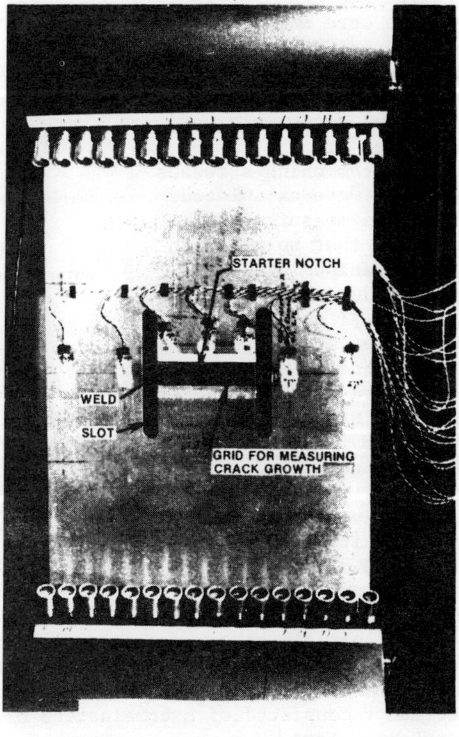

STARTER NOTCH
WELD
SLOT
GRID FOR MEASURING CRACK GROWTH

FIGURE 9 WELDED H—PLATE SPECIMEN

strut, which are balanced by residual compressive stresses in the side struts. The residual stresses in the weld may be relieved merely by sawing through the side struts. Thus, an important advantage of this specimen is that one can measure FCG in similar welds with and without residual stress without resorting to thermal stress relief or mechanical deformation, both of which could result in metallurgical changes in weld-metal structure. The dimensions of the H-plate and the residual stresses in the center and side struts are shown in Fig. 10.

H-plate specimens were fabricated from 12.7 mm (1/2 in.) aluminum alloy 5083-0 plate. Welds were made by the gas metal arc welding process in four passes using alloy 5183 GAD 13 electrode. The H-plate specimens were tested with the weld reinforcement removed. Residual stresses were relieved in Specimen 2 by sawing through the side struts. A 10 mm (0.40 in.) long starter notch at the center of the weld was fatigue precracked prior to FCG testing, as follows:

Spec.	Side Struts	Stress in Center Strut, MPa (ksi) Applied Minimum	Applied Maximum	Effective Minimum	Effective Maximum	Cycles	Avg. Crack Extension, mm (in.)
1	Intact	0	11.7 (1.7)	107.5 (15.6)	119.3 (17.3)	500,000	0
		0	17.2 (2.5)	107.5 (15.6)	124.8 (18.1)	780,000	1.50 (0.06)
2	Separated	0	17.9 (2.6)	0	17.9 (2.6)	1,050,000	0
		107.5 (15.6)	126.2 (18.3)	107.5 (15.6)	126.2 (18.3)	2,100,000	1.21 (0.05)

The applied stresses in Specimen 2 were increased to the same level as the effective combined stresses in Specimen 1 (due to residual and load stresses) in order to initiate a crack.

The FCG tests of the H-plates were made in dry air (R.H. ≤10%) at a frequency of 10 Hz. During the test of Specimen 1 electrical gages were used to monitor strains in the struts, from which stresses and loads were calculated. The strain measurements indicated that no relaxation of residual stresses occurred during precracking.

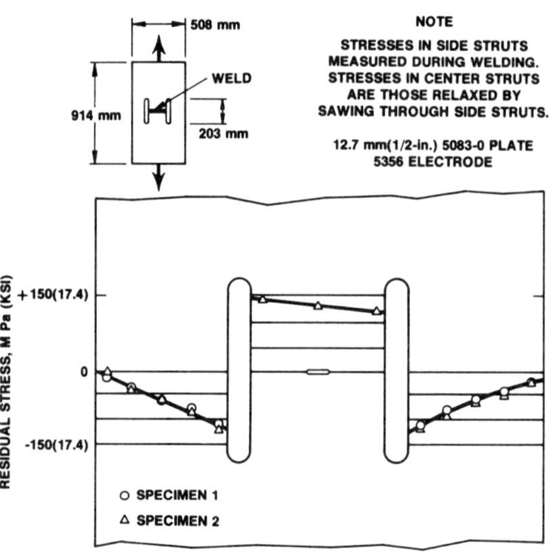

FIGURE 10 RESIDUAL STRESSES IN WELDED H - PLATE SPECIMEN

Fig. 11 shows how the average residual stress and applied stress in the center strut of Specimen 1 decreased with crack growth (Fig. 11). The applied stress range is added to residual stress to produce an effective stress range having, initially, a stress ratio of 0.85. For Specimen 2 the applied loads were adjusted as the crack grew, so that the stress range simulated the effective stress range in the center strut of Specimen 1. Thus, the stress range in Specimen 2 resulted entirely from external loads while the same stress range in Specimen 1 consisted of a combination of residual stress and stress due to external load.

Fig. 12 shows the calculated stress intensity at the crack tip corresponding to the stresses shown in Fig. 11. The FCG data for the two specimens are presented in Fig. 13 in terms of crack growth per cycle, da/dN, versus stress intensity range, ΔK.* These data show that there is no significant difference in FCG resistance of the two specimens. From this it follows that, for the conditions of this investigation, residual welding stresses may be treated the same as those due to external loads as far as FCG is concerned. In other words, residual stresses due to welding are directly additive to load stresses.

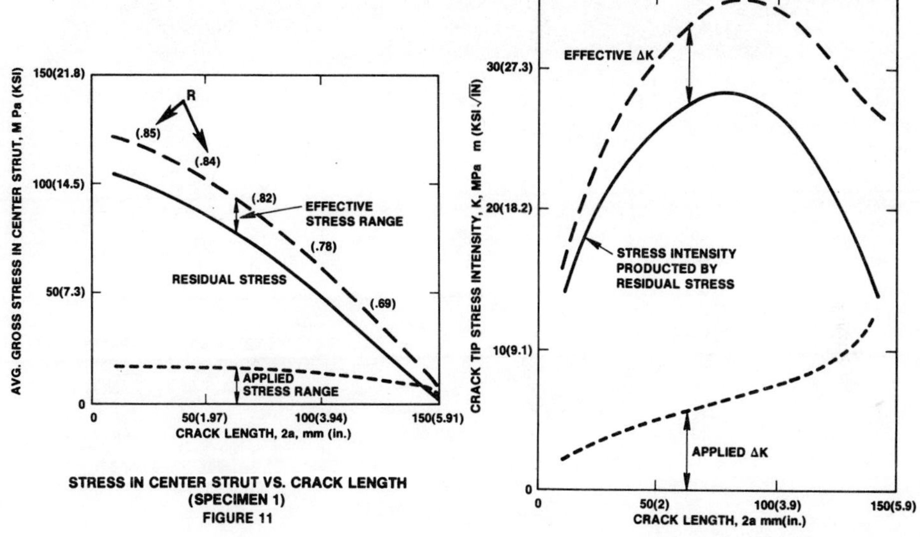

STRESS IN CENTER STRUT VS. CRACK LENGTH
(SPECIMEN 1)
FIGURE 11

FIGURE 12 STRESS INTENSITY VS. CRACK LENGTH
(SPECIMEN 1)

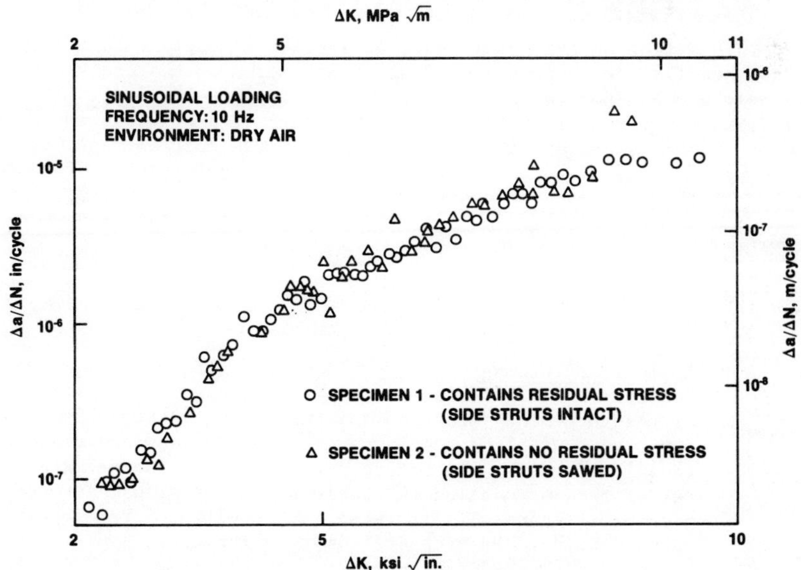

FIGURE 13 FATIGUE CRACK GROWTH RATE
IN H - PLATE SPECIMEN WELDS

The fracture surfaces of two large specimens used to study FCG from a surface flaw in parent material (upper figure) and normal to a longitudinal weld (lower figure)[24] are illustrated in Fig. 14. Except in the early stages of cracking, a longitudinal weld would not be expected to

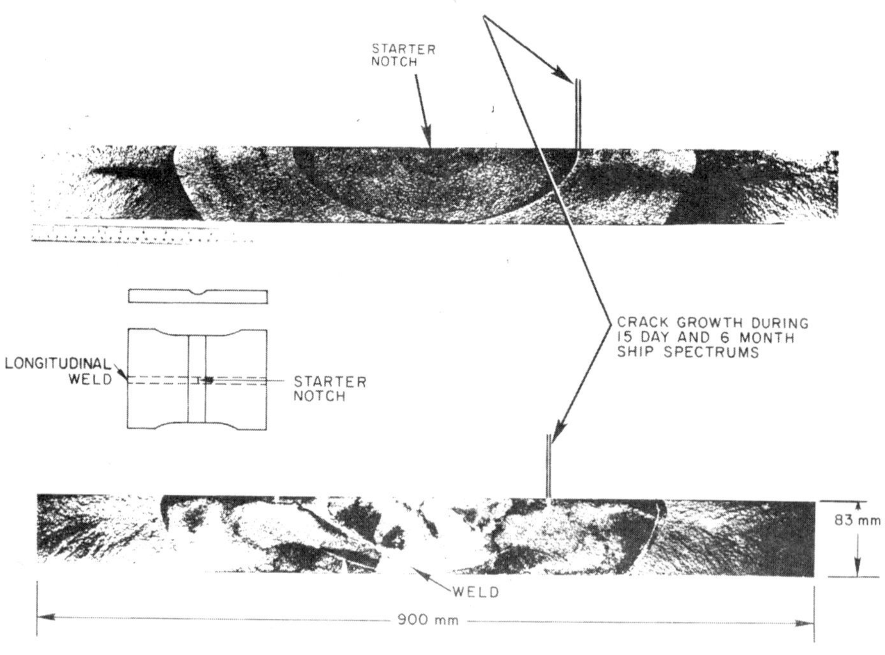

WELDED SPECIMEN

FIGURE 14 FRACTURE SURFACES OF LARGE FATIGUE CRACK GROWTH SPECIMENS

have much effect on the growth rate of a crack in a plane normal to the weld since (1) the residual stresses are relieved as the crack grows and (2) only a small portion of the crack front is in weld metal. This was verified by the results of the investigation. Crack initiation took longer in the welded specimen and propagation was initially slower in the depth direction. Otherwise, there were no significant differences in FCG in the plain and welded specimens.

<u>SUMMARY</u>

1. The effect of residual welding stresses on fatigue crack initiation may be summarized as follows:

   A. <u>Transverse loading</u>
   Residual welding stresses have a minor effect on fatigue life of reinforcement-intact welds; the dominant factor is weld geometry. With the reinforcement removed, surface residual tensile stresses will reduce fatigue life. Thermal stress relief and introducing favorable surface residual stresses by mechanical means will increase fatigue life of reinforcement-removed welds. Peening will improve the fatigue performance of reinforcement-intact welds.

   B. <u>Longitudinal loading</u>
   Residual welding stresses are the dominant factor affecting fatigue life, although weld geometry can also have a significant effect when the reinforcement is left intact.[19] Thermal stress relief or introduction of favorable surface residual stresses by mechanical means will increase fatigue life.

2. The effect of residual welding stresses on fatigue crack growth may be summarized as follows:

   A. Transverse loading
   A through crack in a multi-pass weld will grow faster in the surface region of tensile stresses than in the interior region of compressive stresses. The overall crack growth rate in the presence of such self-balancing transverse stresses, however, is similar to that in welds without residual stresses.

   Residual tensile stresses developed normal to a weld as a result of restraint of the adjacent structure are equivalent to stresses due to external loads in their effect on fatigue crack growth. In other words, this type of stress is directly additive to that produced by external load and therefore has a significant effect on crack growth rate.

   B. Longitudinal loading
   Longitudinal residual welding stresses can affect the early stages of crack growth. As the crack grows into the base metal the residual stresses are relieved and the fatigue crack growth rate becomes similar to that in unwelded metal.

REFERENCES

1. Osgood, W. R., Editor, "Residual Stresses in Metal and Metal Construction," Reinhold Publishing Corp., New York, NY, 1954.
2. Hill, H. N. "Residual Welding Stresses in Aluminum Alloys," Metal Progress, August 1961.
3. Masubuchi, K., "Residual Stresses and Distortion in Welded Aluminum Structures and Their Effects on Service Performance," Welding Research Council Bulletin 174, July 1972.
4. Structural Welding Code, American Welding Society, AWS D1.1-Rev. 2-77.
5. Wirsching, P. H. and Kempert, J. E., "Design Codes that Fight Fatigue," Machine Design, Sept. 23, 1976.
6. Munse, W. H. "Fatigue of Welded Steel Structures," Welding Research Council Monograph, New York, NY, 1964.
7. Gurney, T. R. "Influence of Residual Stresses on Fatigue Strength of Plates with Fillet Welded Attachments," British Welding Journal, Vol. 7, No. 6, June 1960.
8. Nordmark, G. E. "Peening Increases Fatigue Strength of Welded Aluminum," Metal Progress, November 1963.
9. Gurney, T. R. "Fatigue of Welded Structures," Cambridge University Press, London, 1968.
10. Gurney, T. R. and Maddox, S. J., "A Re-analysis of Fatigue Data for Welded Joints in Steel," Welding Research International, Vol. 3, No. 4, 1973.
11. Trufiakow, W. I. and Paricki, "Effect of Residual Stresses on the Fatigue Strength of Welded Joints in Aluminum Alloys," Przeglad, Spawalnictwa, Vol. 29, No. 8, 1977, pp. 176-177.
12. Booth, G. S. "The Effect of Mean Stress on the Fatigue Lives of Ground or Peened Fillet Welded Steel Joints," The Welding Institute 34/1977/E, March 1977.
13. Burk, J. D. "Predicted Effects of Residual Stresses on Fatigue Life of Butt Welds," Ph.D. Thesis, University of Illinois at Urbana-Champaign, 1978.
14. Pisarski, H. G. "Fatigue Crack Propagation in Aluminum Alloy Weldment," Welding Research International, Vol. 6, No. 4, 1976.
15. Kelsey, R. A. "Effect of Heat Input on Welds in Aluminum Alloy 7039," Welding Journal, Welding Research Supplement, Vol. 50, No. 12, December 1971, pp. 507s-514s.

16. Dinsdale, W. O. and Young, J. G., "Significance of Defects in Aluminum Alloy Fusion Welds," British Welding Journal, Vol. 9, 1962, pp. 482-493.
17. Pense, A. W. and Stout, R. D., "Influence of Weld Defects on the Mechanical Properties of Aluminum Alloy Weldments," Welding Research Council Bulletin 152, July 1970.
18. Sanders, W. W., Jr. "Fatigue Behavior of Aluminum Alloy Weldments," Welding Research Council Bulletin 171, April 1972.
19. Sanders, W. W., Jr. and Gannon, S. M., "Fatigue Behavior of Aluminum Alloy 5083 Butt Welds," Welding Research Council Bulletin 199, October 1974.
20. Lawrence, F. V., Jr.; Munse, W. H.; and Burk, J. D., "Effect of Porosity on Fatigue Properties of 5083 Aluminum Alloy Weldments," Welding Research Council Bulletin 206, June 1975.
21. Kelsey, R. A. and Nelson, F. G., "Mechanical Properties of USA/USSR Al-Mg Plate and Welds for LNG Applications," Vol. 24, Advances in Cryogenic Engineering, 1977.
22. Kaufman, J. G. and Kelsey, R. A., "Fracture Toughness and Fatigue Properties of 5083-0 Plate and 5183 Welds for Liquefied Natural Gas Applications," ASTM STP 579 'Properties of Materials for Liquefied Natural Gas Tankage,' 1975, pp. 138-158.
23. Kelsey, R. A.; Nordmark, G. E.; and Clark, J. W., "Fatigue Crack Growth in Aluminum Alloy 5083-0 Thick Plate and Welds for Liquefied Natural Gas Tanks," ASTM STP 556 'Fatigue and Fracture Toughness Behavior--Cryogenic Behavior,' 1974, pp. 159-185.
24. Kelsey, R. A.; Wygonik, R. H.; and Tenge, Per, "Crack Growth and Fracture of Thick 5083-0 Plate Under Liquefied Natural Gas Ship Spectrum Loading," Ibid.
25. Dawes, M. G. "Fatigue Pre-Cracking Weldment Fracture Mechanics Specimens," Metal Construction and British Welding Journal, Vol. 3, No. 2, February 1971.
26. Nelson, F. G. and Kaufman, J. G., "Fracture Toughness of Plain and Welded 3-In.-Thick Aluminum Alloy Plate," ASTM STP 536 'Progress in Flaw Growth and Fracture Toughness Testing,' 1973, pp. 350-376.
27. Kelsey, R. A. "Fatigue and Fracture Characteristics of Al-Mg Butt Welds," Aluminium, October 1977, pp. 600-606.
28. Shumaker, M. B.; Kelsey, R. A.; Sprowls, D. O.; and Williamson, J. G., "Evaluation of Various Techniques for Stress Corrosion Testing Welded Aluminum Alloys," ASTM STP 425, 1967, pp. 317-341.

A STUDY OF VIBRATIONAL RESIDUAL STRESS RELAXATION PARAMETERS IN WELDMENTS

L. H. Burck, R. J. Miklasiewicz, and S. Weiss

Materials Department, University of Wisconsin-Milwaukee
Milwaukee, Wisconsin 53201

ABSTRACT

Although vibrational stress relief of weldments is attractive as an alternative to thermal techniques, there is relatively little understanding of the significant parameters or operating mechanisms. With this lack of understanding effective use and maximum benefits often are not realized. In the present study, specimens machined from A36 base metal and from submerged arc welds in this metal were uniformly loaded in axial tension with various mean strains which simulated residual stresses and with various applied vibratory stresses. Mean stress decay was then measured as a function of number of vibratory cycles for each condition. The extent of mean stress relaxation obtained was found to increase with both increasing mean stress and vibratory stress levels. Stress relaxations of up to 32% were observed in base material for maximum (mean plus vibratory) stresses up to 13% over the monotonic yield strength (8% over the cyclic yield strength). However, some degree of relaxation was also observed for maximum stress levels as low as 75% of the monotonic yield strength (72% of the cyclic yield strength). Vibratory stress relaxation in the higher strength weld metal was found to have the same dependence on mean and vibratory stress as did the base metal when the maximum stresses were normalized by the base and weld metal yield strengths. In all cases the majority of the relief occurred within the first 50,000 vibratory cycles and was completed by 100,000 cycles. These results are discussed with respect to component application and possible fatigue damage.

INTRODUCTION

Residual stresses are unavoidably introduced during manufacturing and fabrication operations. Occasionally these stresses are beneficial but more often they limit performance by accelerating failure from fatigue, brittle fracture, stress corrosion cracking, or dimensional instability. In particular, weldments inherently possess extremely high localized residual stresses which are typically of yield strength magnitude and are especially susceptible to these types of problems.

The most common method of relieving residual stresses prior to service is by thermal treatment during which the material flows plastically when the residual stresses exceed the high-temperature yield stress of the material or by creep deformation in response to the residual stress. Problems associated with such thermal methods include cost, energy consumption, possible microstructural degradation, and surface oxidation.

An alternative to thermal methods is vibrational stress relief. This technique involves the application of mechanical vibrations to the piece at room temperature and thus avoids many of the problems associated with thermal methods. At the present time, however, there is insufficient understanding of the basic mechanisms and relevant parameters involved in the vibrational method to assess accurately its actual or potential effectiveness for many engineering alloys. Furthermore, although welded structures would seem to be principal candidates for vibrational stress relief, very little data obtained directly from weldments has been reported. On the other hand, a number of investigations have shown that stress relaxation can occur as a result of cyclic loading,[1-8] even when the maximum stress is below the yield stress, and several reports of industrial application have appeared.[9-11]

In one study which specifically considered the vibrational stress relief of weld-induced residual stresses, Weiss, Baker and DasGupta[7] found that disc-shaped AISI 1018 steel specimens containing circular mid-radius welds showed significant residual stress relief when vibrated in resonance. However, because of the non-uniform nature of the vibratory stress and the fact that residual stress relief was measured for each welded plate as a whole, it was not possible to assess the relative contributions of the weld metal, the heat affected zone, and the base metal to the overall stress relaxation. Since, in general, these various weld-associated regions have significantly different thermo-mechanical histories, microstructures, and mechanical properties, they might be expected to respond differently to applied vibratory stresses. Such potential differences have significance because of the highly localized nature of many weld residual stresses and the non-uniformity of the vibratory stresses which can be applied in a practical manner to engineering structures.

It was the objective of the presently reported study to compare the responses of simulated residual stresses in a common structural steel (ASTM A36) and in weld metal from a typical submerged arc weld in this material to imposed vibratory loading under controlled conditions. In particular, it was of interest to investigate the relative importance of the various vibratory stress relaxation parameters (alternating, residual, and maximum stress levels and number of vibratory cycles) and to relate the stress relaxation responses of the weld and base metals to their respective mechanical and metallurgical properties.

EXPERIMENTAL PROCEDURE

The base material studied was ASTM A36 steel, the chemical composition of which is given in Table I. Test specimens were machined from welded 5/8 in. (15.9 mm) thick hot-rolled plate in order to contain either all base material or all weld metal in the gage section. A drawing of the test specimen is given in Figure 1.

The weld from which the test specimens were removed was produced by the submerged arc process. A single-V-groove butt joint incorporating a 60° included angle, 0.25 in.(6.35 mm) land and 0.25 in.(6.35 mm) root opening

Table I Chemical Composition of A36 Base Material

Element	Weight Percent
C	0.25
Mn	0.81
P	0.010
S	0.027
Si	0.056

and a steel backing strip was used. Full penetration welds were accomplished using a two pass technique, DCRP, 600 amperes and 29 volts. The electrode and flux materials used were in accordance with the AWS EM12 - K and F72 respectively. Interpass temperature was limited to 300°F (422°K) maximum during welding.

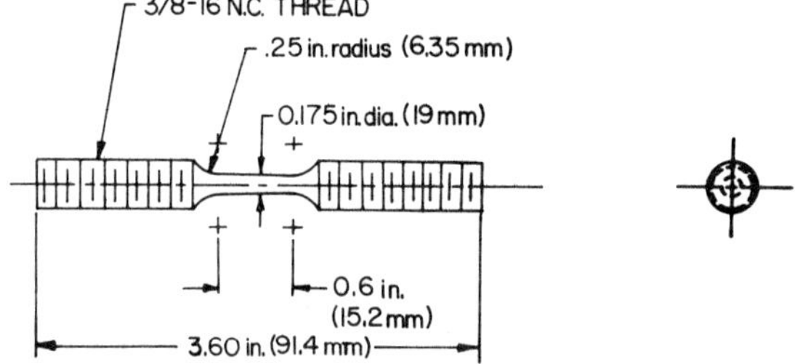

Figure 1 Drawing of the test specimen design.

The specimen design shown in Figure 1 was used both for the vibratory stress relaxation testing and to measure the monotonic tensile properties of the weld and base metals. In addition, a similar specimen was used to determine the cyclic stress-strain properties of the A36 base material by the incremental step technique.

Vibratory testing was accomplished by mounting test specimens in a cage attached to an electromagnetic shaker as shown schematically in Figure 2.

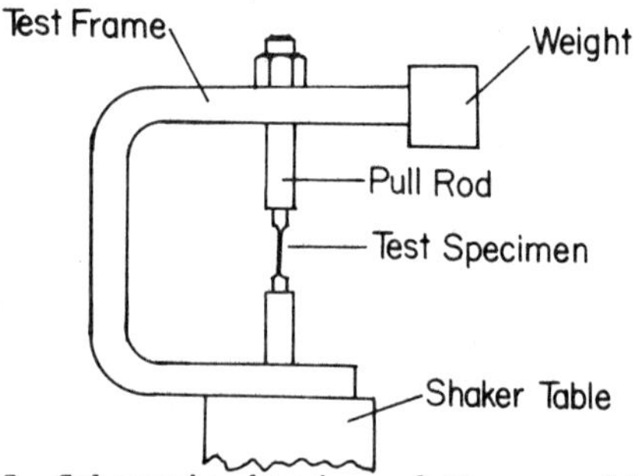

Figure 2 Schematic drawing of the assembly used for vibratory testing.

Static mean strains were applied to the specimens by means of a threaded nut on the top of the specimen pull rod. Vibratory stresses were superimposed by shaking the entire assembly at 44 cycles per second which was eighty percent of the resonant frequency. Loads on the test specimen were monitored by means of a strain-gaged load link in the pull rod assembly and a light-recording oscillograph. In this manner, decay of the static mean stress was measured as a function of number of vibratory cycles for various combinations of initial mean stress and vibratory stress.

In testing the A36 base material, alternating vibratory stress levels of ±2.60 ksi (17.9 MPa), ±4.50 ksi (31.0 MPa), and ±6.50 ksi (44.8 MPa) were utilized. For the weld metal samples the alternating vibratory stress was ±6.5 ksi (44.8 MPa). Mean stress levels were chosen such that the total applied stresses (mean plus alternating) ranged from 67% of the monotonic yield strength in the base metal samples to well above the yield strength and up to 90% of the monotonic yield strength in the weld metal samples.

RESULTS

The mechanical properties of the A36 base material and the weld metal are given in Table II. As shown in the table, the weld metal possessed higher yield and ultimate strengths than did the base material and the base material hardened somewhat under cyclic loading.

Table II Mechanical Properties of the Base Material and Weld Metal

	Monotonic Yield Stress (0.2% Offset)	Cyclic Yield Stress (0.2% Offset)	Ultimate Strength
Base material	45.8 ksi (316 MPa)	47.5 ksi (328 MPa)	73.5 ksi (507 MPa)
Weld Metal	66.2 ksi (456 MPa)	Not measured	84.0 ksi (579 MPa)

During vibratory loading of the test samples, the mean stress values were observed to decrease asympotically to various final values which depended on the levels of initial mean stress and the vibratory stress. In all cases this relaxation occurred within the first 100,000 cycles of vibration and was mostly completed by 50,000 cycles. The final mean stress values, expressed as percentage reductions from the initial values, are shown for the A36 base material in Figure 3 as a function of initial static mean stress for the three values of alternating stress investigated. As shown in Figure 3, both increasing initial mean stress level and increasing alternating stress produced greater stress relaxation effects. Similar behavior was observed in the weld metal tests, the results of which are shown in Figure 4.

As previously noted, the relaxation effects occurred within the first 100,000 cycles of vibration and subsequent cycling produced no further significant changes. This behavior is shown for a representative sample in Figure 5.

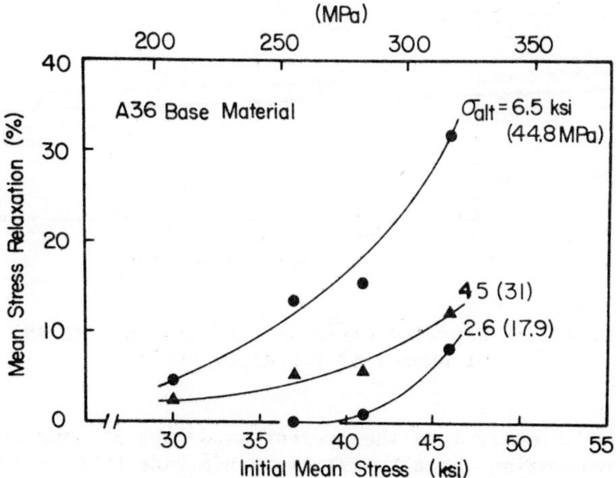

Figure 3 Percentage decrease in mean stress level during vibratory cycling as a function of initial mean stress level for A36 base material.

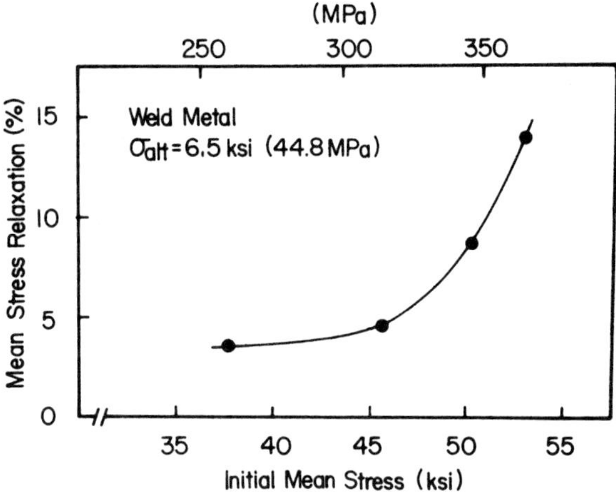

Figure 4 Percentage decrease in mean stress level during vibratory cycling as a function of initial mean stress level for weld metal samples.

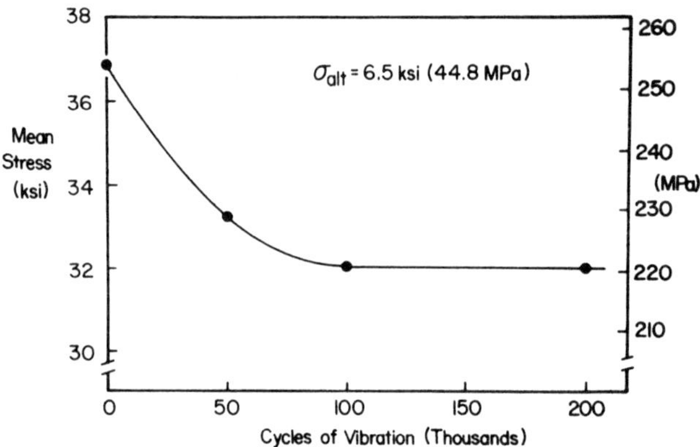

Figure 5 Representative plot of mean stress as a function of number of vibratory cycles.

DISCUSSION

One of the objectives of the present study was a comparison of vibratory stress relaxation characteristics in A36 base metal with those in weld metal. Although it was found that there was less stress relaxation in the weld samples for a given set of conditions the responses of the weld and base metals correlated when compared on the basis of yield strength. This was true even though the respective microstructures and thermomechanical histories differed substantially. For example, Figure 6 shows the weld metal stress relaxation data of Figure 4 together with base metal data for the same alternating stress (Figure 3) plotted as a function of maximum (mean plus alternating) stress to yield stress ratio. It must be noted, however, that the maximum stress was not the only important parameter in determining the extent of stress relaxation realized. For example, significant effects of vibratory stress level are shown in Figure 7 which presents the stress relaxation data of Figures 3 and 4 plotted as a function of maximum (mean plus alternating) stress. Note that increased alternating stress levels produce substantially greater stress relaxation at any given maximum stress.

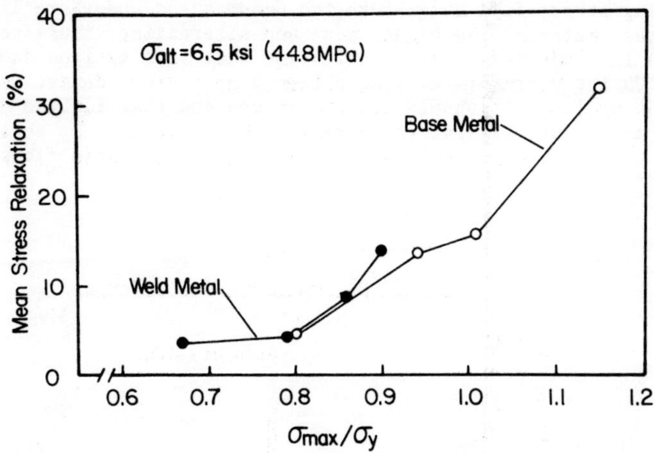

Figure 6 Vibratory stress relaxation data for base metal and weld metal as a function of maximum stress to yield stress ratio, σ Max/σ Yield.

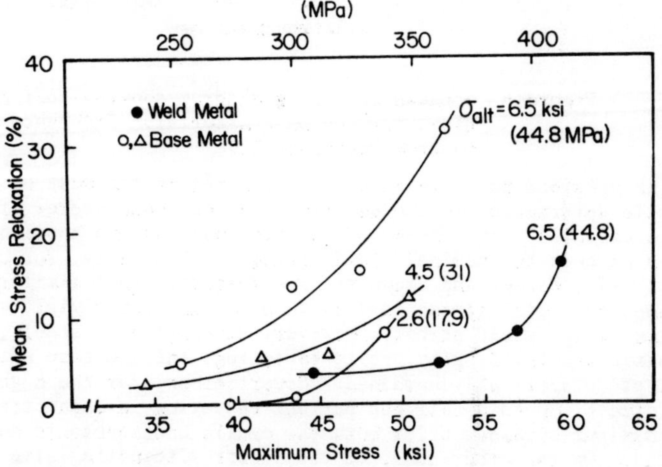

Figure 7 Vibratory stress relaxation data as a function of maximum (mean plus alternating) stress.

A significant factor which must be considered with regard to any vibrational stress relaxation process is the relationship of the alternating stress levels involved to the fatigue strength of the material. Because of the high mean (residual) stresses present, relatively low levels of alternating stress are potentially damaging. Figure 8 shows the mean and alternating stress levels utilized in testing the A36 base material of the present study plotted on a conventional Goodman-Soderberg diagram and Figure 9 shows a similar diagram for the properties and test conditions of the weld metal. For the weld metal specimens, all combinations of mean and alternating stress fall below both the Goodman and Soderberg lines. However, for the base material the higher mean and alternating stress combinations lie above the Soderberg limit, indicating that some fatigue damage may have occurred during vibratory cycling although no obvious damage was observed in 200,000 cycles. It should also be recognized that fatigue damage may occur at much lower alternating stresses than indicated by these diagrams in cases of pre-existing cracks or other significant weld flaws.

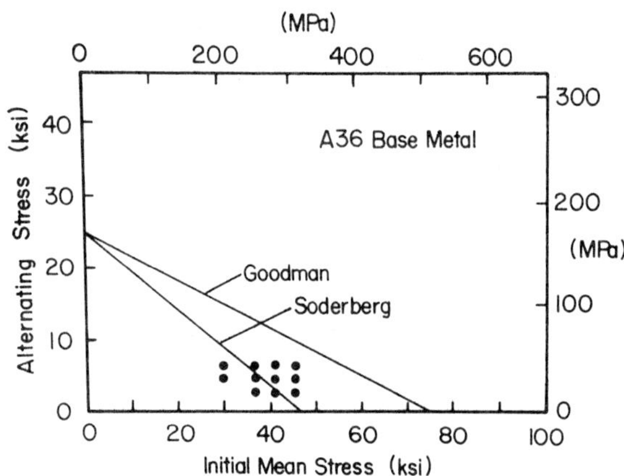

Figure 8 Goodman-Soderberg diagram showing test conditions for the base material. Endurance limit is from Reference 12.

It is apparent that vibrational stress relaxation must be accompanied by plastic deformation in the material and that some degree of relaxation is to be expected when the sum of the residual and the applied vibratory stresses exceeds the nominal yield strength. Similarly, for cyclic softening materials, Wozney and Crawmer[3] have suggested that similar effects can occur when the total stress is less than the monotonic yield strength but above the cyclic yield stress. However, although many steels, particularly martensitic steels, display cyclic softening, the A36 base material of the present study cyclically hardened. Nevertheless, for the highest alternating stress, up to a thirteen percent reduction in mean stress was observed for maximum stresses below both the cyclic and monotonic nominal yield strengths. On the other hand, at the lowest alternating stress level investigated, negligible relaxation effects were observed for total stress levels as high as ninety-five percent of the monotonic yield stress. These results indicate that relaxation can be obtained for total stress levels below the monotonic and cyclic yield strengths, perhaps by the cyclic accumulation of microplastic strain increments or by localized yielding at stress and microstructural inhomogeneities, and that the mechanism involved is strongly dependent on the level of alternating stress.

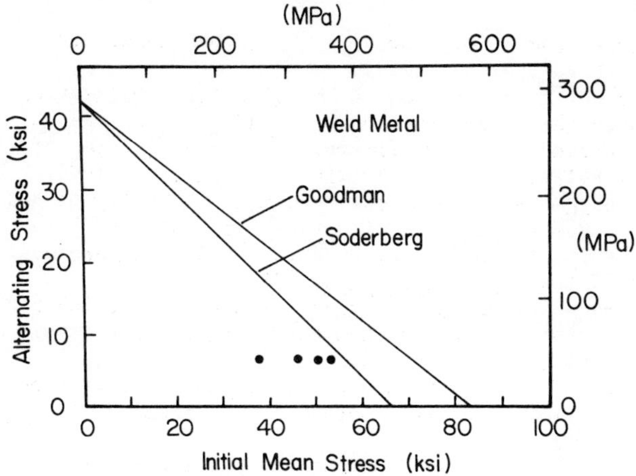

Figure 9 Goodman-Soderberg diagram showing the test conditions for the weld metal. Endurance limit is estimated from the weld metal tensile properties.

In the present tests the specimen fixture shown in Figure 2 had an effective spring constant of 2.12×10^4 lb/in. (3.72 MN/m) at the location of the pull-rod attachment. More rigid fixturing would have produced greater relaxation effects while more compliance would have resulted in less of an effect. Similarly, in practical applications the rigidity of a structure in relation to the portion being stress relieved will affect the degree of vibratory stress relief obtained for a given loading. Also, in this study residual stresses were simulated by the uni-axial loading of test specimens whereas welds in structures are more likely to contain biaxial or triaxial residual stress fields. The significance of the multiaxial nature of such residual stresses with respect to their relief by vibrational methods is yet to be determined.

SUMMARY AND CONCLUSIONS

Specimens machined from A36 steel and from submerged arc welds in this material were loaded in axial tension with various mean strains which simulated residual stresses and with various levels of applied vibratory stress. Mean stress relaxation was measured as a function of these parameters and the number of vibratory cycles.

The extent of mean stress relaxation obtained was observed to increase with both increasing mean stress and with increasing vibratory stress. For the higher alternating stresses investigated, significant stress relaxations were observed in the base material when the maximum (mean plus alternating) stress was well below both the cyclic and monotonic yield strengths. However, for the lowest value of alternating stress, negligible stress relaxation occurred when the maximum stress was below the yield stress. Vibratory stress relaxation behavior in the base material and in the higher strength weld metal were essentially the same at a given ratio of maximum stress to yield stress. In all cases, the majority of relaxation occurred within the first 50,000 vibratory cycles and was essentially completed by 100,000 cycles.

REFERENCES

1. J. Morrow and G. M. Sinclair: Cycle Dependent Stress Relaxation, ASTM STP 237, p. 83 (1958).
2. J. Morrow, A. S. Ross and G. M. Sinclair: Relaxation of Residual Stresses Due to Fatigue Loading, SAE Transactions 68, 40 (1960).
3. G. P. Wozney and G. R. Crawmer: An Investigation of Vibrational Stress Relief in Steel, Welding J. 47, 111-S (1968).
4. B. D. Boggs and J. G. Byrne: Fatigue Stability of Residual Stress in Shot Peened Alloys, Met. Trans. 4, 2153 (1973).
5. E. J. Pattinson and D. S. Dugdale: Fading of Residual Stress Due to Repeated Loading, Metallurgia, 228 (1962).
6. A. L. Esquivel and K. R. Evans: X-Ray Diffraction Study of Residual Microstresses in Shot-Peened and Fatigued 4130 Steel, Exper. Stress Anal., 496 (1968).
7. S. Weiss, G. S. Baker, and R. D. DasGupta: Vibrational Residual Stress Relief in a Plain Carbon Steel Weldment, Welding J. 55, 47-S (1976).
8. A. A. Grudz, et al.: Reducing the Welding Stresses in Plates by Vibration, Automatic Welding, no. 7, 70 (1972). [Avt. Svarka no. 7 (1972].
9. A. Ya. Nedoseka: Vibration Plant for Reducing the Residual Stresses in Welded Structures, Automatic Welding, no. 10, 72 (1973). [Avt. Svarka, no. 10, 74 (1973].
10. O. I. Zubchenko and A. A. Gruzd: Vibrating Loads Used for Relieving The Residual Stresses in Welded Frames, Automatic Welding, no. 9, 59 (1974). [Avt. Svarka 27, no. 9, 64 (1974)].
11. V. M. Sagalevich and A. M. Meister: Elminating Welding Strains and Stresses in Sheet Constructions by Vibration Under Load, Welding Production, no. 9, 1 (1971). [Svar Proiz., no. 9, 1 (1973)].
12. United States Steel Corporation: Design Manual, ADUSS 27-3400-01, (1968).

DISCUSSION

P. Dembowski (General Electric Co.): What were typical specimen to specimen variations in cycles necessary to achieve an equivalent degree of relaxation?

Author: In all cases considered, relaxation was completed by 100,000 cycles with the majority of relaxation occurring within the first 50,000 cycles. In general, the more highly stressed specimens showed more rapid relaxation on a percentage basis; however, all specimens showed the general behavior stated above.

L. W. Sandor (Sun Shipbuilding & Dry Dock Co.): What is your definition of stress relaxation? Would you say that it is a sort of re-distribution of residual stresses? To what extent either the tensile or the compressive residual stress components are relaxed? Are both components relaxed to the same degree? Finally, I would like to ask what are the mechanism(s) involved in this "relaxation process"?

Author: For the purposes of the present study, we define stress relaxation as a decrease in the value of the initially applied static mean stress. In the case of the highly localized, inhomogeneous residual stresses associated with actual welded assemblies, stress relaxation may well involve a redistribution of residual stress since, obviously, the relief of one component of stress (e.g. the peak tensile stress) will also affect the distribution of the equilibrating stress. Overall, there must be an internal balance between the forces associated with the tensile and compressive residual stresses. In the present study, however, the effects observed represent the response of the bulk material because of the uniform nature of the loading.

R. W. Hertzberg (Lehigh University): Could some of the stress relaxation you found be related to crack formation during load cycling?

Author: We do not feel that crack formation during cycling was a significant factor in the stress relaxation which we observed. If cracks were nucleating we would expect a continuous increase in specimen compliance as the cracks grew by fatigue and this would result in a continuously decreasing mean stress in our tests. In fact, this was not observed but rather all relaxation occurred within 100,000 cycles, beyond which no further effects were observed, even for specimens which were further tested for several hundred thousand cycles.

An Advanced State-of-the-Art X-Ray Stress Analyzer

Clayton O. Ruud

University of Denver Research Institute

Denver, Colorado 80208

The objective of this paper is to describe a very rapid, highly accurate, compact portable X-ray instrument for nondestructive stress measurement. As those attending this conference are well aware, a major source of unwanted and unquantified residual stresses is the welding process.

Introduction

X-ray diffraction is the only proven truly nondestructive method that is able to measure these stresses with a reasonable degree of accuracy without making unstressed calibration measurements before the welds are produced. The principle of the X-ray diffraction stress method is illustrated in Figure 1. Here it can be seen that a stress (σ_X) in the plane of the surface would increase the distance between interatomic planes (d). Furthermore, there is a corresponding change in the angle at which X-rays are reflected from those planes, and θ and d are related by the well-known Bragg equation

$$n\lambda = 2d \sin \theta, \tag{1}$$

where n is an integer (usually 1), λ is the wavelength of the monochromatic X-radiation used for analysis, d is the interatomic spacing of a selected set of crystalline planes in the metal and θ is the angle at which the X-rays are reflected from the planes of atoms. Usually 2θ is used in analysis because that angle is equal to the angle between the incident and diffracted beam. The principles of the X-ray diffraction method have been described in a number of texts, of which Barrett or Cullity[1,2] is recommended. Reflections from grains in the metal at two orientations are required for an absolute stress determination. Figure 2 shows how a single direction of incident beam can be reflected in more than one direction. In the single exposure technique which is illustrated in this figure, two reflected X-ray beams are selected which are 180° apart, on opposite sides of the incident beam. The diffracted beam from a single incident X-ray beam is not simply two beams but a cone of reflections.[1,2] The intersection of the cone and a plane perpendicular to the core axis is a circle often referred to as a Debye ring. The single exposure technique (SET) is described in the following equation.

$$\sigma_{SET} = \frac{E}{1+\nu} \frac{(S_2-S_1)}{R_o \sin^2\theta \sin 2\beta} \tag{2}$$

Here σ is the residual and/or applied stress, S_2-S_1 is the relative position of the X-ray peaks with respect to the incident beam, E is the elastic modulus, ν is Poisson's ratio, θ the Bragg angle from an unstressed material, and β is the angle between the incident beam and the normal to the surface of the material. The derivation and a more detailed description of this equation may be found in the SAE Handbook Supplement J784a, published by the Society of Automotive Engineers.[3]

Stress Measuring Instrument

Several years ago, two metallurgists at the University of Denver Research Institute, aware of the lack of application of the X-ray method to stress analysis, set out to determine why applications were so sparse. The conclusion was that the speed, portability and accuracy of available instrumentation severely limited practical applications. We came to the conclusion that the most likely means to improve the X-ray stress method was through the application of an X-ray position sensitive detector. Upon surveying and testing position sensitive detectors available, we developed a new concept and eventually invented a position sensitive scintillation detector. Position sensitive means that the detector is not only capable of determining the number of photons striking its X-ray sensitive surface, but it also can determine the position at which those photons strike the surface. Therefore such a detector is capable of simultaneously determining the intensity of the X-rays and the direction from which they emerge from the surface of a specimen. The detector which we conceived and invented allowed for the development of a stress analyzer of about the size and shape of a quart thermos bottle and weighing less than 10 pounds; figure 3 shows a sketch of the stress head. Upon testing we found that the DRI stress head was more accurate and rapid than the available methods, except for those methods applied under extremely precise laboratory conditions. The device is capable of being fully automated, and the laboratory device which we are presently using is automated except for the positioning of the X-ray head. It is capable of measuring stresses on most materials of practical interest, *i.e.* aluminum alloys, stainless, ferritic and martensitic steels, in 10 to 60 seconds with excellent accuracies.

Results

A CRT oscilloscope display of peaks from the X-ray stress analyzer is shown in Figure 4. This is from 7075-T6 aluminum alloy and it shows the X-ray reflection from the (511,333) peak. The 2θ angle is approximately 162.5° and Cu Kα radiation was used. The beta angle, in this case, was 40°, which gives two psi angles; one at 49° and the other at 31°. Psi is equal to the angle between the normal to the sample and the normal of the planes reflecting a given X-ray beam. You will note that there are two sets of doublets shown. The incident X-ray beam passes between these two doublets and the relative position of the peaks on either side relates to R_2-R_1 in equation (2).

Figure 5 shows results from 6061-T6 aluminum alloy showing peaks from the unstressed condition on the bottom and the stressed condition on top. Stressing was done on a 4-point stress jig and it required 60 seconds to do each set of peaks. Figure 6 shows a CRT oscilloscope photograph from 304 stainless steel. This is a (311) peak reflected at about 147.5°2θ using Cr Kβ radiation. The β angle is 25° which gives psi's of 41 and 9°.

Due to effects of inhomogeneity and anisotropy it is recommended that calibration plots of X-ray peak shift versus applied stress be obtained from material of similar thermo-mechanical conditions to that which is to be tested. Calibration is performed for the X-ray stress head by plotting the peak shift versus applied stress, as read by electrical resistance strain gauges, induced by a 4 point bend stress device similar to that reported by Prevey.[4] Figure 7 shows a calibration curve of stress versus the X-ray shift for 304 stainless steel. A least square fit of the data to a straight line shows a standard deviation of about 1.5 ksi which is the lowest standard deviation for a calibration curve on austenitic stainless steel that to the author's knowledge has been published. The 15 data points here required approximately two hours to obtain including reading four strain gauges each time, which was the most time consuming portion of the experiment.

The described stress analyzer has also been applied to *in situ* residual stress measurements inside of 304 10-inch Schedule 80 stainless steel pipe. Both hoop and longitudinal stress readings inside of the pipe have been obtained.

Conclusion

A new type of X-ray stress anlyzer has been developed as a result of an invention of a position sensitive scintillation X-ray detector. This new stress analyzer provides a method for rapid, accurate and nondestructive stress measurement in structural metallic parts. This highly versatile stress instrument could well be applied to weld parameter development to control residual stresses and for residual and applied stress inspection of existing structures.

Acknowledgement

The work described in this report has been possible through funding with the U.S. Army MERDCOM and the Electric Power Research Institute.

References

1. Barrett, Charles S. and T. B. Massalski, Structure of Metals, 3rd ed., McGraw-Hill Book Company, New York, 1966.

2. Cullity, B. D., Elements of X-Ray Diffraction, 2nd ed., Addison Wesley Publishing Company, Inc., Reading, Mass., 1978.

3. Society of Automotive Engineers, "Residual Stress Measurement by X-Ray Diffraction - SAE J784a," SAE, 400 Commonwealth Dr., Warrendale, PA 15096, 1971.

4. Prevey, P. S., "A Method of Determining the Elastic Properties of Alloys in Selected Crystallographic Directions for X-Ray Diffraction Residual Stress Measurement," Adv. in X-Ray Analysis, Vol. 20, Plenum Publishing Corp., N.Y., p. 345-354, 1976.

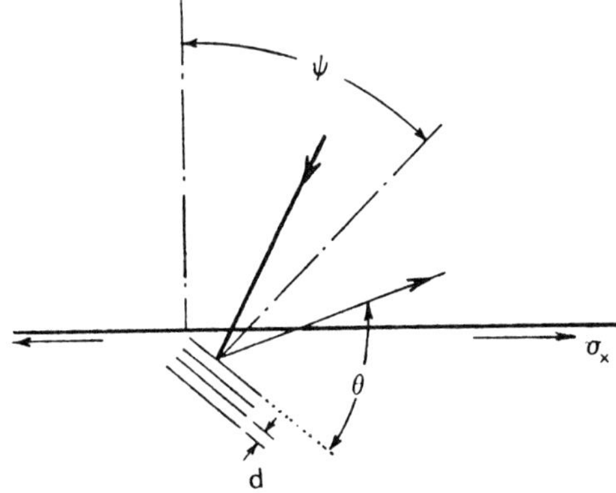

Fig. 1. Surface stress effect upon interplanar spacing (d) and the angle of the diffracted X-ray beam (θ)

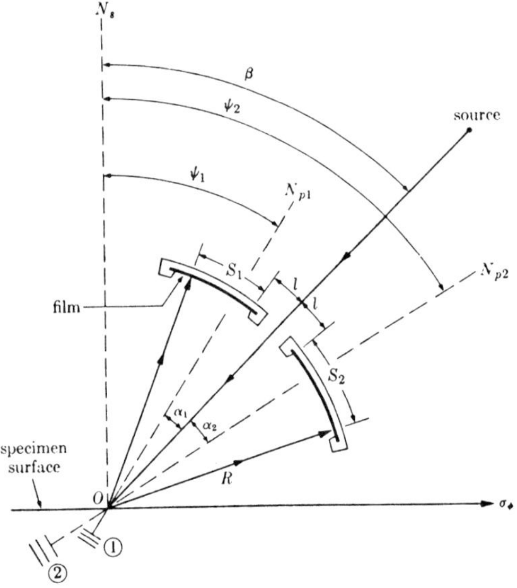

Fig. 2. Conceptual SET

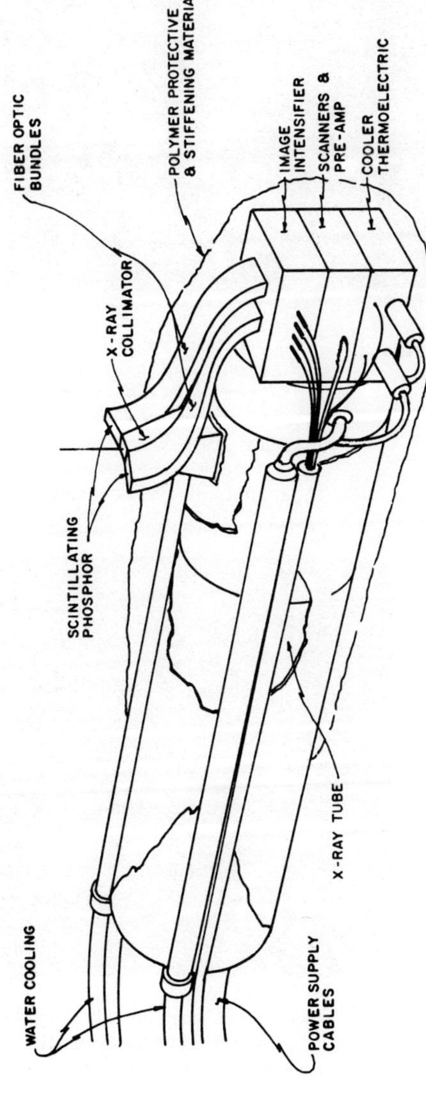

Fig. 3. X-ray stress head with position sensitive scintillation detector and X-ray source

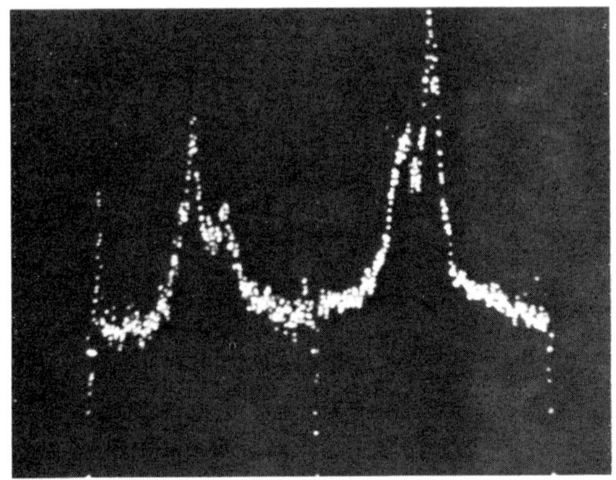

Fig. 4. CRT oscilloscope display from the position sensitive scintillation detector showing two, 180° apart, positions on the Debye rine. The sample is 7075-T6 aluminum alloy and the X-ray peaks represent the Cu K-alpha-1,2 doublet diffracted from the (333,511) set of planes.

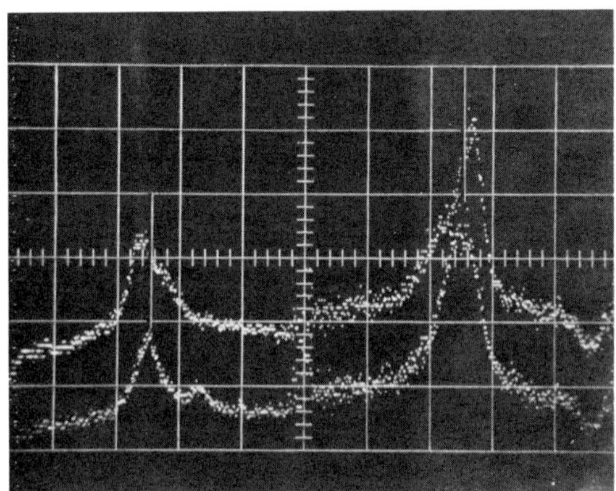

Fig. 5. CRT oscilloscope display from the position sensitive scintillation detector comparing unstressed (bottom set) to stressed (top set). The 4 point bend specimen is 6061-T6 aluminum and the rest of the conditions are the same as in Fig. 4.

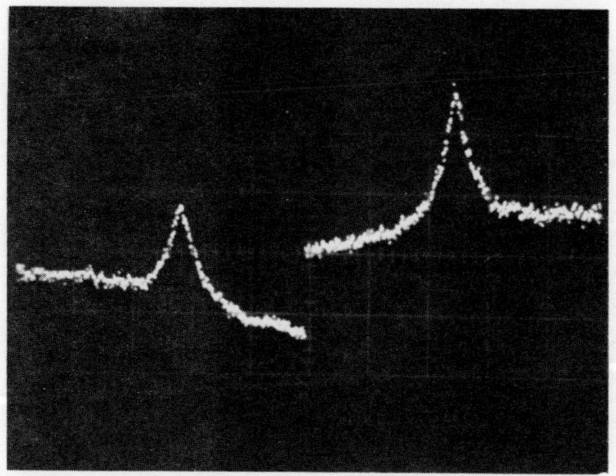

Fig. 6. CRT oscilloscope display from the position sensitive scintillation detector showing two, 180° apart, positions on the Debye ring. The sample is 304 stainless steel sheet and the X-ray peaks represent Cr K-beta diffracted from the (311) set of planes.

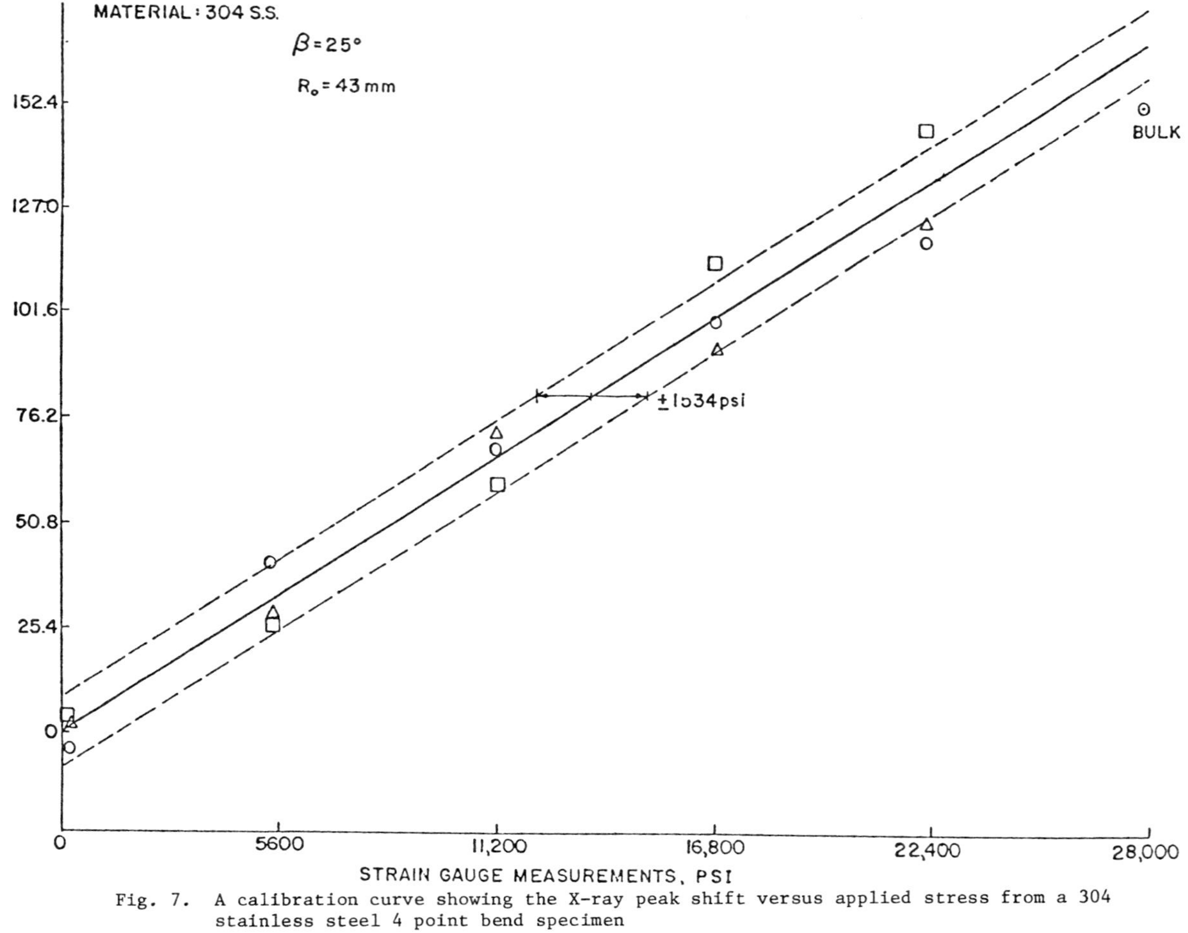

Fig. 7. A calibration curve showing the X-ray peak shift versus applied stress from a 304 stainless steel 4 point bend specimen

J. B. Ballance (Editor, Journal of Metals): What are the projections for this portable X-ray stress analyzer to be manufactured and sold commercially?

Author: At this time the University has no plans to manufacture or sell this device commercially. Presently we are building instruments on a prototype basis for two sponsors and performing stress measurements as a service to the industry.

E. Landerman (Westinghouse PWR): What is the effect of surface finish with the use of the equipment discussed especially since there are needs for evaluation of residual stresses insitu?

Author: The rather low energy X-ray beam used for XRD penetrates to a depth of a few tenths of a thousandth of an inch to as deep as a thousandth. However in many cases where residual stresses are of concern, this shallow depth of penetration is advantageous due to the steep stress gradient. A number of laboratories have used an electro-polishing technique to measure stresses to several tenths of an inch in depth.

VARIABLES INFLUENCING WELD SENSITIZATION OF AUSTENITIC STAINLESS STEEL

H. D. Solomon

General Electric Corporate Research and Development Center
Schenectady, New York 12301

The process of the sensitization of the HAZ of an austenitic stainless steel weld is more complex than that which results from simple isothermal exposures. This paper summarizes the results of recent studies of the variables which influence weld sensitization.

Two sorts of variables are considered; HAZ variables and alloy variables. The former includes the influence of the peak temperature reached in the HAZ during each weld pass, the cooling rate, the strain developed in the HAZ during cooling and the sequence of the thermomechanical cycles experienced by the HAZ. The latter variables include such factors as the influence of any prior annealing treatment, prior cold working and the composition of the alloy.

It is shown that the influence of the cooling rate and sequence of thermo-mechanical events is strongly dependent upon the peak temperature reached as the result of each welding pass. Peak temperatures of approximately 800 - 900°C are more damaging than higher or lower temperatures. Straining during cooling is shown to accelerate sensitization, for a single cycle a strain of about 5% is critical for this acceleration. It is also shown that sensitization can be influenced by prior thermal and mechanical treatments with the <u>least</u> sensitization being experienced by specimens which had been adequately annealed and then rapidly quenched. Lastly, some preliminary data are presented showing that the expected greater resistence to sensitization of 304L over 304 is indeed demonstrated, as is the superiority of 316L over 316, and 316 over 304.

I. INTRODUCTION

Stainless steels can be made sensitive to grain boundary attack in certain media. This sensitization results from the grain boundary chromium depletion that is caused by grain boundary precipitation of chromium carbides. Sensitization can be produced by isothermally heating and holding a stainless steel in the temperature range in which carbide precipitation is both thermodynamically and kinetically possible (i.e., 450 - 850°C for an austenitic stainless steel). It can also be produced by cooling through this sensitization range. It is this latter sort of thermal cycle that a weld heat affected zone (HAZ) experiences. This paper summarizes the results of recent studies of the variables which influence weld sensitization.

Two sorts of variables are considered - HAZ variables and alloy variables. The former includes the influence of the peak temperature reached in the HAZ during each weld pass, the cooling rate, the strain developed in the HAZ during cooling and the sequence of the thermo-mechanical cycles experienced by the HAZ. The latter variables include such factors as the influence of any prior annealing treatment, prior cold working and the composition of the alloy.

II. EXPERIMENTAL PROCEDURES

Two kinds of specimens have been considered. Specimens with a hourglass 1/8" dia. gage section were used in studies where simultaneously controlled thermal and mechanical cycling was desired. These specimens were machined from 1/2" dia. bars of 304 S.S. containing 0.05 wt% C. Induction heating was used in these experiments in conjunction with servo controlled straining which was controlled on the basis of diametrical strain measurements.(1) A dedicated computer was used to control the temperature (by varying the power input) and to subtract out the thermal strains, thus allowing controlled mechanical strains to be applied in any desired manner while independently controlling the temperature.

A more numerous set of experiments employed straight unmachined 1/8" dia. rods which were heated by the passage of a low voltage alternating current through the specimen.(2) These specimens contained 0.08 wt% C. The temperature control was achieved with a chart driven signal generator or more recently with the aid of a microprocessor, both of which varied the power input with reference to a desired temperature cycle and thermocouple reference input. For the most part these experiments were run in a setup which allowed for the free expansion and contraction of the specimen. Some recent tests have also been run in a modified setup in which a microprocessor-controlled vise clamped the specimen when the peak temperature was reached so that on cooling, the thermal strain was converted to a mechanical strain.

The procedure followed in all experiments was to expose a specimen to a thermal or thermo-mechanical cycle and then to use any one of a variety of testing procedures to determine the degree of sensitization. ASTM test A262 practice E(3) (the modified Strauss test) was the most extensively used test. Among the other tests used were ASTM-A262A (the oxalic etch test), the Electro Potentiokinetic Reactivation (EPR) test,(4) and the slow strain rate test in high temperature water.(5) Most of the results presented here will be confined to those obtained with the A262E test; the results obtained with the other tests will be mentioned just briefly.

The compositions of the 1/2" dia. and 1/8" dia. rods are compared in Table I. In addition to the differences in the carbon content and annealed grain sizes listed in Table I, there was also a variation in the as received condition. The 1/8" rods (.08 wt% C) showed considerable metallographic and mechanical evidence of being worked (it was estimated that the degree of deformation was equivalent to about 10% cold working). The 1/2" dia. rods (.05 wt% C) were mill annealed at 1100°C and then air cooled. Reannealing at 1100°C followed by water quenching did not alter the grain size or mechanical properties. Annealing the 1/8" dia. rods (.08 wt% C) at 1100°C recrystallized the rods and eliminated the worked structure. Both sets of rods were tested in their as received and annealed conditions. In addition, some of these annealed rods were also prestrained at room temperature prior to thermo or thermo-mechanical cycling.

TABLE I

COMPOSITION AND GRAIN SIZE OF THE RODS USED IN THIS STUDY

Wt%

	C	Cr	Ni	Mn	Si	S	P	Fe	G.S. ASTM #
1/8" Dia. Rods	0.077	18.2	8.43	1.12	0.40	0.025	0.025	Bal.	5.5 (As received) 7 (Annealed)
1/2" Dia. Rods	0.051	18.6	9.0	1.57	0.73	0.031	0.029	Bal.	8 (Both for the mill annealed and reannealed specimens)

III. RESULTS AND DISCUSSION

A. HAZ Variables

The HAZ of a weld experiences the sort of thermal history shown in Figure 1. These temperature-time profiles are idealized slightly to the extent that linear cooling at elevated temperatures was used. In real weld thermal profiles there is somewhat more of a decrease in cooling rate with decreasing temperatures than is shown in Figure 1.[6-8] Over the range of about 500 - 800°C, however, a linear approximation is quite reasonable. The temperature profiles shown in Figure 1 were obtained from thermocouples attached to test specimens and measure the response during a weld simulation experiment. These cooling rates were patterned after those actually measured in a 26" Sch 80 pipe weld.[8]

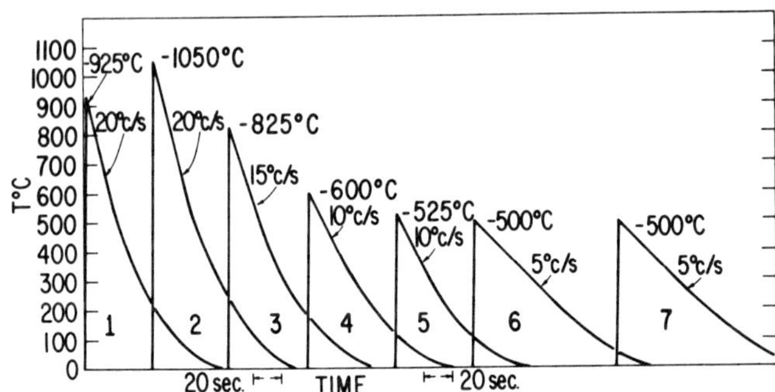

Figure 1: Idealized Weld HAZ thermal cycles patterned after data taken 0.1" from the Fussion Line on the ID of a 26" Sch 80 pipe. (Thermal data taken from Reference 8)

The HAZ experiences a series of heating and cooling cycles with different peak temperatures and cooling rates associated with each cycle. In studying this process, first the sensitization due to a single heating and cooling cycle was considered, then the successive cycles were considered, and finally actual weld thermal histories were considered. The following results and discussion will follow this sequence of tests.

III A.1. Influence of Peak Temperature and Cooling Rate

There are three parts to any of the thermal cycles shown in Figure 1. First there is a heating phase, a peak temperature is reached, and finally there is the cooling. Experiments performed with heating rates in the range of 20 to 135°C/S have failed to show any influence of the heating rate in this range of variation,[2] and therefore no further reference to the heating rate will be made. The cycle variables of interest are the peak temperature of a cycle and the subsequent cooling rate. For simplicity, and because in the sensitization temperature range of 500 - 800°C the cooling rate in a real weld can be approximated as being linear, only linear cooling was considered.

Figures 2 and 3 show the results of quantified A262E tests on specimens cooled at a variety of rates from approximately 1000°C and 800°C. This data was obtained on annealed specimens of the 1/8" dia. rods containing 0.08 wt% C). After a 3 day exposure in the test solution (as described by the ASTM procedures[3]), the specimens were removed and pulled to failure in air at room temperature. The depth of intergranular penetration (IGP) was then measured on the fracture surface. The ultimate tensile strength (max load/original area) is also a measure of the degree of attack. Because the attacked region cannot support as much of a load as unattacked material, the greater the intergranular penetration the lower is the measured value of σ_{UTS}. Both measures of attack are shown in Figures 2 and 3.

Two critical cooling rates are defined in Figures 2 and 3. The zero sensitization rate (ZSR) is the cooling rate at and above which there is no sensitization as measured by A262E. The total sensitization rate (TSR) defines the cooling rate at and below which there is 100% intergranular penetration and failure occurs with no yielding (i.e., σ_{UTS} < the nominal yield stress of 30 Ksi). It is clear from Figures 2 and 3 that ZSR and TSR are a function of the peak temperature of the cycle, being higher (easier sensitization) for a maximum temperature of approximately 800°C.

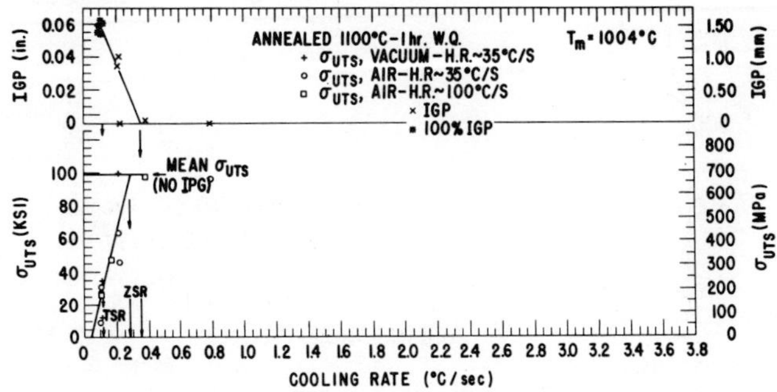

Figure 2: Intergranular Penetration (IGP) and σ_{UTS} vs Cooling Rate for T_m = 1004°C (measurements made on annealed specimens from the 1/8" dia., 0.08 wt% C rods).

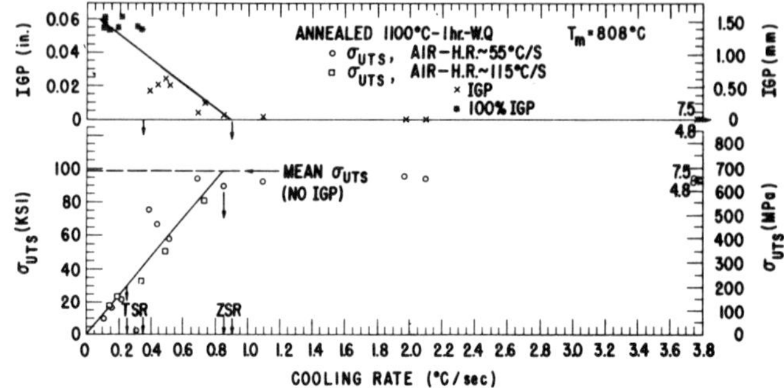

Figure 3: Intergranular Penetration (IGP) and σ_{UTS} vs Cooling Rate for T_m = 808°C (measurements made on annealed specimens from the 1/8" dia., 0.08 wt% C rods).

Peak temperatures for 600 to 1100°C have been studied and the ZSR and TSR values are shown as a function of peak temperature in Figure 4. Also compared in Figure 4 are computer predictions based on isothermal sensitization data.[2] Figure 4 illustrates that the drop of ZSR and TSR at and above 1000°C is not expected from our previous view of sensitization. Heating to 1000°C or even to 1100°C did not alter the structure of these rods, which were previously annealed for one hour at 1100°C. The origin of the difference between specimens cooled from 1000° and 800°C is at present unknown. For both temperatures, the carbides are similar in size and spacing along those grain boundaries which contain carbides[2]. It is the number of grain boundaries which contain carbides which changes with peak temperature. Work is underway to look at Cr, C, and solute segregation to see if any of these factors explains the origin of this effect.

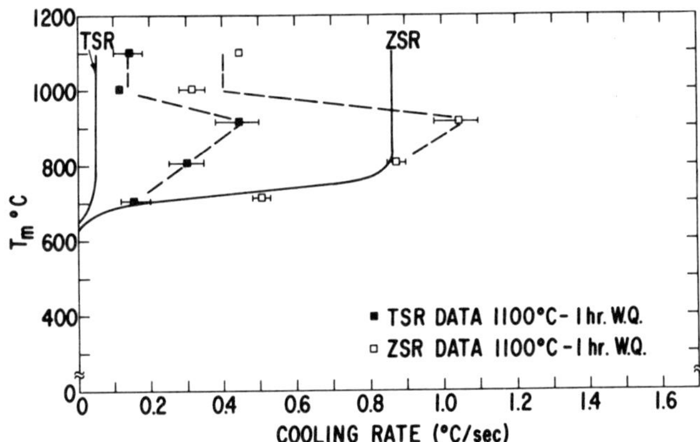

Figure 4: TSR and ZSR vs T_m as experimentally determined and computer predicted measurements (made on annealed (1100°C) 1/8" dia., 0.08 wt% C rods). The computer predictions are defined by the solid lines.

The data of Figures 2, 3, and 4 were from one heat of annealed 304 S.S. using one sensitization detection technique. Similar studies have also been run to date on 7 additional heats of 304 and 316 and they all show this same effect. This effect is also observed when A262A, EPR, and the slow extension rate test are used to evaluate the influence of cooling.[9] Clearly the peak temperature of a thermal cycle plays a significant role in determining the sensitization behavior of a weld HAZ.

III A.2. Multiple Cycling

In multipass welding, the HAZ experiences more than one thermal cycle and thus weld sensitization depends, in general, not on a single temperature-time cycle, but rather on the influence of multiple cycles. Multiple cycles bring a tremendous complication into the problem because the sequence is important as well as the peak temperature and cooling rate associated with each cycle. This is illustrated in Figure 5. This figure shows the post A262E IGP and σ_{UTS} results for specimens experiencing the thermal cycles shown in the inserts and described in the caption (these results were obtained on annealed 1/8" rod specimens containing 0.08 wt% C). As can be seen, the sequence of thermal cycling plays an important role in establishing degree of sensitization. A thermal cycle with a peak temperature of 1000°C can eliminate prior sensitization even though time at this elevated temperature is short, It has also been shown[2] that multiple cycling with Tm = 1000°C is not additive; the sensitization is not enhanced by repetitions of a thermal cycle over that induced by the first cycle.[2] When Tm = 800°C, the prior sensitization is not completely eliminated and thus multiple cycles are to some extent additive (although not 100%).[2]

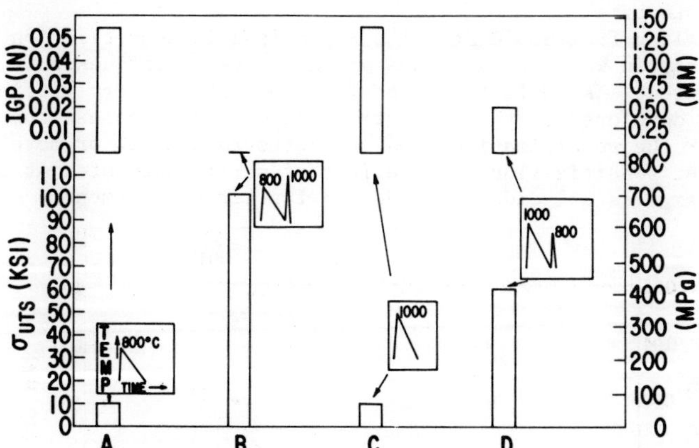

Figure 5: Integranular Penetration Rate (IGP) and σ_{UTS} after multiple cycling:

A - Single cycle, T_m = 800°C, $\dot{C}$ = 0.1 °C/s.

B - Two cycles, first cycle the same as A, then T_m = 1000°C, $\dot{C}$ = 1.0 °C/s.

C - Single cycle, T_m = 1000°C, $\dot{C}$ = 0.1 °C/s.

D - Two cycles, first cycle the same as C, then T_m = 800°C, $\dot{C}$ = 1.0 °C/s.

(measurements made on annealed specimens from the 1/8" dia., 0.08 wt% C rods).

If all the total time at temperature is merely added up with no regard to the sequence of the thermal cycles, then the degree of sensitization is overestimated because resolution effects are disregarded. While such an approach is generally quite inaccurate, it is conservative since the degree of sensitization is overestimated. If an accurate measure of sensitization is required it is necessary to simulate the actual thermal cycles in the proper sequence or to measure the sensitization of actual welds.

III A.3. Influence of Strain During Cooling

The constraint of the cold surrounding base metal causes the weld and HAZ to be strained during the thermal cycle associated with each weld pass. Tests utilizing the application of a tensile or compressive strain during heating and/or cooling were therefore performed to quantitatively determine the influence strain on the sensitization.(1) (These tests were performed on specimens machined from the 1/2" dia. bars containing 0.05 wt%C.) The results of these experiments are summarized in Figure 6 (this data was taken on specimens in the as received air cooled condition). The IGP after an A262E test is shown as a function of the strain applied during cooling at 0.1°C/S or at a higher rate of 0.5 - 0.65°C/S. The strains are mechanical tensile strains developed during cooling, except for 2 specimens where a 5% compressive strain was applied during heating and a 10% tensile strain on cooling (denoted 5 C, 10T). These strains were developed between the peak temperature and room temperature, with the strain per °C being linear. The strain rate was adjusted with the cooling rate so that, for a given strain shown in Figure 6, the strain per °C remained constant.

Cooling from 800°C at 0.1°C/S resulted in sensitization even when the applied strain was zero and thus at this rate an influence of strain was not apparent. At a faster rate of 0.5 - 0.65°C/S very little sensitization was developed in the zero strain case. Strains less than 2.4% did not alter the sensitization behavior but straining 5% or more resulted in significant sensitization. The effect of strain saturated at 5%, strains as large as 25% did not produce a further enhancement.

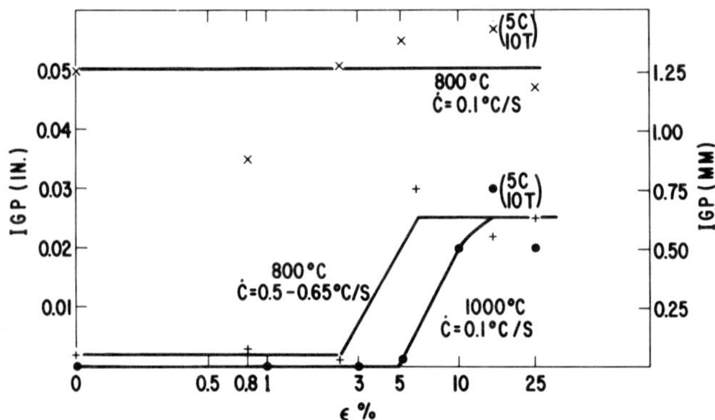

Figure 6: Influence of strain during cooling on sensitization. The sensitization was determined by the degree of intergranular penetration after an A262E test. (Measurements made on mill annealed specimens made from the 1/2" dia., 0.05 wt% C rods).

Cooling from 1000°C is less severe than cooling from 800°C so that at 0.1°C/S no sensitization is observed in the zero strain case. Strains of 5% or more are required for an enhancement of sensitization with saturation of the effect at 10%. With a faster cooling rate from 1000°C, of 0.5° - 0.65°C/S, even strains as large as 25% could not produce sensitization.

Figure 6 illustrates that for the heat studied, relatively large strains are required to enhance sensitization due to a single thermo-mechanical cycle. These strains are in excess of what would be expected in a weld. The exact calculation of the strain developed in the HAZ of a weld is quite complex; an estimate of the magnitude of the strain can, however, be made from the thermal strain developed by the peak temperature. If the peak temperature of the HAZ is 1000°C above the base metal, then for an average coefficient of expansion of 2×10^{-5} per °C[10], the thermal strain ε ($\varepsilon = \alpha\Delta T$) is 2%. The thermal strain can be relieved, in part, by displacements in the free surface of the weld, local plastic distortions, and long range elastic deflections. All of the thermal strain is not, therefore, converted into plastic strain in the HAZ. Figure 6 predicts that even if all the thermal strain were converted into plastic strain, the strain level would not be sufficient to cause an enhancement of sensitization. While this is true for the heat under discussion, it is not necessarily true for all heats of 304 S.S. The 1/2" dia. rods had a lower carbon content than the 1/8" dia. rods and are therefore less easily sensitized (0.05 wt% vs. 0.08 wt%). Preliminary evidence currently being gathered indeed shows that this low carbon content heat is less susceptible to the influence of strain on cooling. Some of this evidence will be discussed in the next section. It should also be kept in mind that for thermal cycles with $T_m \leq 800$°C, the influence of strain is partially additive. Multiple cycling of specimens from the 1/2" dia. rod, cooled from 800°C with only 0.8% strain per cycle, have shown that even with this low a strain there can be an influence of strain if there is multiple cycling.[11]

III.A.4. <u>Weld HAZ Simulation</u>

The HAZ of a weld experiences a series of thermomechanical cycles which were simulated on specimens machined from annealed 1/8" dia. rods containing 0.08 wt% C. The microprocessor controlled temperature controller working in conjunction with AC heating of the specimen caused the specimen to experience the thermal cycle shown in Figure 1. A pneumatic vise also controlled by the microprocessor could be made to lock the specimen in a fixed position at any point in the thermal cycle. Tests were run with the specimen free to expand and contract or in other tests with the vise locking the specimen at the peak temperature, thus causing the thermal strain to be converted into plastic strain as the specimen cooled. In this latter type of test, the specimen was unlocked after it cooled and then allowed to heat unconstrained during the next thermal cycle. At the peak temperature, the specimen was again locked and more strain was imposed upon the specimen as it again cooled. This thermomechanical cycle is not exactly like that which exists in a HAZ since the specimen is heated without constraint (this is necessary to avoid buckling). While not exactly duplicating the thermomechanical straining in the HAZ, this technique does provide a simple way of putting some constraint on a simulation specimen.

The weld HAZ thermal history being simulated was established in the welding of a 26" dia. pipe[8]. The first (root) pass produces less of a peak temperature than the second pass because less heat input was used so as not to burn out the Grinnel insert. The first, second, and third passes were made by GTAW with bare filler wire. Subsequent passes were done by SMAW with coated electrodes. The third and subsequent passes produced successively lower peak temperatures at the ID (.1" from the fusion line) where the temperature was monitored. This is because as the weld is built up, the weld puddle gets further away from the point of temperature measurement. Test specimens were subjected to a simulated version of this history. These cycles are shown in Figure 1.

The results of experiments utilizing weld simulation with and without constraint are illustrated in Figure 7 and Table II. Figure 7 compares the sensitization of annealed specimens of the 1/8" dia. rod sensitized with and without constraint. The degree of sensitization is illustrated by the A262 practice A(3) oxalic acid etch test. As can be seen in Figure 1 and Table II, the above weld thermal cycle is insufficient to cause sensitization in this alloy; constraint during cooling is required. Table II compares the fraction of grain boundaries attacked (as measured with the aid of grid counting)(9) as a function of the prior history. As can be seen, constraint accelerates the sensitization even after the first cycle. The less than 2% strain on cooling ($\alpha\Delta T = 2 \times 10^{-5} \times 900 = 1.8\%$) was enough to enhance sensitization. This result cannot be directly compared to the single thermomechanical cycling results cited in the previous section because different tests were used to resolve the sensitization. A262A tests have, however, also been run on the 1/2" dia. rods used in the strain controlled experiments described previously, and these tests confirm that this alloy shows less of an influence of strain during cooling than the higher carbon 1/8" dia. rods discussed here.

The second cycle, where peak temperature is 1050°C, removes the sensitization of the prior cycle in exactly the same was as was illustrated in Figure 5. The third pass, with the 825°C peak temperature, resensitizes the steel. Here there is an influence of the strain imposed during cooling

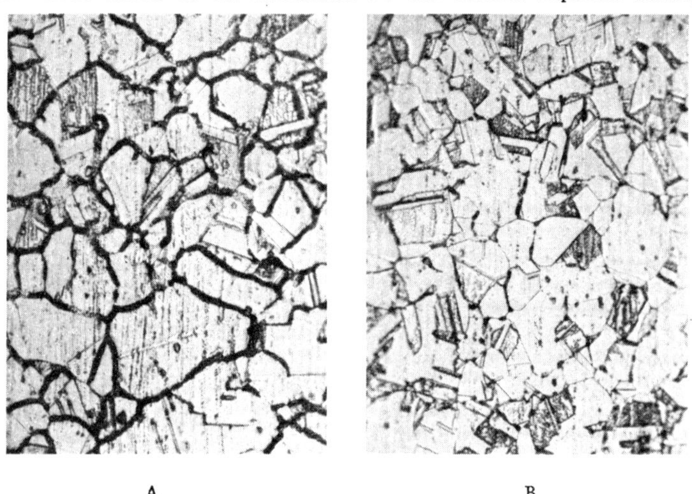

A B

Figure 7: Sensitization revealed by A262A (measurements made on annealed specimens from the 1/8" dia., 0.08 wt% C rods).

A - Specimen exposed to the complete weld simulation (7 cycles) with constraint, 300X

B - Specimen exposed to the complete weld simulation (7 cycles) without constraint, 300X and that imposed during the prior cycle. Further cycles do not seem to have further increased the sensitization, but this observation may merely reflect the fact that the A262A test cannot discriminate between degrees of sensitization when they all cause most of the grain boundaries to be attacked.[9] The A262A test saturates at a lower level of sensitization than the A262E test.[9]

TABLE II

SENSITIZATION AFTER WELD SIMULATION CYCLES. RESULTS OF A262A TESTS

(The cycle numbers refer to the cycles shown in Figure 1. The numbers denote which cycles were experienced).

With Constraint

	1	1,2	1,2,3	1,2,3,4	1,2,3,4,5	All 7
% of Attack in A262A	62	0	79	66	68	68

Without Constraint

	1	1,2	1,2,3	1,2,3,4	1,2,3,4,5	All 7
% of Attack in A262A	33	0	29	-	-	23

B. Alloy Variables

The preceding was concerned with the influence of the thermo-mechanical history of a HAZ on sensitization. Unfortunately, the problem of weld sensitization is not completely understood merely with this information. There is considerable heat to heat variation in the response to a weld thermal cycle. Some of the origins of this variation are discussed in this section. First the influence of treatments prior to welding will be considered and then there will be a discussion of the influence of alloy composition.

III.B.1. Influence of Prior Thermal and Mechanical Treatments

The condition of the steel prior to welding can influence the severity of the weld sensitization. This is illustrated in Figures 8 and 9. Figure 8 shows results obtained on the 0.08 wt% C 1/8" dia. rod specimens that have seen different treatments prior to sensitizing. There is clear metallographic evidence that the as received specimens were worked, but the exact thermomechanical history is unknown.[11] Whatever the exact history, reheating to 1000°C eliminates its influence. Heating to less than 1000°C does not. At these lower temperatures the zero sensitization cooling rate (ZSR) and total sensitization cooling rates are higher for the as received specimens than for the annealed ones (1100°C - 1 Hr water quenched). Tests run on annealed specimens which were prestrained 10% in tension prior to the thermal cycling have shown that most of the difference between the as received and annealed specimens is due to the prior working.[11] The 1/2" dia. rods with their lower carbon content (.05 wt%) also show an influence of prestrain, but it is much less than that exhibited by these higher carbon content (.08 wt%) 1/8" dia. rods.[11] As expected, heating to 1000°C removes the influence of the prior working in both the 1/8" dia. and 1/2" dia. rods.

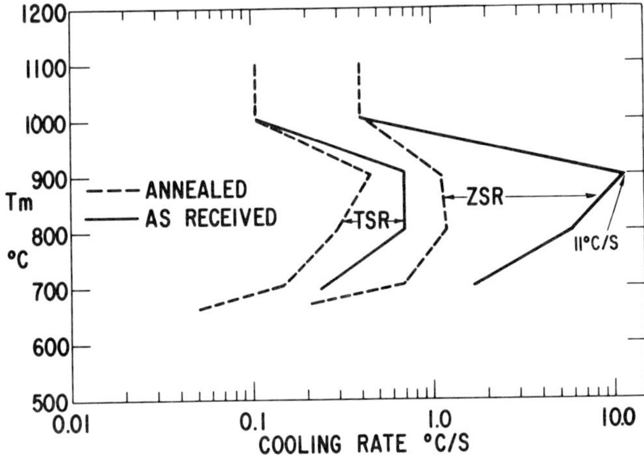

Figure 8: Influence of Prior Working on ZSR and TSR as a function of T_m. (Measurements made on the 1/8" dia. rods, 0.08 wt% C)

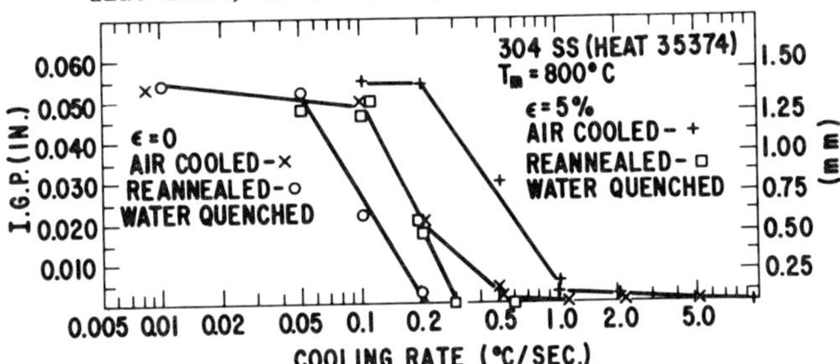

Figure 9: Influence of Air cooling vs Water Quenching on subsequent sensitization by cooling from 800°C at different rates. The sensitization was measured by A262E (measurements made on specimens from the 1/2" dia., 0.05 wt% C rods). Some specimens were strained 5% during cooling.

Prior deformation is not the only pretreatment which can influence subsequent sensitization. Figure 9 shows the difference between the sensitization of specimens of the 1/2" dia. rods (.05 wt% C) which were air cooled after an 1100°C anneal or reannealed at 1100°C followed by water quenching. The results show that the air cooled specimens are more easily sensitized from 800°C than the water quenched specimens. Straining 5% during cooling enhances the sensitization of both sets of specimens. Figure 9 shows that the influence of air cooling after annealing on the subsequent sensitization produced by cooling from 800°C is about the same as that produced by straining during cooling. Figure 9 also shows the way in which several factors may combine to cause sensitization. A water quenched specimen cooled from 800°C at 0.3°C/S is not sensitized. An air cooled specimen which is strained during cooling will be sensitized at 2.0°C/S. Similar experiments to those shown in Figure 9 were also performed on specimens cooled from 1000°C. In agreement with the behavior illustrated in Figure 8, there was no difference in the sensitization behavior of air cooled and water cooled specimens when they were subsequently sensitized by cooling from 1000°C.

III.B.2. Influence of Composition

It has long been known that sensitization can be controlled by varying the alloy chemistry. This is accomplished by removing the carbon and/or adding other elements such as Mo, Nb or Ti. Figure 10 shows that these remedies work. In this figure, the cooling rate required to just start sensitization (ZSR) when one cools from 800°C is displayed as a function of the carbon content for 304 and 316 SS. This is preliminary data but it clearly shows the benefit of 304L and 316L (which contain a maximum of 0.03 wt% C) as compared to alloys containing more than 0.03 wt% C. This figure also shows that for 0.05 wt% carbon content, 316 S.S. is superior to 304 S.S. This result is not surprising and is in agreement with many previous studies.

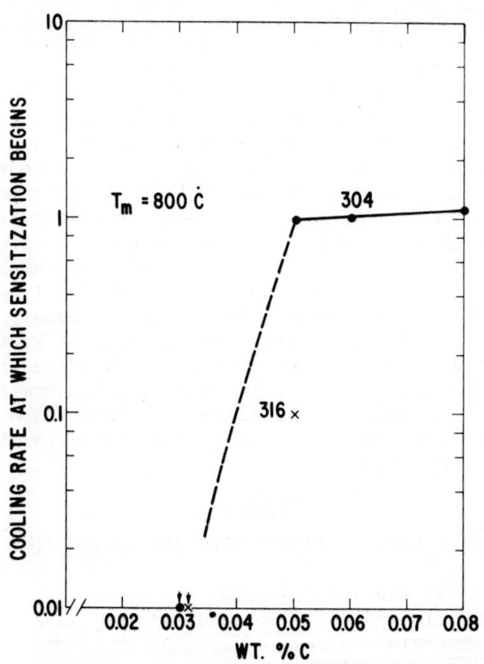

Figure 10: Influence of Wt% C on the critical cooling rate for the start of sensitization from 800°C, as revealed by A262E.

The preliminary evidence is that the carbon content also interacts with prior deformation and straining during cooling. As has been noted, the lower carbon content (.05 wt%C) 1/2" dia. rods show less of an influence of both prestrain and strain during cooling than the 1/8" dia. rods which contain 0.08 wt% C. This is a reasonable behavior. Dislocation theory predicts that prior strain and strain during cooling probably act to increase the carbide nucleation rate and the diffusion rate. The more carbon present the greater the amount of carbide precipitated and the greater will be the influence of strain. There can be no acceleration of precipitation if there is no carbon present. Thus in the limit of zero carbon in the alloy, there should be no influence of deformation. Much more work is needed, however, to develop the quantitative interplay between the carbon content and the influence of deformation.

IV. SUMMARY

Weld sensitization is a complex phenomenon. It is influenced by the alloy chemistry, the thermomechanical history prior to welding, as well as the exact thermomechanical history established during welding. Because the history is so important, it is not possible to describe the process in terms of a simple state function which might be determined from information such as the power input during welding. It is necessary to know not only such variables as the peak temperature and cooling rate at the point of interest in the HAZ, but also the sequence of the thermomechanical cycles and the prior history.

Table III summarizes the influence of the variables considered here on the sensitization produced by cooling through the sensitization region and that experienced by the HAZ. Table III clearly shows the direction one must go to minimize sensitization.

Many of the recommendations of Table III (such as using low carbon content alloys) have been known for more than 40 years; others have heretofore been unknown. Of particular interest is the influence of the peak temperature. It is at the points in the HAZ where the peak temperature is only 800°C that is the most sensitized. This fact is of particular interest in single pass welds or where a local heat treatment is used to solution-anneal the HAZ. If there is a temperature gradient (which will always be the case unless one uses a furnace solution anneal of the whole part) then there will be some point where Tm = 800°C. This is where sensitization can occur. To prevent this, it is necessary to have taken precautions with respect to the prior condition of the steel and to cool rapidly. Some guidance can be taken from the data presented here but this guidance must be taken with the caution that many factors influence the critical cooling rate at which sensitization occurs. Data generated on one heat of 304 is, therefore not directly applicable to others, although the general trends still apply.

TABLE III

APPROACHES TO REDUCE WELD SENSITIZATION

- Solution anneal and water quench.
- Avoid deformation which is not annealed out prior to welding.
- Minimize constraint to reduce deformation during welding.
- Avoid multiple thermal cycles where possible.
- Reduce the amount of area which sees a peak temperature of 800 - 900°C.
- Use low carbon alloys.

ACKNOWLEDGEMENTS

The author would like to acknowledge D. Lord as the designer of the hardware and software for the combined strain-temperature cycling tests and his performing these tests. The author would also like to acknowledge his stimulating discussions with C. L. Briant, R. E. Carter, W. L. Clarke, T. M. Devine, R. E. Hanneman, M. F. Henry, D. Lord and M. J. Povich. This work was supported in part by the Electric Power Research Institute under contract RP-1072-1.

REFERENCES

1. H. D. Solomon and C. L. Lord, in Preparation.
2. H. D. Solomon, Corrosion, V34, PP 183-193, 1978.
3. American Society for Testing and Materials, Standard Recommended Practices for Detecting Susceptibility to Intergranular attack in Stainless Steels/Designation A262-70 Practice E.
4. W. L. Clarke, R. L. Cowan, and W. L. Walker, General Electric Company Report NEDO-12669, May 1977, and to be published in ASTM Special Technical Publication, ASTM Symposium on Evaluation Criteria for Determining the Susceptibility of Stainless Steels to Intergranular Corrosion, Toronto, Ontario, May 2 and 3, 1977.
5. W. L. Clarke, R. L. Cowan, and J. C. Danko, General Electric Company Report NEDO-12670, May 1977, and to be published in ASTM STP, G-1 Symposium on Stress Corrosion Cracking - The Constant Strain Rate Technique, Toronto, Ontario, May 2 - 4, 1977.
6. E. F. Nippes, L. L. Merrill and W. F. Savage, Welding J. Research Supplement, V28, PP 5565 - 5645, 1949.
7. H. Zitter, Archiv für das Eisenhuttenwesen, V28, PP 401 - 416, 1957.
8. A. J. Giannuzzi, G. E. Report for EPRI, NEDO-20985-5, August 1976.
9. H. D. Solomon, W. L. Clarke and T. M. Devine, In Preparation.
10. Metals Handbook, V1 1961, P 422.
11. H. D. Solomon, In Preparation.

DISCUSSION

J. Ballance (Journal of Metals): Could you comment further on your findings on the effects of grain-boundary solutes on sensitization?

Author: Preliminary findings point to boron segregation to grain boundaries as being responsible for the difference between specimens cooled from 1000°C and 800°C. Boron seems to segregate when one cools from 1000°C but not when the cooling is from 800°C. This segregation effects the fraction of the grain boundaries which become sensitized by carbide formation. If these preliminary results are upheld, then Clyde Briant and myself plan to publish them.

S. Matthews (Stellite Division): A great deal of data has been shown on the effect of cooling rates, in which values of 0.1 to 1.0°C/sec were used in the experiments. If conventional welding processes produce cooling rates of typically 10 to 20°C/sec, could you comment on your rationale behind using admittedly unrealistic HAZ cooling rates?

Author: Continuous cooling sensitization experiments are a way of characterizing the relative sensitization behavior of different alloys. An alloy which requires a cooling rate of 0.01°C/S is much less sensitive to sensitization than one which will sensitize at 1.0°C/S. Hence, these experiments can be used to rank alloys, which was one of the objectives of the program.

G. A. Knorovsky (Sandia Labs): Have you and Dick Chrenko put your heads together to come up with an explanation for the lesser sensitivity to stress corrosion cracking of larger diameter piping?

Author: Stress corrosion tests in high temperature water have shown laser welds to be superior to conventional welds in their stress corrosion cracking resistance.

The heat input is very localized in a laser weld. This enables one to make thick single pass welds where conventional processes require multi pass welds. The localized heating and single pass nature of the welds minimize sensitization and give rise to the good SCC resistance in high temperature water. These factors, however, also give rise to a localized but high residual stress.

H. W. Kerr (University of Waterloo): K. Aust appeared to show sensitization without carbides. What is the explanation for his results vs. the present results which show carbide formation?

Author: Aust showed that solute segregation to grain boundaries could cause intergranular attack in certain highly oxidizing media. When there was carbide precipitation there was no attack. In contrast in these present studies less oxidizing solutions have been used and carbide precipitation (and Cr depletion) is required for attack.

Scanning Auger Electron Spectroscopy Studies of Grain-boundary Segregation in Type 304 Stainless Steel*

J. Y. Park, S. Danyluk, and D. E. Busch

Materials Science Division
ARGONNE NATIONAL LABORATORY
Argonne, Illinois 60439

Abstract

Scanning Auger electron spectroscopy studies have been conducted on grain-boundary surfaces of Type 304 stainless steel that were fractured in situ. To enhance the probability of intergranular fracture, the specimens were first subjected to creep deformation for 1000 h at 700°C. A semi-quantitative surface chemical composition was calculated from the peak heights of Auger electron spectra. The concentration of Cr at the fracture surface was not different from the bulk value. This indicates that the long-term heat treatment caused healing of the sensitization. The concentrations of S, C, and Si at the fracture surface were at least an order of magnitude higher than the bulk values. Chemical composition profiles obtained by ion-sputtering indicated that segregation of S, C, P, and Si occurred within a depth of several atomic monolayers from the grain-boundary surface. Mo, Mn and Cu were not detected. The concentrations of Ni and Fe are in good agreement with the bulk chemical analysis.

---

*Work supported by the Electric Power Research Institute.

Scanning Auger Electron Spectroscopy Studies of Grain-boundary Segregation in Type 304 Stainless Steel*

J. Y. Park, S. Danyluk, and D. E. Busch

Materials Science Division
ARGONNE NATIONAL LABORATORY
Argonne, Illinois 60439

Introduction

Intergranular stress-corrosion cracking (ISCC) of Types 316 and 304 stainless steel has occurred in furnace-sensitized sections and heat-affected zones (HAZ's) of weldments in Boiling-Water-Reactor (BWR) components.[1,2] These studies confirmed that sensitization of an alloy (i.e., the formation of carbide precipitates and depletion of chromium at the grain boundaries during welding) is one important contributing factor to ISCC; thus, sensitization has been viewed as the cause for the property changes in the HAZ's and their increased susceptibility to ISCC. However, some heats of material are more resistant to sensitization and ISCC than others, and this heat-to-heat variation cannot be explained simply by differences in bulk chromium or carbon concentration. Impurity segregation or other alloying elements at the grain boundaries may also influence the kinetics of sensitization and contribute to ISCC.

With the advent of surface-sensitive analytical techniques such as Auger electron spectroscopy, it has become possible to examine the alloy chemistry near grain boundaries. Chromium depletion or other segregation phenomena at grain boundaries can therefore be directly analyzed. Using this technique, it was found that grain-boundary segregation (GBS) in ferrous alloys is a strong influencing factor in temper brittleness.[3-5] The study of GBS in austenitic stainless steel has been hampered, however, by experimental difficulties in exposing contamination-free grain-boundary surfaces at which the segregation may be measured. It was recently suggested that a creep-impact fracture technique[6] or slow straining in an H_2 atmosphere[7] may be used in situ in the AES chamber to expose grain boundaries of austenitic stainless steel. The present paper reports the re-

*Work supported by the Electric Power Research Institute.

sults of AES analysis on intergranularly fractured surfaces of Type 304 SS using the creep-impact fracture technique.

Experiment and Results

Type 304 stainless steel (chemical analysis given in Table I) was cold rolled to 0.0375-cm-thick flat stock, solution annealed for 1.8 ks in argon at 1025°C, and water quenched. The grain size of the material was ∿25 μm. The uniaxial creep-rupture specimens of a gauge length of 2.22 cm and a width of 0.559 cm were prepared. The creep test was conducted on a conventional constant-load creep machine in an argon environment at 700°C at 55 MPa for 4 Ms. After creep deformation, a 0.1-cm-thick slice of the specimen was mounted on an AES impact-fracture stage (part of a scanning AES system manufactured by Physical Electronic Industries, Inc.) in an ultrahigh-vacuum AES chamber (1.3×10^{-8} Pa). The specimen was then fractured by a hammer blow with an impact force of ∿2.7 J. AES analysis was accomplished on the fractured grain-boundary surface using a 3-μm-diameter primary electron beam at 5 KeV and 1 μA. After the AES analysis, the fracture surface was reexamined with a high-resolution scanning electron microscope to confirm the intergranular fracture mode and identify other microstructural features, e.g., creep cavities and microvoids. Figure 1 is a scanning electron micrograph of a fracture surface. The fracture mode was predominantly intergranular. AES analysis was performed on the grain faces in areas A and B. Some creep cavities and voids were present in area B. Figure 2 is an AES spectrum showing [dN/dE] vs E, where N is number density of Auger electrons and E is energy level. Energies of Auger electron transitions, shown by peaks in the AES spectrum, are used to identify elements. AES peaks for P, S, and C are evident in the spectrum in addition to peaks for Fe, Cr, and Ni. The peak representing the oxygen transition (510 eV) is absent. This is evidence that the fracture surface is not contaminated. The chemical composition of the surface was calculated from the peak-to-peak height for each element, using the procedure developed by Davis et al.:[10]

$$C_x = \left(\frac{H_x}{S_x} \Big/ \sum_i \frac{H_i}{S_i} \right) \qquad (1)$$

where C_x is the concentration of element x, H_x and H_i are the peak-to-peak heights of elements x and i, respectively, and S_x and S_i are the sensitivity factors of elements x and i, respectively, relative to a standard material. This formalism has been used previously to analyze stainless steel and good correlation has been obtained between the calculated concentrations and the results of bulk chemical analysis. Details of the calculation procedure are given in Ref. 10.

The results of calculations using the data shown in Fig. 2 are given in Table I along with the bulk chemical analysis for comparison. The concentrations of C, S, and P at the grain boundaries were at least an order of magnitude higher than the values obtained by bulk chemical analysis. The concentrations of the major alloy elements, Cr, Ni and Fe, were approximately equal to those obtained by bulk chemical analysis. Two C peaks were observed (at 253 and 260 eV) in addition to the main peak at the 272-eV transition. This indicates that some of the C has formed a carbide phase. Peaks for Mn, Mo and Cu were not large enough to be clearly identified. Figure 3 is an AES spectrum obtained at location B (Fig. 1). The peak-to-peak height for S (152 eV) is larger than that in the spectrum of Fig. 2. The concentration of S at location B is calcu-

TABLE I. Grain-boundary and Bulk Chemical Composition (wt%) of Type 304 Stainless Steel Sample

	Fe	Ni	Cr	C	S	P	Mo	Mn	Si	Cu
Grain Boundary (AES Analysis)	64.2	9.2	22.8	1.7	0.1	0.5	ND[a]	ND	1.4	ND
Bulk (Chemical Analysis)	-	9.3	17.7	0.046	0.012	0.026	0.33	1.17	0.47	0.20

[a]ND = not detected.

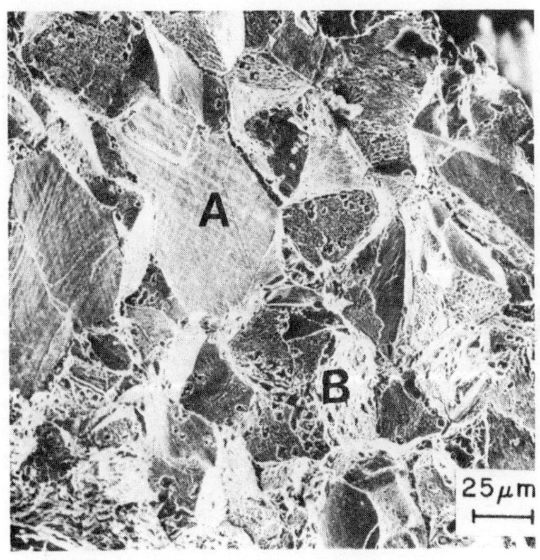

Fig. 1. Scanning Electron Micrograph of a Fracture Surface of Type 304 Stainless Steel. ANL Neg. No. 306-76-136.

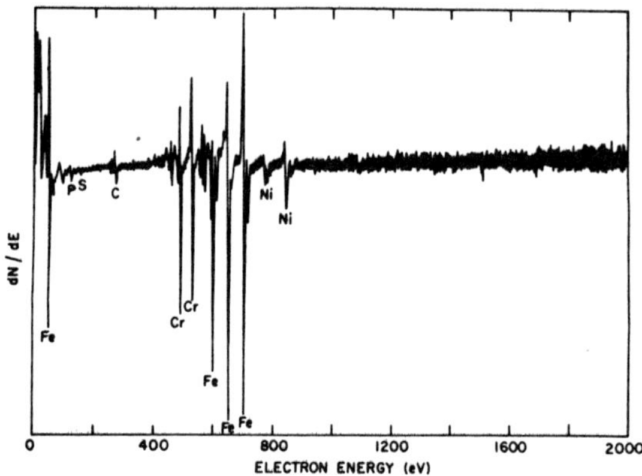

Fig. 2. Auger Electron Spectrum of Fracture Surface at Location A of Fig. 1. Neg. No. MSD-63512.

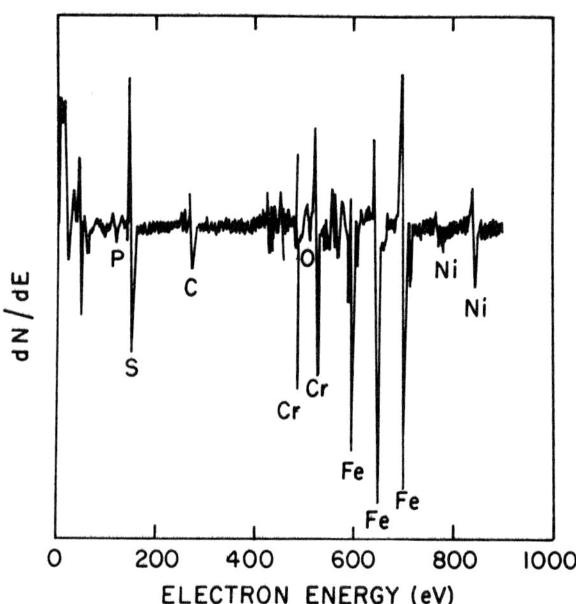

Fig. 3. Auger Electron Spectrum of Fracture Surface at Location B of Fig. 1. Neg. No. MSD-65507.

lated from Eq. (1) to be as much as 6 wt%. This indicates that local variation in S segregation is large. The O peak seen in Fig. 3 at 510 eV may be due to surface absorption after 7.2×10^4 s (20 h) of operation of the AES system.

After the analysis of the fresh fracture surface was completed, the fracture surface was sputtered with 1-KeV argon ions at a current of 60–130 μA (measured at the sample) and in-depth concentration profile was obtained at location B (Fig. 1). Figure 4 shows the concentration of Ni,

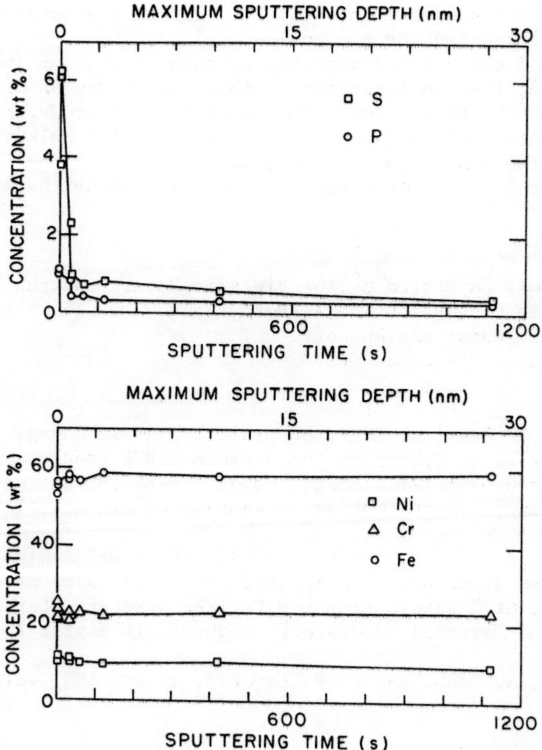

Fig. 4. In-depth Concentration Profiles Near Grain-boundary Surface at Location B of Fig. 2. Neg. No. MSD-63936.

Cr, P, and S plotted versus sputtering time and depth. The depth was calculated using the average sputtering rate, 2.5×10^{-2} nm/s, which was obtained by measuring the actual loss of material from the surface of

Type 304 SS under the same sputtering conditions. Variation in the sputtering rate is expected, however, because of variations in the angle of sputtering and the different sputter rates of segregated species. Figure 4 shows that S and P have steep concentration gradients within a few atomic monolayers from the fracture surface. Similar results were observed for C and Si. The concentration of Cr at the fracture surface was not different from the bulk value, and no appreciable concentration change was observed for Cr, Ni or Fe throughout the in-depth concentration profile. This indicates that the long-term (3.6-Ms or 1000-h) heat treatment caused healing of the sensitization, which should have occurred during the early part of creep treatment at 700°C.

AES studies of GBS in sensitized Type 304 SS are in progress. Two heats of material, which have the same bulk chemical composition but different susceptibility to ISCC, are being used. The results will be examined for possible correlations between GBS and susceptibility to ISCC.

Conclusion

Grain-boundary segregation of S, P, C and Si was observed in Type 304 SS that was heat treated for 3.6 Ms (1000 h) at 700°C. The segregation occurred within a depth of several atomic monolayers from the grain-boundary surface; the concentration in this region was an order of magnitude higher than the bulk value. The concentration of Cr, Ni and Fe at the grain-boundary surface was not different from the bulk value. This indicates that the long-term heat treatment caused healing of the sensitization. Mo, Mn, and Cu were not detected clearly by the AES technique.

Acknowledgments

This work was supported by the Electric Power Research Institute und under Contract No. RP449-1. The authors thank W. J. Shack and R. W. Weeks for their encouragement and support.

References

1. J. Y. Park, S. Danyluk, R. B. Poeppel, and C. F. Cheng, "Metallurgical Examination of Cracks in the Dresden-2 BWR Emergency Core-spray System 10-inch Diameter Piping, "Report prepared for the Commonwealth Edison Company, Argonne National Laboratory, Argonne, IL (April 1976).

2. S. Danyluk, J. Y. Park, and C. F. Cheng, "Failure Analysis of the Type 304 Stainless Steel 4-in. Recirculation Bypass Lines of the Monticello BWR Power Plant," Report prepared for the Northern States Power Company, Argonne National Laboratory, Argonne, IL (April 1976).

3. D. F. Stein, A. Joshi and R. P. LaForce, Trans. ASM, Vol. 62, p. 776 (1969).

4. P. W. Palmberg and H. L. Marcus, Trans. ASM, Vol. 62, p. 1016 (1969).

5. R. Viswanathan, Met. Trans., Vol. 2, p. 809 (1971).

6. J. Y. Park and S. Danyluk, Corrosion, Vol. 33, p. 304 (1977).

7. C. L. Briant, Met. Trans. Vol. 9A, p. 731 (1978).

8. L. E. Davis, N. C. MacDonald, P. W. Palmberg, G. E. Riach, and R. E. Weber, <u>Handbook of Auger Electron Spectroscopy</u>, Second Edition (Physical Electronic Industries, Inc., Eden Prairie, MN (1976).

DISCUSSION

D. J. Gooch (Central Electricity Research Laboratories): Could not the observed segregation have occurred during the creep deformation for 1000h at 700°C?

Author: Yes, the observed segregation is definitely a result of the creep-deformation treatment.

J. J. Pepe (General Electric Company): Would you describe the Creep Method used to expose grain boundaries? How is this related to stress corrosion cracking?

Author: The creep method is described in the manuscript. It is used to expose grain boundaries for measurement of chemical composition by Auger electron spectroscopy. Variation in the chemical composition at grain boundaries, caused, for example, by impurity segregation or chromium depletion, are known to be an important causative factor in intergranular stress-corrosion cracking.

ESTIMATION OF DEGREE OF SENSITIZATION OF TYPE 304 STAINLESS STEEL WELDMENTS BY ELECTROCHEMICAL TECHNIQUE

J. Fukakura, H. Kashiwaya, T. Mori
Heavy Apparatus Engineering Laboratory
S. Iwasaki and M. Arii
Nuclear Energy Group
Toshiba Corporation, Japan

This paper describes the preliminary test results utilizing a non destructive method to quantitatively estimate the degree of sensitization established by isothermally heat treating type 304 stainless steel. The Electrochemical Potentiokinetic Reactivation (EPR) measurement was adopted for this quantitative estimate of sensitization.

Solution heat treated 304 stainless steel specimens were isothermally heat treated at 620°C for 1 hr. to produce a first stage of sensitization. EPR measurements were then made prior to a second stage heat treatment. This second stage treatment was performed in the range of 500°C to 1100°C for from 30 sec. to 600 sec. EPR measurements were then made after this second stage heat treatment in order to determine the increase in sensitization that occurred below 800°C and the recovery that occurred at this and higher temperatures.

INTRODUCTION

The occurrence of stress corrosion cracking (SCC) in Boiling Water Reactor Plants has prompted a considerable effort to search for the origin of this phenomenon and to establish the best countermeasures. Our attention has been concentrated on the examination of the influence of the three main factors responsible for SCC. They are sensitization, weld residual stress and the environment. This paper is concerned with only the first factor.

The degree of sensitization(DOS)was determined by the Electrochemical Potentiokinetic Reactivation Technique (EPR).[1,2,3] This technique was used to quantify the sensitization produced by the first stage sensitization treatment and any changes produced by the second stage treatment. This study was aimed at quantifying the addition sensitization that occurs below about 800°C and the thermal conditions required to eliminate the effects of the first stage thermal exposure. It is also hoped that data generated in this way can be used to predict the DOS from a knowledge of the thermal history.

EXPERIMENTAL PROCEDURES

Materials and Test Specimens

Test specimens of the geometry shown in Figure 1 were cut from a 4 in. Sch. 80 type 304 pipe. The composition of this pipe is shown in Table 1.

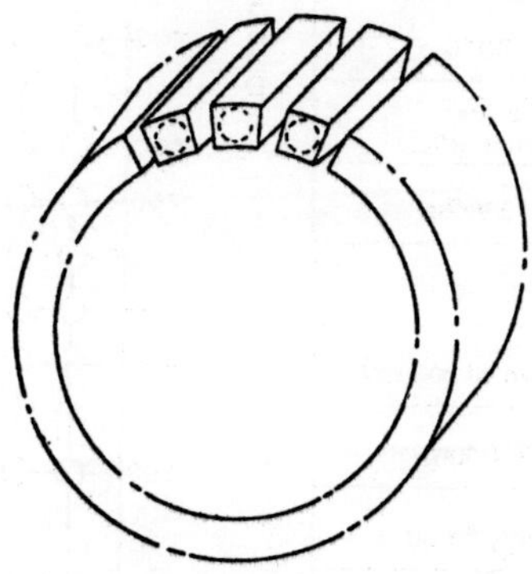

Fig. 1(a). Specimen Preparation

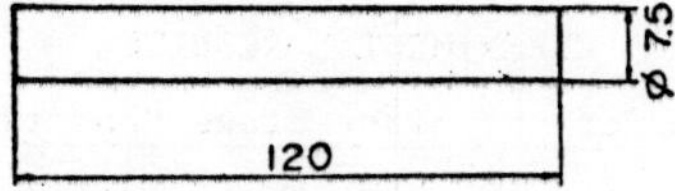

Fig. 1(b). Specimen Geometry

Table 1. Chemical Composition (%)

	C	Si	Mn	P	S	Ni	Cr
Analysis	0.05	0.47	1.49	0.029	0.007	9.30	18.40

Sensitization Heat Treatment

After machining all the test specimens were annealed at 1050°C for 30 min. The specimens were given the first stage isothermal sensitization at 620°C for 1 hr. EPR measurements were then performed according to the conditions listed in Table 2, with the peak reactivation current chosen as the parameters defining the DOS. The specimens were then given the second stage isothermal heat treatment. The matrix of conditions is shown in Figure 2. After exposure due to one of the conditions listed in this matrix, the specimens were again EPR tested.

Table 2. Condition of EPR Measurement

Temperature	30°C
Solution	0.5M H_2SO_4 +0.01M KSCN
Scanning Rate	6 V/hr
Standard Electrode	S.C.E.
Potential Holding	2 min at 200 mV
Counter Electrode	Plutinum
Surface Finish	Emery # 800
Deaeration	5min Prior

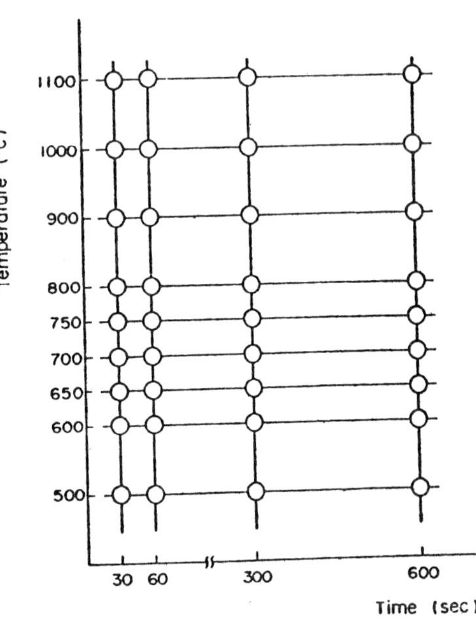

Fig.2. Matrix of Sensitization Treatment

EXPERIMENTAL RESULTS

Degree of Sensitization of the First Stage Sensitization

All of the specimens (about 40) were heat treated in a single bundle, none the less, there was a considerable fluxuation in the peak reactivation current (Phi). This current varied from 3.74 mA/cm^2 to 6.62 mA/cm^2 for the group of specimens exposed to the same sensitization cycle. It was concluded from multiple measurements performed on the same sample that the variation in Phi was not due to variations inherent in the measurement but represented sample to sample variations.

Degree of Sensitization Due to the Second Stage Sensitization

Each of the 40 specimens which were exposed to the first stage sensitization were given a second isothermal sensitization (one of the conditions in the matrix of Figure 2). The peak current in the reactivation test obtained after this second and final heat treatment is defined as Phf and is shown as a function of heat treatment in Figure 3. The variation in peak current between the first and second sensitization treatments $\Delta Ph = Phf - Phi$ is shown in Figure 4.

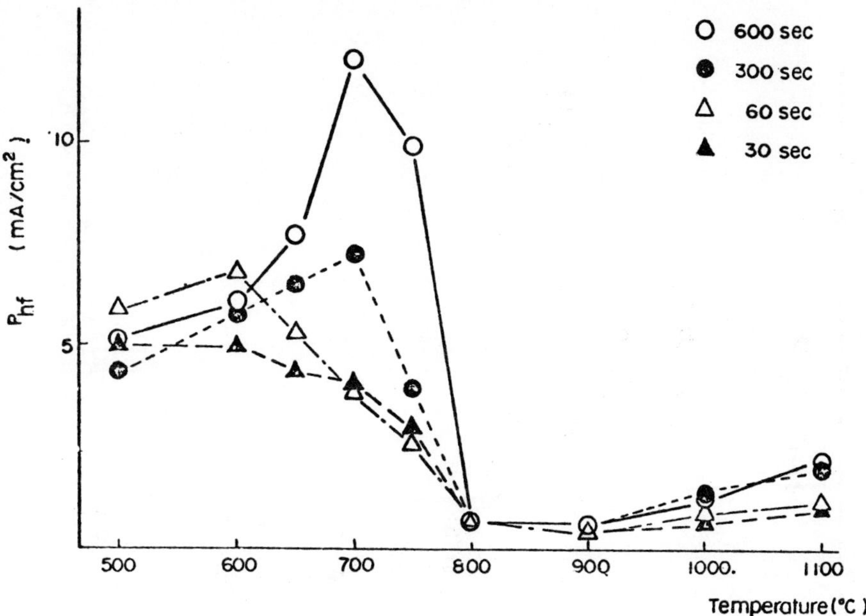

Fig. 3. P_{hf} after Second Stage Sensitization

Observation of the Surface of EPR Specimens

The surface of several specimens after the EPR test are shown in Figure 5. These photographs were taken with an SEM. Figure 5a shows the dissolution of the grain boundary region in a specimen which experienced only the first (620°C - 1 hr.) sensitization treatment. The EPR test produced severe wide ditches at some but not all of the grain boundaries. A second stage 700°C - 600 sec. heat treatment caused all of the grain boundaries to be attacked. This 700°C - 600 sec. heat treatment was the most severe i.e., ΔPh was the greatest. For other conditions where ΔPh was less, there was no change in the structure over that observed after cooling the first thermal treatment. At elevated temperatures where ΔPh was negative, the degree of grain boundary attack was decreased by this high temperature thermal treatment. This is shown in Figure 5c. The 800°C - 600 sec. heat treatment completely eliminated any grain boundary attack in a subsequent EPR test.

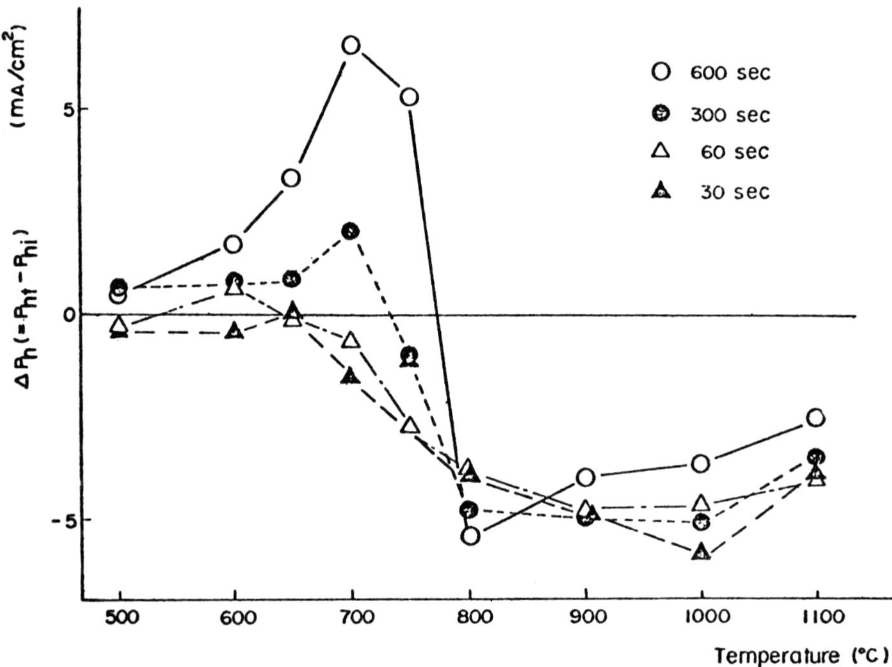

Fig.4. ΔP_h after Second Stage Sensitization

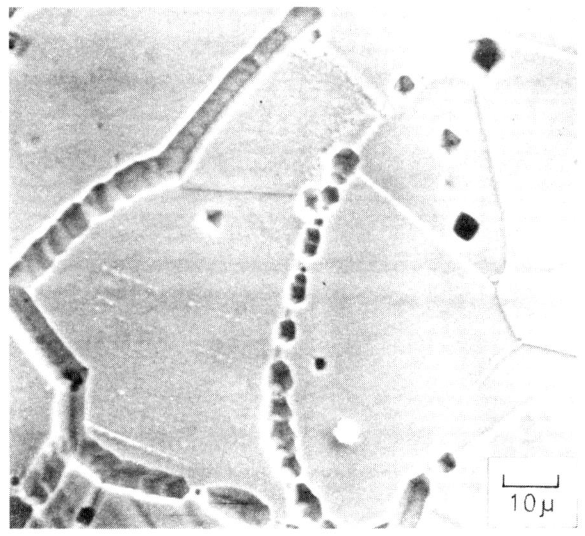

Fig. 5a 620°C for 1 hr.

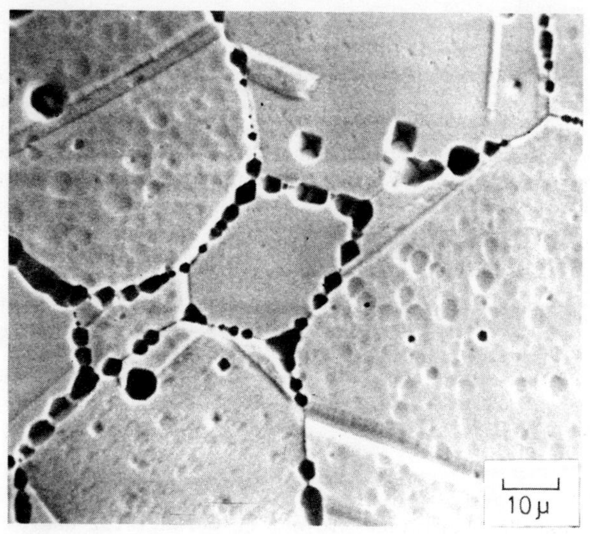

Fig. 5b 700°C for 600 sec.

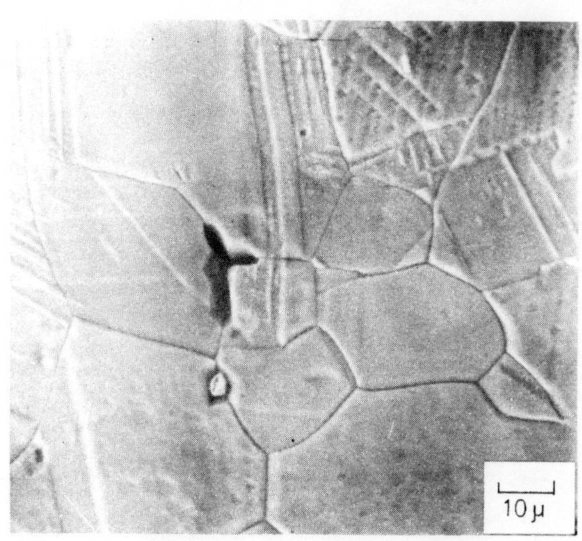

Fig. 5c 800°C for 600 sec.

DISCUSSION

There are two significant findings of this study. Firstly, that the EPR test is a sensitive measure of sensitization. The second finding was that below 800°C a thermal treatment will increase the degree of sensitization with the maximum increase occurring at 700°C. At 800°C and above sensitization is eliminated. This resolution of carbides is rapid and can occur in as little as 30 sec. at 800°C.

CONCLUSIONS

A. Degree of sensitization increases dominantly at temperatures adjacent to 700°C.

B. Recovery proceeds so quickly at temperatures higher than 800°C that short time heat treatment even of 30 sec. is sufficient to cancel the previous sensitization.

ACKNOWLEDGEMENT

We wish to express deep gratitude to our co-workers for useful discussions and especially to Mr. Y. Nishikawa for careful performance of this experiment.

REFERENCES

1. P. Novak, et.al., "Testing the Susceptibility of Stainless Steel to Intergranular Corrosion by a Reactivation Method", Corrosion; Vol. 31, P. 344.

2. W. L. Clarke, et.al., "Detection of Sensitization in Stainless Steel using Electrochemical Techniques", Proceedings of NACE Meeting 1977, San Francisco, Paper No. 180.

3. F. Umemura, "Electrochemical Reactivation Method for Evaluating IGC and IGSCC Susceptibility of Sensitized Stainless Steels" to be published in 7th International Congress on Metallic Corrosion, Rio de Janeiro, October 1978.

Intergranular Corrosion Behavior of Wrought and Weld Deposited 308 Stainless Steel

T. M. Devine

General Electric Company

The influence of ferrite morphology and carbon content on the intergranular corrosion behavior of 308 stainless steel was investigated using 4 wrought alloys and 6 weld deposited alloys. The 4 wrought alloys were heat treated at 4 different annealing temperatures to introduce 4 different ferrite levels. The annealed samples along with the weld deposited alloys were aged at temperatures ranging from 550°C to 700°C for times varying between 1/4 and 400 hours and then tested for intergranular corrosion susceptibility in acidified copper-copper sulfate solution. It is shown that samples possessing a critical amount and distribution of ferrite have a greatly reduced susceptibility to intergranular corrosion attack and a model is presented to calculate the critical amount and distribution of ferrite required as a function of carbon content.

Introduction

Type 308 stainless steel is perhaps the most widely used two-phase austenitic-ferritic stainless steel. It is employed extensively in a variety of forms in applications where corrosion resistance is of importance. This includes uses as weld filler for joining sections of austenitic 304 stainless steel, as a weld deposited, corrosion resistant cladding on many structural components such as pressure vessels, nozzles, etc., and as a casting (CF3, CF8) alloy. The remarkable resistance of austenitic-ferritic stainless steels to intergranular corrosion associated with carbide precipitation has been known for some time.[1-3] In previous publications we have determined the mechanism of intergranular corrosion of duplex stainless steels.[4-6] The model states that the critical amount of ferrite required to impart a significant reduction in susceptibility to intergranular corrosion be expressed not in terms of the amount of ferrite present but rather in terms of the amount of austenite-ferrite boundary area. In this investigation we examine the intergranular corrosion behavior of duplex 308 stainless steel with different ferrite morphologies in an attempt to determine the appropriate manner in which to express the critical amount and distribution of ferrite required to impart resistance to intergranular corrosion.

Procedure

One commercially obtained heat, #L-B7, and 3 experimental heats of 308 stainless steel whose compositions are listed in Table I were used to evaluate the effects of aging treatment on the intergranular corrosion behavior of wrought 308 stainless steel. Heat L-B7 was obtained in the form of as cold drawn .045 inch diameter wire. The 3 experimental heats were used to study the effects of carbon content on intergranular corrosion behavior. The three heats were cast as 30 pound tapered squares, 3 inch square x 2-1/4 inch square x 11 inches long, hot forged at 1200°C to 1-1/2 inch x 2-1/2 inch, rectangular cross section, hot rolled to 0.090 inch thickness at 1175°C and cold rolled to 0.050 inch thick sheet. The sheet was then annealed at 1100°C, water quenched and sheared into test coupons 7 inch x 1/2 inch. Samples were electropolished in a solution of 60% H_3PO_4 + 40% H_2SO_4 at 40°-60°C using $\sim$ 2 amps/cm^2 for 5 minutes. Following electropolishing samples were ultrasonically cleaned in a detergent solution and then distilled water, dried and encapsulated in quartz tubes which were first evacuated and then back-filled with 1/4 atmosphere of argon. Following heat treatment specimens were water quenched by plunging the tubes into water and breaking them. The samples were then electropolished and cleaned as described above, wrapped in zirconium foil, encapsulated in evacuated quartz tubes and aged at temperatures ranging from 480°C to 700°C for times of 1/4 hour to 1000 hours. Following aging the samples were again electropolished and cleaned and tested for susceptibility to intergranular corrosion by immersing the samples wrapped in copper wire in acidified copper-copper sulfate solution for 72 hours. The latter was conducted according to ASTM specification A262E.[7] Upon removal from the solution, the samples were bent open and examined with an optical microscope for evidence of intergranular corrosion.

Identical intergranular corrosion tests were conducted on 308 weld metal deposited onto fully annealed 1/2 inch thick x 4 inch wide x 6 inch long plate of 304L stainless steel. The compositions of the 5 stick electrodes and the uncoated wire are listed in Table II. The latter TIG weld was deposited at a nominal rate of 5 inches/minute using 105 amps and 16 volts. The stick electrodes were manually deposited at a nominal rate of 6 inches/minute using 95 amps and 23 volts. The last column in Table II lists the Ferrite Number measured for each weld deposit using a Magne Gauge. The Ferrite Number measurements were made using the Draw File procedure recommended by the Welding Research Council.[8] The weld deposits were tested for intergranular corrosion resistance in both the as-welded and as-aged conditions. Prior to corrosion testing the welds were electropolished and cleaned as described above.

Results

Annealing 308 stainless steel at sufficiently high temperatures introduces ferrite into the microstructure. The amount of ferrite produced increases with increasing anneal temperature. As shown in Figure 1, a one-hour anneal of heat L-B7 at 1350°C followed by a water quench results in a duplex microstructure consisting of approximately 20 v/o ferrite as measured by area fraction. Annealing at lower temperatures produces less and less ferrite until at 1200°C the structure in 100% austenitic as shown in Figure 2.

The duplex structure exhibits a much finer grain size than the austenitic structure even though it was created at a higher temperature. Apparently the austenite and ferrite phases restrict each other's grain growth. Table III describes the variation of ferrite volume fraction found in each alloy investigated in this program as a function of annealing temperature along with the amount of austenite-ferrite boundary area. As the carbon content of the alloy decreases the amount of ferrite increases for each annealing temperature. Ferrite volume fractions were measured by area fraction. In comparing the wrought alloys with the weld deposited alloys, the ferrite phase in the latter has a much finer distribution and for a given volume fraction of ferrite the weld deposited material has a much larger amount of austenite-ferrite boundary area than the wrought material.

Figures 3 and 4 summarize the effect of aging treatment on the intergranular corrosion behavior in A262E of 308 stainless steel, heat L-B7, containing varying amounts of ferrite. Solid points indicate complete through-the-thickness intergranular fracture of the specimen during bending after the 72-hour immersion period in the boiling acidified copper-copper sulfate solution. Half-filled points indicate that bending produced macroscopically visible cracks. Points with a cross (+) indicate that extensive microscopically visible cracking occurred due to bending. However, the intergranular corrosion penetration was not sufficient to result in deep intergranular cracking which would be macroscopically visible. Open points indicate the presence of isolated intergranular cracks one or two grains in length and X's indicate no cracking.

The results obtained for the fully austenitic version of heat L-B7 produced by annealing at 1200°C for 1 hour and presented in Figure 3 are similar to those obtained on other austenitic grades of stainless steel such as 304. A "C"-shaped curve can be drawn through the data separating the aging treatments which result in severe intergranular corrosion from those which give rise to only isolated intergranular cracks one or two grains in length. Although samples were aged for up to 400 hours at 600°C, no healing effects were observed. Such samples suffered 100% intergranular penetration.

The A262E results as a function of heat treatment obtained for duplex material produced by annealing heat L-B7 at 1250°C for 1 hour indicate that this material was at least as susceptible to intergranular corrosion as the all-austenitic material. The alloy contained 4 v/o ferrite and corrosion occurred along both the austenite-ferrite and austenite-austenite grain boundaries. No healing occurred after aging the samples for over 400 hours at 550°C.

Figure 4 illustrates the effect of aging on the intergranular corrosion behavior of the 10 v/o ferrite alloy of heat L-B7. The results were nearly identical for the 20 v/o ferrite alloy. The latter material developed a sensitized microstructure earlier and exhibited immunity to intergranular corrosion sooner at each aging temperature than the 10 v/o ferrite alloy. In contrast to the all-austenitic material and the 4 v/o ferrite material, both of which exhibited a large zone of susceptibility to severe intergranular corrosion, the 10 and 20 v/o ferrite alloys possessed only a small zone of susceptibility to severe intergranular corrosion. Immunity of the 10 and 20 v/o ferrite alloys to intergranular corrosion could be restored at each temperature by aging for sufficient time. Aging the austenitic and 4 v/o ferrite alloys for as long as 400 hours failed to restore immunity to intergranular corrosion. In further contrast to the all-austenitic material and the 4 v/o ferrite duplex alloy both of which locally corroded along austenite-austenite grain boundaries when sensitized, the 10 and 20 v/o ferrite alloys were locally attacked only along austenite-ferrite boundaries when sensitized at 600°C. No intergranular corrosion occurred along austenite-austenite grain boundaries in the 10 and 20 v/o ferrite alloys sensitized at 600°C.

The results obtained for the 3 laboratory heats of 308 stainless steel parallel those of heat L-B7 and are summarized in Figures 5-7. The curves on the TTT diagrams enclose aging treatments which resulted in extensive microscopically visible cracking during bending following A262E testing. A minimum of 18 aging times were employed at each aging temperature to define these curves. As shown in Figure 5, there is sufficient ferrite present in the lowest carbon containing alloy that, regardless of initial annealing temperature, the material behaves as a duplex alloy and exhibits a healing phenomenon at each aging temperautre. For the .038% carbon containing alloy, there was sufficient ferrite present even in the material annealed at 1200°C to affect a healing effect at 600°C. The .055% carbon alloy had to be annealed at 1300°C or higher before sufficient ferrite was introduced to produce a healing effect at 600°C.

Table IV summarizes the effect of aging treatment on the intergranular corrosion behavior of the 6 weld deposited alloys of 308 stainless steel. Three alloys possessed sufficient ferrite to either be immune to intergranular corrosion or to at least undergo a healing phenomenon at 600°C. The remaining alloys did not indicate a healing phenomenon within 100 hrs. at 600°C.

Discussion

In contrasting the intergranular corrosion behavior due to intergranular carbide precipitation in austenitic stainless steel with that of high ferrite ($\gtrsim$8v/o) duplex stainless steel, a sensitized microstructure develops in the latter material after shorter aging times than for the austenitic alloys. With continued aging, immunity to intergranular corrosion is subsequently restored in both alloy systems. However, immunity occurs much earlier in the duplex alloys than in the austenitic alloys, (e.g., 6-10 hrs. vs. >600 hrs. at 600°C for heat L-B7). And, in the duplex alloys containing the requisite amount and distribution of ferrite, localized attack occurs only along austenite-ferrite boundaries. The austenite-austenite grain boundaries are immune to attack. In previous publications we have determined the mechanism of intergranular corrosion in two-phase austenitic-ferritic stainless steels.[4-6] In brief, the rate determining step in $M_{23}C_6$ precipitation is chromium mobility. The diffusivity of chromium in the ferrite phase is approximately 1000 times greater than that in austenite phase at 600°C. Thus, after very short aging times, $M_{23}C_6$ precipitates along austenite-ferrite boundaries (there are no ferrite-ferrite boundaries in the ferrite levels investigated) due to chromium and iron in the ferrite reacting with carbon. The available carbon content of the ferrite is quickly exhausted and subsequent carbon comes almost exclusively from the austenite. This depletes the austenite of carbon and precludes the possibility of carbide precipitation along austenite-austenite grain boundaries. Since the austenite grain boundaries are free of $M_{23}C_6$ precipitation and their accompanying chromium depleted zones, they are immune to intergranular corrosion in A262E. Although the chromium in the $M_{23}C_6$ precipitate lying along the austenite-ferrite boundaries is supplied principally by the ferrite phase, a small but significant amount is contributed by the austenite phase. In fact, the small amount of chromium contributed by the austenite phase creates a narrow chromium-depleted zone in the austenite at the original austenite-ferrite interface. It is this narrow chromium-depleted zone in the austenite phase that causes intergranular corrosion along the austenite-ferrite boundaries in A262E. Immunity to intergranular corrosion in the duplex alloys occurs after longer aging times (e.g. ∿7 hrs. at 600°C) when the narrow chromium-depleted zone in the austenite phase at the original austenite-ferrite interface, is replenished with a critical amount of chromium required to restore the corrosion resistance of the area. This healing of the chromium-depleted zone in such a short time is made possible by virtue of the unusually small width of the zone.

The above mechanism of the intergranular corrosion of duplex stainless steels suggests that in order to limit the aging treatments which can result in susceptibility to intergranular corrosion by having healing occur at very short aging times, it is necessary to have a critical amount and distribution of austenite-ferrite boundary area.[4-6] Given this critical amount and distribution, the available carbon supply of the austenite will be tied up as $M_{23}C_6$ precipitates along austenite-ferrite boundaries. This will preclude precipitation of $M_{23}C_6$ along austenite-austenite grain boundaries and thereby prevent intergranular corrosion along these grain boundaries. The austenite phase will only contribute a small amount of chromium to the carbides resulting in the formation of a very narrow chromium depleted zone at the austenite-carbide interface (located at the original austenite-ferrite boundary) which is replenished with chromium very quickly. Thus, while previous investigators have sought to determine one critical volume fraction of ferrite required to limit intergranular corrosion susceptibility of duplex alloys, our model requires a critical amount and distribution of austenite-ferrite boundary area. The inapplicability of a criteria for intergranular corrosion resistance based on the volume percent ferrite present is best illustrated by considering the results obtained for the two wrought alloys heat L-B7/1250°C - 1 hour, and heat 294/1250°C - 1 hour, and the weld deposit heat 5410. All three had approximately 4 v/o ferrite. Since both heat L-B7/1250°C - 1 hour which contained .040 carbon and heat 294/1250°C - 1 hour which contained .055 carbon behaved like fully austenitic stainless steel and did not show a healing phenomenon at 600°C or 550°C, the weld deposit heat 5410 which contained .050 carbon should also not show a healing phenomenon. However, the results indicate that this weld deposit does in fact exhibit a healing effect at 600°C. As shown below all of these results are completely consistent with our model which requires a critical amount and distribution of austenite-ferrite boundary area for resistance to intergranular corrosion.

There are two analytical approaches we can take to develop expressions for the required amount and distribution of austenite-ferrite boundary area. In the first, we simply require that there be sufficient austenite-ferrite boundary area to accomodate the available carbon supply as a 1600 Å-thick carbide layer covering the austenite-ferrite boundaries. Earlier work indicated this to be the average carbide thickness along the austenite-ferrite boundaries.[6] This leads to the following expression for the amount of austenite-ferrite boundary area as a function of carbon content of the alloy.

(1) $S_V^{\alpha-\gamma} = 1.14 \times 10^4 \times \% C$

The second approach to developing an expression for the critical amount of austenite-ferrite boundary area is to require that the ferrite phase be distributed throughout the austenite phase in such a manner that each carbon atom within the austenite phase has easy access to an austenite-ferrite boundary. This is analogous to requiring that the mean free path of the austenite-phase is sufficient to deplete the carbon content of the austenite matrix to some low value $\bar{c}$ within a specified time t. We therefore equate the austenite mean free path length to a diffusion distance over which the average carbon content of the austenite phase is reduced to $\bar{c}$ in time t.[10]

$$(2) \quad \lambda_\gamma = \left[\frac{\pi^2 Dt}{\ln\left(\frac{8}{\pi^2}\ \frac{Co}{\bar{c}}\right)}\right]^{1/2}$$

where D = diffusivity of carbon in austenite
t = aging time
Co = initial carbon content of austenite
$\bar{c}$ = mean carbon content of austenite after aging for time t

The amount of austenite-ferrite boundary area per unit volume can easily be determined by metallographic techniques. It is simply twice the number of intercepts per unit length of a random test line with austenite-ferrite boundaries.[9]

$$(3) \quad S_V^{\alpha-\gamma} = 2N_L$$

where $S_V^{\alpha-\gamma}$ = amount of austenite-ferrite boundary area per unit volume
N_L = number of times a random test line intercepts an austenite-ferrite boundary per unit length of test line.

Using (3), therefore, we can rewrite (1) as

$$(4) \quad N_L^{\alpha-\gamma} = 5.7 \times 10^3 \times \%\ C$$

and using the additional metallographic relationship

$$(5) \quad \lambda_\gamma = \frac{1 - L_L^\alpha}{1/2\ N_L^{\alpha-\gamma}}$$

we can rewrite (2) as

$$(6) \quad N_L^{\alpha-\gamma} = 2V_V^\gamma \left[\frac{\pi^2\ Dt}{\ln \frac{8}{\pi^2} \frac{Co}{\bar{c}}}\right]^{-1/2}$$

Thus we can express each criteria in terms of N_L and carbon content of the alloy.

Figure 8 is a plot of equation (4) and describes the amount of austenite-ferrite boundary area required to accomodate all of the available carbon as a 1600 Å thick $M_{23}C_6$ precipitate. The second curve in this figure plots twice this amount of austenite-ferrite boundary area. The latter may be a more realistic criteria since we have shown in a separate investigation that only one-half of the austenite-ferrite boundaries contain $M_{23}C_6$ precipitates.[6]

Figure 9 is a plot of equation (6) for three different values of $\bar{c}$ and two different ferrite levels.

Figure 10 plots the data obtained for the 16 wrought alloys and 6 weld deposits considered in this investigation. The solid points indicate the material was immune to intergranular corrosion when aged at 600°C for longer than 24 hours. The open points indicate that in spite of the amount of ferrite present, the alloy behaved like a fully austenitic stainless steel and did not exhibit a healing phenomenon with continued aging at 600°C. As shown, the data is reasonably well separated by the two curves drawn. The straight line requires twice the amount of austenite-ferrite boundary area than that required to tie up all of the available carbon as $M_{23}C_6$. The curved line expresses the distribution of ferrite required in a 10 v/o ferrite alloy to have an austenite-ferrite boundary within easy carbon diffusion distance throughout the austenite so as to be able to reduce the mean bulk austenite carbon content to .020 after 5 hours at 600°C. That the data is described well by the criteria requiring twice the amount of austenite-ferrite boundary area than that required to tie up all of the carbon as $M_{23}C_6$ along austenite-ferrite boundaries is consistent with the mechanism of intergranular corrosion in duplex stainless steels described elsewhere[4-6] and the observation that only 50% of all the austenite-ferrite boundaries are covered with carbides.[6] That the data is described well by the criteria requiring that the austenite-ferrite boundaries be distributed in such a way to permit reduction of the austenite mean carbon content to .020 after 5 hours at 600°C is consistent with the mechanism of intergranular corrosion and the work of Bleton, *et al*[12] who reported that fully austenitic 18-8 type stainless steels with .020% carbon or less are extremely resistant to intergranular corrosion. Both criteria are equally successful at describing the behavior of duplex stainless steels with widely varying ferrite-morphologies - wrought and welded. This is because each criteria is based on the amount of austenite-ferrite boundary area on which the mechanism of intergranular corrosion depends and not on the volume fraction of ferrite which can vary widely for the same amount of austenite-ferrite boundary area.

Acknowledgements

Miss B. J. Drummond conducted the intergranular corrosion tests. T. E. Douglas supervised the processing of the wrought alloys and Messrs. G. Pagnotta and L. Wojcik deposited the welds.

References

1) P. Payson, *Transations* AIME, 100, p.306, 1932.
2) L. B. Pfeil and D. G. Jones, *Journal of the Iron and Steel Institute*, p. 337, 1933.
3) V. R. Scherer, G. Riedrich, and G. Hoch, *Archiv für das Eisenhüttenwesen*, p. 1, 1939.
4) T. M. Devine, "The Effect of Microstructure on the Stress Corrosion Cracking Susceptibility of Austenitic and Duplex Stainless Steels," *Corrosion 77*, NACE, San Francisco, March 14-16, 1977.
5) T. M. Devine, "The Mechanism of Intergranular Stress Corrosion Cracking in Duplex 308 Stainless Steel", *Corrosion 78*, NACE, Houston, Texas, March 6-10, 1978.
6) T. M. Devine, "The Mechanism of Intergranular Corrosion and Pitting Corrosion in Austenitic and Duplex 308 Stainless Steel", accepted for publication by *The Journal of the Electrochemical Society*.
7) ASTM Designation: A262-70 practise E.
8) "WRC Development Program for Standard Weld Metal Sample Preparation Procedures", Advisory Subcommittee for Welding Stainless Steels, High Alloys Committee, Welding Research Council, December, 1972.
9) E. E. Underwood, "Surface Area and Length in Volume", p. 77, in *Quantitative Microscopy*, Eds. R. T. DeHoff and F. N. Rhines, McGraw-Hill Book Company, New York, 1968.
10) P. G. Shewmon, *Diffusion in Solids*, McGraw-Hill, New York, 1963.
11) J. Gurland, "Spatial Distribution of Discrete Particles", p. 278 in *Quantitative Microscopy*, *Op. Cit.*.
12. J. Bleton, J. Blenot and P. Bastien, *Rev. Met.*, *48*, p. 525, 1935.

TABLE 1

COMPOSITION OF WROUGHT 308 STAINLESS STEEL HEATS

	Cr	Ni	Mn	Si	S	P	C
Commercial							
Heat L-B7	20.95	9.82	1.76	0.41	.008	.016	.040
Experimental							
Heat 272	20.5	10.0	1.7	0.5	.02	.02	.025
Heat 293	↓	↓	↓	↓	↓	↓	.038
Heat 294	↓	↓	↓	↓	↓	↓	.055

TABLE II

	Cr	Ni	Mn	Si	S	P	C
Manual Metal Arc							
IN10B	18.92	10.30	1.71	0.32	0.018	0.020	0.056
5410	20.97	9.52	1.65	0.33	0.015	0.027	0.050
G6976	20.65	9.83	1.70	0.20	0.017	0.019	0.053
2E7L	20.23	9.32	1.69	0.31	0.019	0.015	0.044
2E11L	20.03	8.88	1.84	1.05	0.019	0.019	0.036
Tungsten Inert Gas							
4S983	20.95	9.82	1.76	0.41	0.008	0.016	0.040

TABLE III

	Heat No.	Heat Treatment	v/o Ferrite Ferrite No.	Austenite-Ferrite Boundary Area per Unit Vol.(cm^{-1})
Wrought				
	L-B7	1200°C/1	0	0
		1250°C/1	4	237
		1300°C/1	10	465
		1350°C/1	20	535
	292	1200°C/1	14	350
		1250°C/1	17	355
		1300°C/1	21	428
		1350°C/1	31	433
	293	1200°C/1	6	155
		1250°C/1	9	285
		1300°C/1	17	310
		1350°C/1	25	388
	294	1200°C/1	2	123
		1250°C/1	4	153
		1300°C/1	9	255
		1350°C/1	20	410
Weld Deposits				
	IN10B	As Deposited	2.25	145
	5410	"	4.50	476
	G6976	"	2.95	279
	2E7L	"	4.15	308
	2E11L	"	5.14	1061
	45983	"	8.28	1241

TABLE IV
INFLUENCE OF AGING TREATMENT ON THE INTERGRANULAR CORROSION BEHAVIOR OF WELD DEPOSITED 308 STAINLESS STEEL

Heat	As Deposited	550°C/24 hrs.	600°C/2 hrs.	600°C/48hrs.	600°C/96 hrs.
IN10B	X	◐	◐	◐	◐
G6976	X	◐	◐	◐	⊕
2E7L	X	◐	◐	◐	⊕
5410	X	◐	◐	○	X
2E11	X	◐	⊕	○	X
45983	X	X	X	X	X

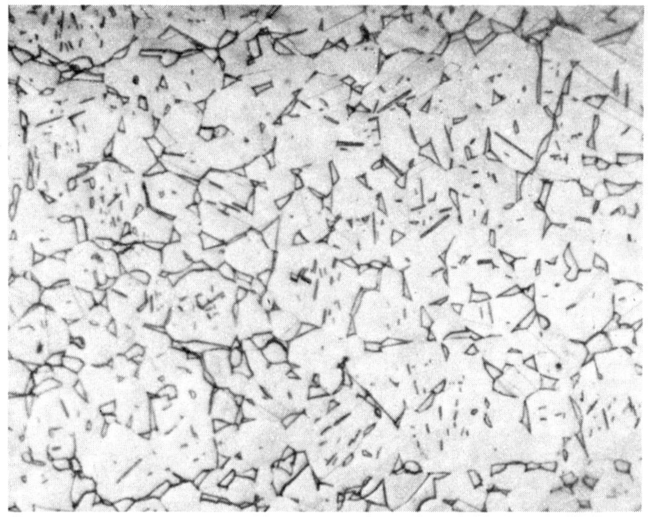

FIG 1

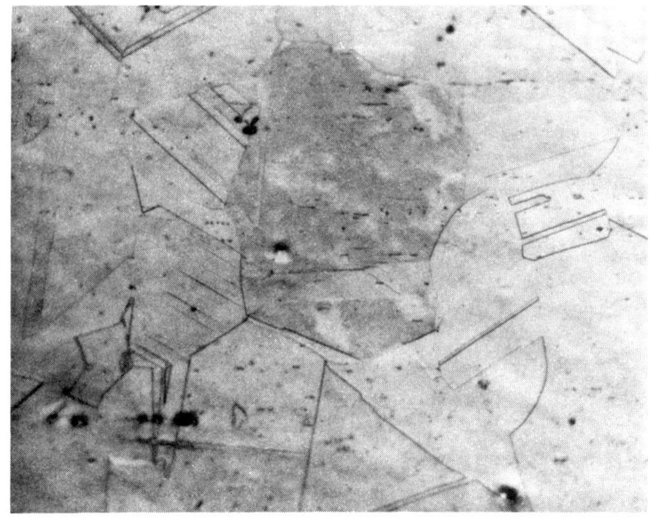

FIG 2

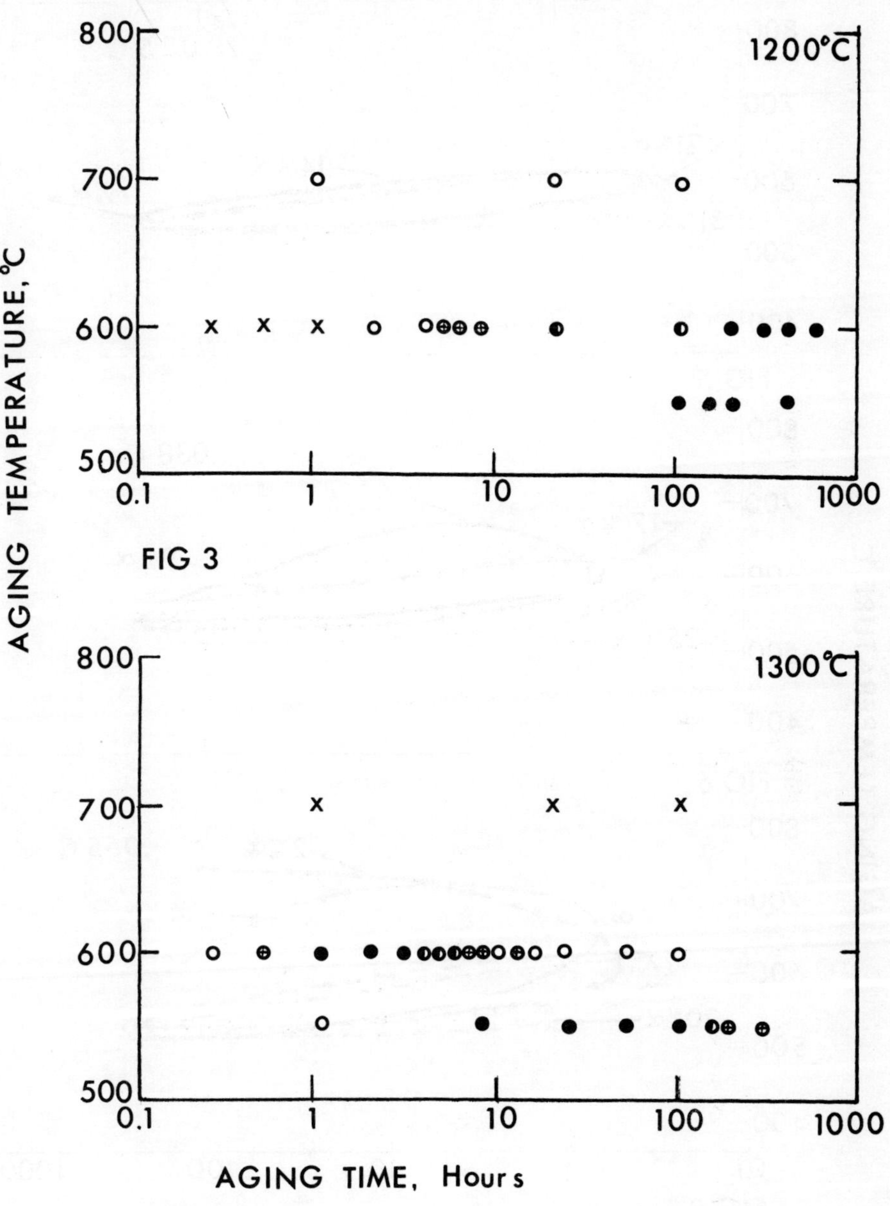
800
700
600
500
1200°C
0.1
1
10
100
1000
AGING TEMPERATURE, °C
800
700
600
500
1300°C
0.1
1
10
100
1000
AGING TIME, Hours

FIG 3

FIG 4

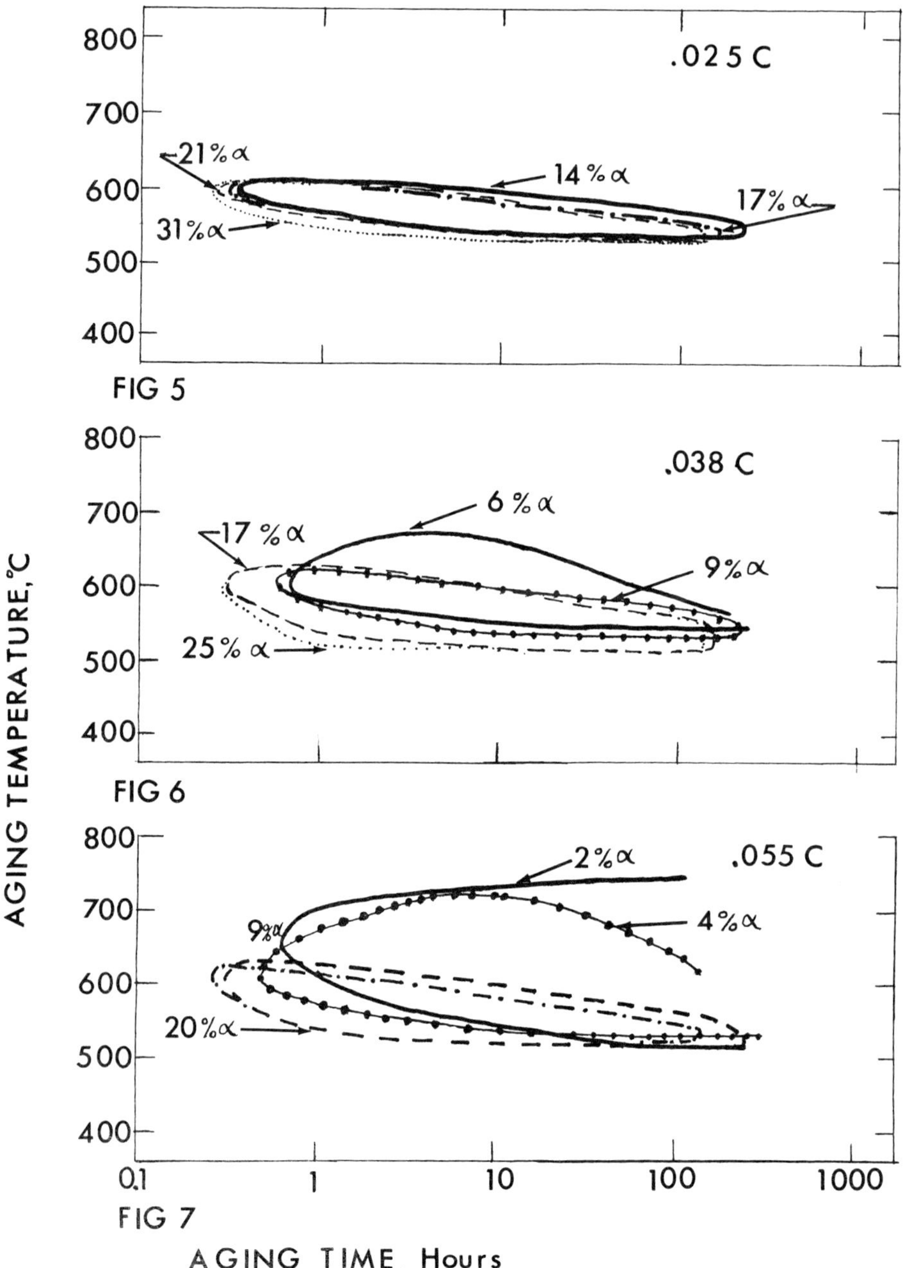
800
700
600
500
400
.025 C
21%α
14%α
17%α
31%α
FIG 5
800
700
600
500
400
.038 C
6%α
17%α
9%α
25%α
FIG 6
AGING TEMPERATURE,°C
800
700
600
500
400
2%α
.055 C
9%α
4%α
20%α
0.1
1
10
100
1000
FIG 7
AGING TIME, Hours

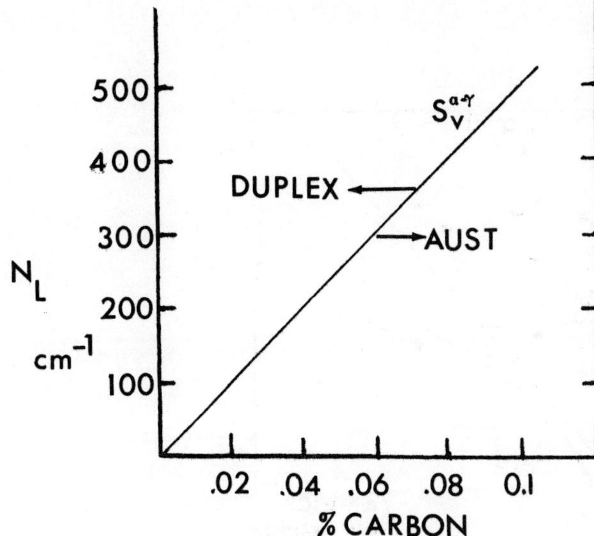
500
400
300
200
100
N
L
cm-1
DUPLEX
AUST
$S_V^{\alpha-\gamma}$
.02
.04
.06
.08
0.1
% CARBON

FIG 8

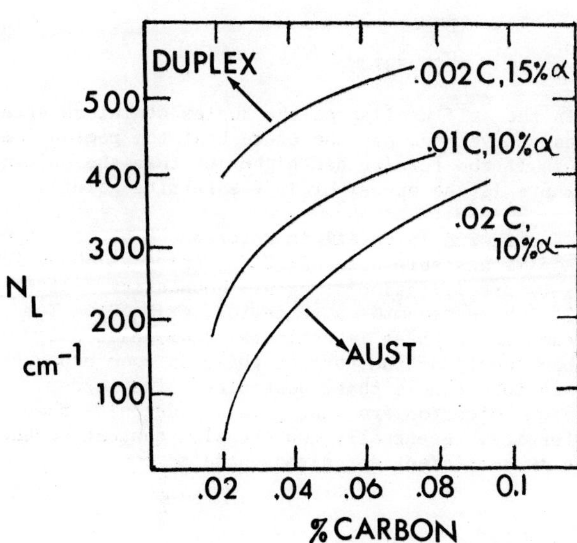
DUPLEX
.002 C, 15% α
.01 C, 10% α
.02 C,
10% α
AUST
500
400
300
200
100
N
L
cm-1
.02
.04
.06
.08
0.1
% CARBON

FIG 9

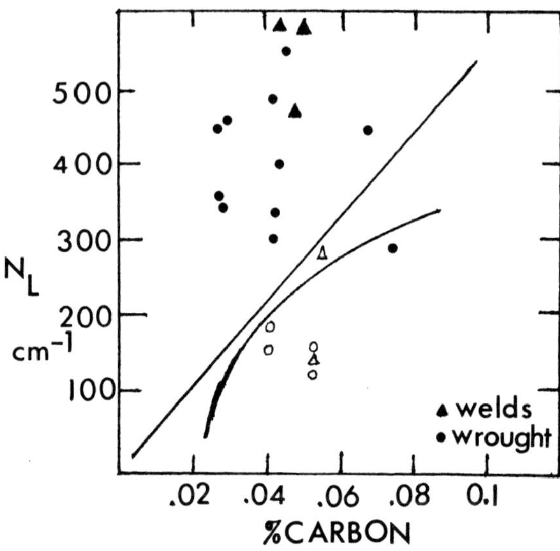

FIG 10

DISCUSSION

T. W. Eager (M.I.T.): In the self healing of the duplex stainless steel, the ferrite is being consumed yet you say the Cr to heal the region comes from the austenite grain. If the ferrite has higher Cr than the austenite and is being consumed, where is the excess Cr in the ferrite going?

Author: This question is answered in detail in reference 6 of the paper. Briefly, the ferrite phase is unstable below 1200°C. Aging at 600°C for 30 minutes results in $M_{23}C_6$ precipitation along α-γ boundaries. After 2 hours at 600°C the ferrite phase decomposes to cellular γ+$M_{23}C_6$. The chromium content of the austenite phase is estimated from phase diagrams to be ≈14 w/o. The carbon supply of the ferrite phase is soon exhausted and the ferrite decomposes into single phase austenite. The latter reaction occurs by chromium rejection from the growing austenite phase into the untransformed ferrite. Eventually the chromium content of the ferrite at the interface is sufficient for sigma phase to form.

WELD RESIDUAL STRESS MEASUREMENTS ON AUSTENITIC STAINLESS STEEL PIPES*

R. M. Chrenko

General Electric Corporate Research and Development Center
Schenectady, New York 12301

The purpose of this study was to measure the residual stress near welds in austenitic stainless steel pipes of different diameters and in pipes joined using different welding techniques. The welding techniques include standard, laser and plasma welding.

The axial ID residual stresses near welds in Schedule 80 stainless steel pipes are usually tensile. The stresses can be made appreciably less tensile or even compressive by heat sink welding, whereby the ID is water cooled after the first pass while the rest of the weld is completed. The tensile stresses present in pipes already welded with no water cooling can be reduced by applying heat to the OD near the finished weld while water cooling the ID. One means of applying the OD heat is to weld on an OD overlay, called here a backlay.

The residual stress measurements were made using x-ray diffraction methods with strain gauge stress relief corrections included where appropriate.

* The work on 304 SS pipes of different diameter was partially supported by a contract from the Electric Power Research Institute. RP-449-2

INTRODUCTION

Welded austenitic stainless steel pipes occasionally crack via an intergranular stress corrosion cracking (IGSCC) mechanism. In order for IGSCC to occur it is generally accepted that three factors must be present at the same location: a sensitized microstructure, a certain environmental condition, and a critical stress or strain condition. The degree to which any one factor must be present for IGSCC to occur depends on the intensity and overlap of the other two factors. These three factors are being studied in our laboratory in an attempt to alleviate or eliminate further IGSCC in these austenitic stainless steel welded pipes.

The stress to which the material is subjected while in use is a combination of the applied stress and the residual stress. The applied stress can usually be calculated with reasonable accuracy but this is not always true with respect to the residual stress. The residual stresses present in welded austenitic stainless steel pipes will be described here. The residual stresses of primary interest, and those that will be reported, are those on the ID near welds in the axial or longitudinal direction (i.e., those in the direction perpendicular to the weld). Only the surface residual stresses will be described.

The study consists of:

(1) Measurement of residual stresses in standard welded type 304 SS pipes of different diameters.

(2) A comparison of welding stresses due to laser, plasma, and standard welding processes for 4" diameter type 304 SS pipes.

(3) A comparison of the welding stresses in thick type 316 SS spool pieces welded with ID argon cooling and with ID water cooling.

(4) Measurement of the welding stresses present in an already welded 4" diameter type 304 SS pipe after a backlay had been applied while the ID was water cooled.

Each of these four studies will be described in more detail and then a summary will be presented. The residual stress measurements were made using x-ray diffraction methods with strain gauge stress relief corrections included where appropriate.

EXPERIMENTAL

The experimental part of this study consisted of preparing the welded specimens and measuring the residual stresses. Most of the welds were standard butt welds made with a double J weld preparation configuration. The pipes were machined in the counterbore region prior to welding. The Grinnel insert fusion and the first layer, consisting of two passes, were made using the gas tungsten arc welding (GTAW) process. The remaining groove was filled using the shielded metal arc welding (SMAW) process. Normal ranges of heat inputs were used, usually between 18000 and 40000 joules/inch (0.7 and 1.6 kJ/mm). Pipes of nominal 4", 10", and 26" (10, 25.4, and 66 cm) diameter were welded using the same general procedure. The type 304 stainless steel pipes examined were schedule 80. The one 4" diameter pipe of type 316 SS had a wall thickness approximately twice that of schedule 80 4" pipe. The non-standard welds will be described in the section that discusses the stresses.

The x-ray residual stress data were obtained on a Rigaku Strainflex mobile residual stress analyzer. In most cases the two exposure method was used to obtain the stresses and the 2θ values were determined by the half value breadth method. The equation for calculating the residual stresses was:

$$\sigma = (2\theta_{\Psi = 0} - 2\theta_{\Psi = \Psi}) \cdot \frac{\cot\theta}{2} \cdot \frac{E}{1+\nu} \cdot \frac{1}{\sin^2\Psi} \cdot \frac{\pi}{180} \qquad (1)$$

where 2θ is in degrees, θ is the average value of $\theta_{\Psi = 0}$ and $\theta_{\Psi = \Psi}$ and is approximately constant. The final equation for the stress in terms of measured 2θ becomes:

$$\sigma = K \, \Delta 2\theta \qquad (2)$$

In most cases the Ψ value used in equation (1) was $\Psi = 45^o$.

Some of the stresses were calculated from the plot of 2θ versus $\sin^2\Psi$, where the Ψ values were usually between $\Psi = 0$ and $\Psi = 45^o$. In this case the residual stress is determined from:

$$\sigma = K' \frac{\partial\, 2\theta}{\partial \sin^2\Psi}$$

CrKα radiation was used for all measurements and the diffraction was from the {220} austenite planes giving a $2\theta \sim 128^o$. For 304 SS the x-ray value of $\frac{E}{1+\nu}$ was determined in a four point bend apparatus on a piece of lightly ground, as received, material. The value obtained was $\frac{E}{(1+\nu)} = 25.2 \times 10^3$ KSI. Assuming $\nu = 0.29$, the x-ray value for E for the {220} planes is $E = 32.5 \times 10^3$ KSI or approximately 15% higher than the reported bulk value. The same value of $\frac{E}{1+\nu}$ was also used for the type 316 SS.

RESULTS

Residual Stresses in Welded Pipes of Different Diameters

Welding stresses were determined on schedule 80 type 304 SS pipes of 4", 10", and 26" diameter. Of interest were the residual stress variations on any one pipe and among pipes of different diameters. A typical ID surface axial residual stress profile with distance from the weld is shown in Figure 1. Such profiles were found for all the azimuths for all the pipes, but with differences in the vertical displacement of the whole curve.

The residual stresses can vary appreciably around the circumference of a pipe. An example is given in Figure 2 which shows the residual stress 0.2" from the weld center line for a 26" pipe. Included are the ID surface axial x-ray stresses and the total stresses (x-ray stresses corrected for stress relief due to cutting the x-ray specimens out of the pipe). This 0.2" distance was chosen since this location is in the weld heat affected zone (HAZ), it is the area closest to the weld on which reliable strain gage and x-ray measurements can be made, and the stresses in this location are usually tensile or at least more tensile than the stresses further from the weld.

The azimuthal (circumferential) stress variations seen in Figure 2 for the 26" pipe have also been found in the welded 4" and 10" pipes. The azimuthal variations are smaller the larger the pipe diameter. It is suggested that this trend is due to the larger thermal heat sink of the thicker walled larger diameter pipes making thermal variations around the azimuth during welding less severe as compared to the smaller diameter pipes.

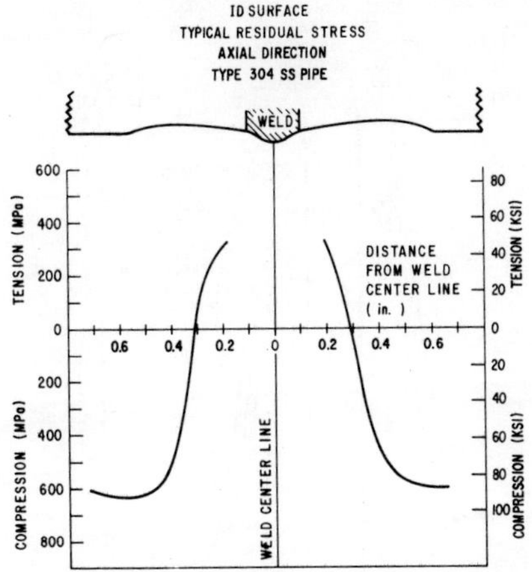

Figure 1: Typical ID surface x-ray residual stress for 4 in., 10 in., and 26 in. diameter welded schedule 80 type 304 SS pipes. Stress as a function of distance from weld center line.

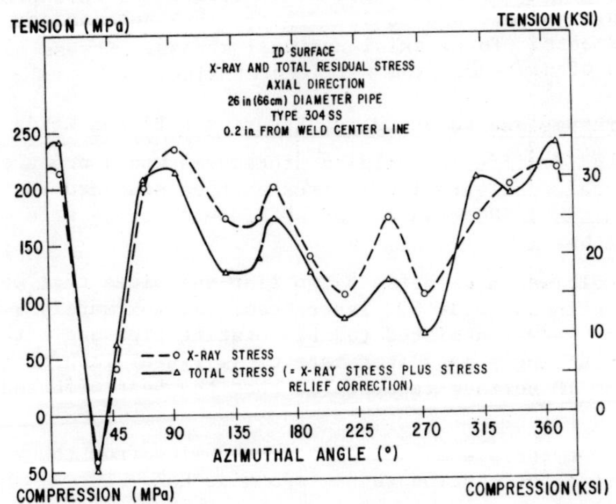

Figure 2: Azimuthal (circumferential) variation of x-ray and total ID surface residual stress in axial direction for 26 in. diameter welded 304 SS pipe. Stress location is 0.2 in. (0.5 cm) from weld center line.

Of more interest than azimuthal stress variations for IGSCC purposes is the magnitude of the axial residual stress at the most tensile azimuth for the three different diameter pipes. The total axial residual stresses for the most tensile azimuths for the three different diameter pipes are plotted in Figure 3. In the heat affected zone these most tensile stresses are in the order: $\sigma_{4"} > \sigma_{10"} > \sigma_{26"}$. Hence, all other factors being equal, the effect of the residual stress on IGSCC would be expected to be more severe in the 4" diameter pipes. This is the trend found in actual field experience, where 4" diameter pipes have had the most indications of IGSCC.

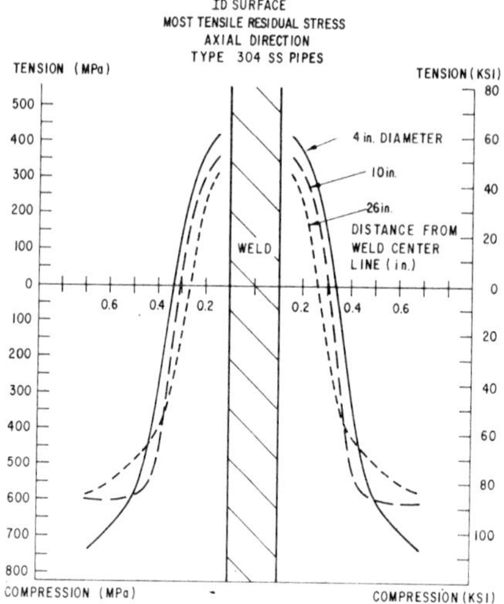

Figure 3: Most tensile azimuth ID surface residual stress profiles for 4 in., 10 in., and 26 in. diameter schedule 80 304 SS pipe weldments. Total axial residual stress. Stress as a function of distance from weld center line.

Welding Stresses due to Standard, Laser, and Plasma Welds

To determine if different welding procedures would produce weld joints having less tensile residual stresses than standard butt welded pipes, several schedule 80 type 304 SS 4" diameter pipes were welded using laser and plasma heating.

The laser weld was made between two flat-end pipes that were placed under a 12 kw continuous mode CO_2 laser beam. Approximately 9 kw of the 10.6 μm radiation was transmitted to the rotating pipe and a through wall weld was made in a single pass at a rate of ∿1 linear inch/sec. From the coloration on the ID surface it appeared that the heat affected zone was very narrow.

The details of the plasma weld are forthcoming from the vendor. However, the ID surface coloration suggests a wide HAZ.

A comparison of the most tensile ID surface axial residual stresses for standard, laser and plasma welded pipes is shown in Figure 4. Near the weld the highest tensile stress for the laser welded pipe is greater than in standard butt welded pipes but the region of tensile stresses lies closer to the weld. For the plasma welded pipe the highest tensile stress is less than in standard butt welded pipes but the region of tensile stresses extends further from the weld center line. The oscillations in the surface stress for the plasma welded pipe have been confirmed by repeated measurements and have not been seen in laser or standard welded 4" diameter type 304 SS schedule 80 pipes.

The data and the appearance of the HAZ suggest that the more localized the welding heat the more tensile are the axial (longitudinal) ID surface residual stresses near the weld and the region of tensile stresses lies closer to the weld. Whereas these three welding processes differ, the tensile areas under the curves starting 0.1" from the weld center line (this is the weld fusion line for the standard and plasma welded pipes) are the same within 10%.

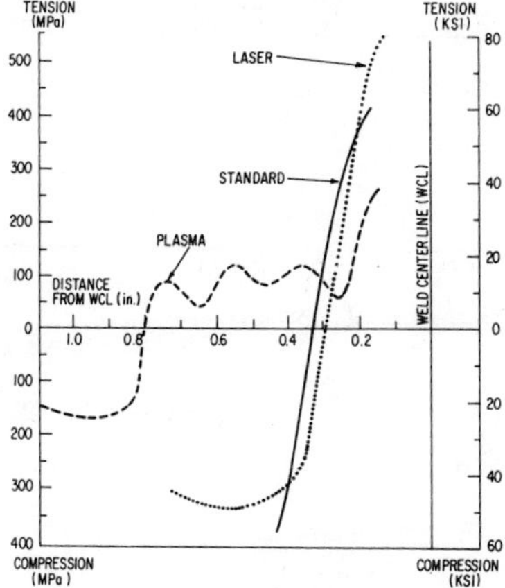

Figure 4: ID surface residual stress in the axial direction for 4 in. diameter type 304 SS laser, standard, and plasma welded pipes. Stress as a function of distance from weld center line.

Welding Stresses in Heat Sink Welded Type 316 SS Spool Pieces

Residual stress measurements have been made on the pipe configuration shown in Figure 5. In practice the thicker 316 SS spool pieces could also be 304 SS. In manufacturing, the 304 - 316 subassemblies would be first welded and then solution annealed to remove the sensitization in the heat affected zone of the 304 SS. The subassemblies would then be taken to the building site and the final 316 - 316 weld made. This procedure should reduce IGSCC susceptibility because the sensitization in the heat affected zone of the 304 SS is removed by the solution anneal and the 316 SS spool pieces are thicker than the rest of the pipe with the result that less stress and strain occurs near the weld for an applied load. In addition, and of more importance to this residual stress study, if ID water cooling is used after the fusion pass or first layer, then compressive axial residual stresses should be induced on the ID surface near the weld. This water cooling method to induce ID compressive stresses can also be used for welding pipes of standard configurations.

Several 304-316-316-304 assemblies have been welded according to the above procedure, namely; (1) the 304-316 subassemblies were welded; (2) the subassemblies were solution annealed (1100°C, 1 hour, argon, water quench); (3) the 316-316 welds were made using argon gas backing or water spray cooling on the ID.

The ID surface residual stress measurements for 316-316 spool piece welds made with argon and water spray cooling (after the first pass) are shown in Figure 6, which also shows the shape of the weld and adjacent areas. These axial ID surface stresses for the pipe welded with argon gas backing are tensile near the weld whereas the stresses for the pipe cooled with a water spray on the ID are compressive. The compressive stresses are induced on the ID surface by the axial contraction of the weld metal at or near the OD surface while the ID surface and subsurface metal is kept cool by the water spraying. This welding process whereby the ID is water cooled after the fusion pass or first layer has been given the name of heat sink welding (HSW).

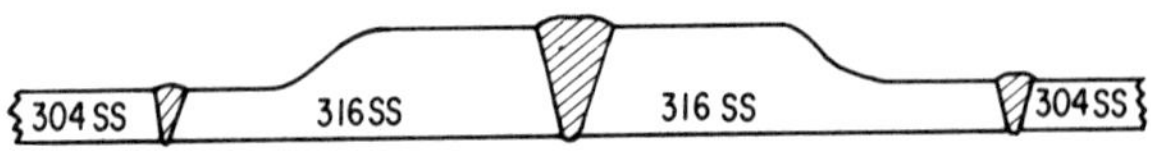

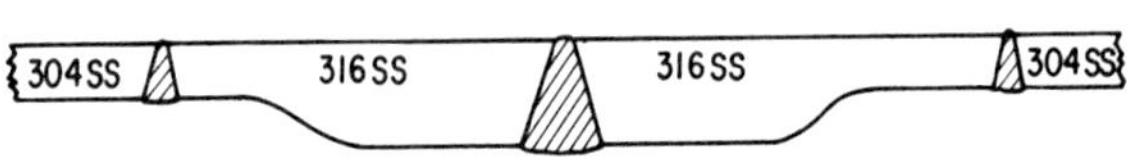

Figure 5: Configuration of pipes for using type 316 SS spool pieces with type 304 SS pipes.

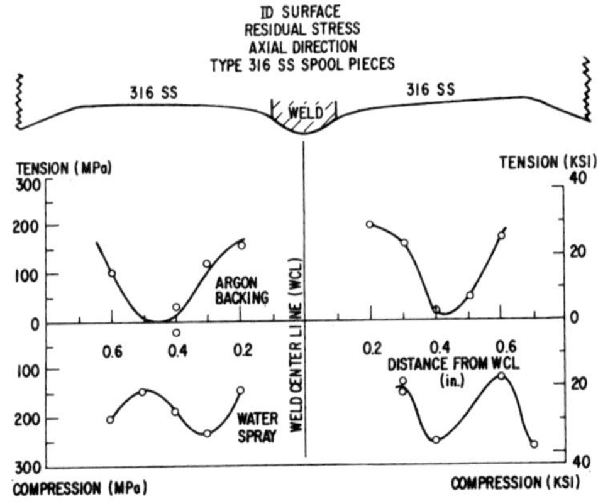

Figure 6: ID surface residual stress in the axial direction for 4 in. diameter type 316 SS spool pieces welded with ID argon gas backing and ID water spray cooling. Stress as a function of distance from weld center line.

The stress pattern for the argon backed pipe is somewhat different from patterns obtained from similarly welded schedule 80 type 304 SS 4 in. diameter pipes, which show tensile stresses near the weld out to ∿0.3 - 0.4 inches from the weld center line and then compressive stresses. (See Figure 1). Perhaps the difference is due to the greater wall thickness of the type 316 SS spool pieces near the weld (thickness > 2X thickness of schedule 80 4 in. diameter pipe). The significance of the residual stress oscillations with distance from the weld is not known.

Welding Stresses in a Type 304 SS Pipe with a Backlay

A backlay consists of welding a type 308 SS overlay on the OD of an already welded pipe while cooling the ID with water. This backlay is placed over a standard circumferential weld and the typical appearance of a weld covered by a backlay is shown in Figure 7.

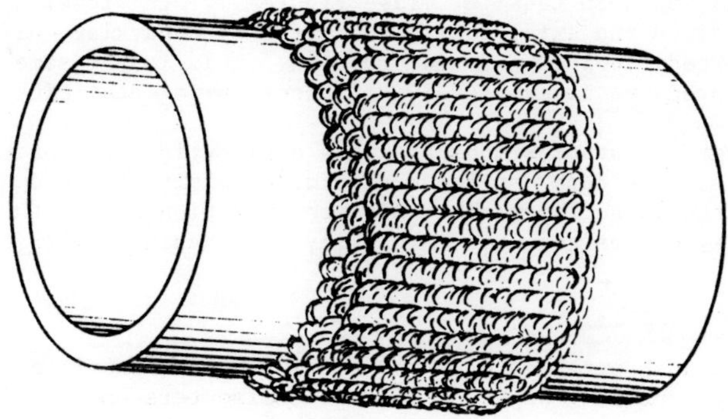

Figure 7: View of a 3 layer backlay over a normal circumferential weld in a 4 in. diameter type 304 SS pipe. The ID is water spray cooled while the backlay is being applied.

The benefits of the backlay are: (1) the area near the weld is thicker with the result that for a given applied load the stress and strain in the ID heat affected zone are reduced, (2) the use of ID water cooling while applying the backlay to the OD should induce compressive residual stresses near the weld on the ID; (3) if the ID is adequately water cooled, then the already sensitized heat affected zone near the original weld should not become more sensitized. Also, the remainder of the ID should not become sensitized due to the application of the backlay.

The ID surface residual stress distribution (axial or longitudinal direction of stresses) for a backlay is shown in Figure 8. The axial

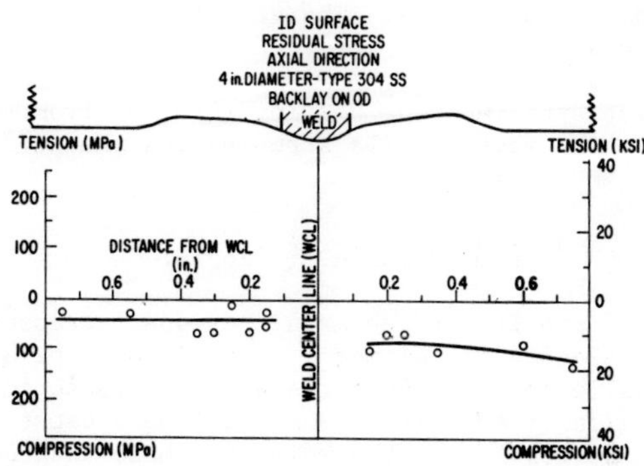

Figure 8: ID surface residual stress in the axial direction for 4" diameter type 304 SS welded pipe after application of a backlay. The circumferential weld was first made and then the backlay was applied while the ID was water spray cooled. Stress as a function of distance from weld center line.

stresses are compressive near the weld, whereas these stresses for a standard weld would be tensile. (See Figure 1). Compressive stresses are induced since the axial contraction of the OD backlay while cooling pulls the already cool ID into compression. This is the same basic effect that produces ID compressive stresses during heat sink welding.

Metallography studies indicate that application of a backlay on a 4 in. diameter pipe does not cause carbide precipitation and hence sensitization of the ID surface. The only sensitization due to the application of the backlay occurs near the OD backlay to a depth of ∿1/3 the wall thickness.

SUMMARY

Weld residual stresses have been measured on austenitic stainless steel (mostly type 304) pipes of different diameters and on pipes welded using different processes. All pipes had been machined in the counterbore area prior to welding and received no post-weld surface treatment. Only surface residual stresses on the ID in the axial or longitudinal direction are discussed.

The axial ID surface residual stresses measured on standard butt welded schedule 80 type 304 SS pipes are summarized below:

1. Bell shaped residual stress distributions centered at the weld have been found for 4 in., 10 in., and 26 in. (10, 25.4, and 66 cm) diameter welded pipe.

2. The stresses are usually tensile in the heat affected zone 0.2 in. (0.5 cm) from weld centerline.

3. The stresses are highly compressive 0.5 in. (1.2 cm) from weld centerline.

4. At a constant distance from weld centerline in the heat affected zone, e.g., 0.2 in. (0.5 cm), the stresses vary appreciably around the circumference. The variations are less the larger the pipe diameter.

5. In the heat affected zone 0.2 in. (0.5 cm) from the weld centerline the most tensile stresses for the different diameter pipes are in order:

$$\sigma_{4\ in.} > \sigma_{10\ in.} > \sigma_{26\ in.}$$

A comparison of the ID surface axial residual stresses for standard, laser, and plasma welded 4 in. diameter schedule 80 type 304 SS pipes shows the most tensile stresses near the weld are those induced by laser welding but the region of tensile stresses is the closest to the weld. The lowest tensile stresses near the weld are those induced by plasma welding but the region of tensile stresses extends furthest from the weld. The tensile stresses induced by a standard weld are intermediate in terms of their magnitude near the weld and the distance they extend from the weld.

ID surface residual stress measurements on type 316 SS spool pieces indicate that, by water cooling the ID while welding, compressive ID axial stresses are induced near the weld. The same weld made by using argon gas backing has tensile stresses near the weld. The process of cooling the ID with water after the fusion or first pass is called heat sink welding (HSW).

ID surface residual stress measurements were made on an already welded 4 in. diameter schedule 80 type 304 SS pipe after a backlay had been applied to the OD while the ID was water cooled. The axial stresses near the weld were slightly compressive, and hence, much less tensile than stresses measured in similar pipes without backlays.

ACKNOWLEDGMENTS

The author thanks R. A. Thompson for strain gauge stress relief corrections and R. E. Hanneman for stimulating discussions on various aspects of these studies.

DISCUSSION

L. W. Sandor (Sun Shipbuilding & Dry Dock Co.): You seemed to have indicated that the larger the pipe diameter, the greater the residual stress. I wonder, if you could comment on the whys and wherefores of that?

Author: Actually, it's the other way around. The maximum ID axial tensile stress for the larger diameter pipe is less tensile than for smaller diameter pipes. The different diameter pipes all have residual stress variations around the azimuth. However, if one compares the most tensile stress measured on each pipe, then this most tensile stress for the larger diameter pipe is less tensile than for the smaller diameter pipes.

One factor responsible for the tensile axial stresses near the ID weld is the circumferential weld shrinkage. The resultant inward bend induces this stress. The larger diameter pipes have a thicker wall section and we hypothesize this thicker wall leads to less circumferential shrinkage near the weld and therefore less tensile axial stress.

S. S. Glickstein (Westinghouse-Bettis Lab): Can you explain the reason for the large tensile stress measured for the laser weld relative to the plasma arc weld?

Author: The residual stress data, in conjunction with the appearance of the ID heat affected zone, suggest that the more localized the source of welding heat, the higher the axial tensile stresses and the steeper the stress gradient near the weld. The extended hot zone that occurred during the plasma welding allowed the contracting weld metal to produce contraction and tensile stresses over a wide area but the resultant tensile stresses were averaged out to a lower value.

A. Glover (Canadian Welding Inst.): In the data you presented on the residual stress measurement after laser welding you stated that this was the peak value - did this correspond to the stop position (particularly considering the rate of welding)?

Author: For the laser welded pipe we examined a number of azimuths, none of which corresponded to start or stop positions, so I cannot tell if these start or stop positions gave higher tensile stresses than those shown. The stresses reported are for the azimuth that gave the highest tensile stresses.

I should say that in several normal butt welds (double J configuration, GTAW, SMAW) we attempted to correlate variations in residual stress with various start and stop positions, but could find no correlation.

L. M. Petrick (Ebasco Services, Inc.): The literature has not reported a threshold stress for stress corrosion cracking, even in a sensitized region. Does your work suggest any hope for threshold stress possibility?

Author: The threshold stress at which cracking occurs depends on the environment as well as the microstructure of the material and such factors as surface condition. However, one test that is used to find regions of tensile stresses on austenitic stainless steel is boiling $MgCl_2$ at 155°C (ASTM G36-73). Work at Electricitie de France indicates that cracks occur in type 316 SS when the stress is $\gtrsim$ 20 KSI when immersed in $MgCl_2$. For type 304 SS I have heard the number 10 KSI, but I do not know the original source of the information.

We have found a qualitative correlation between $MgCl_2$ cracking of 304 SS and our residual stress measurements in that cracking only occurred in tensile regions. We have not determined the minimum stress needed for cracking 304 SS in $MgCl_2$. It should be remembered that the $MgCl_2$ test indicates tensile regions and may not simulate the cracking mode (transgranular, intergranular, or a combination) that may occur in actual practice.

A Correlation Between Fractography, Metallography and

K_{Iscc} Values of [illegible] Steel Weldments

[illegible] and [illegible]

Material Science and Technology Division
Naval Research Laboratory
Washington, D.C. [illegible]

ABSTRACT

[illegible] weldments of [illegible] were fabricated with electrodes of matching composition. Two of the weldments of differing yield strengths were tested for SCC [illegible] and [illegible] were [illegible] after [illegible] and at [illegible] (538 °C) [illegible] the other [illegible]. In the as-welded condition the specimen with the higher yield strength fractured by microvoid coalescence (MVC) while the specimen of lower yield strength fractured by cleavage. The cleavage facets were fine and regular in shape. Although the fracture modes of the two specimens [illegible], the K_{Iscc} values [illegible] for the weldment that fractured by MVC [illegible] as compared to the as-welded [illegible] analysis showed that a [illegible]

INTRODUCTION

[illegible] (SCC) [illegible] such as [illegible] (K) [illegible] used to evaluate [illegible] performance [illegible] of [illegible] behavior [illegible] on the microstructure of the [illegible] and on [illegible]

These [illegible] correlate fracto- [illegible] and [illegible] values of [illegible] high-strength steel weldments.

A Correlation Between Fractography, Metallography and K_{Iscc} Values of HY-180 Steel Weldments

F. W. Fraser and E. A. Metzbower

Material Science and Technology Division
Naval Research Laboratory
Washington, DC 20375

ABSTRACT

Four GTA weldments of HY-180(10Ni-8Co) steel were fabricated with electrodes of matching composition. Two of the weldments, of differing yield strengths, were tested for SCC in the as welded condition and two were tested after tempering, one at 1000°F(538°C) and the other at 1150°F(621°C). In the as welded condition the specimen with the higher yield strength fractured by microvoid coalescence (MVC) while the specimen of lower yield strength fractured by cleavage. The cleavage facets were fine and regular in shape. Although the fracture modes in the two specimens differed, the K_{Iscc} values were nominally the same. Tempering the weldments led to fracture by MVC and resulted in improved K_{Iscc} values as compared to the as welded specimens. Metallographic analysis showed that a correlation exists between the fracture mode, K_{Iscc} value and the microstructure.

INTRODUCTION

Stress corrosion cracking (SCC) parameters such as the critical stress intensity factor for crack initiation, K_{Iscc} can be used to evaluate the performance of welded steels subjected to aggressive environments. However, SCC behavior depends ultimately on the microstructure of the weld metal and on the fracture separation paths.

This investigation was undertaken to correlate fracture separation processes, related microstructures and the K_{Iscc} values of four high strength steel weldments.

MATERIALS

Four GTA weldments of HY-180(10Ni-8Co) steel (specimens I,J,Q and H), fabricated using matching electrodes and tested for SCC,[1] were studied in this investigation.

Specimens J,Q and H were fine bead weldments, all fabricated at the same wire feed rate and travel speed, whereas specimen I was a coarse bead weldment fabricated at a higher wire feed rate, a reduced travel speed and a higher voltage and amperage. The double V welds were 38mm (1.5 in.) thick. The SCC characteristics were determined by the cantilever beam method on a single-edge-notched, fatigue precracked specimen.[2] Specimens I and J were tested in the as welded condition, whereas specimens H and Q were subjected to post weld heat treatments before testing. The heat treatments, yield strengths and K_{Iscc} values for the four weldments are given in Table 1.

Table I

Specimen	Heat treatment	ys ksi	ys MPa	K_{Iscc} ksi$\sqrt{in}$	K_{Iscc} MPa$\sqrt{m}$
I	as welded, coarse bead	165	1138	80	88
J	as welded, fine bead	188	1296	81	89
Q	tempered, 1000°F (538°C), 1 hr water quenched fine bead	193	1331	95	104
H	tempered, 1150°F (621°C), 1 hr water quenched fine bead	125	862	118	130

PROCEDURE

One fracture surface of each specimen was stripped with replicating tape and then cleaned ultrasonically. After preliminary macroscopic examination, fractographic analysis was carried out in the scanning electron microscope (SEM). Each specimen encompassed a fatigue, SCC and fast fracture area. (Figs. 1,2,3, and 4 show macrographs of the four fracture surfaces). The SCC zone lies between the fatigue and the fast fracture areas, as indicated by the arrows.

Upon completion of the fractographic analysis, metallographic specimens were taken by sectioning each fracture specimen in both the longitudinal and the transverse directions. The longitudinal sections were taken through the smooth SCC zone of each specimen and intersected fatigue, SCC and fast fracture areas. The transverse sections were taken through the SCC zones exclusively and included the rough textured weldment center, the smooth SCC zone and the embrittled edge. The sections were then prepared for metallographic analysis. The etchant used was Kalling's reagent.[3]

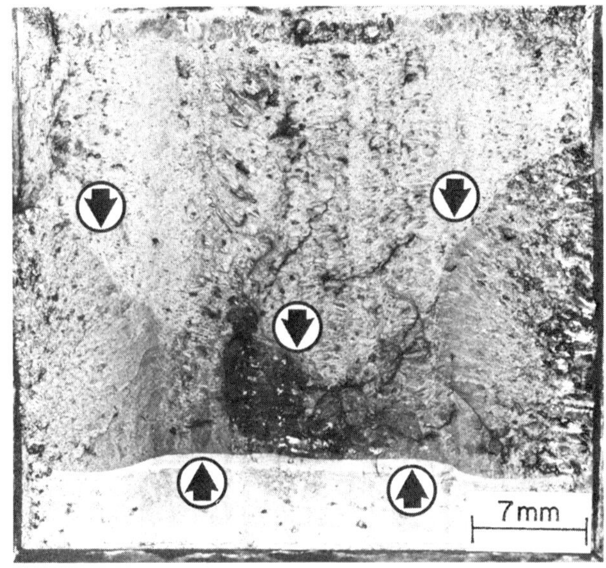

Fig. 1. Specimen I. Fracture surface. Arrows indicate the location of the SCC zone.

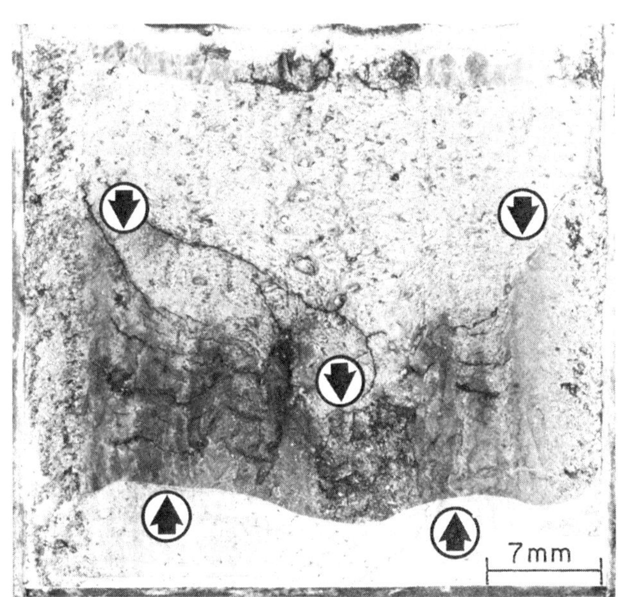

Fig. 2. Specimen J. Fracture surface. Arrows indicate location of SCC zone.

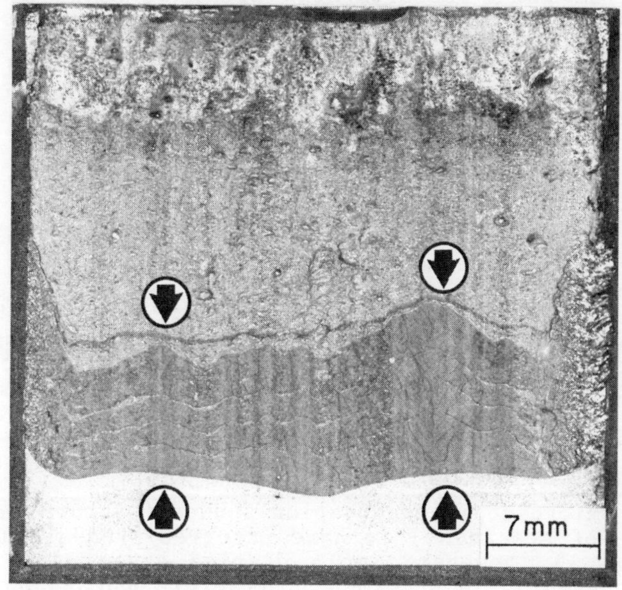

Fig. 3. Specimen Q. Fracture surface. Arrows indicate location of the SCC zone.

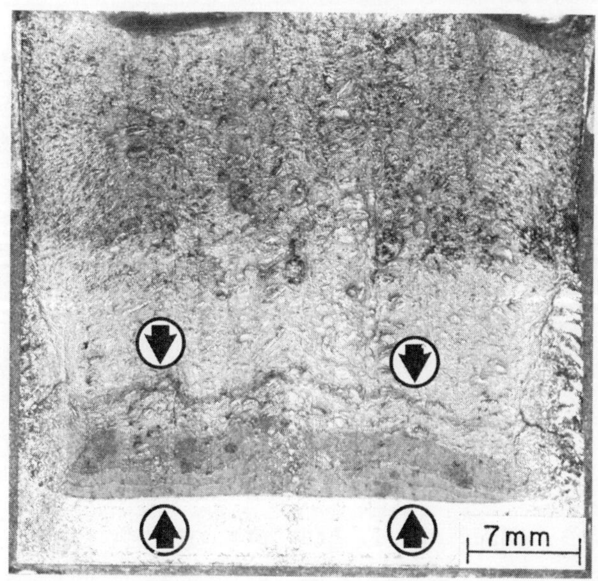

Fig. 4. Specimen H. Fracture surface. Arrows indicate location of SCC zone.

Hardness measurements were made using a Vicker's microhardness testor under a 300g load. All measurements were converted to the Rockwell C scale (R_c).

To investigate microsegregation of the alloying elements within the weld metal, x-ray analysis was carried out using an electron beam microprobe.

RESULTS

FRACTOGRAPHIC ANALYSIS

Macroscopic examination showed the SCC area in specimens I and J to be concave (Figs. 1 and 2), whereas in specimens Q and H this area was of relatively uniform depth (Figs. 3 and 4). A rough, coarse textured fracture occurred at the side edge of each specimen, reflecting the absence of thermal cycling on the last weld bead, and at the center of each specimen indicating that this area had been subjected to less thermal cycling than had the remaining weld metal. These coarse areas were most pronounced in the as welded specimens I and J. The remainder of the SCC zones in all four specimens appeared smooth and fine textured.

Microscopic examination showed fracture in the smooth SCC zone of specimen I to have occurred predominantly by cleavage (Fig. 5), however, small dimpled areas were found scattered throughout. In specimen J, fracture had occurred by MVC (Fig. 6). The cleavage facets in specimen I were very fine and regular in shape. In both I and J, relatively large cleavage facets were found at the embrittled edges in the SCC areas, and also at the coarse textured centers of these specimens, with extensive secondary cracking occurring in this latter area in specimen I.

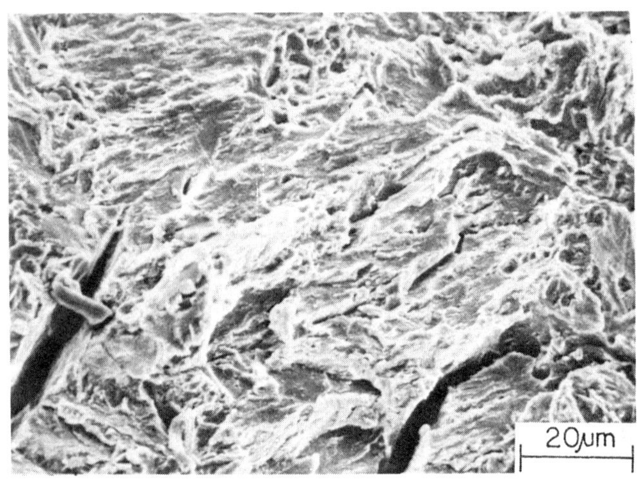

Fig. 5. Specimen I. Fracture by cleavage in the SCC zone.

Fast fracture in both specimens had occurred by MVC with dimples ranging from fine, equiaxed to large, conical formations in specimen I, while equiaxed dimples of relatively uniform size were found throughout the fast fracture area in specimen J.

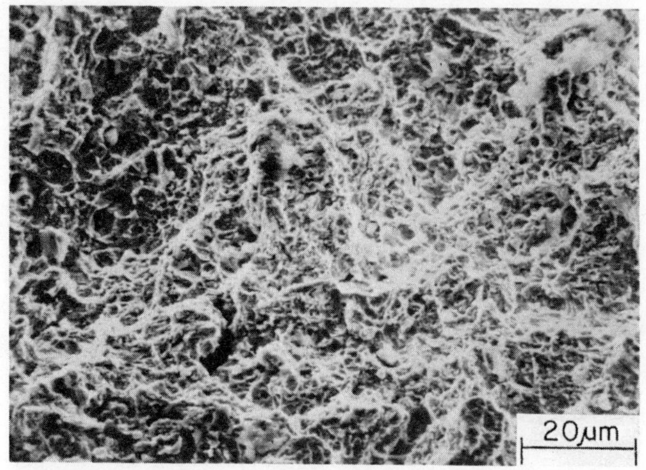

Fig. 6. Specimen J. Fracture by MVC in the SCC zone.

The tempered specimens, Q and H, showed similar fracture modes in both the SCC and the fast fracture areas. In the smooth SCC areas, fracture had occurred by MVC whereas in the coarse regions, at the centers and sides of the specimens, fracture had occurred by MVC and cleavage. The dimples formed in specimen Q were coarser than those of specimen H.

No evidence of intergranular separation was found in any of the specimens analyzed.

MICROSTRUCTURAL ANALYSIS

Metallographic examination of the microstructure of the longitudinal cross section, in the vicinity of the fracture edge in specimen I, showed a highly acicular structure of tempered martensite with a second phase at the grain boundaries (Figs. 7 and 8). This second phase appeared as a light etched band of leaflike structure. The hardness of the grain boundary phase was 49.5 Rc whereas that of the acicular structure within the grain, was 47.8 Rc.

Microprobe analysis showed a slight increase in the concentrations of Ni and Cr within the grain boundary phase.

A uniform microstructure of tempered martensite with a hardness of 46.6 Rc, was found along the fracture edge in the longitudinal section of specimen J (Fig. 9). The microstructures of the tempered specimens Q and H were similar in appearance to that of the grain boundary phase in specimen I. Specimen Q (Fig. 10) contained light and dark etched areas with hardnesses of 49.3 Rc and 48.1 Rc respectively.

The structure of specimen H (Fig. 11) was finer than that of Q and considerably softer, with a hardness of 39.1 Rc. The illustrated microstructures of specimens Q and H are also from the longitudinal sections.

No significant secondary cracks were found in either the fatigue or fast fracture areas.

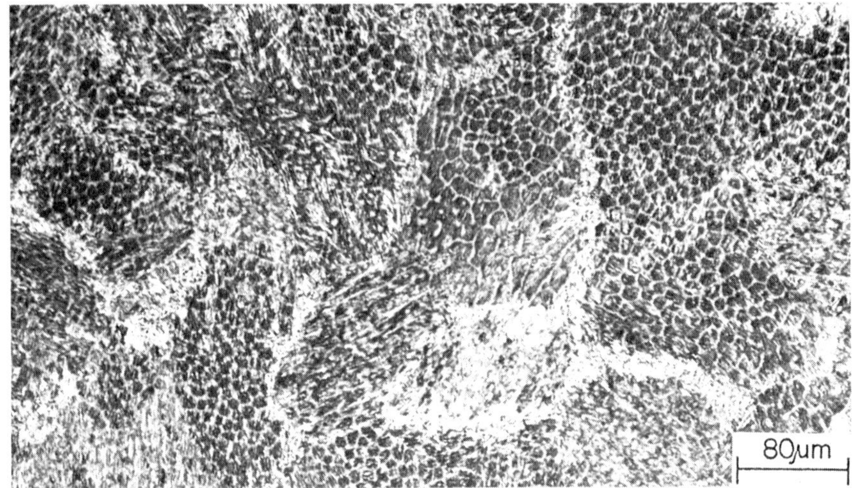

Fig. 7. Specimen I. Light etched second phase at the grain boundaries. Etched with Kalling's Reagent

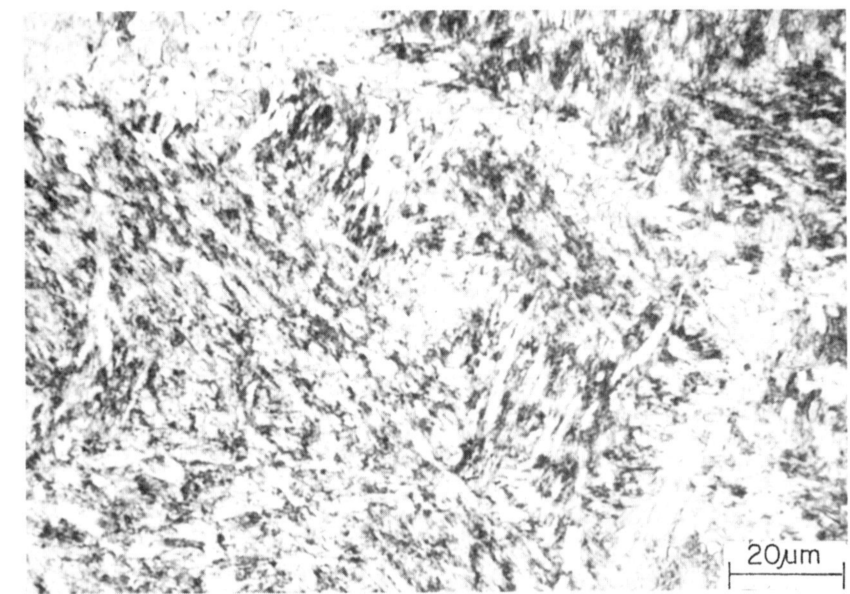

Fig. 8. Specimen I. Tempered acicular structure within the grain and leaf like structure at the grain boundary

In both of the as welded specimens, the center of the weldment seen in the transverse section consisted of a relatively coarse columnar structure. The structure in the smooth SCC zone was similar to that seen in the longitudinal sections but showed varying degrees of tempering depending on the extent

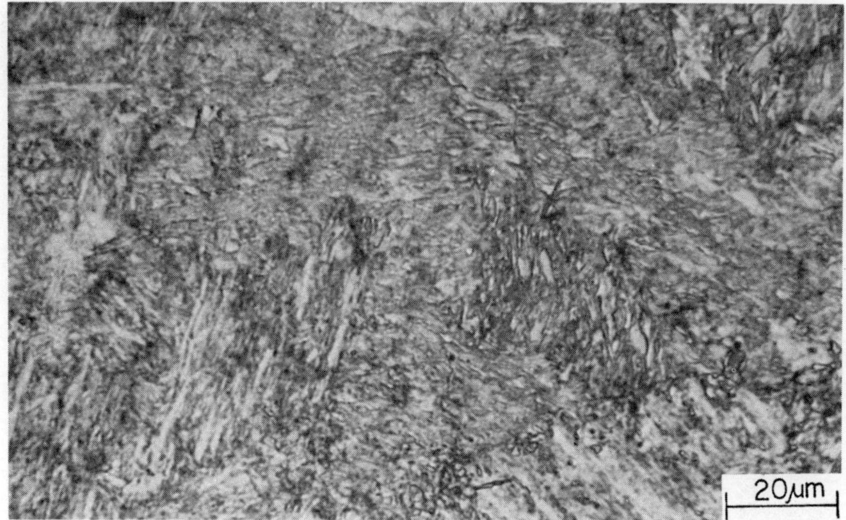

Fig. 9. Specimen J. Tempered, martensitic microstructure.

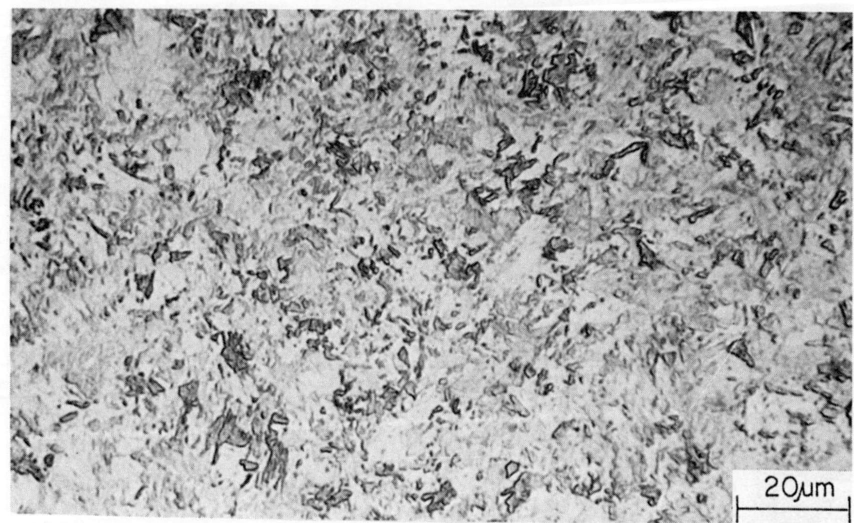

Fig. 10. Specimen Q. Tempered microstructure, similar to that of the grain boundary phase in Specimen I.

of thermal cycling. As the number of thermal cycles decreased, the microstructure became increasingly coarse. The coarsest structure was found in the final uncycled weld beads at the edge of the specimen. All four specimens did however contain secondary cracks in the SCC zones. In addition to those originating at the fracture edge and extending vertically into the microstructure, specimen I contained short, broad, blunt micro-

Fig. 11. Specimen H. Highly tempered microstructure.

cracks associated with the cell walls of the solidification substructure. In specimen J, the microcracks were finer, and appeared to be associated with chain like formations of precipitated second phase particles throughout the microstructure.

Relatively few secondary cracks were found in the tempered specimens Q and H. Those that had occurred were short, sharp cracks, all of which had originated at the fracture edge.

DISCUSSION

The residual stresses introduced during welding give rise to an SCC zone of irregular shape, as seen in the as welded specimens I and J. The shape of the crack front is also influenced by available hydrogen and the susceptibility of the microstructure to SCC.

The coarse grained, untempered structure developed in the weld metal deposited in the final passes during fabrication of the weldment, and its subsequent exposure to the corrosive environment during SCC testing, result in embrittlement as evidenced by the large cleavage facets seen on the fracture surface. The crack front advances more rapidly in this area than in the remaining weld metal where the microstructure is refined, due to repeated thermal cycling, and where exposure of a fresh surface to the corrosive medium is restricted by the crack opening.

Tempering of the weldment tends to relieve the residual stresses, to some extent, and to produce a refined and relatively uniform microstructure, thereby reducing the susceptibility of the weld metal to embrittlement, and resulting in a more uniform SCC zone as found in specimens Q and H.

As welding parameters and subsequent heat treatments are varied, K_{Iscc} values afford a measure of the relative sensitivity of the weld metal to SCC. As seen in Table 1, tempering increases the K_{Iscc} value relative to the as deposited weld metal. The highest value is attained in weld metal overtempering to a lower yield strength.

In the smooth SCC areas in all three fine bead weldments, fracture occurred by MVC and differed only in the degree of uniformity and refinement of the microvoids. Repeated thermal cycling during deposition of the fine weld beads results in a relatively refined microstructure which is further refined by tempering, as shown in Figs. 10 and 11.

A comparison of the properties of the fine and coarse bead as deposited weld metal in Table 1, shows similar K_{Iscc} values for these two weldments although the fracture modes differed. The coarse bead weldment I fractured by cleavage. In a prior investigation of the HY-130 system,[1] fracture by cleavage was generally associated with lowered K_{Iscc} values. The fracture surfaces of these specimens were characterized by relatively large, irregular cleavage facets. On the other hand, the surface of specimen I showed mainly fine, uniform cleavage facets. In respect to the K_{Iscc} values, therefore, the degree of uniformity and refinement of the fracture mode appears to be of greater significance than the fracture mode, per se.

The cleavage facets which characterized specimen I were reflected in the fine, acicular, martensitic microstructure within the grains at the fracture edge of this specimen. Kalling's reagent, the etchant used to develop the microstructure, revealed a white etched phase at the grain boundaries. Austenite which has at least partially transformed to martensite will appear as a white etched structure, when etched with Kalling's reagent, while tempered martensite will appear dark. Although some austenite may be retained in the grain boundary phase, evidence of transformation to martensite is seen in the higher hardness of this phase relative to the tempered grain interior.

The only evidence of alloy segregation was found in the grain boundary phase. During thermal cycling, diffusion of Ni and Cr to the grain boundary resulted in a microstructure similar in appearance and hardness to that produced by tempering specimen Q. As discussed previously, specimen Q fractured by MVC. It seems reasonable, therefore, to attribute the small areas of MVC, scattered randomly across the fracture surface of specimen I, to fracture of the grain boundary phase. This phase undoubtedly contributed to the relatively high resistance of specimen I to SCC.

CONCLUSIONS

The susceptibility of HY-180 weld metal to SCC is directly related to the microstructure.

Resistance of the weld metal to SCC increases as the microstructure becomes increasingly refined with tempering.

There is a direct correlation between the microstructure, the fracture mode and the texture of fracture in the SCC zone.

Similar microstructures, whether attained through variations in composition or by heat treatment, exhibit similar microfracture characteristics.

REFERENCES

1. C. T. Fujii, F. W. Fraser, and E. A. Metzbower, Stress-Corrosion Cracking Characterization of High-Strength Steels-Base Metals and Weldments. NRL Report 8230, May 1978.

2. B. F. Brown, A New Stress-Corrosion Cracking Test for High Strength Alloys. Materials Research and Standards, Vol. 6, No. 3, 129 (1966).

3. ASM STDS, Part 11. Metallography; Non-destructive Testing, p. 57 (1977).

DISCUSSION

L. W. Sandor (Sun Shipbuilding and Dry Dock Co.): Could you please give me your definition of "fine bead" technique and how you did it?

Author: With the fine bead welding technique, weldments are fabricated at a higher travel speed, reduced wire feed rate and lower voltage and amperage as compared to normal welding practices which produce a coarser weld bead.

In fabricating the coarse bead specimens, 23 weld beads were deposited whereas 70-100 weld beads were deposited in the fine bead specimen.

The welding parameters for the two techniques are given below.

	Fine Bead	Coarse Bead
Travel Speed	6"/min.	4"/min.
Wire Feed Rate	19"/min.	70"/min.
Voltage	13V	16V
Amperage	230 Amps	360 Amps
Wire Diam.	1/16"	1/16"

L. M. Petrick (Ebasco Services, Inc.): Was consideration given to a second $1000^{\circ}F$ temperature, to overcome or at least reduce, the amount of untempered martensite which may be present in the $1000^{\circ}F$ tempered specimens?

(There is a possibility that M_f is below ambient temperature for the HY180 alloy, particularly when subjected to the rapid weld cool and segregation).

Author: No further heat treatment was considered for either of the tempered specimens.

H. C. Rogers (Drexel University): You referred to material at the grain boundary as a "phase". Do you think it is a different phase or does it merely vary in composition and/or microstructure?

Author: I referred to the white etched material at the grain boundary as a phase in that it differs from the grain interior in composition, microstructure and hardness. I use as my definition of phase that given by L. H. Van Vlack in the text Elements of Materials Science and Engineering, "---- that part of a material which is distinct from others in structure and/or composition.

Microstructure and Stress Corrosion Cracking of Low-Carbon Alloy Steel Welds

Part I - The Effect of Welding Variables on Microstructure of HY-130 Steels

C. Chen, A. W. Thompson and I. M. Bernstein
Department of Metallurgy & Materials Science
Carnegie-Mellon University
Pittsburgh, PA 15213

Evaluation of the environmental performance of welded HY-130 steel can be performed in terms of stress corrosion cracking (SCC) parameters such as the threshold stress intensity, K_{ISCC}; but since such parameters by themselves provide no insight into reasons for good or poor performance, little guidance toward improvements is obtained. Understanding of SCC behavior must ultimately be stated in terms of microstructural nuclei and paths for fracture events, the interaction of which determines an overall parameter such as K_{ISCC}.

The microstructure, particularly in multi-layer welding, is quite complicated due to thermal effects of the welding cycle; in the case of HY-130, only limited information on microstructural details is available. As part of a larger study on stress corrosion of HY-130 welds, a thorough investigation of welded microstructure was made. Part I of this study describes the effect of welding variables such as welding process, heat inputs, bead size, and the base plate thickness on microstructure of HY-130 steels. Part II of this study relates the microstructure to the level of SCC resistance, through the relation between microstructure and the fracture separation process.

INTRODUCTION

HY-130 is a 5% Ni steel also containing Cr, Mo, and V, with low carbon (∿0.1%); it exhibits the medium strength (∿900 MPa), high fracture toughness, and good weldability needed for structural materials. As part of a larger U.S. Navy program on high-strength steel welds, it was found that weldments of HY-130 steel were more susceptible to stress corrosion cracking (SCC) in seawater than was HY-130 base metal[1]. The weld structure, solute segregation during solidification, and hydrogen content during welding could lower resistance to SCC in the weld.

It is believed that SCC in high strength steels is predominantly a hydrogen embrittlement process, dependent on metallurgical variables in particular environments. Bernstein and Thompson[2a, 2b] reviewed environmental fracture of steels and concluded that metallurgical variables such as composition, grain size and texture, microstructural features, and thermal treatment could be related to the degree of environmental embrittlement. Final microstructure of the steel might be the most important variable to affect the environmental-induced cracking. The nature and character of the final microstructure are also affected by other variables; for example, a given composition of steel can often be prepared with several different microstructures and corresponding strength levels, and these will differ in the extent to which they are hydrogen embrittled. The weld microstructure is greatly affected by the cooling rates in welding, which are a complex function of heat input, the mass of base metal, preheat and interpass temperature, and type of electrode, etc. The focus of this part, Part I, is on microstructure because it is the material characteristic which is amenable to modification and control through welding practice.

EXPERIMENTAL PROCEDURE

Six welded specimens (Weldment Code C, D, F, EZI-24, EZE-9, V8-1) were used to study the microstructure as a function of welding variables. The material for this investigation was supplied by the Naval Ship Research and Development Center (NSRDC). The weldments were produced in HY-130 base plate. The type of joint design was single V for specimen EZI-24 and EZE-9 and double V for the others. The chemical compositions of the filler metals, McKay 14018 SMAW and Linde 140S GMAW electrodes, are given in Table II.

The microstructure of the welds was studied primarily by transmission electron microscopy (TEM). Specimens were cut from the fusion zone for examination in TEM. Thin foils were prepared by final jet-polishing in a 66% methanol-34% nitric acid electrolyte after preliminary grinding and chemical thinning. The electrolyte was maintained at a temperature of -35°C, and a potential in the range of 12 to 15 V was used for thinning.

RESULTS

C, D, and F Welds

Figure 1a - 1d show the typical microstructural appearance of the weld metals, C, D and F in the light microscope. The microstructure of specimen D (Fig. 1a) is evidently a mixture of martensite and bainite. The structure of specimen C appears to contain mainly bainite with small amounts of martensite, as shown in Fig. 1b. While the predominant microstructure of specimen F appears to be ferrite and bainite (Fig. 1c), Figure 1d shows the martensite which is produced in the uppermost layer (surface) weld beads of the specimen. However, the structures of these welds contain considerably more complexity than can be recognized from optical micrographs. According, detailed structural analyses were carried out by TEM using the thin-foil technique to identify microstructural features in detail.

Figure 2a shows the bainitic structure (Bainite I) [3, 4] in specimen D, in which the structure consists of bainite (ferrite laths) with the untransformed austenite (marked A's in the figure) between them. This appeared to have developed by the growth of parallel ferrite laths of the same orientation. The untransformed austenite could be due to segregation of carbon and other solutes during solidification. Austenite has been identified by dark-field using an austenite reflection, (002). The corresponding diffraction pattern of Fig. 2a, and its schematic representation, are shown in Figures 2b and 2c, respectively. It is obvious that the (002) austenite reflections have a relatively large separation in reciprocal space from the b.c.c. reflections and are therefore the preferred reflections to identify the presence of austenite. Figure 2d also shows bainitic structure with austenite in the dark area. Dislocated, twinned and auto-tempered (in small amount) martensite were observed in specimen D as shown in Fig. 2e, f, and g, respectively.

Figure 3a is the bainitic structure (Bainite I) in specimen C. Some polygonal ferrite (Fig. 3b), twinned and auto-tempered martensite (Fig. 3c) have also been observed in small amount in this specimen. The relative amount of twinned martensite found in specimen C is less than that in specimen D. Normally, this type of martensite would not be expected to form in a steel of this composition, and presumably it occurred as a result of solidification segregation.

Figure 4a is the typical polygonal ferrite structure in specimen F. The dislocation substructure in this region was fairly high and occasionally a small amount of austenite, which is marked A in the figure, was located at the nodes of grain boundaries. The selected area diffraction pattern also indicates there are (002) austenite reflections, which can be distinguished from the (011) ferrite reflections as shown in Fig. 4b. Figure 4c is the microstructure of bainite with some carbides precipitated at the boundaries.

EZI-24, EZE-9 and V8-1 Welds

The microstructure of specimen EZI-24 appears to contain mainly bainite of different forms and size, and martensite. Figure 5a is a bright-field (BF) image which shows the structure often called "granular bainite."[5] The structure is oriented; this particular area, which is not typical of the material, contains elongated regions of stabilized austenite in a ferrite matrix. The dark-field (DF) image using the $(02\bar{2})$ reflection indicates there is untransformed austenite present as shown in Fig. 5b. Figure 5c is a much more typical bainitic structure. The morphology of carbide precipitation is similar to that of lower bainite in higher carbon steels. Figure 5d shows bainite in which carbides have precipitated in both lath boundaries and lath interiors. Figure 5e illustrates the bainitic structure with wide laths ($\sim$ 1.5 μm). The martensite in this specimen is predominantly of the twinned martensite type, with plates of several orientations.

Figure 6a is a bright-field photograph which shows the bainitic structure with retained austenite (marked A's in the figure) in specimen EZE-9. Figure 6b shows the typical "Bainite I" microstructure in this specimen. Fig. 6c illustrates another variant of bainite; the structure comprises untransformed austenite at lath boundaries and some small islands which could be austenite or martensite inside the lath. Figure 6d shows the martensitic structure in specimen EZE-9; a few internal twins are also observed.

There is a significant difference in microstructural features between specimen V8-1 and the other specimens in this study. The main structure in specimen V8-1 is auto-tempered martensite as shown in Fig. 7a. Carbide-free bainite, which is often called acicular ferrite, was observed, in which bainites, developed by the repeated nucleation of the substructural unit as suggested by Oblak and Hehemann,[6] can be seen in the lower portion of Fig. 7b. Twinned martensite and lower bainite were also found in small amounts in specimen V8-1.

DISCUSSION

Since the microstructure in a weld is greatly affected by the cooling rate in welding, one would expect that the particular thermal history, including preheat/interpass temperature, the welding method utilized, the thickness of base metal, and type of electrode, would normally control the cooling rate of the molten pool and thereby the microstructure of the weld. These welding variables were separated into four groups for meaningful comparison, as follows: (i) heat input and preheat/interpass temperature; (ii) welding process and electrode; (iii) thickness of base plate; and (iv) bead size.

Heat Input and Preheat/Interpass Temperature

High heat inputs are generally needed to deposit larger beads and thus increase the rate of welding. The higher heat input (32 kJ/cm) and higher preheat/interpass temperature (191-204°C) during welding of specimen C would have resulted in longer cooling times, which favor high temperature transformation products. In specimen C, these products are predominantly bainite, with martensite and ferrite in small amounts. The lower heat input (18 kJ/cm) and lower preheat/interpass temperature (135-149°C) of specimen D resulted in a mixed structure of martensite and bainite. Not only is this structure finer in scale than that of specimen C, but it also contains considerably more of the stronger constituent, martensite.

Welding Process and Electrode

Linde 140S gas metal-arc and McKay 14018 shielded metal-arc welding electrodes were used to make weldments of specimens EZ1-24 and EZE-9, respectively. It has been reported that the Linde 140S GMAW metal exhibited higher hardness values than the McKay 14018 SMAW metal under identical simulated weld cooling condition.[7] The Ms temperatures are 415°to 450°C for the McKay 14018 SMAW metal and 410° to 430°C for the Linde 140S GMAW metal.[7] Since these martensitic transformations occur at relatively high temperature, it would be expected that only the high temperature products (e.g. Bainite I and granular bainite) in the bainitic transformation region, as well as martensite, can form upon cooling. These types of microstructure were all observed in the weld metals (EZI-24 and EZE-9).

But in the GMAW metal, EZI-24, lower bainite, which is a low temperature product in the bainitic transformation region, was also present in large amounts, as shown in Fig. 5c. Decreasing the transformation temperature appeared to narrow the bainitic laths. As a result, carbon diffusion distances are shorter, and the precipitation of carbide could occur at lath boundaries (Fig. 5d). Figure 5c and 5d also show that the transformation to lower bainite was not accompanied by untransformed austenite. This lower bainite structure was not found in the SMAW EZE-9 specimen. Another significant microstructural feature of GMAW metal (EZI-24) is that it contains more twinned martensite (Fig. 5f) than SMAW metal (EZE-9).

The controlling factors which cause the microstructure of GMAW metal to differ from SMAW metal are not known. One possibility is that the wider range of preheat/interpass temperature, 65-150°C (the recommended preheat and interpass temperatures for both processes are 105 to 150°C), and the lower Ms of GMAW metal (EZI-24, might contribute to the various forms of microstructure in the weldment. It is clear that the welding process and type of electrode utilized have great influence on the microstructure of the welds. It is also noted that although heat input for a given weld provides useful information in making comparsions of welds by different processes, there is also a difference in the normal heat intensity of the two processes.

Thickness of Base Plate

The thickness (or the thermal mass) of the base plate is also an important factor in determining the cooling rate of the weld. Specimens EZI-24 and V8-1 have similar welding variables except for plate thickness

(2.5 cm vs. 5.1 cm, in Table I). The microstructure changed from a bainite and martensite (not tempered) structure in specimen EZI-24, to a tempered-martensite structure in specimen V8-1. Evidently the increased cooling rate due to the larger heat sink in the thick specimen, V8-1, favors martensitic transformation rather than bainite, and also precipitation of carbides during cooling from M_s ($\sim 400^{\circ}C$) to room temperature.

Bead Size

The size of individual weld beads could also be an important aspect of thermal history, since it is related to other variables such as heat input, preheat/interpass temperature, and welding process, etc. For instance, the coarser beads of specimen C required higher heat input and preheat/interpass temperature than that of finer beads in specimen D in the GMAW materials. In GTA-welded specimen F, the preheat/interpass temperature is the same as specimen D and the heat input is lower than that in specimen D. As a result, the number of beads necessary to make the weld increased. The ultrafine beads of specimen F resulted in a structure which has been refined by the multiple thermal cycling. The structure of this weld is a mixture of fine-grained ferrite and bainite; these are still higher temperature transformation products than either specimen D or specimen C.

CONCLUSIONS

Detailed metallographic study of weld microstructures using light microscopy and, in more detail, transmission electron microscopy, has permitted determination of the constituents in a number of HY-130 welds. These can be systematically correlated with the thermal history imposed in the welding process by such variables as heat input, preheat/interpass temperature, welding method, plate thickness, and bead size. The resulting microstructural features should strongly influence the strength, ductility, toughness and other mechanical properties of the welds, as well as environmental fracture resistance. These properties form the subject matter of Part II of this study.

ACKNOWLEDGEMENTS

We appreciate helpful discussions with C. A. Zanis, C. T. Fujii, and C. M. Adams, and experimental assistance from E. Danielson. This work was supported by the Naval Sea Systems Command, SEA 035, through the Office of Naval Research.

REFERENCES

1. C. A. Zanis, SAMPE Quarterly, January 1978, p. 8.

2a. I. M. Bernstein and A. W. Thompson, International Metals Reviews, Vol. 21, 1976, p. 269.

2b. A. W. Thompson and I. M. Bernstein, Advances in Corrosion Science and Technology, Vol. 7, Plenum, New York, in press.

3. Y. Ohmori, Trans. ISIJ, Vol. 13, 1973, p. 55.

4. H. Ohtani, F. Terasaki, and T. Kunitake, Trans. ISIJ, Vol. 12, 1972, p. 117.

5. L. Habraken and J. L. DeBrouwer, Fundamental of Metallography, Presses Academiques Européenes, Bruxelles, 1971, pp. 210 and 342-48.

6. J. M. Oblak and R. F. Hehemann, Symposium on Transformation and Hardenability in Steels, Inst. Metals, London, 1967, p. 15

7. R. D. Wyckoff, R. T. Brenna and A. Pollack, Continuous Cooling Transformations in HY-130 Steel Weld Metal, Report 4446, Naval Ship R & D Center, Annapolis, February 1975.

Table I. Welding Variables

Weldment Code*	Process	Heat Input (kJ/cm)	Preheat/ Interpass Temp (^{o}C)	Electrode Type & Dia. (mm)	Plate Thickness (cm)	Remark
C	GMA	32	191-204	Linde 140S 1.59	3.8	Coarse bead (∿15 beads)
D	GMA	18	135-149	Linde 140S 1.59	3.8	Fine bead (∿35 beads)
F	GTA	11	135-149	Linde 140S 1.59	3.8	Ultrafine bead (>100 beads)
EZI-24	GMA	16	65-135	Linde 140S 1.59	2.5	-
EZE-9	SMA	14-16	135-150	McKay 14018 3.97	2.5	-
V8-1	GMA	14	107-135	Linde 140S 1.59	5.1	-

*Post-weld heat treatment was not performed to all weldments.

Table II. Typical Chemical Composition (Wt%) of Electrodes

SMAW McKay 14018	GMAW Linde 140S	
C - 0.044	C - 0.098	
Ni - 3.43	Ni - 2.57	Zr - 0.008
Mn - 0.89	Mn - 1.61	V - 0.008
Mo - 0.73	Mo - 0.92	P - 0.005
Cr - 0.45	Cr - 0.75	S - 0.004
Si - 0.27	Si - 0.35	H_2 - 0.6 ppm
V - 0.01	Cu - 0.056	
S - 0.006	Ti - 0.023	
P - 0.005	Al - 0.01	

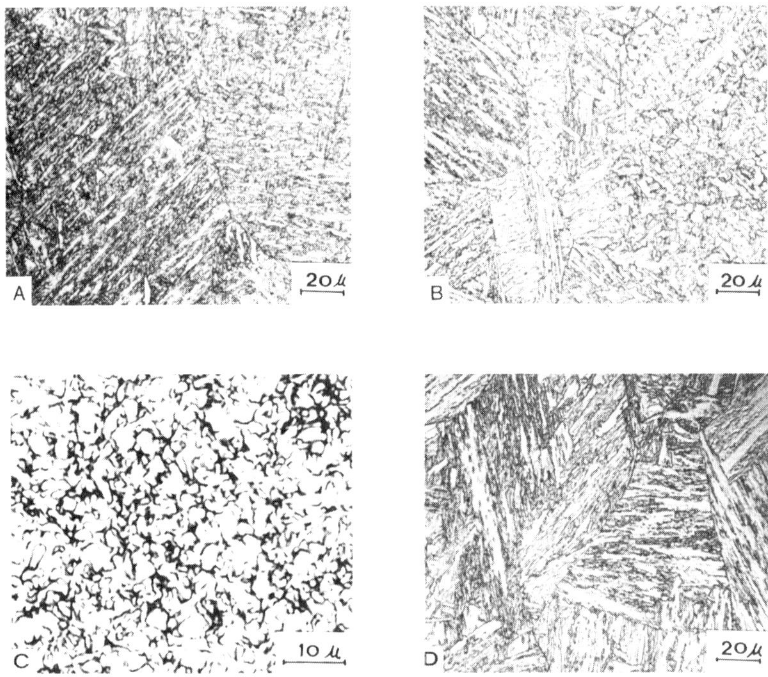

Figure 1. Optical microstructural appearance of the weld metals: (a) specimen D. (b) specimen C. (c) specimen F, weld interior. (d) specimen F, weld surface.

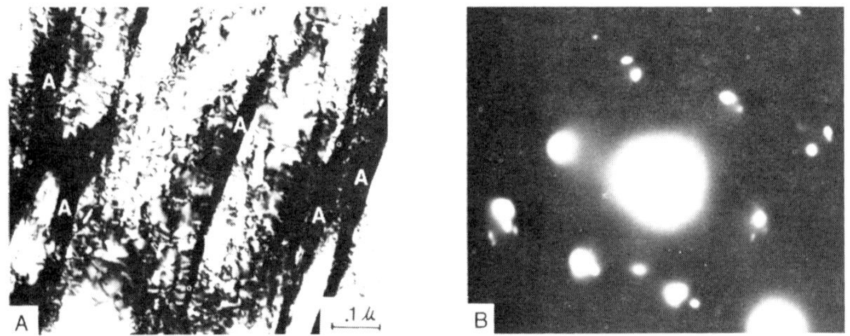

Figure 2. Figure captions on next page.

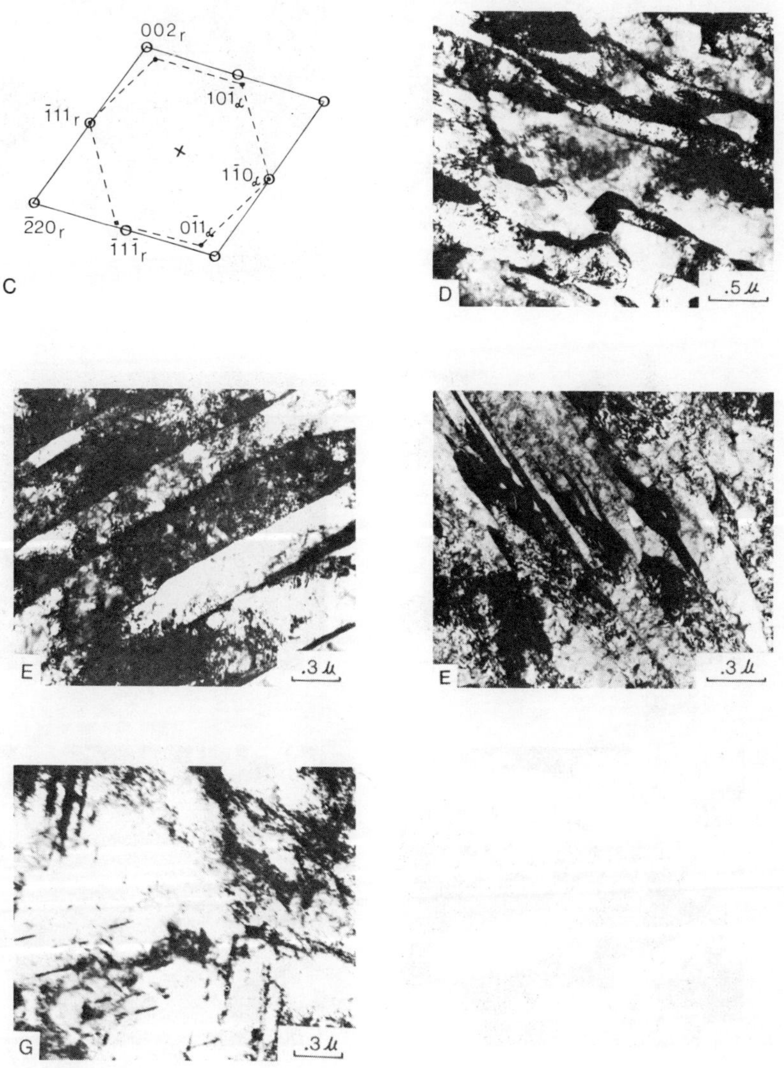

Figure 2. Transmission electron micrographs of specimen D. (a) Bainite I. (b) and (c) The corresponding diffraction pattern and its schematic representation of Fig. 2a, respectively. (d) Bainitic structure with austenite in dark areas. (e) Dislocated martensite. (f) Twinned martensite. (g) Auto-tempered martensite.

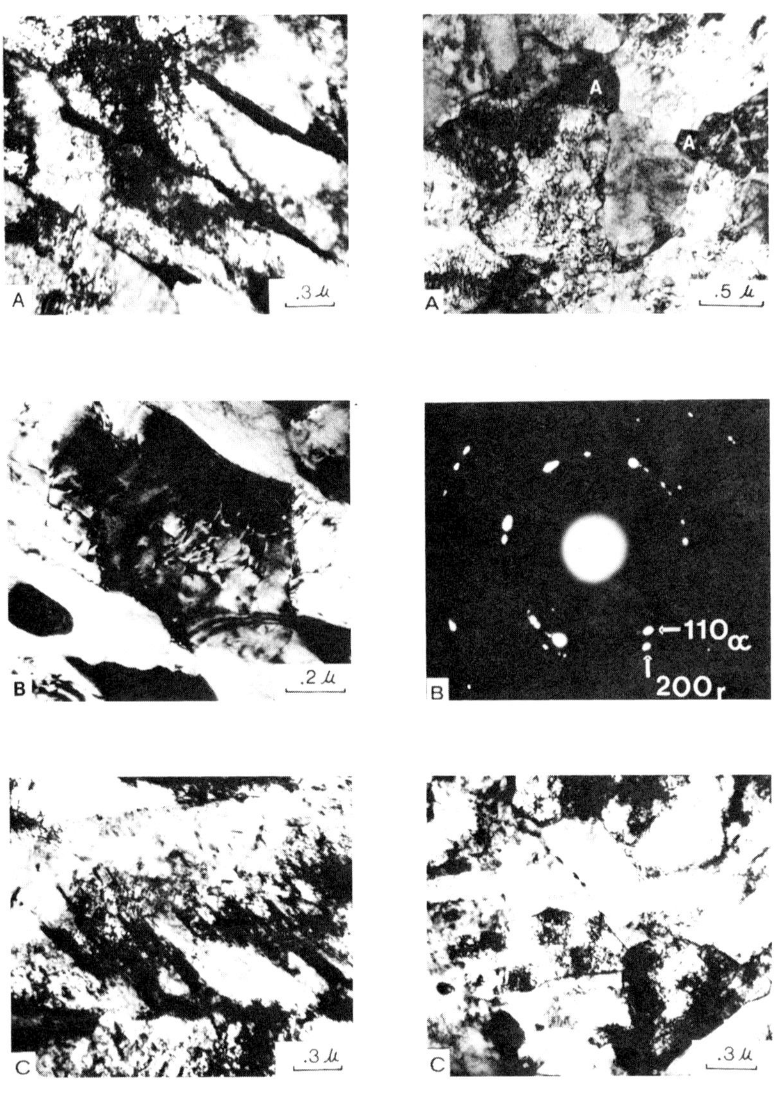

Figure 3. Transmission electron micrographs of specimen C. (a) Bainite I. (b) Polygonal ferrite. (c) Twinned and auto-tempered martensite.

Figure 4. Transmission electron micrographs of specimen D. (a) Typical polygonal ferrite. (b) The selected area diffraction pattern of (a). (c) Bainite with some carbides precipitated at boundaries.

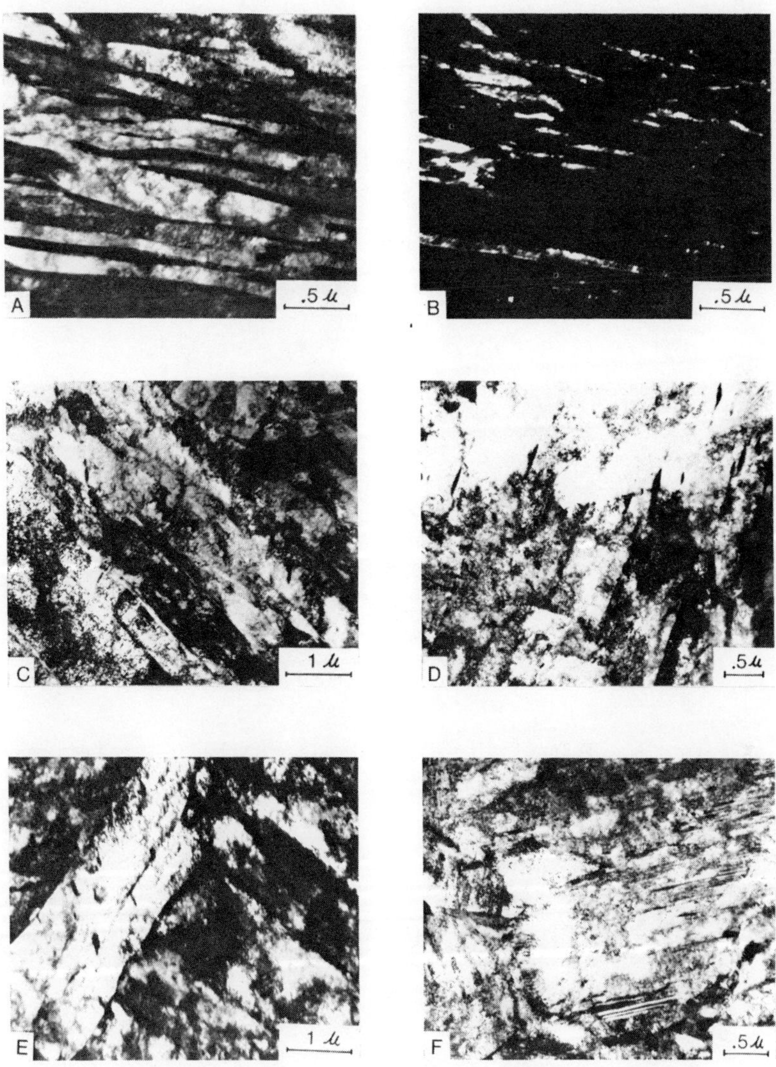

Figure 5. Transmission electron micrographs of specimen EZI-24. (a) and (b) Bright- and dark-field image of "granular bainite," respectively. (c) Lower bainite. (d) Bainite with carbides precipitated in both lath boundaries and lath interiors. (e) Bainite with wide laths. (f) Twinned martensite.

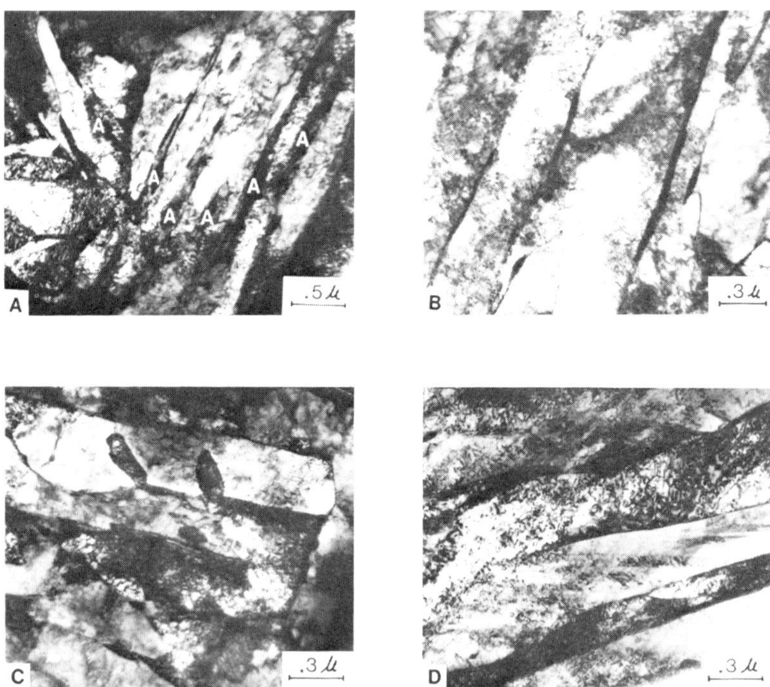

Figure 6. Transmission electron micrographs of specimen EZE-9. (a) Bainite with retained austenite. (b) Bainite I. (c) Another variant of bainite. (d) Martensite.

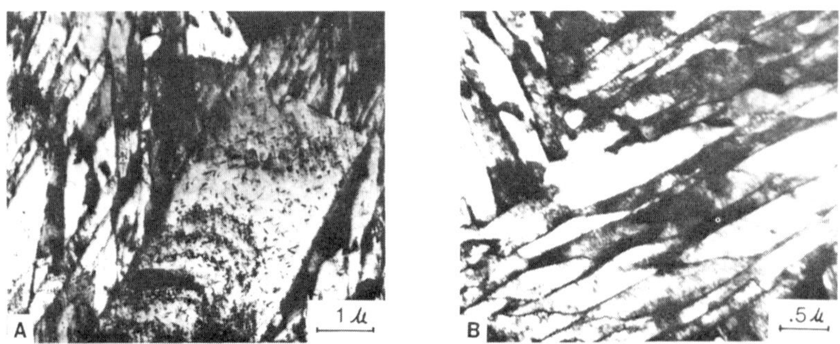

Figure 7. Transmission electron micrographs of specimen V8-1. (a) Auto-tempered martensite. (b) Carbide-free bainite or accicular ferrite.

Microstructure and Stress Corrosion Cracking of Low-Carbon Alloy Steel Welds

Part II - Fractographic Studies of Stress Corrosion Cracking of HY-130 Steel Weldments

C. Chen, A. W. Thompson and I. M. Bernstein
Department of Metallurgy & Materials Science
Carnegie-Mellon University
Pittsburgh, PA 15213

ABSTRACT

The intent of this study was to characterize stress corrosion failures of HY-130 weld metals, produced in both wedge opening load (WOL) and cantilever bend (CB) specimens, and to relate the level of cracking resistance through the fracture separation process to the microstructure of the weld. Since the environmentally-induced fracture of high strength steels is believed to occur predominantly by a hydrogen-related mechanism, hydrogen permeation experiments were also performed on material from CB specimens. The mechanical properties, K_{ISCC} values, results of permeation experiments, and the fracture-microstructure relationship are discussed.

INTRODUCTION

The fracture surface appearance in hydrogen-bearing environments observed at moderately high magnification often is altered by hydrogen and is usually complex.[1] Fracture details effectively represent a spectrum of relative local ductility, ranging from very ductile modes like microvoid coalescence, through poorly-defined transgranular modes like quasi-cleavage, to relatively brittle modes, such as intergranular fracture and cleavage. Hydrogen can induce any of these from one of the others, or it may leave the mode unchanged.[2] Bernstein and Thompson[3] observed that the environmentally-induced cracking of steels results either from the direct action of hydrogen or is mainly controlled by the response of the material to hydrogen. They suggested that stress corrosion cracking (SCC) and hydrogen embrittlement in iron and steel are overlapping processes containing common elements of behavior. Beachem[4] showed that fracture modes operate according to the microstructure, stress intensity at crack tip, and hydrogen concentration under SCC and HAC (hydrogen-assisted cracking) conditions. Cain and Troiano,[5] working on AISI 4620 steel, indicated that the microstructure might have a greater influence on hydrogen susceptibility than strength level. Gooch studied[6] the SCC susceptibility of high strength welds and found that for optimum SCC resistance, the combinations of weld metal composition and heat treatment which produce twinned martensite should be avoided. It is generally agreed that untempered martensite is the most deleterious microstructure, while the preferred substructure combines a well-tempered martensite or bainite with an ausworking process to produce a refined packet size and a uniform dispersion of carbides.[7] The foregoing suggests that the various microstructures formed as a result of differing welding conditions could be responsible for changes in SCC performance of the welds, which could also be related to the fracture modes.

Hydrogen is introduced into material during exposure to aqueous environments by a series of electrochemical reactions. It is possible that hydrogen permeation experiments could provide some useful information concerning the relative resistance to SCC of the material, although hydrogen which diffuses into a straining crack tip region may not move in the same way or at the same rate as observed in a permeation study.

EXPERIMENTAL PROCEDURE

The welded materials which were used to study the microstructure in the companion paper, Part I, (except V8-1) were also used in this investigation. SCC tests were performed by using either 1T-WOL specimens in natural seawater or cantilever bend (CB) specimens in 3.5% NaCl solution. The mechanical properties, K_{ISCC} values, SCC test environment, and specimen type for these welds were determined by NSRDC and are listed in Table I. The length of exposure of 1T-WOL specimens was more than 2,000 hours, while in CB specimens, which were performed under laboratory screening conditions, the time was no more than 500 hours. Susceptible material (EZI-24) is denoted by a "low K" designation and the more resistant material (EZE-9) is given the "high K" designation in 1T-WOL specimens.

Fracture surfaces were carefully cleaned either with replicating tape or with an inhibited acid etch. These surfaces were then examined by scanning electron microscopy (SEM). In addition a dual technique was used to simultaneously reveal the fracture surface and the microstructure in 1T-WOL specimens.

Specimens (2.54 x 2.54 x 0.04 cm) used for hydrogen permeation experiments were sectioned in a direction parallel to the fracture surface and within the fusion zone. They were polished with 600 grit silicon carbide paper, spot welded at one corner to an iron wire, ultrasonically cleaned in acetone, and then plated with a thin layer of Pd. A cell developed by Devanathan and Stachursky[8] was used to determine the permeability of hydrogen through these membranes. 0.1 N NaOH with addition of 33 μg/mℓ $NaAsO_2$ solution was used and a charging current of 2mA/cm^2 was applied on the cathode side of the membrane. The cathode area was 0.97 cm^2 and the temperature was maintained at room temperature. The anode side was filled with 0. 1 N NaOH and a potential of +200 mV (vs. SCE) was applied by a potentiostat, and a chart recorder was used to record the anodic current.

RESULTS

A. 1T-WOL SPECIMENS

The fracture separation processes were examined with SEM to determine the fracture surface topography of specimens. Fracture surface "mapping" of these specimens was also done using the same technique. Macrographs and fractographs of the fracture surface of these weld metals are shown in Figures 1-3. In the low-K, EZI-24 sample (Fig. 1), fracture was mainly intergranular or interphase in nature with evidence of secondary cracking, microvoid coalescence, and some quasi-cleavage in the region of SCC (Fig. 2a-d). Figure 2a shows the SCC region near the fatigue pre-crack. Figure 2b is a high-magnification view of the area outlined by the rectangle in Fig. 2a, revealing a fine structure of dimples, some secondary cracks, and corrosion pits. In the high-K sample, EZE-9, (Fig. 3), failures were primarily by microvoid coalescence in the regions of SCC.

Large voids and cracks (Fig. 3a) were found in specimen EZE-9. Deformation markings (ripples) on dimple walls were observed as shown in Fig. 3b, which is also a high-magnification view of the area outlined by the rectangle in Fig. 3a. However, in the region of fast fracture, microvoid coalescence is the dominant separation process of the welds (e.g. Fig. 3d). Fracture by intergranular separation has been observed either with columnar or coarse equiaxed grains. The fracture in these regions most likely occurred by cracking along columnar grain or prior austenite grain boundaries, as shown in Fig. 4a and 4b, respectively. There is also a strong tendency for secondary cracks to follow boundaries. For examples, in Fig. 4a and 4c secondary cracks are associated with a prior austenite grain boundary. In the case of fracture by microvoid coalescence, the fracture-microstructure relationship is much less obvious, as shown in Fig. 4d for specimen EZE-9.

B. CANTILEVER BEND SPECIMENS

The low magnification appearance of the fracture surfaces of specimens C, D, and F, produced in SCC tests, is shown in Fig. 5. SEM fractographs taken from the SCC region which is indicated A, B, C and D in Fig. 5 are shown in Figures 6a - d. The same arrangement for specimens D and F are given in Fig. 7a - d and Fig. 8a - d, respectively. The mechanism of fracture of specimen C is a mixture of microvoid coalescence and quasi-cleavage, with some secondary cracks as shown in Fig. 6. The fractographic observation (Fig. 7) shows that the fracture separation process of specimen D involves intergranular cracks, quasi-cleavage, microvoid coalescence, and secondary cracks. Figure 7d shows a secondary crack at a grain boundary triple point (left in the figure) on the intergranular fracture surface. The amount of brittle fracture (intergranular cracks and quasi-cleavage) contributed more than 50% of the total fracture surface.

The fracture surface of specimen F exhibits fine dimples (Fig. 8a and 8c) and some local areas of quasi-cleavage (Fig. 8b). Figure 4d is the typical fracture appearance observed in the outer (near-surface) layers of specimen F. It is characterized by quasi-cleavage fracture with some ductile tearing, which may occur along martensite plate boundaries. It is possible that the quasi-cleavage facet in Fig. 8d is related to the sub-packet size of martensite as shown in Fig. 1d in Part I of the study. The microstructure observed in the localized region of quasi-cleavage fracture (Fig. 8b) in specimen F is martensite. The reason for this martensite structure is not known, but it could have arisen in the root pass.

The results of hydrogen permeation studies of the welds C, D, and F are shown in Fig. 9, which is a plot of log (time in min.) versus permeation current (anodic current), a direct measure of the flux of hydrogen appearing on the output side of the membrane. J_∞ is the permeation current at the steady state. It is noted that the J_∞ for specimen F is lower than either specimen C or specimen D cathodically charged in the 0. 1 N NaOH solution.

DISCUSSION

Examination of the results in Part I has shown that the "high K" sample, EZE-9, is generally associated with a relatively fine bainitic structure (not including lower bainite), with islands of retained austenite and some martensite. The "low K" sample, EZI-24, is associated with a mixed bainitic structure in which broad laths of lower bainite, Figs. 5c and 5d in Part I, predominate (with carbides precipitated at both lath boundary and lath interior); twinned martensite is also present, with retained austenite being rare. The details of the fracture morphology (Fig. 2 versus Fig. 4) suggest that the smaller bainite plate dimensions, with retained austenite and no coarse carbides present in the SMAW sample, EZE-9, lead to more consistently ductile fracture, while the coarser carbides, occasional strip-like austenite, and apparently segregation-induced twinned martensite (probably stronger and less ductile, see Table I) in the GMAW weld, EZI-24, led to more brittle-appearing fracture. Note the secondary cracking in Fig. 2b. Another important microstructural feature in welded samples is the cellular substructure formed during the solidification process and delineating regions of compositional differences due to segregation of solutes. This substructure was revealed as illustrated in Fig. 10: a dual micrograph of the fracture surface and the solidification structure. There appears to be no general tendency for the crack to interact with the solidification substructure.

The amount of brittle fracture observed in the fracture surfaces in the SCC areas of specimen C is less than that of specimen D. The presence of large amounts of hard martensite could contribute to the more brittle fracture behavior of specimen D. Although specimen C has a lower yield strength (σ_Y), it has a higher K_{ISCC}/σ_Y ratio, $0.66\sqrt{in.}$, than that of specimen D (0.57). A higher K_{ISCC}/σ_Y ratio suggests an ability to sustain the presence of larger flaws at a given design stress (usually a specified fraction of σ_Y) before succumbing to SCC.

Generally, the fracture appearance of specimen F (except near the surfaces) is more ductile than either specimen C or specimen D. It is intriguing that despite the absence of the strongest constituent, martensite (except at the surface), the yield strength was the highest of the three specimens. This could be due to the fine-grained and high dislocation density structure, which is probably a result of the repeated thermal stresses of the larger number of weld passes. The K_{ISCC}/σ_Y ratio for specimen F is 0.68, which is similar to the ratio of specimen C, despite the higher yield strength of F. It has been found that if the surface layers of martensite were removed, the value of K_{ISCC} increased.[9] Bernstein and Thompson[3] suggested that a refined substructure gave the best resistance to environmental embrittlement, which could be the case for specimen F (fine-grained ferrite and bainite).

In the study of GMA and SMA weld metals, the most important factors affecting SCC susceptibility appear to be retained austenite and the amount of untempered and/or twinned martensite present. Untempered and twinned martensites are known to lower the resistance to SCC.[3] The role of retained austenite continues to be controversial; Ritchie et al[10] have shown it to be beneficial in reducing SCC crack velocity (but not K_{ISCC}), while Webster[11] claimed it also improves toughness. Cracks propagating through martensite are thought to be stopped upon intersecting the hydrogen-resistant retained austenite. Whether the same process could occur in bainite is not clear, but Naylor and Krahe[12] have claimed that bainite size and morphology affect toughness, so that these factors alone in the SMAW welds (EZE-9) should prove beneficial. It seems reasonable to expect the retained austenite to contribute to this effect, but the large differences in K_{ISCC}, Table I, evidently cannot be ascribed to the retained austenite.[10]

By definition, the permeability is proportional to the product of hydrogen solubility and diffusivity.[13] Wang[14] tested AISI 4340 and 18% Ni (250) maraging steel, and concluded that the excellent SCC resistance of maraging steel could be attributed to the slow diffusion of hydrogen in it. On the other hand, the lowest SCC resistance in untempered martensite exhibited the highest hydrogen absorbed at steady state, compared to the other microstructures in AISI 4340 steel. Relative hydrogen permeability has also been used to rationalize SCC crack velocities in the presence of retained austenite.[10] Accordingly, the results in Fig. 9 help understand the relative performances of the C, D and F welds.

The absorption of hydrogen in twinned martensite (Fig. 4e) could promote SCC susceptibility and might assist the brittle fracture observed in specimen D. On the other hand, the J_∞ for specimen F is lower than either specimen C or specimen D, as can be seen in Fig. 9. It is possible that slower diffusion and/or lower hydrogen solubility could be beneficial to SCC resistance. Further studies, however, would be necessary to characterize the trapping sites and processes in order to obtain a better understanding of the environmentally-induced cracking of materials like the HY-130 steel of this study.

SUMMARY AND CONCLUSIONS

Study of SCC and microstructure of HY-130 welds has shown that welding variables are important factors in determining the microstructure, mechanical properties, and SCC performance of the weld. In general, specimens which are less susceptible to SCC exhibit a primary fracture mode of microvoid coalescence; this is true for both WOL and CB specimens. In the more susceptible materials, the fracture mode is much more brittle in appearance (intergranular cracks and quasi-cleavage). The solidification substructure does not appear, with the information available, to exercise a major effect on the operating crack path. A refined structure of GTA weld metal is resistant to SCC; in the GMA or SMA weld metals, the most important factors affecting the SCC susceptibility appear to be retained austenite and the amount of untempered and/or twinned martensite. It is not certain that retained austenite in a bainitic structure can be beneficial to SCC properties of low-carbon alloy steel welds. But the present study indicates austenite might play a critical role. More work on a wide range of microstructure would be needed to prove this, together with studies on the effect of bainite and carbide size. Finally, hydrogen permeation behavior provides an interesting correlation with the relative resistance to SCC of the welds. The main trend is that the material with the lowest J_∞ (permeation current at steady state) had the highest K_{ISCC} value of the welds studied.

ACKNOWLEDGEMENTS

We gratefully acknowledge the support of the Naval Sea Systems Command, SEA 035, through the Office of Naval Research.

REFERENCES

1. A. W. Thompson, Environment-Sensitive Fracture of Engineering Materials, TMS-AIME, in press.

2. A. W. Thompson, Environmental Degradation of Engineering Materials, (M. R. Louthan and R. P. McNitt, eds.), pp. 3-17, V.P.I. Press, Blacksburg, VA, 1977.

3. I. M. Bernstein and A. W. Thompson, International Metals Reviews, Vol. 21, 1976, p. 269.

4. C. D. Beachem, Met. Trans., Vol. 3, 1972, p. 437.

5. W. M. Cain and A. R. Troiano, Petr. Eng., May 1965, p. 78.

6. T. G. Gooch, Welding J. (Research Supplement), Vol. 53, 1974, p. 287S.

7. I. M. Bernstein and A. W. Thompson, Strengthening Mechanisms and Alloy Design, Chapter IX, Academic Press, New York, pp. 303-347.

8. M. A. Devanathan and Z. Stachurski, Proc. Roy. Soc., Vol. 270, 1962, p. 90.

9. C. A. Zanis, NSRDC, private communication, 1978.

10. R. O. Ritchie, M. H. Castro Cedeño, V. F. Zackay, and E. R. Parker, Met. Trans., Vol. 9A, 1978, p. 35.

11. D. Webster, Met. Trans., Vol. 2, 1971, p. 2097.

12. J. P. Naylor and P. R. Krahe, Met. Trans., Vol. 5, 1974, p. 1699.

13. For example, C. Sykes, H. Burton, and C. C. Gregg, J. Iron Steel Inst., Vol. 156, 1947, p. 155.

14. M. T. Wang, "Hydrogen Permeation Behavior of High-Strength Steels," Technical Report AFML-72-102, Part I (Corrosion Cracking of Metallic Materials), August, 1972, p. 289.

Table I. Mechanical Properties and SCC Data of the Welds.

Material	Yield Strength, MPa	Elong., %	K_{ISCC}, MPa$\sqrt{m}$	SCC Test Environment	Specimen Type*
EZI-24	920	13	34 (low K)	Seawater + applied -1500 mV**	IT-WOL
EZE-9	940	16	114 (high K)	Seawater + Zn Coupled	IT-WOL
C	869	22	91	3.5% NaCl + Zn Coupled	3.8 x 5.1 x 30 cm Cantilever beam
D	993	19	90	3.5% NaCl + Zn Coupled	3.8 x 5.1 x 30 cm Cantilever beam
F	1041 (Upper Y. P.)	25	113	3.5% NaCl + Zn Coupled	3.8 x 5.1 x 30 cm Cantilever beam

*All specimens were fatigue pre-cracked.

**Fracture behavior similar to Zn-coupled specimens.

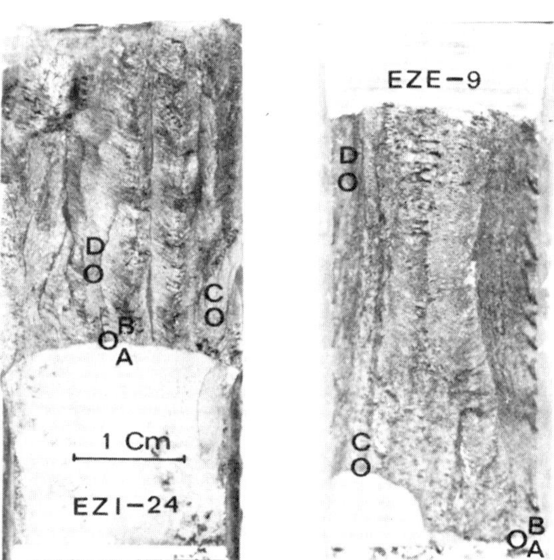

Figure 1. Fracture surface of WOL specimens tested in seawater.

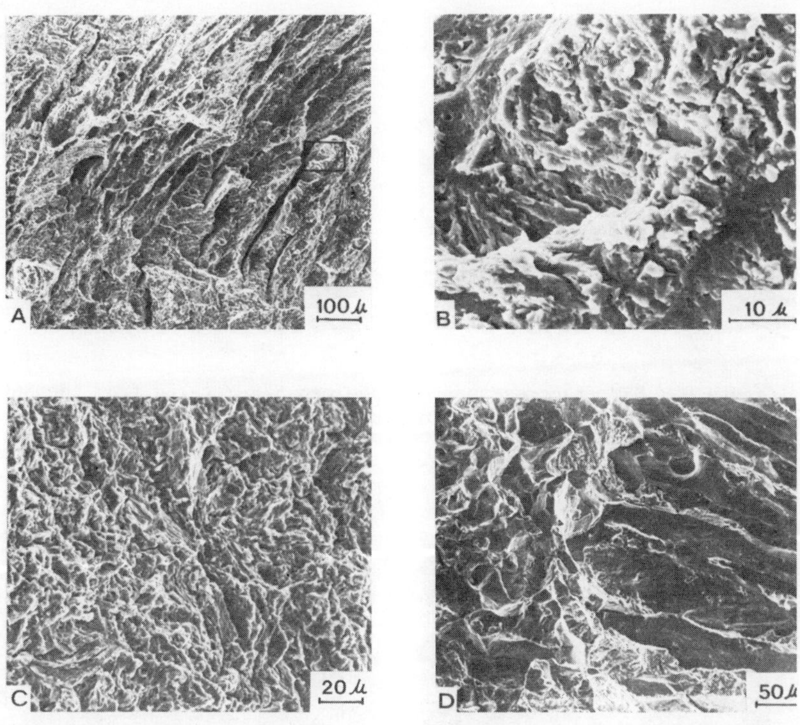

Figure 2. SEM fractographs of specimen EZI-24 taken from the SCC regions as indicated in Fig. 1.

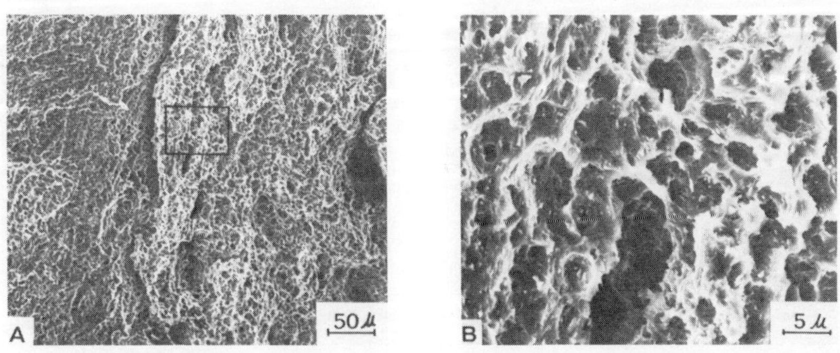

Figure 3. Figure captions on next page.

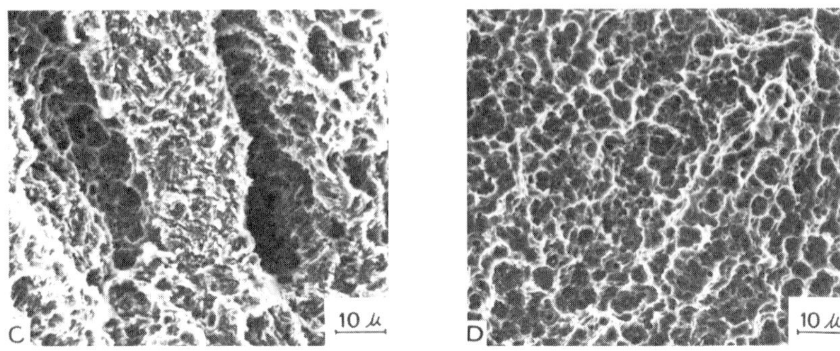

Figure 3. SEM fractrographs of specimen EZE-9 taken from the SCC regions as indicated in Fig. 1.

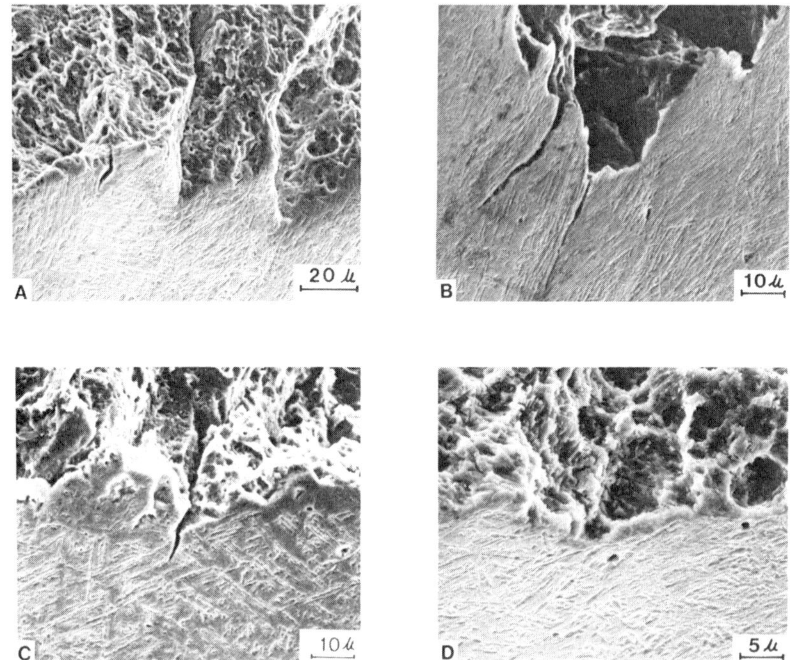

Figure 4. Dual photographs (fractrograph-microstructure) of WOL specimens. (a), (b), and (c) specimen EZ1-24. (d) specimen EZE-9.

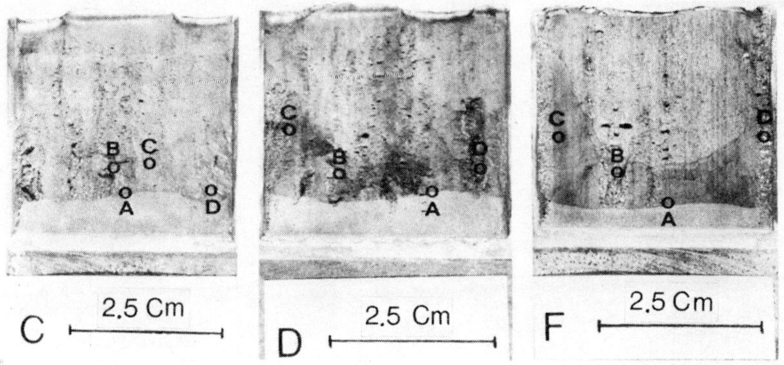

Figure 5. Fracture surface of CB specimens tested in 3.5% NaCl.

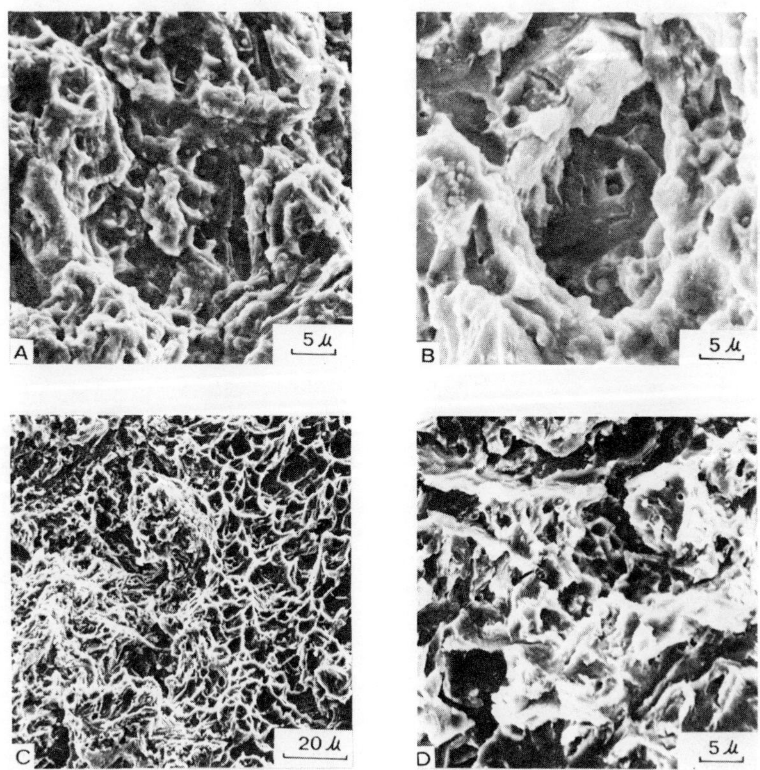

Figure 6. SEM fractographs of specimen C taken from the SCC regions as indicated in Fig. 5.

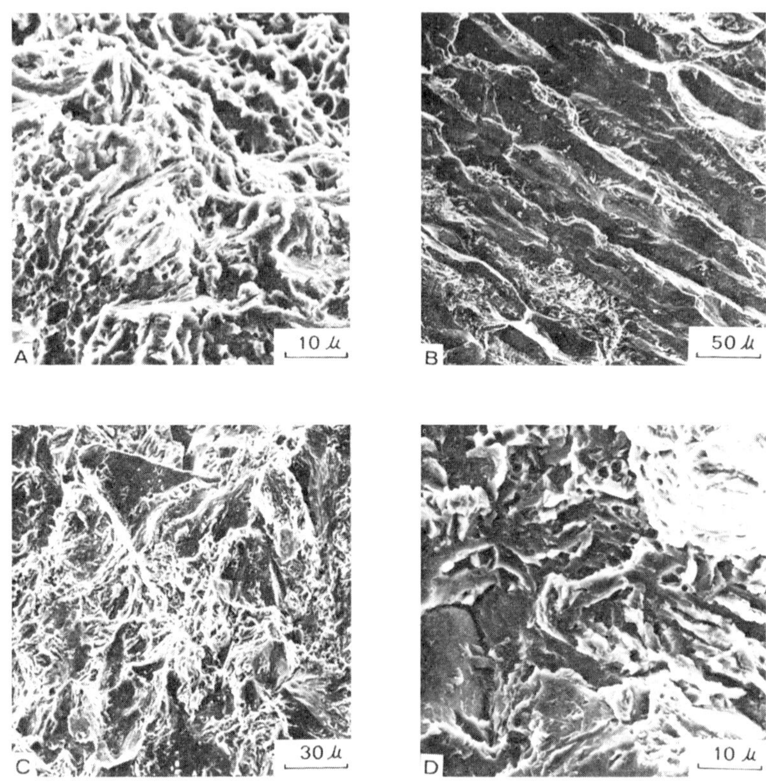

Figure 7. SEM fractographs of specimen D taken from the SCC regions as indicated in Fig. 5.

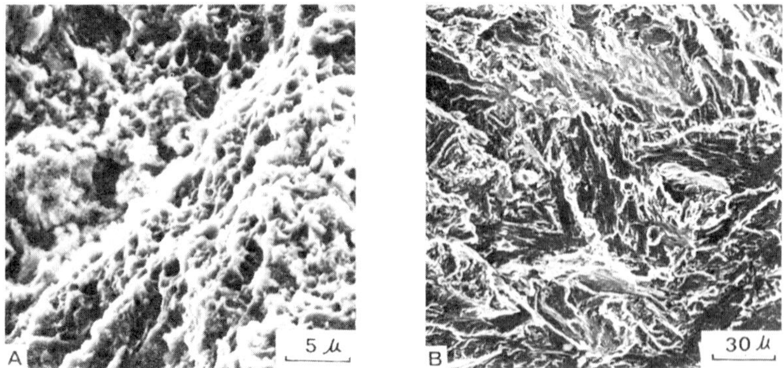

Figure 8. Figure Captions on next page.

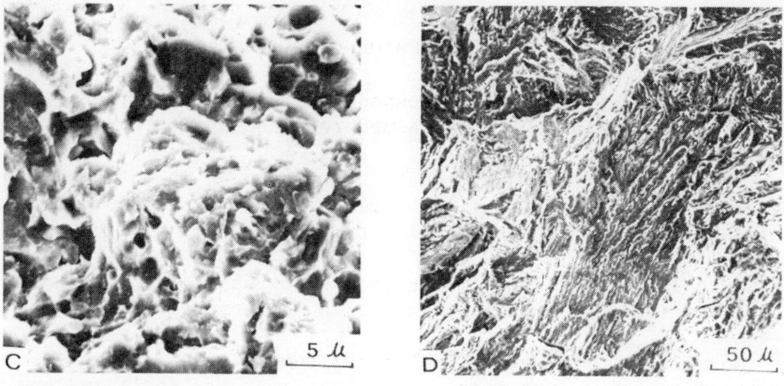

Figure 8. SEM fractographs of specimen F taken from the SCC regions as indicated in Fig. 5..

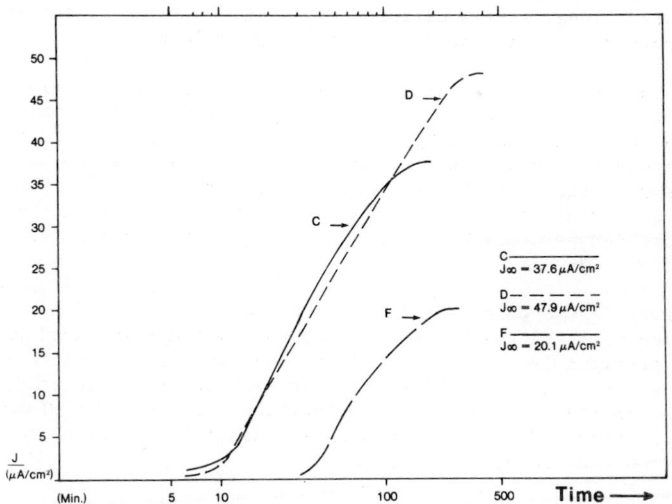

Figure 9. Hydrogen permeation-time curves of specimens, C, D and F.

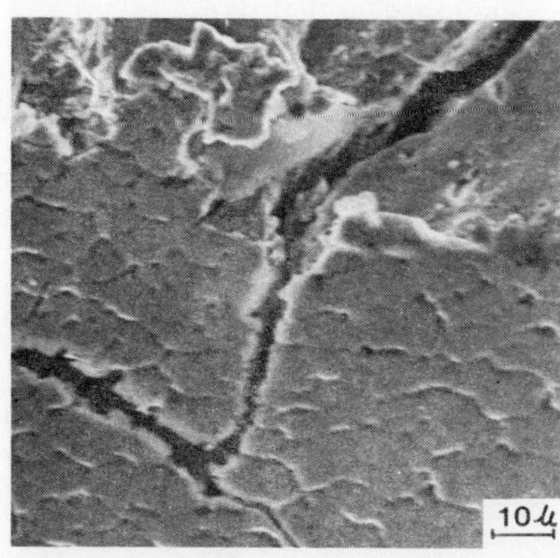

Figure 10. Solidification microstructure and fracture surface of specimen EZI-24.

DISCUSSION

P. V. Dembowski (General Electric Company): Is there a difference in crack growth rate in specimens which contained retained austenite?

Author: We did not do crack growth rate measurement in our specimens. But Ritchie et.al. (Ref. 10 in part AI) have shown that retained austenite is beneficial in reducing SCC crack velocity (not K_{ISCC}).

J. T. McGrath (Dept. of Energy Mines and Resources): Was the amount of retained austenite more prevalent in the low carbon weld metal?

Author: Not really. The overall chemical compositions are also important. The amount of retained austenite in the specimens was small and could be due to inhomogenity of filler metal or/and segregation during solidification.

H. C. Rogers (Drexel University): The H_2 diffusivity difference found to exist in different steels was relatively small. Did you consider possible effects of residual stress differences existing in the different weld on the observed diffusivity and/or permeability differences?

Author: The hydrogen diffusivities were calculated from permeation transients (either rise or decay transient) on specimens which contained residual stress in the welds. Since stress has been shown not to affect diffusivity, while increasing the hydrogen solubility in the lattice (W. Beck et.al., Proceedings of the Royal Society of London, Series A, Vol. 290, 1965, p. 191). The possible effects of residual stress might affect the permeability (e.g. increased permeation current at steady state), but presumably not the diffusivity.

D. W. Harvey (ARCOS Corp.): (1) One problem associated with metallographic examination of multi-run joints is that the examiner is faced with a composition gradient from the root area to the cap area and microstructural variations due to the reheating cycle of each applied weld bead. It is therefore important to identify the location that electron microscopy specimens were extracted from. Unless this precaution is taken comparative metallography has no absolute value.

Would the authors please indicate how this precaution was allowed for?

(2) The majority of the microstructures consisted of highly dislocated non-polygonal acicular ferrite as expected from very low-carbon low-alloy materials. They could not correctly be described as bainite.

Author: (1) For IT-WOL specimens, the joint design was single-V with root opening of 12.7 mm and incline angle of 60°. Therefore, one should have little or no base metal dilution in the regions close to fracture surface. For CB specimens, the joint design was modified double-V with root opening of 6.4 mm and incline angle of 75°. I expect the regions other than the middle-thickness region should also have little or no dilution. All the micrographs were taken from the areas which were very close to the fracture plane of the SCC specimens and there is believed to be no significant base metal dilution.

(2) This comment may relate to a matter of terminology. "Dislocated acicular ferrite" is an excellent way of describing the appearance of low-carbon bainite.

PROPERTIES AND STRUCTURES OF LASER WELDS OF HIGH STRENGTH ALLOYS

E. A. Metzbower and D. W. Moon

Material Science and Technology Division
Naval Research Laboratory
Washington, DC 20375

ABSTRACT

A series of laser welds of 12-mm (0.5-in) thick plates of high strength steel, aluminum and titanium were fabricated and tested. The welds had a simple butt weld geometry and the heat inputs varied from 0.8 kj/mm (21 kj/in) for the steel and titanium alloy to 0.3 kj/mm (7 kj/in) for the aluminum alloy. Mechanical properties (yield and ultimate strengths, elongation and reduction-in-area) were measured and calculated. The fracture resistances of the welds were measured by a subsize dynamic tear (DT) test. Fractographic observations were made on both tensile and DT test specimens. Metallographic examinations were made in the base plate, fusion and heat-affected zones. Correlations between structures and properties will be made. Comparisons between laser weld properties and properties of welds fabricated by other techniques will be made.

INTRODUCTION

A focused, high power, coherent, monochromatic light beam or laser beam can be utilized to vaporize the surface of a metal and produce a deep penetration column of vapor extending into the metal. This vapor column is surrounded by a liquid pool which, if it is moved along the path to be joined, produces welds with large depth-to-width ratios. The development of multi-kilowatt, continuous carbon dioxide gas laser systems has provided us with a deep penetration laser welding capability.

Laser beam welding has been utilized to join a variety of metals and alloys including low alloy and stainless steel; aluminum alloys;[1] titanium alloys; refractory and high temperature alloys. Porosity-free, ductile welds can be attained with tensile strengths near that of base plate. Hardness increases are obtained in alloys having microstructures related to the cooling rates. Laser beam welds have low heat inputs, consequently the cooling rates are high. The impact strength[1,2] of laser beam welds in several high strength low alloy steels has been attributed to a lowering of the impurity inclusion content and a change in the inclusion size distribution.

Deep penetration welds produced with high power CO_2 lasers are similar to electron beam welds, however, laser beam welding offers several advantages: a) a vacuum environment is not required for the work piece since the laser beam is easily transmitted through the air; b) x-rays are not generated in the laser beam/workpiece interaction; c) the laser beam may be readily focused, aligned and redirected by optical elements.

This study was undertaken to measure mechanical properties and fracture toughnesses of laser beam welds in a variety of high strength alloys and to correlate these properties with their microstructure, hardness, and fractography. The welds were fabricated by United Technologies Research Center and represent "a best effort" type of approach rather than an extensive research program aimed at developing optimum properties.

PROCEDURE

The materials that were laser beam welded are listed in Table 1 along with their nominal compositions. The plates that were welded were about 12-mm (0.5-in) thick and were 100-mm (4.0-in) x 200-mm (8.0-in). These plates were welded to form a 200-mm x 200-mm (8-in x 8-in) weldment. All welds were butt welds. Welding parameters are given in Table 2. Helium was used in all cases to provide an inert gas shield over the reactive, hot welds and also to assist in plasma control.

Table 1

Nominal Compositions of Base Plates

Material	Fe	C	Cr	Mo	V	Ni	Al	Ti	Nb	Mg	Mn
HY-80	bal	0.18	1.68	0.45	-	3.0	-	-	-	-	-
HY-130	bal	0.11	0.55	0.55	0.04	5.0	-	-	-	-	0.8
Aluminum	-	-	0.12	-	-	-	bal	-	-	5.10	0.8
Titanium	-	-	-	0.80	1.00	-	6.0	bal	2.0	-	-

Table 2

Laser Beam Welding Parameters

Material	Power (kw)	Speed mm/sec	(ipm)	Heat Input kj/mm	(kj/in)
HY-80	10.6	12.7	(30)	0.83	(21.2)
HY-130	11.0	12.7	(30)	0.87	(22.0)
Al alloy	8.0	27.5	(65)	0.29	(7.4)
Ti alloy	11.0	14.8	(35)	0.74	(19.0)

The weldments were given visual and radiographic examinations. Some of the weldments especially the aluminum alloy showed excessive porosity.

The weldments were then cut into blanks for mechanical property and fracture toughness specimens. Mechanical properties were measured on a standard ASTM round specimen 6.4-mm (0.252–in) in diameter. The weld was transverse to the loading direction in the mechanical property measurements. Fracture toughness was measured on a subsize, 12-mm (0.5-in) thick, dynamic tear (DT) test specimen that was notched in the weld. Fractographic observations were made in both the tensile and DT fracture surfaces.

Microstructures were observed and recorded in the fusion zone, heat-affected zone, and the base material for all specimens. Hardness traverses were made across these zones utilizing a diamond pyramid indenter under a 300 gram load. The Vickers hardnesses were then converted to the pertinent Rockwell scale except for the Al alloy.

RESULTS

Mechanical Properties

The mechanical properties were determined on an ASTM standard 6.4-mm (0.252-in) round tensile bar specimen. The yield strength, ultimate strength, per cent elongation and per cent reduction in area were determined for each specimen. Per cent elongation was measured on a 25-mm (1.0-in) gauge length. The average values of these properties are given in Table 3. Eight laser beam welds of HY-130 were tested and all fractured in the base metal. Sixteen specimens of HY-80 were tested, twelve of these fractured in the base plate, whereas four fractured in the weld. The four test specimens that fractured in the weld showed average yield and ultimate strengths but a significant reduction in ductility. Considerable evidence for porosity was seen in these fracture surfaces. Sixteen laser beam welds of the titanium alloy were tested, eleven failed in the base plate, whereas five failed in the weld. For these five there was a loss not only in ductility but also in yield strength. Again, considerable porosity was seen on these fracture surfaces. All of the aluminum alloys failed in the weld and all had considerable porosity in the fracture surface. The ductility of the aluminum laser beam welds was low.

Table 3

Average Tensile Properties

Material	0.2% YS (MPa)	(ksi)	UTS (MPa)	(ksi)	% Elongation	% Reduction-in–Area
HY-80	627	91	752	109	23.5	75.8
HY-130	924	134	958	139	19.4	77.0
Al alloy	193	28	289	42	9.2	16.9
Ti alloy	758	110	855	124	11.5	31.7

Fracture Toughness

The fracture resistance of the laser beam welds were evaluated at room temperature in a double pendulum machine using a subsize DT specimen. The thickness of the specimen was 12-mm (0.5-in) instead of the 16-mm (0.625-in) thickness normally used. The values of the DT energies for the laser beam welds are given in Table 4. Evidence for porosity was seen in the fracture surfaces of the HY-130 steel, and both the titanium and aluminum alloys. The HY-80 steel fracture surface had a coarse grain appearance as if fusion was incomplete and hot cracking had occurred.

Table 4

DT Energies of Laser Welds

Material	AU DT Energy Nm	(ft.lb)	Range of DT Energy Nm	(ft.lb)
HY-80	76	(56)	51- 176	(38-130)
HY-130	955	(705)	817-1266	(603-934)
Titanium alloy	249	(184)	176- 311	(130-230)
Aluminum alloy	116	(86)	101- 128	(75- 95)

Fractography

Several fracture surfaces of laser beam welds for each material were examined in a scanning electron microscope. Except for the aluminum alloy, the tensile fractures occurred in the base metal. The fracture separation process in these base plate fractures was microvoid coalescence. Porosity was evident on the fracture surface when failure occurred in the laser beam welds as seen in Fig. 1. With the exception of HY-80, the fracture separation process (between the pores) in all of the DT specimens and the aluminum alloy tensile specimens was microvoid coalescence. The number, size and distribution of pores varied not only from alloy to alloy but also from specimen to specimen with the alloy. The HY—80 DT fracture surfaces showed indications of hot cracking (see Fig. 2) and extensive secondary cracking which most likely was present prior to testing.

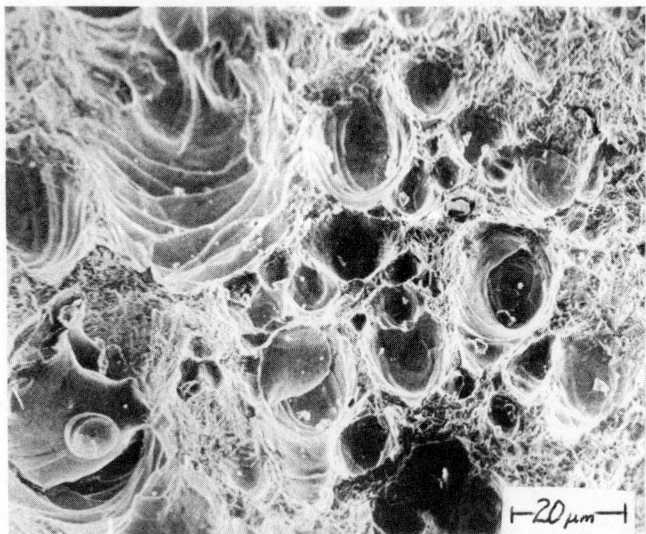

Fig. 1. Voids on the fracture surface of the DT specimen of the titanium alloy.

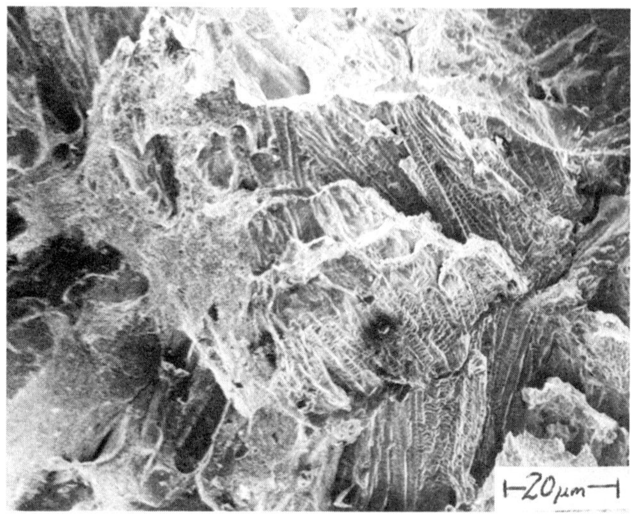

Fig. 2. Hot cracking and secondary cracking in HY-80.

Metallography

The transverse cross sections of the weldments were prepared for microstructural and solidification structural examination. The microstructures of each material were examined in the regions of the weld, the heat affected zone (HAZ), and the base metal.

In the HY-130 laser beam welds, the microstructure of the fusion zone is composed mostly of martensite with some bainite as shown in Fig. 3. In the medium grain size region of the HAZ adjacent to the fusion boundary, martensite and bainite were found. The microstructure of the finer grain region of the HAZ adjacent to the base metal consists of martensite and some ferrite. The base metal shows a quenched and tempered martensitic structure. Figure 4 shows the very fine solidification structure of the fusion zone in HY–130 weldment. The microstructures of the HY–80 and the HY–130 laser beam weldments are very similar.

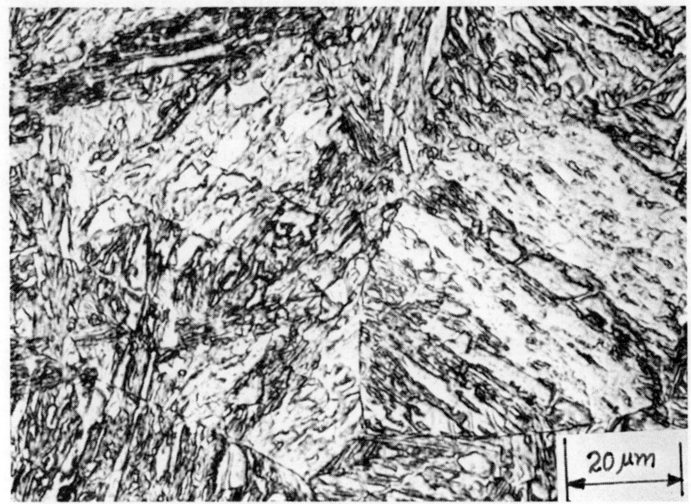

Fig. 3. Martensite and some bainite in weld metal of LB HY-130 weldment.

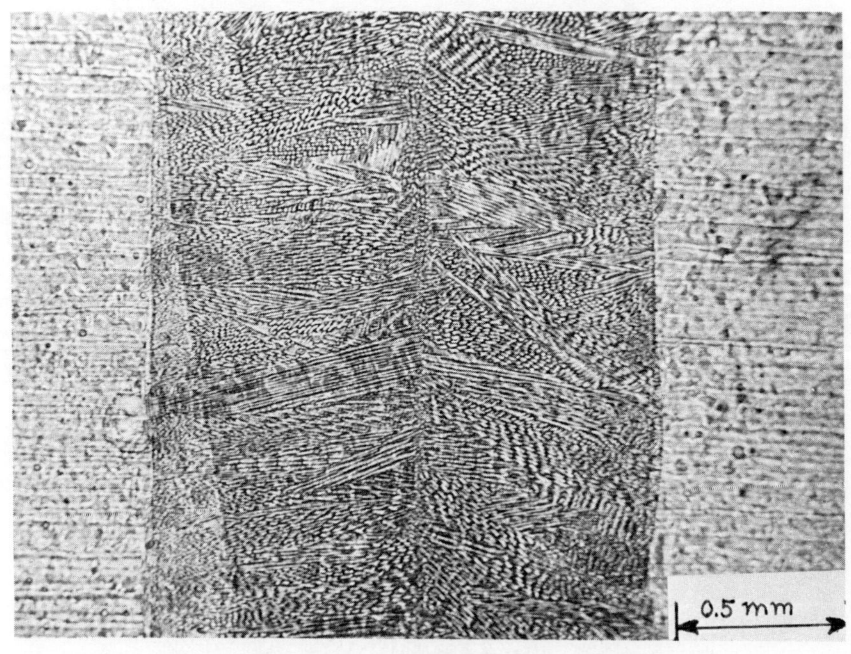

Fig. 4. Solidification structure of fusion zone in HY-130.

The microstructures of the fusion zone in the Ti alloy consist of $\alpha'$and prior β grain boundaries as seen in Fig. 5. Figure 6 shows the microstructure of the Ti base metal consisting of elongated and equiaxed primary α grains in β and transformed β matrix containing Widmanstätten α. The HAZ exhibits essentially the same microstructure as that of the base metal, i.e., primary α, Widmanstatten structure and β. The size and number of the primary α in the HAZ decreases and the β phase increases as the weld-HAZ interface is approached.

Fig. 5. α' and prior β grain boundaries in fusion zone of titanium alloy.

Fig. 6. Primary α, Widmanstätten α and β in base metal of titanium alloy.

In the Al 5456 laser beam weldment, very fine grain α Aluminum, fine Mg_2Al_3 precipitates at and within the grains, and large insoluble constituents mostly of Mg_2Si were observed in the fusion zone micrograph as shown in Fig. 7.

The HAZ microstructure consists of elongated grains, Mg_2Al_3 precipitates and insoluble constituents similar to those in the base metal. Figure 8 shows the base metal comprised of strain hardened, elongated α Al, β Al (Mg_2Al_3) at the grain boundaries and in the grains, and large insoluble particles such as black Mg_2Si and gray $(Fe,Mn)Al_6$.

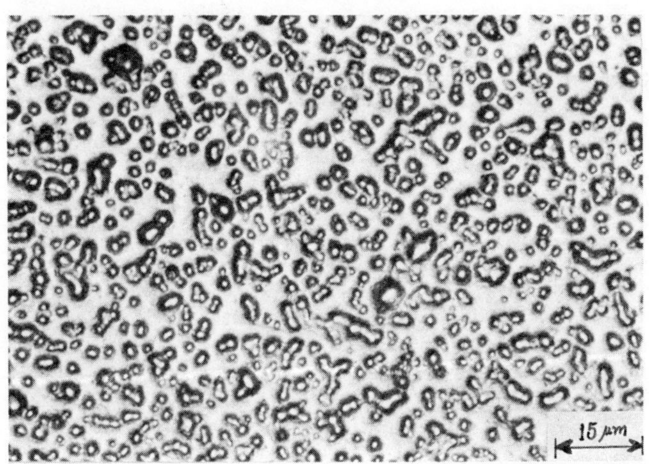

Fig. 7. Fine grained α aluminum in fusion zone of Al 5456 weldment.

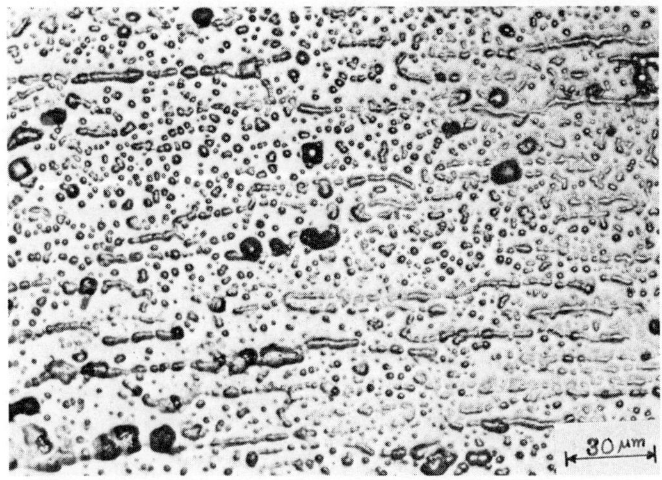

Fig. 8. Elongated α aluminum and dispersed fine Mg_2Al_3 particles in base metal of Al 5456 weldment.

Hardness

Transverse cross sections of the weldments were cut for microhardness testing. Using a diamond pyramid indenter with a 300 gm load, the microhardnesses were measured from one side of the base metal across the weld metal to the other side of the base metal.

The hardness of the weld and the HAZ in steels are much higher than that of the base metal as shown in Fig. 9 and 10. Average hardness values of the weld are 44 Rc for HY-80 and 48 Rc for HY-130 steel. The hardness of the HAZ varied from 27 Rc to 46 Rc in HY–80 and from 37 Rc to 39 Rc in HY-130 steel. The average hardness of the base metals is 18 Rc for HY-80 and 32 Rc for HY-130 steel.

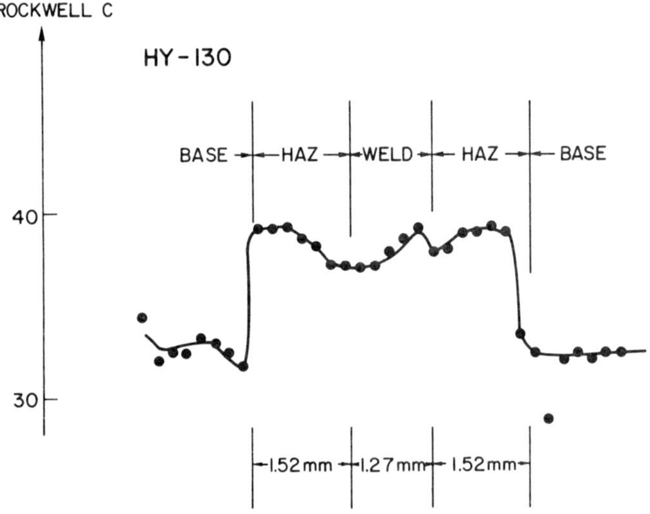

Fig. 9. Hardness traverse of HY-130

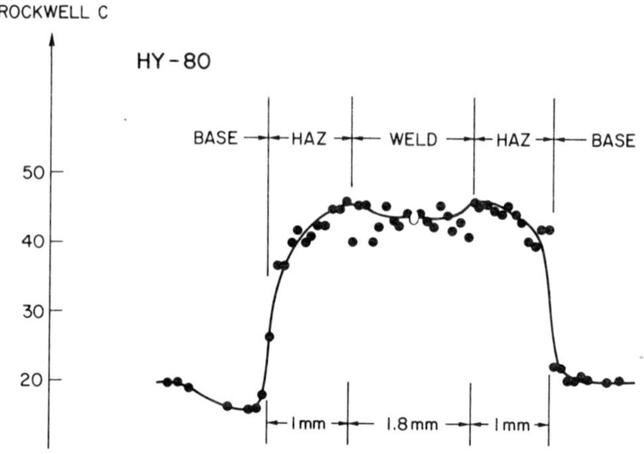

Fig. 10. Hardness traverse of HY-80

The hardness transverse of the Ti alloy exhibits little change across the three different zones with average value of 36 kc as seen in Fig. 11.

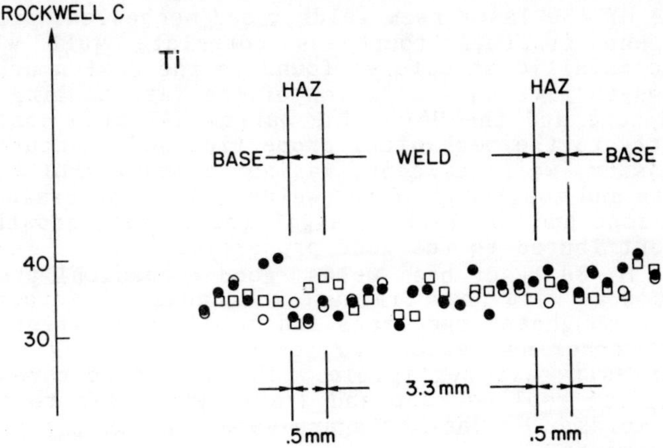

Fig. 11. Hardness traverse of Ti alloy

The hardness behavior of the Al alloy is presented in Fig. 12. The Vickers number varied from 87.0 in the base metal to 77.5 in the HAZ adjacent to the weld-HAZ interface. The highest Vickers number of the weld metal is 88.0.

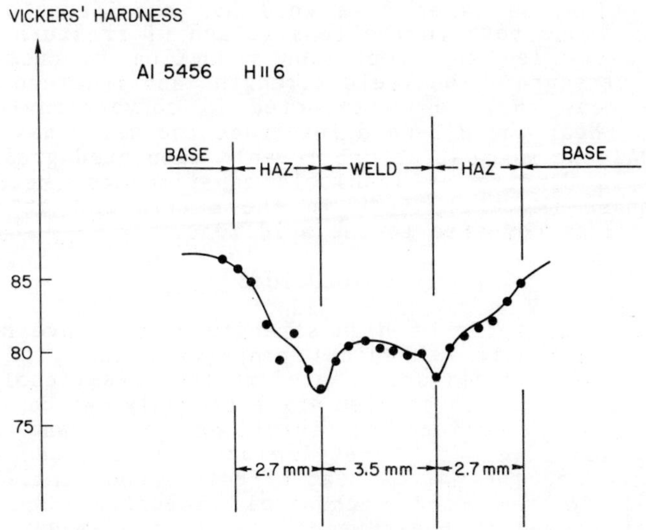

Fig. 12. Hardness traverse of Al alloy

DISCUSSION

For the purposes of discussion, each alloy will be commented upon in turn.

In the HY-130 laser beam welds, good mechanical properties and excellent fracture toughness correlate well with the martensitic/bainitic structures found in the fusion area. The high hardnesses give an indication of the fast cooling both in the fusion zone and the HAZ. The narrow HAZ also contributes significantly to the mechanical properties and fracture toughness. Porosity, while evident, was not a meaningful factor in the strength and toughness of the welds. The fine grain size of the fusion zone and the lack of significant grain growth in the HAZ also contributed to the good properties.

In the HY-80 laser beam welds, good mechanical properties were measured as were poor fracture toughness. Fractography of the fracture toughness specimens indicated an apparent lack of fusion, hot cracking, extensive secondary cracking and grain growth. Porosity was negligible. The microstructures correspond to a fast cooling rate and are very similar to those of the HY-130 welds. Hardnesses again were that of a fast transformation product.

In the titanium alloy, the mechanical properties of the laser weld compare favorably to those of the base plate and gas metal arc welds.[3] The fracture toughness is excellent. Fractography revealed considerable small porosity in the fracture surfaces of the DT specimens. This is attributable to titanium high affinity to oxygen. A variety of transformed and untransformed phases (α', β' and α, β) were observed in the metallographic surfaces. The fluctuations in hardness are probably due to the different hardness values of these phases.

The aluminum laser beam weld had considerable porosity which was found both in the tensile and DT fracture surfaces. This porosity led to an obvious reduction in ductility and ultimate strength. The yield strengths and fracture toughnesses were near that to be expected in conventional welds of aluminum. Near the HAZ-weld interface the alloy has softened. In the HAZ the microstructure reveals elongated grains and an increase in density of insoluble constituents (Mg_2Si) and a second phase (Mg_2Al_3), both in the matrix and at the grain boundaries, as compared to the weld zone.

CONCLUSIONS

Laser beam welds of high strength alloys have been shown to possess adequate mechanical properties and good fracture toughness. Hardnesses are related to their fast cooling rate. Porosity is a definite problem which probably can be alleviated by judicious selections of shielding gases and effective shielding gas coverage. The fine grain size produced in the fusion zone and the narrow heat affected zone contribute significantly to the good mechanical properties and fracture toughnesses found in these welds. A certain amount of second phase vaporization lowers the number of inclusions and changes the inclusion distribution especially in the lower melting point alloys.

REFERENCES

1. Welding Handbook, Volume 1, Seventh Edition, The American Welding Society, Miami, FL, 1976, pp. 21-22.

2. E. M. Breinan and C. M. Banas, "Fusion Purification During Welding with High Power CO_2 Lasers," Proceedings of the Second International Symposium of the Japan Welding Society, Osaka, Japan, Aug 24-28, 1975.

3. Welding Handbook, Section 4, Sixth Edition, The American Welding Society, Miami, FL, 1972, p. 69.9.

HEAT-AFFECTED ZONE TOUGHNESS IN HIGH-STRENGTH PIPELINE STEELS

A.B. Rothwell[1], A.G. Glover[2], J.T. McGrath[2] and G.C. Weatherly[3]

[1] Noranda Research Centre, Pointe Claire, Quebec, Canada

[2] Canadian Welding Development Institute, Toronto, Ont., Canada

[3] University of Toronto, Toronto, Ont., Canada

SUMMARY

This publication describes the results of crack-opening displacement (COD) tests on, and optical and electron metallographic examination of, a series of low-carbon steels formulated to cover a wide range of compositions currently applied in high-strength linepipe. All steels were subjected to a thermal cycle corresponding to a high heat-input, submerged arc seam weld. At 0.09%C, increasing alloy additions (Mn,Mo,Ni) led to a progressive deterioration in toughness: a reduction in carbon content to 0.04% improved toughness considerably, and at this level of carbon, COD transition temperature appeared to pass through a maximum as alloying was increased.

All microstructures fell into the category loosely classified as "coarse bainitic", but the carbon-rich regions at the interstices of the bainitic ferrite grains, which transformed to ferrite-carbide aggregates in the less-hardenable steels, were replaced by partially-transformed regions of martensite and retained austenite (MA) as alloying elements were added.

Fractographic examination showed that brittle fracture facet size, and that of non-propagating micro-cracks, was of the same order as, though somewhat larger than, the austenite grain-size, suggesting that the controlling factor in determining toughness may be the ease of crack propagation across a high-angle boundary; variations in lath width or carbide (MA) size may thus only be important inasmuch as they affect the yield strength.

1. Introduction

During the last ten years, the extension of pipeline operations to more remote and inhospitable environments has led to major developments in the metallurgy of linepipe materials. In particular, the need for economies of scale, in order to contain project costs, has created a demand for larger-diameter, thicker-wall pipes which satisfy increasingly stringent low-temperature toughness requirements, and whose weldability must also be guaranteed under the most exacting field conditions. Traditional carbon-manganese, micro-alloyed steels have not been able to fulfil this need, and many pipemakers have responded by reducing the carbon content to 0.1% or below, while making small additions of molybdenum, sometimes together with increased manganese contents, nickel or other elements. The overall effect of the increased level of alloying is to modify yielding and initial work-hardening behaviour in such a way as to offset the drop in yield strength normally associated with pipe-forming, expansion and the extraction of flattened-strap tensile testpieces, thus allowing yield strengths of over 500 MPa (70 ksi) to be reached in pipe 19mm (0.75 in) or more in thickness.

In terms of weldability in the field, such steels, as a result of their low carbon content, may be expected to be at least the equal of lower-strength steels, and laboratory studies and field experience show that this is generally so[1-3]. Some studies of seam-weld properties have also been carried out[1,4-9], but a full comprehension of the toughness-microstructure relationship in the weld metal and heat-affected zone (HAZ) has not yet been reached. In particular, as a result of peculiarities in testing geometry and/or the limited range of chemical composition or of heat-input examined, the metallurgy of the HAZ is not fully understood. Accordingly, the present study aimed to cover a wide range of chemical composition, within the class of steels under discussion, and to extend the range of cooling rate to the slowest likely to be encountered in the seam-welding of mainline pipe. A simulation technique was adopted in order to overcome problems of test geometry and notch location, and to provide the fullest possible metallurgical background. The merits and limitations of simulation methods have been discussed in detail elsewhere[10], and will not be entered into here; it is generally accepted that they provide a powerful technique for the comparison of different materials and for the study of fundamental metallurgical principles.

2. Experimental

The experimental materials comprised two groups of steels, having nominal carbon contents of 0.04 and 0.09%, respectively (Table I).

STEEL	Chemical Composition, wt %					Transformation Temperature, °C/°F			HV_5	$T_{0.2}$, °C/°F
	C	Mn	Mo	Ni	C.E*	Start	50%	Finish		
A	0.09	1.38	0.015	--	0.323	647/1197	601/1114	518/964	203	-30/-22
B	0.09	1.37	0.305	--	0.379	608/1126	576/1069	500/932	219	+25/+77
C	0.085	1.62	0.40	--	0.435	587/1089	550/1022	495/923	236	+35/+95
D	0.09	1.56	0.34	0.45	0.448	579/1074	543/1009	493/919	239	+35/+95
E	0.04	1.29	0.345	--	0.324	655/1211	608/1126	521/970	201	-10/+14
F	0.05	1.48	0.455	--	0.388	629/1164	590/1094	520/968	216	+ 5/+41
G	0.04	1.60	0.355	0.47	0.409	611/1132	577/1071	500/932	214	+ 5/+41
H	0.04	2.10	0.38	0.47	0.497	595/1103	554/1029	483/901	229	- 5/+23

$$* \; C + \frac{Mn}{6} + \frac{Cr + Mo + V}{5} + \frac{Ni + Cu}{15}$$

All steels contain 0.12 - 0.33% Si, 0.02 - 0.03% Al, ≤0.005% S, 0.007 - 0.019% P, 0.055 - 0.07% Cb.

TABLE I - Chemical composition, transformation temperatures, and simulated HAZ hardness and COD transition temperature of experimental steels.

All steels contained nominally 0.06% Cb, a level which has been shown capable of causing significant embrittlement at slow cooling rates[11]. Within each group, hardenability was progressively increased by additions of manganese, molybdenum or nickel. Levels of other elements were typical of high-quality, high-strength linepipe steel.

The steels were rolled to 19mm (0.75 in) plate by a schedule devised to simulate two-stage controlled rolling, with re-heating at 1150°C (2100°F), 50% deformation below 900°C (1650°F) and a finish rolling temperature of 800°C (1470°F).

Gleeble simulation blanks 10.4 x 10.4 x 100mm (0.409 x 0.409 x 4 in) were prepared, the axis of the blanks being parallel to the principal rolling direction. In addition, three blanks for each steel were machined to 10mm (0.394 in) diameter over their central portion, to facilitate dilatometric measurements during thermal cycling.

All blanks were cycled with a peak temperature of 1320°C (2410°F); the thermal cycle, which is shown in Fig. 1, corresponds to a submerged-arc weld with a heat-input of 4.05 kJ/mm (103 kJ/in) between 18mm (0.72 in) plates. This heat input represents an upper limit for the seam-welding of plate of such a thickness, and gives a cooling time between 800 and 500°C (1470 and 930°F) of ~80s.

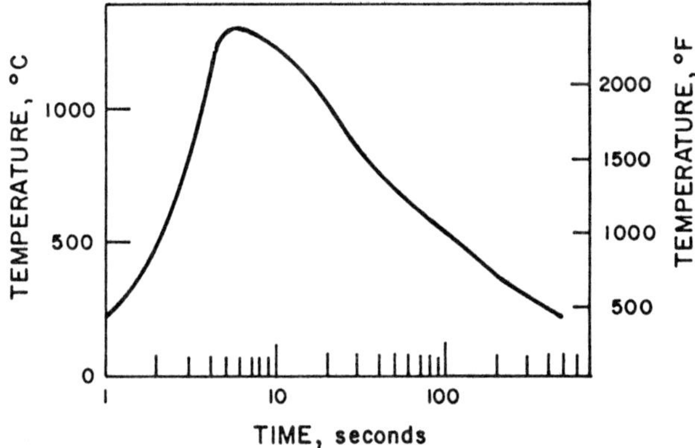

Fig. 1 - Weld thermal cycle used in simulation.

Transformation start, 50% completion, and finish temperatures were determined by dilatometry, during cooling, on triplicate samples. The start and finish temperatures were measured in correspondence with the smallest, detectable deviation from the linear dilatation traces of the austenitic and fully-transformed microstructures, respectively, and were subject to some dispersion as a result of the variable quality of the dilatometric traces produced by the Gleeble. The 50% transformation temperature could be determined with much greater accuracy, since the rate of transformation is at a maximum.

Metallographic sections were prepared from dilatometric testpieces, in correspondence with the position of the control thermocouple. General microstructure was determined after etching in nital, while the structure of the carbon-rich regions was examined at high magnification after etching in saturated picral. Diamond pyramid hardnesses were determined, with a 5kg load, as the mean of five impressions on each section. Blanks for thin foils were also cut from the centre of dilatometric samples, and thinned by conventional techniques in such a way that the final foil lay in the cross-sectional plane containing the thermocouple position.

Blanks for COD testing were ground to 10mm (0.394 in) square, to remove all traces of oxidation and decarburisation; a 0.15mm (0.006 in)-wide slot, 1.25mm (0.05 in) deep, was ground in a face normal to the original plate surface, at the point corresponding to the control thermocouple position. The notch was extended in fatigue, at a ΔK of $\sim$15 MPa$\sqrt{m}$ ($\sim$13.5 ksi$\sqrt{in}$), to a total depth of 2mm (0.08 in). Specimens were tested in three-point bending, over an appropriate range of temperature, surface displacement being detected by a clip-gauge. Crack-tip opening displacement was calculated according to the proposed British Standard procedure, with a rotational factor of 2.5.

Selected brittle fracture surfaces were nickel-plated, sectioned and examined in the optical microscope. Others were examined in the SEM, in some instances after a light electropolish on part of the surface to reveal the underlying microstructure, according to the technique of Chesnutt and Spurling[12].

3. Results

3.1 Transformation behaviour

As may be seen in Table I, transformation temperatures showed a progressive decrease with increasing level of alloying: start and 50% temperatures fell by some 50-70°C (90-125°F), at each level of carbon, on passing from the lowest to the highest alloy content, while the change in transformation finish temperature was more limited. Within each group, transformation temperature followed an approximately linear, inverse relationship with the conventional carbon equivalent. However, the transformation temperatures of the low-carbon group were higher, at a given level of carbon equivalent, than those of the high-carbon group. Similarly, the hardness increased more rapidly with increasing carbon equivalent in the high-carbon group than in the low-carbon group. These two observations are clearly linked, and Fig. 2 shows that, if hardness is plotted against 50% transformation temperature, results for both groups fall within a single band. The present results also lie within a band determined in earlier work on Gleeble-simulated HAZs covering a wide range of microalloying additions and cooling rates[2].

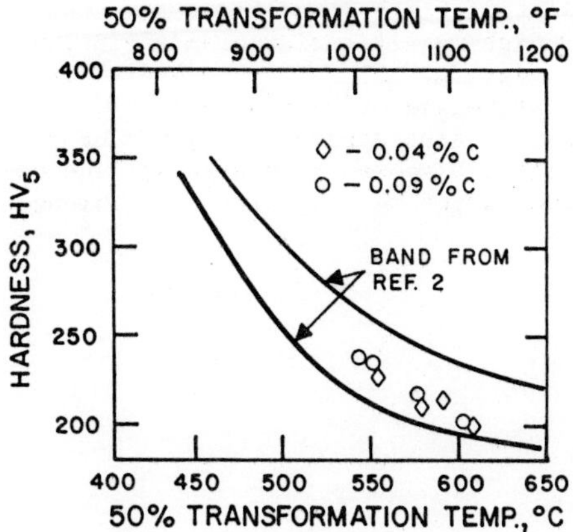

Fig. 2 - Variation of simulated HAZ hardness with transformation temperature

It is worth re-emphasizing, perhaps, that carbon equivalent is not a satisfactory predictor of hardness in these low-carbon steels, even at a constant, relatively slow cooling rate and a constant level of micro-alloying; carbon content itself assumes an increasingly dominant role in a number of aspects of weldability as it falls below ∿0.1%.

3.2 Microstructure

Typical gross microstructures of the higher-carbon group of steels are shown in Fig. 3. All microstructures may be classified as predominantly coarse bainite. As shown in Fig. 3a), the bainitic ferrite laths in steel A were generally separated by elongated, dark-etching regions, apparently carbide aggregates. Steel D, which had the lowest transformation temperature, appeared to have slightly finer ferrite laths, while the regions between the laths were predominantly light-etching (Fig. 3b).

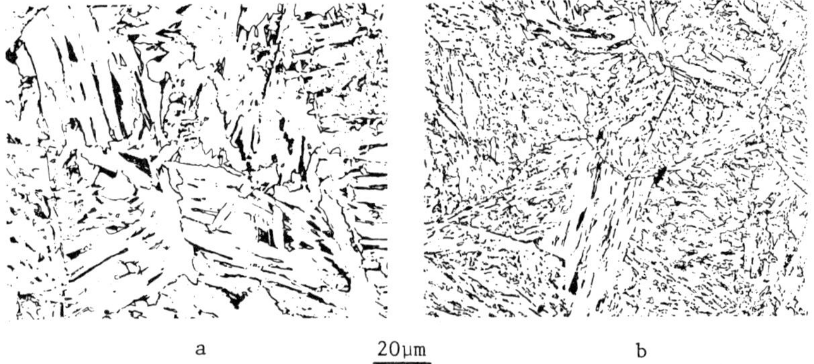

a 20μm b

Fig. 3 - Optical micrographs, nital etch. Original magnification x475, a) Steel A; b) Steel D.

The electron microscopic examination, described below, established that these regions consisted of martensite and retained austenite, and they will be referred to as MA in the ensuing discussion. The MA particles in steel D. were considerably finer than the carbide regions in steel A.

The carbon-rich regions of the bainitic microstructure are more clearly differentiated by high-magnification examination, after etching in saturated picral. Fig. 4a) shows that each dark-etching region in steel A consists of an aggregate of carbide particles. A few, lighter-etching MA regions are also apparent. As may be seen in Fig. 4b), the MA particles in steel D showed both angular, equiaxed and elongated shapes, with some larger MA particles lying along the prior austenite grain boundaries.

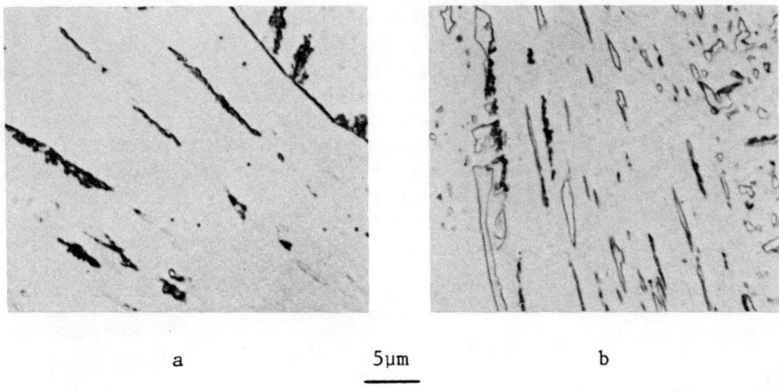

Fig. 4 - Optical micrographs, picral etch. Original magnification x1900. a) Steel A; b) Steel D.

The microstructures of the low-carbon steels were broadly similar to those described above. Clearly, the density of MA or carbide particles was lower in these steels. Steel E showed a mixture of MA and carbide; in some cases a single carbon-rich region would be partially-transformed to carbides, the remainder consisting of MA (Fig. 5). Again, as the hardenability was increased, the proportion of MA to carbide progressively increased and the lath width decreased slightly. Little or no carbide was apparent in steels G or H.

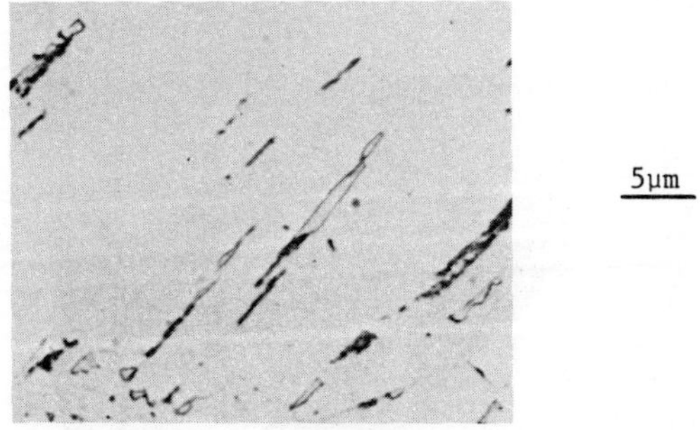

Fig. 5 - Steel E, picral etch. Original magnification x1900.

Electron microscopy on thin foils taken from the simulated heat-affected zones confirmed the basic microstructural features described above. The microstructure of steel A showed elongated, bainitic ferrite grains together with areas made up of aggregates of carbide lamellae (Fig. 6a). Figure 6b), also from steel A, shows an elongated region, part of which had transformed to carbides, and part of which consisted of MA (dark portion). Electron diffraction from the carbide region gave the expected, multiple-spot, Fe_3C pattern, while the dark portion gave few diffraction spots, which were not related to the Fe_3C structure. In the higher-carbon group, the MA particles always appeared opaque, and no internal structural detail could be resolved. This effect is not well-understood, but has been noted by a number of authors, and may indicate that the degree of transformation to martensite is relatively limited. By contrast, structural detail was often visible within MA particles in the low-carbon group; Fig. 6c) shows MA regions in steel G, in which there is clear evidence of internal twinning. Diffraction patterns from such regions showed characteristic streaking and splitting of spots; in some regions, more than one orientation of twinning was apparent.

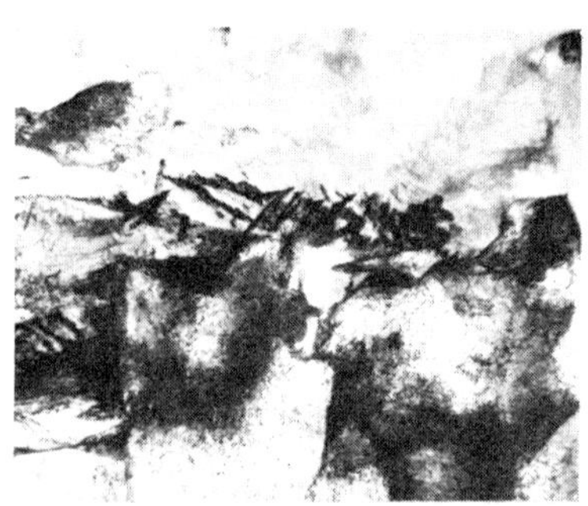

a 1μm b

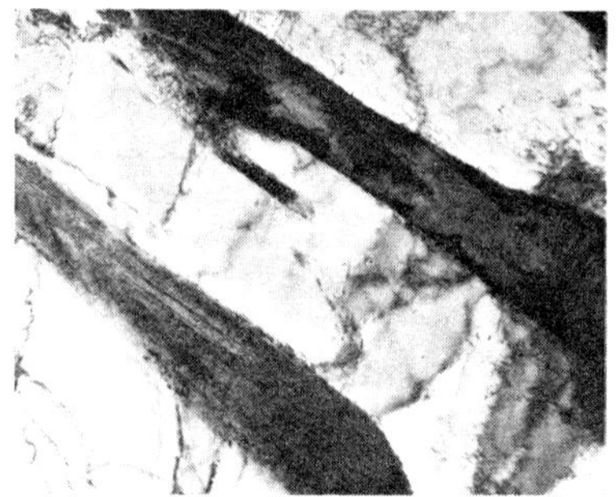

0.5μm

c

Fig. 6 - Thin-foil electron micrographs. a) and b) Steel A, original magnification x10,000; c) Steel G, original magnification x21,000

As a result of the rather high dislocation density within the bainitic ferrite grains, no clear and unambiguous evidence of columbium carbide precipitation was available from the thin foil study. The reason for the generally-observed embrittling effect of columbium in slowly-cooled welds has been the subject of controversy, but a parallel, comprehensive study of HAZ microstructure had indicated that solution and reprecipitation of Cb (C,N)were definitely taking place in the course of a thermal cycle similar to that of the present work[13]. A carbon extraction replica from steel B (Fig. 7), in fact, shows a fine dispersion of particles typical of carbides taken into solution near the peak temperature and re-precipitated during cooling.

0.2μm

Fig. 7 - Steel B. Carbon extraction replica: original magnification x43,000

3.3 Toughness

0.2mm COD transition temperatures for the eight steels are shown in Table I; in most cases, some degree of scatter was apparent within the transition region, and the temperatures quoted are for curves drawn through the lower-bound COD values.

In the higher-carbon group of steels, toughness deteriorated as alloying elements were added and the transformation temperature was depressed; the most notable deterioration took place between steel A and B. There was no significant, intrinsic effect of nickel.

A reduction in carbon content to 0.04% led to a considerable improvement in toughness. In the low-carbon group, the transition temperature appeared to pass through a maximum: while the improvement between steels G and H may appear insignificant, it must be viewed in relation to the large increase in carbon equivalent between these steels, and the concomitant fall in transformation temperature and increase in hardness. These would have been accompanied by a considerable deterioration in toughness, had the trend followed by the other steels been continued.

4. Discussion

Previous attempts to explain the variations of weld-zone fracture toughness with chemical composition have addressed both microstructural and mechanical aspects of the problem. In the former area, austenite grain-size, gross microstructural category (martensite, bainite, acicular ferrite) and precipitation effects are known to be significant[13,14].

Within the mechanical area, an increase in hardness (or strength) is expected to lead to a deterioration in toughness[15], though it must be emphasized that this will only be generally true within the same microstructural category; for example, the replacement of a bainitic microstructure by martensite, though it is accompanied by an increase in hardness, generally improves toughness in low-carbon steels[2].

It is clearly rather a complex task to identify the precise, metallurgical factors which are controlling toughness in the present instance. Austenite grain-size, which is certainly of paramount importance[13], did not vary significantly from steel to steel. All values measured were in the range 75-90 μm; the average fracture facet size appeared to be in the range 110-120 μm, or slightly larger than the prior austenite grain size, and corresponded typically to about two bainite colonies. Secondary, non-propagating microcracks observed in the SEM (Fig. 8) and on fracture profiles in the optical microscope appeared to be of similar dimensions; in general, then, cracks propagated readily across ferrite-ferrite boundaries within a bainitic colony, as has been noted previously[16].

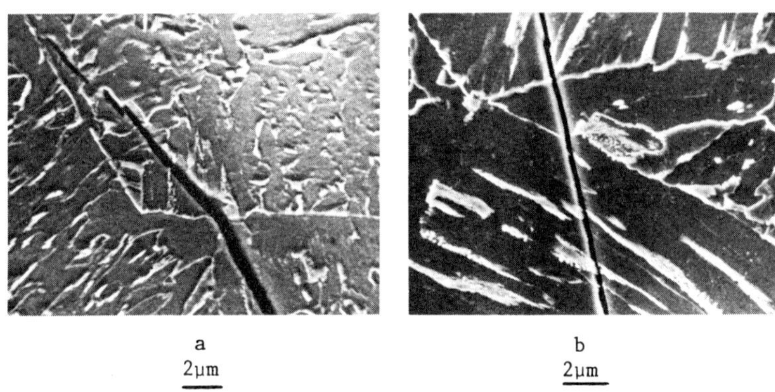

Fig. 8 - Non-propagating microcracks on lightly-polished and etched fracture surface, Steel A . a) original magnification x3200 b) original magnification x3900.

The critical fracture initiation event would thus be the extension of a microcrack across a high-angle boundary. The tendency of the MA particles to be "trapped" in linear arrays at such boundaries (Fig. 9) may be significant in this context, and has already been discussed in reference to weld metal[17].

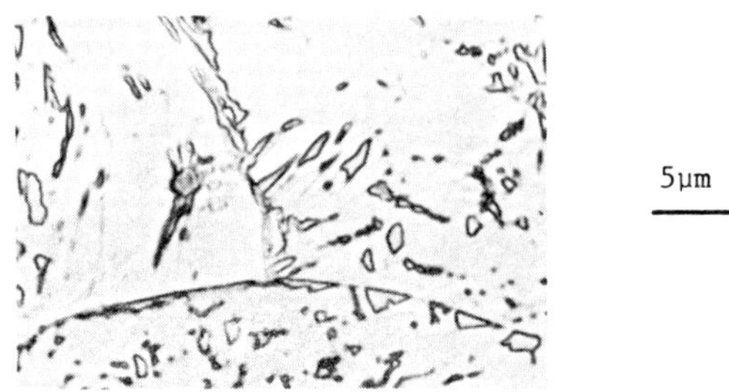

Fig. 9 - Steel D, picral etch. Alignments of MA particles at prior austenite grain boundaries. Original magnification x1900 .

It is not suggested that cracks are likely to propagate preferentially along such arrays, and indeed, this was never observed in fractographic studies. However, the presence of a brittle microconstituent at high angle boundaries, generally with associated transformation stresses, may be expected to aid the propagation of a crack from one grain to another in much the same way as grain boundary carbides are found to do in polygonal ferrite structures. It may also be significant that, while considerable numbers of non-propagating microcracks were observed in the fracture initiation zone of the tougher steels, these were in general few in the less-tough materials. This may indicate that in the steels containing a greater proportion of MA, microcracks have a high probability of finding an easy mode of extension across the first one or two high-angle boundaries and thus of becoming hyper-critical. The probability of finding stable microcracks is thus limited.

As regards the mechanism of formation of MA, rather than carbide aggregates or the discrete, lath-boundary carbides typical of classical, upper bainite, much remains to be learnt. In general, within the range of conditions typical of welding, a reduction in transformation temperature, whether as a result of increased cooling rate or alloying, will tend to promote MA. In the case of alloying, it appears that, while interstitial diffusion is fast enough for the austenite to become increasingly enriched as transformation progresses, there is little partition of substitutional alloying elements. Transformation of the austenite is thus progressively retarded, and nucleation and growth of carbides is suppressed. It also appears that a lower carbon content, at a given transformation start temperature, favours the formation of MA. This is probably the result of a reduction in the rate of carbon enrichment, so that there is less probability of the necessary carbon content for the nucleation of carbides being established before the temperature has fallen below the carbide formation range.

A final factor which may be involved in the observed variations in toughness is a change in the fine dispersion of columbium carbide. Such a change could be brought about by interactions between columbium and other carbide formers such as molybdenum, or could simply result from the fall in transformation temperature with increasing alloying. Little can be said concerning the former possibility, but as regards the latter, a reduction in transformation temperature, within the range studied, would be expected to lead to a decrease in precipitation hardening[13]. This, in turn, should improve toughness.

In summary, while the variations in toughness of the higher-carbon steels could, in principle, be explained in terms of hardness alone (Table I), it is more likely that the observed results express the summation of a number of effects. In particular, the increasing hardness and tendency to form alignments of MA would both tend to lead to a deterioration in toughness, which outweighs any improvement due to the suppression of columbium carbide precipitation. The situation in the low-carbon group is a little more complex, in particular as regards the improvement in toughness from steel G to steel H, despite the higher hardness of the latter. The transformation finish temperature of steel H excludes the possibility of the formation of significant proportions of martensite from un-enriched austenite, and in fact the microstructure still appeared to be fully bainitic. The only plausible explanation would thus appear to be the suppression of columbium carbide precipitation; it is logical that this should be more pronounced, at a given transformation temperature, in the lower carbon steels.

5. Conclusions

1. As the level of alloying was increased in a series of 0.09% C pipeline steels, transformation temperature progressively fell, hardness rose, and toughness deteriorated. A reduction in carbon content to 0.04% brought about a considerable improvement in toughness, and there was some indication that the COD transition temperature passed through a maximum, with increasing alloy content.

2. As transformation temperature decreased, the bainitic ferrite lath width became somewhat smaller, and the carbide aggregates formed at the interstices of the laths were progressively replaced by MA. The latter was generally rather opaque to electrons in the higher-carbon steels, and no structural detail could be resolved, but in the lower-carbon series, fine, internal twinning was clearly apparent. There was some tendency for MA particles to form alignments at high angle boundaries, possibly facilitating the extension of microcracks from one prior austenite grain or bainite colony to another.

3. No significant variations in prior austenite grain size, bainitic colony size or gross microstructure were apparent. The observed effects of transformation temperature on toughness must thus be accounted for by a balance between matrix strength, precipitation hardening and its suppression at lower temperatures, and the presence of alignments of MA particles at high-angle boundaries.

Acknowledgements:

The technical assistance of R. Dufresne, J.-P. Fachinetti, and I.D. Montgomery in the execution of this work is gratefully acknowledged. Thanks are also due to the management of Noranda Research Centre for permission to publish.

References:

1. J.M. Sawhill, Jr., and T. Wada: Weld. J., 1975, 54 No. 1, p. 1s.

2. A.B. Rothwell and F. Bonomo: "Welding of Linepipe Steels", pp. 118-146. Welding Research Council, New York, 1977.

3. C. Düren and H. Müsch: Paper presented at Joint ASM/AIM Symposium. "Welding of HSLA (Microalloyed) Steels", Rome, November, 1976. ASM, Metals Park, in press.

4. G. Bernard, F. Faure and P. Maitrepierre: ibid.

5. A. Poli, C. Santafé, A. Pozzi and L. Bertuzzi: ibid.

6. J.M. Sawhill, Jr., P. Boussel and J.W. Morrow: ibid.

7. C. Düren, G. Hieber and W.W. Wiedenhoff: "Welding of Linepipe Steels", pp. 3-35. Welding Research Council, New York, 1977.

8. S.F. Baumann, J.M. Sawhill, Jr., and M. Nakabayashi: ibid, pp.56-85.

9. H. Gondoh, H. Nakasugi and H. Masui: Paper presented to 16th Annual Conference of Metallurgists, CIM, Vancouver, August, 1977.

10. "Weld Thermal Simulators for Research and Problem Solving": Welding Institute Research Application Seminar, The Institute, Abington, 1972.

11. N.E. Hannerz: Weld. J., 1975, 54 No. 5, p. 162s.

12. J.C. Chesnutt and R.A. Spurling: Met. Trans. A., 1977, 8A, p. 216.

13. A.B. Rothwell, J.T. McGrath, A.G. Glover, B.A. Graville and G.C. Weatherly: Final Report to Dept. of Energy, Mines and Resources: Research Contract OSQ77-00055 - Heat-Affected Zone Toughness of Welded Joints in Micro-Alloy Steels, May, 1978.

14. R.E. Dolby and J.F. Knott: JISI, 1972, 210, p. 857.

15. R.E. Dolby: Paper presented at Joint ASM/AIM Symposium "Welding of HSLA (Microalloyed) Steels", Rome, November, 1976. ASM, Metals Park, in press.

16. T. Gladman, D. Dulieu and I.D. McIvor: "Microalloying 75", p. 32. Union Carbide Corp., New York, 1977.

17. J.G. Garland and P. Kirkwood: Metal Construction, 1975, 7 No. 5, p. 275.

DISCUSSION

M. Bibby (Carleton University): What is the current thinking on the toughness level in the HAZ necessary for large diameter pipeline practice?

Author: As I am sure you are aware, considerable debate and discussion is currently going on, in North America and elsewhere, on this topic. In the weld zone, where structure and properties can vary widely over quite small distances, there may be considerable philosophical difficulties in applying a straightforward fracture mechanics approach to defect tolerance, as the first instability does not necessarily lead to fracture. Nonetheless, I feel sure that regulatory authorities (and others) would be reluctant, at the present time, to accept the possibility that regions of significant extent show little or no ductility (say, critical COD <0.1mm) in static tests at the minimum service temperature. How this problem can be realistically dealt with, under production conditons, is not easy to see; perhaps rather elaborate testing will be necessary at the qualification stage, with quality control measures during production being of a more conventional nature.

Fatigue Behavior of HSLA Steel and Associated Weldments

P. E. Bretz (1), B. L. Braglia (2), and R. W. Hertzberg (1)

(1) Department of Metallurgy and Materials Engineering
Lehigh University
Bethlehem, PA 18015

(2) Electro-Motive Division
General Motors Corporation
La Grange, IL 60525

Fatigue crack propagation (FCP) studies were conducted on VAN-80, an inclusion shape-controlled, high strength, low alloy (HSLA) steel plate, and on weldments in this steel. Base metal (BM) crack growth characteristics were evaluated for the hot-rolled plate. Weld metal (WM) and heat-affected zone (HAZ) fatigue properties were studied by introducing a slotted notch in the required regions and allowing for crack extension parallel to the direction of welding. The gas metal-arc (GMA) welding process was used, with the filler metal conforming to AWS E70S-1b. All welds were stress-relieved prior to fatigue testing.

Fatigue crack propagation rates were found to be comparable in the WM and HAZ regions and were significantly lower than the base metal results at the same applied stress intensity range levels. Such behavior suggests that the weld metal and heat affected regions experienced lower ΔK values than in base metal samples and/or that different micromechanisms were responsible for the fatigue cracking process in these three microstructures.

In an effort to determine the mechanism responsible for the lower effective ΔK levels in the WM and HAZ relative to the base plate, fractographic examinations of all specimens were conducted. Observation of fracture surface replicas in the transmission electron microscope revealed that fatigue striation spacings in the three regions followed the same pattern exhibited by the respective FCP results; i.e., striation spacings were smaller for WM and HAZ tests than for BM tests. This behavior is consistent with a lower prevailing effective ΔK level at crack tips in the WM and HAZ microstructures, though the reason for such behavior remained unspecified. However, metallographic profiles revealed that the fatigue cracks within the WM and HAZ were often deflected by their respective acicular microstructures and generated rougher surfaces as compared with the relatively unperturbed nature of fatigue crack advance in the base metal. Based on these fractographic results, it is concluded that the acicular microstructures of the HAZ and WM retarded fatigue crack growth by allowing for repeated deflection and branching of the advancing fatigue crack, thereby reducing the effective stress intensity conditions at the crack tip. The lower effective ΔK level is consistent with the observed lower fatigue crack growth rates measured in these two regions.

1. Introduction

In recent years, the cyclic response of engineering materials has been evaluated through measurement of the fatigue crack propagation (FCP) rate as a function of the stress intensity factor range. Such data are usually described in terms of the Paris power relationship

$$\frac{da}{dn} = C\ \Delta K^{m} \qquad (1)$$

where

$\frac{da}{dn}$ = the incremental crack growth rate,

ΔK = stress intensity factor range,

C, m = material constants.

The effect of welding on FCP behavior has received some attention. Parry, et al., determined room temperature crack growth rates in ASTM A 514 base plate and weldments, and found comparable growth rates at the same ΔK range for the base metal (BM), weld metal (WM), and heat-affected (HAZ) regions in the stress-relieved condition, and in two plates in the as-welded condition.[1] WM and HAZ tests in two other plates in the as-welded condition revealed growth rates that were clearly lower than that found in the BM tests. These differences in growth rates were rationalized in terms of residual stress considerations. The elevated temperature FCP properties of weldments in AISI Type 304 stainless steel were examined by James who noted somewhat lower growth rates in the WM as compared to the BM for tests conducted at 538 C (1000 F).[2,3] The reasons for the lower WM crack growth rates were tentatively theorized to be a result of this fine duplex WM microstructure and a stronger tendency for crack branching in the WM.

Significantly lower crack growth rates in weldments of 5% Ni steel as compared to the base metal FCP behavior were determined also by Bucci and co-workers in the regime $\Delta K \leq 55$ MPa $\sqrt{m}$ (50 ksi $\sqrt{in.}$).[4] In this instance, the superior fatigue behavior of the WM was attributed to crack closure effects as a result of the higher ductility of the WM as compared to the BM. Similar FCP behavior for 5% Ni steel was found by Sarno, et al.[5]

The objective of this work was to compare FCP behavior in the BM, WM, and HAZ of a high strength low alloy (HSLA) steel, VAN-80, to determine whether any differences exist in FCP response associated with these three weld regions. In addition, fractographic analyses were planned to determine the fatigue fracture micromechanisms responsible for any variations in crack growth behavior among the three tested regions. This HSLA steel (conforming to ASTM A656, Gr.1), was chosen for the present FCP study because its FCP behavior has previously been reported,[6] and the alloy had been shown to be highly weldable.[7] VAN-80 derives its high minimum yield strength of 550 MPa (80 ksi) from a very fine grain size (ASTM No. 12) along with vanadium carbonitride precipitation strengthening.

2. Experimental Procedures

2.1 Welding and Fatigue Testing

The VAN-80 base material was obtained with a nominal plate thickness of 4.83 mm (0.190 in.). All welding was done using the gas metal-arc (GMA) process according to the procedure outlined in Table I, with welds being prepared both longitudinally and transversely with respect to the rolling direction. Following welding, a stress relief treatment of one hour at 593 C (1100 F) was performed in order to reduce or eliminate any effect of residual stresses on the FCP process. Chemical and mechanical property data for both base metal and electrode are given in Tables II and III.

Fatigue specimens were prepared by grinding off the weld reinforcement and machining 76 x 76 mm (3 x 3 in.) compact tension samples. The weld was etched prior to machining so that the starter notch could be placed accurately within the WM or HAZ, as required. The actual plate thickness in the welded test specimens varied from 3.18 to 3.68 mm (0.125 to 0.145 in.), depending on machining requirements.

Room temperature fatigue testing was performed in laboratory air on an MTS electrohydraulic closed loop test machine at a load ratio of R=0.1 ($R = \frac{\sigma_{min}}{\sigma_{max}}$). The initial test frequency was 50 Hz and was reduced as specimen compliance increased. Crack tip positions were monitored using a traveling microscope with a vernier dial. Computer analysis of the FCP data included a crack tip plastic zone size correction. The correction algorithm used the stress intensity factor range calculated for each crack tip reading,

$$\Delta K_i = Y_i \, \Delta \sigma \sqrt{a_i} \tag{2}$$

to estimate the plastic zone size $(r_y)_i$ through the use of the equation

$$(r_y)_i = \frac{1}{2\pi} \left(\frac{\Delta K_i}{\sigma_{ys}}\right)^2 \tag{3}$$

This value was then added to the instantaneous crack length a_i to calculate an effective stress intensity factor for each datum; i.e.,

$$(\Delta K_{eff})_i = Y_i \, \Delta \sigma \sqrt{a_i + (r_y)_i}$$

2.2 Fractography and Metallography

An electron fractographic analysis of two-stage platinum/carbon fracture surface replicas was conducted on a Philips EM 300 electron microscope at an accelerating potential of 60 kV. Each fatigue striation spacing datum represented an average of at least ten different sets of readings from a given replica; each replica constituted a one mm long section of the fracture surface.

Metallographic specimens were prepared by conventional mechanical polishing techniques with a final 0.06 μm alumina slurry polish being followed by etching in a 2% nital solution. Fracture profiles were plated with an electroless Ni solution before polishing to insure edge retention.

3. Presentation and Discussion of Results

3.1 Microstructures

The microstructures of the three crack path zones are shown in Figure 1. The fine-grained ferrite/pearlite structure of the VAN-80 base metal is in marked contrast to the acicular structures of the WM and HAZ regions. The weld metal consists mainly of bainite and proeutectoid ferrite while the large-grained HAZ contains Widmanstatten ferrite and some bainite.

3.2 Fatigue Crack Propagation Results

The longitudinal and transverse growth rates for the base metal as a function of stress intensity factor range are shown in Figure 2. The data indicate that the FCP behavior of this steel is essentially isotropic, in that the maximum difference in growth rates is less than a factor of two.[8] This finding is an agreement with previous FCP data for VAN-80.[6] The crack growth rate data for the WM and HAZ regions are presented in Figures 3 and 4, respectively, and again shown little variation between longitudinal and transverse tests.

A comparison of crack growth behavior in the three weld zones is shown in Figure 5, using least-squares linear regression lines of the data. It can be seen that the crack growth rates are highest for the BM and lowest for the WM at all stress intensity factor values examined. The superiority of WM crack propagation behavior is consistant with previously reported FCP findings.[2-5]

As discussed previously, three mechanisms have been proposed to account for the observed decrease in crack growth rates in weld metal as compared to the base plate: a favorable residual stress pattern due to welding; crack closure effects; and microstructurally induced crack branching. At this point, it is appropriate to examine the applicability of these arguments to the FCP results presented here. While the influence of a favorable residual stress pattern cannot be entirely discounted in this study, it would seem that the stress relief treatment was sufficient to reduce the residual stresses to a significant degree. Hence, the large decrease in crack growth rates in both WM and HAZ fatigue tests would not appear to be the result of a favorable residual stress pattern and its consequent reduction in the effective crack tip stress intensity condition. Secondly, fatigue crack closure effects have been found to reduce crack growth rates in WM tests where the weld metal has a greater ductility than the base metal. When crack closure occurs, the crack is open for only a part of the applied load cycle, even if the minimum loading is greater than zero; consquently, the stress intensity range at the crack tip is not defined by the applied stress range $\sigma_{max} - \sigma_{min}$, but rather by the stress range over which the crack is fully open. This effective stress range may be given by $\sigma_{max} - \sigma_{op}$, where σ_{op} is the stress at which the crack opens ($\sigma_{op} > \sigma_{min}$). This closure effect reduces the effective magnitude of ΔK at the crack tip and decreases the resulting crack growth rates.[9] Since the degree of the closure effect was believed to be governed by the ductility of the metal, one would predict that the greater the ductility of a weld deposit relative to the base metal, the lower would be the WM FCP rates. A logical extension of this argument is that materials with the lowest strength and highest ductility should have the lowest FCP rates due to a high σ_{op}. However, this behavior is not generally observed.[10]

A third possible explanation for the reduced FCP rates in welded specimens concerns the beneficial role of microstructure in attenuating the fatigue crack process.[2,3] For example, it has been observed that a duplex weld structure can lead to significantly more crack branching than that found in a more uniform base metal microstructure. Such repeated crack branching would be expected to dissipate a portion of the strain energy at the tip of the propagating crack and lead to slower FCP rates. It is worth noting that crack branching has led to reduced FCP rates in titanium alloys,[11,12] nickel-base eutectic superalloys,[13,14] and several grades of forging steel.[15] It seems appropriate, therefore, to examine this HSLA alloy steel to determine whether the superior FCP response of the HAZ and WM relative to the base plate was related to preferential crack branching in the HAZ and WM zones.

3.3 Fatigue Fracture Micromechanisms

A TEM examination of the fatigue fracture surfaces was conducted to compare present results with those previously reported by Hertzberg and Goodenow[6] for the same material and for several other ferrite-pearlite steels.[16-20] It was found that these ferrite-pearlite steels exhibited a transition from a structure sensitive propagation mechanism (characterized by intergranular failure) at low ΔK levels to a structure insensitive mechanism (fatigue striations) at intermediate ΔK levels. The results of the present investigation indicate that VAN-80 BM undergoes a similar fatigue fracture mechanism transition in the range 14-20 MPa$\sqrt{m}$ (12.7-18 ksi$\sqrt{in.}$).[21] Above this range, only fatigue striations are observed (up to at least ΔK = 35 MPa $\sqrt{m}$ (31.8 ksi$\sqrt{in.}$))(Figure 6). Between 10-14 MPa$\sqrt{m}$ (9.1-12.7 ksi $\sqrt{in.}$) distinctive intergranular fracture surface markings are observed and reflect fatigue crack growth along ferrite-ferrite boundaries and along pearlite boundaries. An example of pearlite boundary failure is shown in Figure 7. As the stress intensity range increased from 14 to 20 MPa$\sqrt{m}$(12.7 to 18 ksi$\sqrt{in.}$) the amount of intergranular fatigue fracture decreased while the percentage of striated area increased. These fracture surface markings were observed at comparable ΔK levels for both the present BM specimens and those tested by Hertzberg and Goodenow.

Additional fractographic studies were conducted on the HAZ and WM fatigue fracture surfaces. For the HAZ specimens, intergranular features were observed at ΔK levels as high as 23.7 MPa$\sqrt{m}$ (21.6 ksi$\sqrt{in.}$) while striations were seen on replicas from regions experiencing ΔK levels $\geq$ 15 MPa$\sqrt{m}$ (13.6 ksi$\sqrt{in.}$). However, no evidence of pearlite or ferrite boundary failure was noted on the HAZ fracture surfaces. For the WM fracture surfaces, no intergranular fatigue fracture micromechanism was observed for $\Delta K \geq 19$ MPa$\sqrt{m}$ (17.3 ksi$\sqrt{in.}$).

These base metal fractographic results suggest that the overall range of the stress intensity factor that can be inferred from fractographic information is broader than that associated with the striation formation regime. For example, the presence of intergranular failure alone implies a stress intensity range of 10-14 MPa$\sqrt{m}$ (13-18 ksi$\sqrt{in.}$) in VAN-80 steel.

A quantitative analysis of the fatigue fracture surfaces was conducted by plotting striation spacing measurements versus the corresponding ΔK levels in each sample. The results of these measurements appear in Figure 8; the height of each data box is $\pm$ one standard deviation about the mean spacing value, and the width represents the range of ΔK values for each section of the fracture surface replica examined. It can be seen that these microscopic data follow a trend similar to the macroscopic FCP data; i.e., the striation spacings at any ΔK levels are greatest for the BM tests and smallest for the WM fatigue tests.

These data would suggest that the difference in FCP rates for the three weld zones was due to the respective microstructures. In themselves, the striation results cannot distinguish between residual stress and crack branching as the cause for the presumed difference in ΔK_{eff}.

It would seem reasonable to conclude that residual stress effects were minimal in this investigation since all samples had been stress relieved prior to testing. Therefore, attention was given toward establishing whether or not crack path perturbations had caused the observed difference in FCP rates in the three weld zones. To this end, metallographic fracture surface profiles were examined from each fracture region. The BM profiles indicate that the fatigue crack grew in a relatively unperturbed manner across this region (Figure 9). By contrast, the HAZ and WM profiles show significant crack branching and crack deflection within the acicular microstructure. Such crack branching should redistribute the crack tip stress and lead to a lower effective crack tip stress intensity factor. We conclude from these metallographic profiles, therefore, that crack branching was instrumental in retarding fatigue crack growth rates in the WM and HAZ regions as compared to the base metal.

4. Conclusions

1. Fatigue crack propagation rates in weld metal and heat-affected zone regions were significantly lower than in the hot-rolled VAN-80 base metal.

2. Fatigue striation spacings on weld metal and heat-affected zone fracture surfaces were also smaller than those found on the base metal fracture surfaces.

3. The acicular ferrite and bainite in the weld metal and heat affected zone regions caused repeated deflection and branching of the propagating fatigue crack, thereby reducing the effective stress intensity conditions at the crack tip and retarding crack growth rates in these regions.

4. A fatigue fracture mechanism transition was identified for the VAN-80 base metal in the range of 10-20 MPa$\sqrt{m}$ (9-18 ksi$\sqrt{in.}$). A maximum amount of intergranular failure along ferrite and pearlite colony

boundaries was identified in the middle of this range (approximately 14-15 MPa $\sqrt{m}$). Above this level, striation formation was found in increasing amounts; below 15 MPa $\sqrt{m}$, a non-discript fracture micromorphology was seen to gradually replace the intergranular fracture regions. This fracture mechanism transition in the VAN-80 material supply is similar to that reported by others in higher alloy steels.[22,23]

Acknowledgements

The authors wish to express their appreciation to the Pennsylvania Science and Engineering Foundation for their support of this work. In addition, we would like to thank the Jones and Laughlin Steel Corporation for providing the VAN-80 steel plate, Mr. B. R. Somers for assistance in preparing the welded specimens, and the Electro-Motive Division, General Motors Corporation, for chemical analyses of the test material supply.

References

1. Parry, M., Nordberg, H., and Hertzberg, R. W., Welding Journal, 51, no. 10 (October, 1972) p. 485-s.
2. James, L. A., Journal of Testing and Evaluation, 1, no. 1 (January, 1973), p. 52.
3. James, L. A., Welding Journal Research Supplement, 52, no. 4 (April, 1973), p. 173-s.
4. Bucci, R. J., Greene, B. N., and Paris, P. C., Progress in Flaw Growth and Fracture Toughness Testing, ASTM STP 536, American Society for Testing and Materials, 1973, p. 206.
5. Sarno, D. A., Bruner, J. P., and Kampschaefer, G. E., Welding Journal, 53, no. 11 (November, 1974), p. 486-s.
6. Hertzberg, R. W. and Goodenow, R. H., Proceedings of Microalloying 75, Union Carbide Corporation (1977), p. 503.
7. Pollard, B. and Aronson, A. H., Welding Journal, 49, no. 12 (1970) p. 559-s.
8. Clark, W. G., Jr. and Hudak, S. J., Jr., Journal of Testing and Evaluation, JTEVA, 3, no. 6 (1975), p. 454.
9. Elber, W., Damage Tolerance in Aircraft Structure, ASTM STP 486, American Society for Testing and Materials (1971), p. 230.
10. Barsom, J. M., J. Eng. Ind., ASME, Series B, 93, no. 4 (1971), p.1190.
11. Thompson, A. W., et al., Proceedings, 3rd International Titanium Conference, Moscow, 1976.
12. Bania, P. J. and Eylon, D., Met. Trans., 9A, no. 6 (1978), p. 847.
13. Stoloff, N. S., et al., Technical Report No. 2 on Contract N00014-75-C-0503, NR031-745, Rensselaer Polytechnic Institute, Troy, New York, August 1976.
14. Bretz, P. E. and Hertzberg, R. W. submitted for publication in Journal of Materials Science.
15. Tu, L. K. L. and Seth, B. B., Journal of Testing and Evaluation, JTEVA, 6, No. 1 (January, 1978), p. 66.
16. Richards, C. E. and Lindley, T. C., Eng.Fract. Mech. 4 (1972), p. 951.
17. Birkbeck, G., et al., J. Mat. Sci., 6 (1971), p. 319.
18. Cooke, R. J., et al., Eng.Fract. Mech., 7, (1975), p. 69.
19. Ritchie, R. O. and Knott, J. F., Acta Met., 21, (1973), p. 639.
20. Waldron, G. W. J., et al., Proceedings of the 3rd Annual SEM Symposium, 1970, p. 297.
21. Braglia, B. L., M. S. Thesis, Lehigh University, 1978.
22. Cooke, R. J., et al., Eng. Fract. Mech., 7 (1975), p. 187.
23. Clark, G., Pickard, A. C., and Knott, J. F., Eng. Fract. Mech., 8 (1976), p. 449.

TABLE I

Welding Parameters

Process: GMA
Electrode and Diameter: Airco A608 (AWS E70S-1b.), 0.89 mm (0.035 in.)
Shielding Gas and Flow Rate: Ar-2%O_2, 0.43 ℓ/sec (55 cf/hr.)
Polarity: DCRP
Voltage: 26 V
Amperage: 170 A
Travel Speed: 5.5-5.9 mm/sec (13-14 in./min.)
Joint Detail: Butt Joint, 2.6 mm (0.085 in.) gap, Cu backup bar for first pass. One pass each side.

TABLE II

CHEMICAL COMPOSITION (wt.%)

	Electrode			VAN-80	
	SPEC AWS E70S-1b.	AIRCO A608 NOMINAL	ACTUAL	SPEC A-656 GRADE 1	ACTUAL
C	.07-.12	.10	*NA	.18 MAX	.12
Mn	1.60-2.10	1.91	1.66	1.60 MAX	1.43
P	.025 MAX	.015	NA	.040 MAX	.030
S	.035 MAX	.016	NA	.050 MAX	.007
Si	.50-.80	.68	NA	.60 MAX	.48
V	--	--	--	.05-.15	.125
Al	--	--	--	.02 MIN	.02
N	--	--	--	.005-.030	NA
Cr	--	--	--	--	.03
Mo	.40-.60	.51	.48	--	.016
Ni	.15 MAX	.056	NA	--	.057

*Not Available

TABLE III

MECHANICAL PROPERTIES

	ELECTRODE		VAN-80		
	SPEC	ACTUAL	SPEC	ACTUAL	
				Longitudinal	Transverse
Tensile Strength, MPa (ksi)	496 min. (72)	NA	655-793 (95-115)	758 (110)	758 (110)
Yield Strength, MPa (ksi)	414 min. (60)	NA	552 min. (80)	634 (92)	655 (95)
Total % Elongation	17 min.	NA	12 min.	25	21.5
Uniform % Elongation	--	NA	--	14.5	13.5

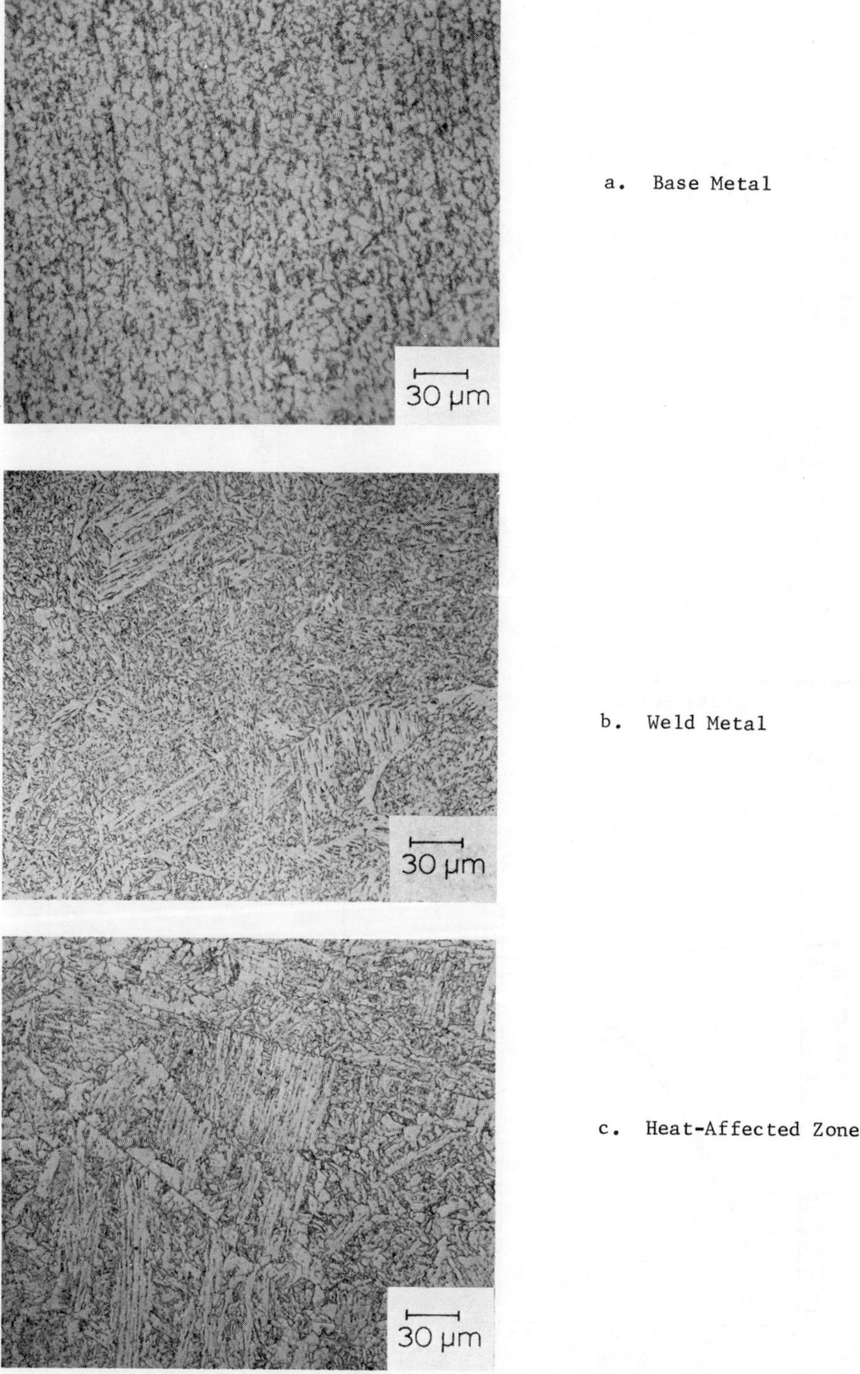

a. Base Metal

b. Weld Metal

c. Heat-Affected Zone

Figure 1. Representative Microstructures of the Three Weld Regions

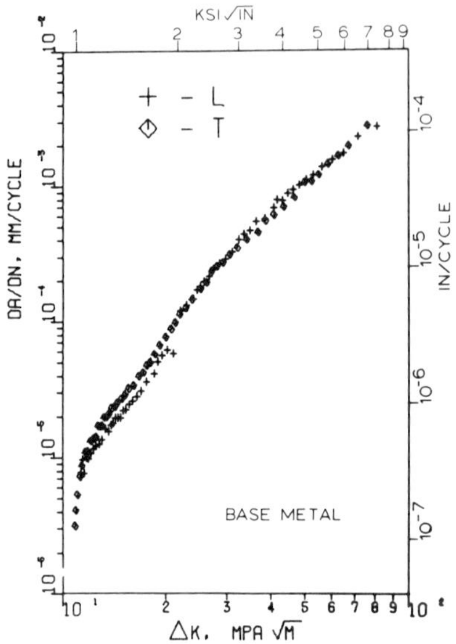

Figure 2. FCP Data for Van-80 Base Metal

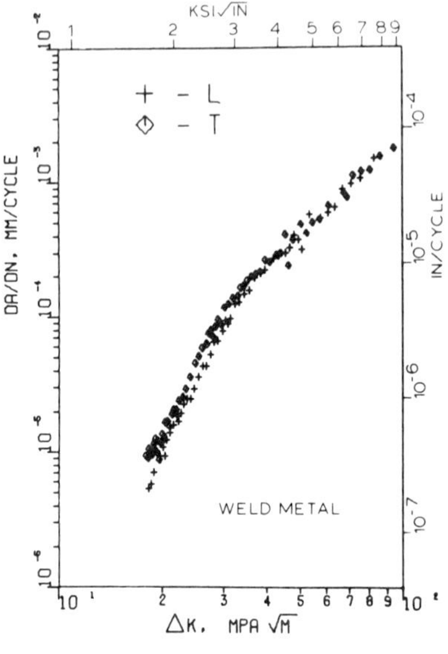

Figure 3. FCP Data for Weld Metal

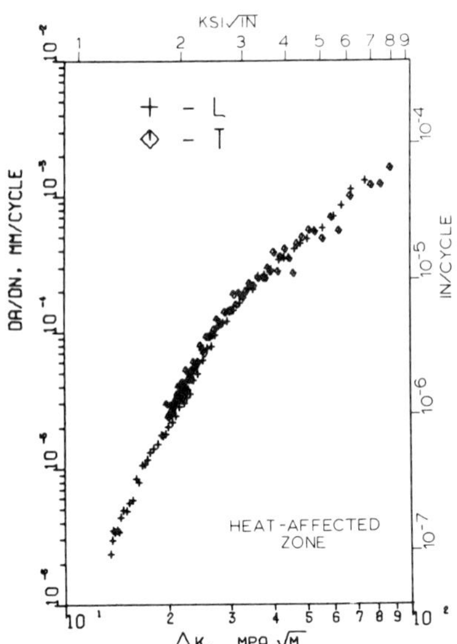

Figure 4. FCP Data for Heat-Affected Zone

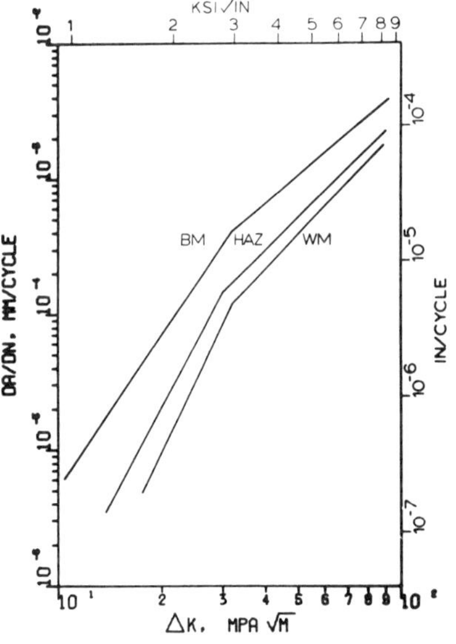

Figure 5. Comparison of FCP Behavior in the Three Weld Regions

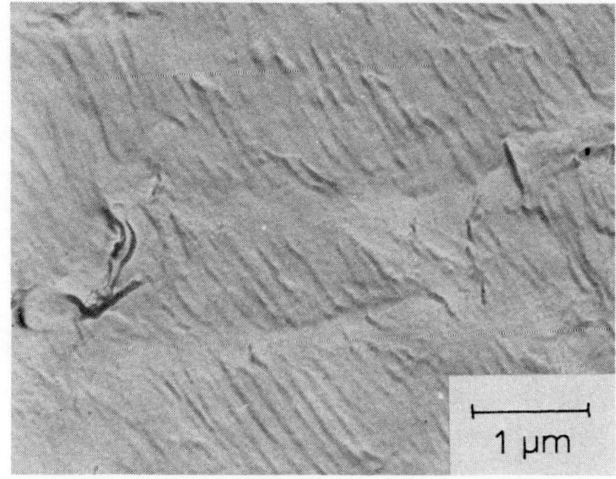

Figure 6. Typical Fatigue Striations.
TEM Fractograph

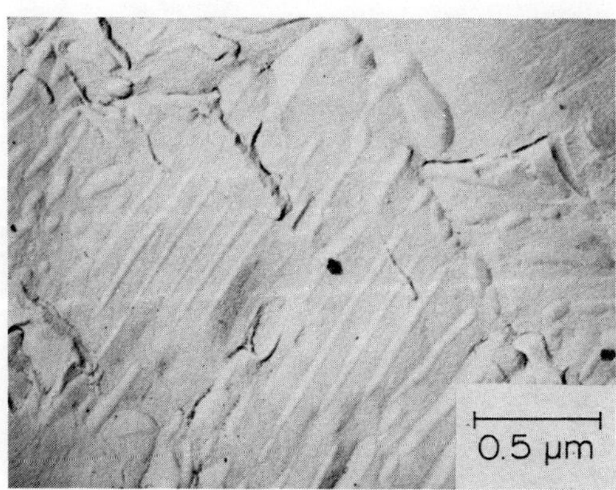

Figure 7. Pearlite Boundary Failure.
TEM Fractograph.

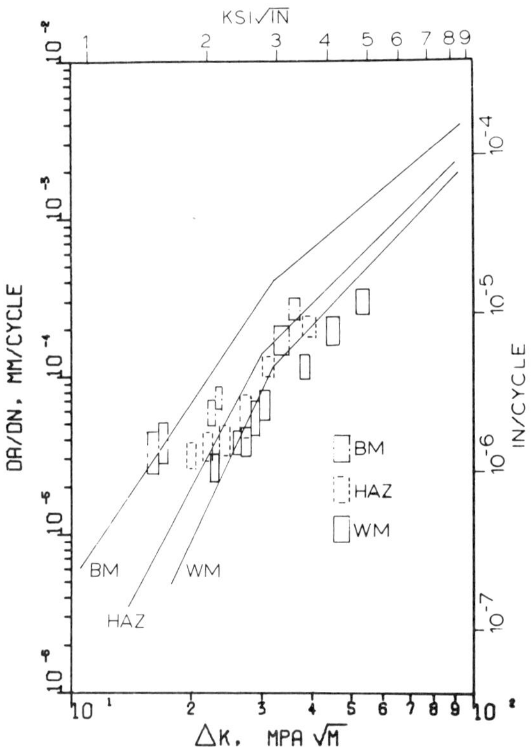

Figure 8. Comparison of Microscopic (Striation Spacing) Data with Measured Crack Growth Rates in the Three Weld Regions.

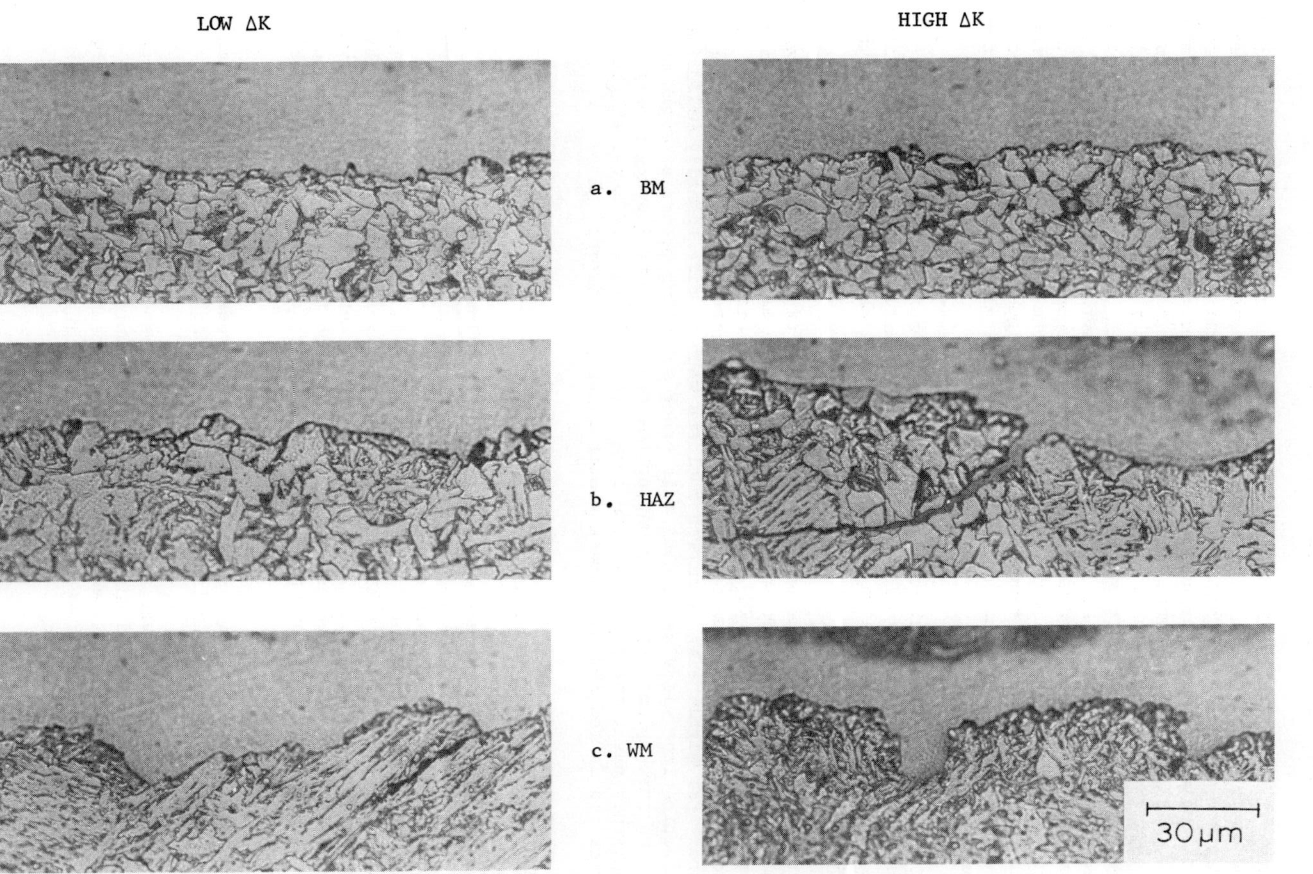

Figure 9. Fracture Profiles in the Three Weld Regions.

DISCUSSION

R. A. Kelsey (Alcoa Labs): Could the orientation-fracture surface texture be related to variation in grain size?

Author: A fatigue crack which is deflected from the nominal crack plane by the grain boundary in a coarse-grained metal ought to result in a larger fracture surface perturbation than in a fine-grained metal. In this sense, the fracture surface roughness could be considered to be related to grain size variations. However, this argument suggests that coarse-grained metals should exhibit lower fatigue crack growth rates than fine-grained metals because of the more tortuous crack paths in the former. Generally, though, fatigue behavior has been found to be indeptndent of grain size except at very low crack growth rates in the threshold regime. (None of our tests were conducted in this region.) We feel that the rougher fracture surfaces and lower crack growth rates in WM and HAZ fatigue tests are a result of the crack branching produced by these asymmetrical microstructures, and not the grain size.

H. C. Rogers (Drexel University): Term "crack bifurcation" is somewhat confusing. One slide very clearly shows crack blunting by transverse splitting in a number of places along crack path, causing the main crack path to be shifted. Comments?

Author: A more accurate but cumbersome phrase might be "microstructurally-induced crack growth retardation." In the case of coarse asymmetric microstructures such as the WM and HAZ regions, the crack frequently splits with each segment being deflected from the nominal crack plane by the microstructure. For highly aligned microstructures such as directionally solidified eutectics (the slide mentioned) this crack splitting occurs at right angles to the crack plane, thereby blunting the crack tip and slowing propagation. In either case, the result is the same - the effective ΔK level at the crack tip is reduced, and crack growth rates are lowered.

L. W. Sandor (Sun Shipbuilding & Dry Dock Co.): What causes tip bifurcation? Could you explain the mechanism involved?

Author: The path which a propagating crack follows is a result of the competition between intra- and intergranular crack growth mechanisms. When the crack reaches a grain boundary, it must choose to propagate either through or around the next grain. This choice is influenced by the relative strengths of the local grain boundary and the specific grain about to be traversed. For example, if such a grain is stronger than the surrounding grain boundary, then the crack would be expected to propagate around one or both sides of the grain; this would result in crack tip deflection or bifurcation.

J. T. McGrath (Dept. of Energy Mines and Resources): In HAZ propagation studies did the crack deviate into weld metal and base metal regions?

Author: We cannot answer this question with absolute certainty. The test specimens were machined such that the starter notch was located in the center of the HAZ or WM, as required. In all tests, the crack grew parallel to the direction of welding; hence, we concluded that the HAZ cracks remained in this region.

QUANTITATIVE MICROSTRUCTURAL STUDIES OF SUBMERGED ARC WELDS IN HSLA STEELS

A.J. Pacey and H.W. Kerr

Department of Mechanical Engineering
University of Waterloo
Waterloo, Ontario, Canada
N2L 3G1

ABSTRACT

A quantitative study was made of the effects of microalloying elements Cb, V and Ti on weld metal microstructure in several weld series. The quantitative microstructural analysis included volume per cent proeutectoid ferrite, acicular ferrite, lath structure, "other", inclusions, and acicular ferrite grain length. This study has emphasized that a statistical variation of microstructure occurs along the length of a weld. Cb was found to reduce the proeutectoid ferrite content and Ti reduced the acicular ferrite grain length in the welds studied. Statistical analysis was employed in an attempt to determine the effects of individual alloying elements on the microstructure. A GTA remelt experiment showed that upon cooling, the following phases occurred:

(1) Proeutectoid ferrite nucleates on γ grain boundaries.
(2) The planar interface between proeutectoid ferrite and γ breaks down, forming ferrite side platelets (lath structure).
(3) Intergranular acicular ferrite completes the gross solid state transformations.

INTRODUCTION

The development of HSLA "microalloy" steels has resulted in a need for understanding the structure and properties of welds made in these materials. For example, it is expected that future Arctic pipelines will utilize low carbon (0.15 wt %) steels having yield strengths of 70 ksi (480 MPa). These large diameter pipes are presently fabricated by submerged arc welding.

Several studies on the properties of the submerged arc welds in candidate steels have been completed. Although it is generally accepted that a fine grained "acicular ferrite" weld metal microstructure is desirable, phases such as proeutectoid (or grain boundary) ferrite, Widmanstatten ferrite sideplates, and upper bainite are also present. In addition, smaller fractions of other phases, such as martensite-austenite complexes, carbides, and inclusions are produced in these welds. The steels often contain 9 or 10 alloying elements. The effects of individual elements on fractions of various phases has only been studied quantitatively in a few cases. Thus it is difficult to compare the structures, and properties, of welds made in different laboratories, even when the welding conditions are similar. This lack of quantitative metallographic data stems first from the difficulty, and expense, of making a series of welds in which the percentage of only one element is changed in order to clarify its effects. Secondly, even when such series are made, the comparison of properties with composition often employs only qualitative studies of microstructural changes.[1-5] In other cases where quantitative amounts of different phases have been recorded, the interpretation is limited by a small number of welds in which only one element is varied, with a given welding heat input.[6-7] Recent work[8] has demonstrated non-linear effects for different elements, so that detailed analysis requires several percentages of a given element, with everything else constant.

This paper reports quantitative microstructural studies of several series of welds in which only one element was varied. The detrimental effects of proeutectoid ferrite, lath structures and "other" phases have been noted, but their relative importance is controversial.[9] Hence, for a given fraction of tough acicular ferrite, the overall toughness may change due to changes in fractions of the remaining constituents. Acicular ferrite toughness is expected to be related to the ferrite grain size within it, but the grain size is usually not measured. Both the effects of composition on structure, and the structure-property relations can only be clarified by quantitative microstructural studies.

EXPERIMENTAL PROCEUDURE

Materials

Weld metal specimens were kindly supplied by two sources. Welds made at Armco Steel, designated "A" series, were sent by R.J. Jesseman. In these welds, only Cb was varied for 4 different Cb steel plates and 3 different alloy wires.[10] Less complete A Series with varying Ti, V, and Cb were also supplied. Welding conditions for the A series welds are fully

TABLE 1 - CHEMICAL COMPOSITION OF SECOND-PASS DEPOSIT IN "A" SERIES SUBMERGED-ARC WELDS

Series	Weld-Ment No.	Carbon	Manga-nese	Sulfur	Silicon	Chrom-ium	Nickel	Molyb-denum	Colum-bium	Titan-ium	Vana-dium	Total Nitro-gen	Oxygen
		CHEMICAL COMPOSITION											
	1	0.11%	1.10%	0.015%	0.32%	0.052%	0.057%	<0.06%	<0.003%	<0.009%	-	0.0069%	0.042%
	2	0.10%	1.05%	0.016%	0.32%	0.050%	0.055%	<0.06%	0.011%	<0.009%	-	0.0059%	0.027%
A-1	3	0.10%	1.06%	0.016%	0.31%	0.049%	0.058%	<0.06%	0.037%	<0.009%	-	0.0073%	0.037%
	4	0.11%	1.01%	0.016%	0.29%	0.076%	0.068%	<0.06%	0.053%	<0.009%	-	0.0093%	0.041%
(Ti)	5	0.060%	1.26%	0.016%	0.36%	0.055%	0.051%	<0.06%	0.045%	<0.009%	-	0.0073%	0.040%
	6	0.060%	1.42%	0.015%	0.35%	0.057%	-	<0.06%	<0.003%	0.054%	-	0.0065%	0.046%
	7	0.11%	1.08%	0.015%	0.34%	0.056%	0.081%	0.065%	0.048%	0.039%	<0.006%	0.0055%	0.037%
	8	0.099%	0.88%	0.015%	0.31%	0.050%	0.080%	0.060%	0.047%	<0.006%	0.059%	0.0057%	0.038%
	9	0.11%	1.13%	0.012%	0.27%	0.063%	0.58%	0.08%	<0.003%	<0.009%	-	0.0066%	0.038%
	10	0.10%	1.07%	0.012%	0.26%	0.063%	0.58%	0.08%	0.012%	<0.009%	-	0.0061%	0.042%
A-2	11	0.10%	1.09%	0.013%	0.29%	0.062%	0.68%	0.09%	0.033%	<0.009%	-	0.0073%	0.028%
	12	0.10%	1.03%	0.012%	0.27%	0.093%	0.72%	0.09%	0.044%	<0.009%	-	0.0067%	0.044%
(Ti)	13	0.062%	1.29%	0.012%	0.34%	0.065%	0.64%	0.09%	0.040%	<0.009%	-	0.0066%	0.049%
	14	0.060%	1.38%	0.012%	0.34%	0.058%	0.60%	0.09%	<0.003%	0.054%	-	0.0070%	0.047%
	15	0.11%	1.32%	0.010%	0.34%	0.16%	0.70%	0.17%	0.004%	<0.009%	-	0.0083%	0.040%
	16	0.10%	1.24%	0.011%	0.32%	0.16%	0.66%	0.17%	0.012%	<0.009%	-	0.0076%	0.035%
A-3	17	0.10%	1.29%	0.011%	0.33%	0.16%	0.70%	0.16%	0.033%	<0.009%	-	0.0076%	0.041%
	18	0.11%	1.18%	0.011%	0.31%	0.18%	0.60%	0.15%	0.053%	<0.009%	-	0.0080%	0.036%
(Ti)	19	0.058%	1.47%	0.012%	0.40%	0.16%	0.69%	0.19%	0.049%	<0.009%	-	0.0085%	0.038%
	20	0.058%	1.60%	0.010%	0.38%	0.16%	0.67%	0.16%	<0.003%	0.043%	-	0.0087%	0.037%
	21	0.10%	1.27%	0.010%	0.34%	0.15%	0.65%	0.17%	0.051%	0.044%	<0.006%	0.0073%	0.037%
	22	0.10%	1.15%	0.010%	0.34%	0.17%	0.75%	0.19%	0.047%	<0.006%	0.062%	0.0076%	0.039%

described elsewhere.[10] The effect of Ti and Cb has been investigated individually in weld series supplied to us by the Steel Company of Canada[11] and are designated as "S" series. These welds were made using AC tandem arc, using 5/32" (4mm) wire, 750-1000 amps at 28-36 volts for the lead wire and 1/8" (3.2mm) wire, 500-750 amps at 32-38 volts for the trailing wire, and a travel speed of 20-30 in/min (51-76 cm/min). The compositions of the A and S welds are given in Tables 1 and 2 respectively.

TABLE 2

CHEMICAL COMPOSITION OF SECOND PASS DEPOSITS IN "S" SERIES

Series	Weld-Ment No.	C	P	S	Mn	Si	Cu	Ni	Cr	Mo	V	Cb	Ti	Acid Soluble Al	Ce	O_2	N_2
S-1	23	0.13	0.009	0.012	1.36	0.24	0.136	0.230	0.048	0.149	<0.005	<0.005	0.0	<0.005		0.0413	0.009
	24	0.14	0.007	0.011	1.34	0.25	0.133	0.225	0.048	0.151	<0.005	0.018	0.0	<0.005		0.0283	0.010
	25	0.13	0.007	0.010	1.33	0.23	0.142	0.225	0.046	0.146	<0.005	0.03	0.0	<0.005		0.0302	0.010
S-2	26	0.110	0.009	0.005	1.41	0.42	0.211	0.104	0.07	0.217	0.052	0.023	0.007	0.014	.021	0.051	0.008
	27	0.105	0.012	0.005	1.41	0.36	0.190	0.110	0.064	0.236	0.045	0.019	0.016	0.015	.020	0.065	0.008
	28	0.101	0.010	0.005	1.45	0.39	0.195	0.108	0.067	0.231	0.048	0.021	0.022	0.014	.021	0.064	0.005
S-3	29	0.053	0.009	0.007	1.71	0.31	0.062	0.235	0.058	0.478	0.006	0.03	<0.005	0.014	.004	0.041	0.004
	30	0.053	0.009	0.006	1.68	0.33	0.061	0.230	0.057	0.468	<0.005	0.031	0.007	0.013	.004	0.059	0.006
	31	0.054	0.01	0.007	1.60	0.32	0.056	0.233	0.056	0.477	<0.005	0.031	0.015	0.013	.004	0.043	0.002
S-7	32	0.056	0.008	0.014	1.66	0.41	0.132	0.361	0.073	0.324	<0.003	0.031	0.005	0.011	<.002	0.038	0.006

Micro-Structural Analysis

Polished weld specimens were deeply etched in 2% nital to emphasize the microstructural phases. The specimens were viewed on a ground glass screen of a Zeiss optical microscope equipped with a 5 x 5 cross-hair grid. The microstructural phase at a cross-hair was identified at 1000 locations for each specimen, using a 20 x 50 array along the centre line of the 2nd weld bead, approximately 2mm below the surface of the plate.

Four phases were identified; proeutectoid ferrite (PF), lath structure (LS), acicular ferrite (AF), and "other". "Other" consists of blocky carbides and martensite-austenite complexes. Similar classifications have been suggested by other authors.[12,13] Reproducibility of the microstructure along a weld was investigated for the A series containing columbium by examining 3 sections along each weld: opposite sides of a section approximately 18 mm thick, and another, stress-relieved, section. The resulting error bars are shown in Figs. 1-4. The absolute error was generally greater for phases whose volume fraction was greater: Table 3. This exercise was not repeated for the other welds, but similar errors can be expected in welds made in other laboratories. When the volume fraction of one phase approaches 100%, the absolute errors are expected to decrease; i.e. maximum error is expected for phase fractions of about 50%.

TABLE 3

SUMMARY OF STANDARD ERRORS IN THE MICROSTRUCTURAL ANALYSIS

Microstructural Phase	Range of Volume %	Range of Error (Volume %)
PF	1.5 - 20	0.6 - 2.8
AF	35 - 55	4.1 - 8.2
LS	30 - 46	3.8 - 11.3
Other	6 - 8	0.3 - 1.6

The effect of magnification was also examined in two weld sections by doing point counts at 352X and 800X magnification. The volume percentages recorded were within ± 2 volume percent. Hence, at least for this range of magnification the observations on a given specimen are fairly reproducible. The S series welds were examined at 800X, and the A series welds at 352X and/or 800X.

Acicular Ferrite Grain Length (AFGL)

Optical Micrographs were taken of acicular ferrite grains in all of the welds at 600X. The length of at least 20 of the longest grains were measured and averaged.

Inclusion Analysis

Transverse sections of the welds were polished, using standard procedures, to a 0.06μm finish. The as-polished specimens were viewed on the centreline of the 2nd pass of each weld using a Zeiss optical microscope coupled to a Q.T.M. Model 720. The inclusion size distributions and area fractions were recorded over an area of 15,500 square microns. Earlier unpublished results employing negatives rather than specimens on the Q.T.M. were found to consistently underestimate the inclusion content.

RESULTS

Quantitative Metallography

The full results of the quantitative metallography are given in Table 4 for the A series welds and Table 5 for the S series welds. In order to examine the effects of individual elements, the effects of Cb and Ti

TABLE 4

SUMMARY OF MICROSTRUCTURAL ANALYSIS FOR THE "A" SERIES SUBMERGED ARC WELDS

Specimen Number	Microstructural Phase (Volume %) PF Avg.	PF Error*	AF Avg.	AF Error*	LS Avg.	LS Error*	OTHER Avg.	OTHER Error*
1	19.8	2.4	34.9	4.5	37.7	6.4	7.6	0.49
2	14.4	2.2	38.9	4.8	39.3	7.0	7.4	0.6
3	8.6	2.8	50.8	4.2	37.9	7.3	6.1	0.8
4	9.7	0.7	40.7	5.5	42.2	6.9	7.1	1.0
5	9.0		54.9		31.2		4.9	
6	3.7		46.5		44.0		5.8	
7	11.4		64.1		19.8		4.7	
8	14.1		32.7		46.1		7.1	
9	7.7	0.4	39.2	8.2	46.1	7.6	7.0	1.0
10	8.2	1.8	54.6	5.0	30.3	3.8	6.5	0.65
11	7.1	4.7	55.6	8.2	30.9	11.3	6.4	1.6
12	3.6	0.9	47.7	4.1	40.5	4.0	8.2	0.64
13	4.8		48.4		41.3		5.5	
14	6.6		70.5		18.3		4.6	
15	5.7	3.8	41.6	4.5	45.3	7.6	7.6	0.7
16	2.4	0.6	48.5	5.0	41.3	5.2	7.8	0.5
17	1.4	0.8	51.3	5.0	39.1	6.1	8.2	0.3
18	1.5	1.0	50.0	6.2	45.0	7.5	7.5	0.3
19	1.0		46.4		47.0		5.6	
20	5.2		81.7		9.9		4.2	
21	10.4		74.2		10.2		5.2	
22	1.2		44.3		48.1		6.4	

* No error indicates only one measurement.

on the fractions of various constituents are plotted in Figs. 1-5 and 8-13 respectively. The acicular ferrite grain lengths (AFGL) are plotted for the different weld series as a function of Cb and Ti in Figs. 6 and 12 respectively.

1) Effects of Columbium

The effect of columbium on the % proeutectoid ferrite (PF) is shown in Fig. 1. Several authors[2-4,10,13,14] have stated that increasing Cb decreases PF. Fig. 1 confirms this observation for series A-1. However, with a higher content of other alloying elements, the effect was reduced, and became insignificant in some series. Large scatter in PF content was observed in some series A welds. This emphasizes the unreliability of using a single section for determining the effects of composition on microstructure.

TABLE 5

SUMMARY OF MICROSTRUCTURAL ANALYSIS FOR THE "S" SERIES SUBMERGED-ARC WELDS

Weldment No.	Microstructural Phase (Volume%)			
	PF	AF	LS	OTHER
23	7.2	70.3	10.6	11.9
24	6.9	72.5	12.2	8.4
25	3.6	70.8	14.8	10.8
26	6.2	79.2	13.3	1.3
27	4.8	84.5	8.3	2.4
28	6.3	80 5	8.9	4.3
29	1.5	85.6	7.3	5.6
30	2.4	82.8	8.3	6.5
31	3.9	79.0	10.5	6.6
32	4.9	57.9	30.6	6.6

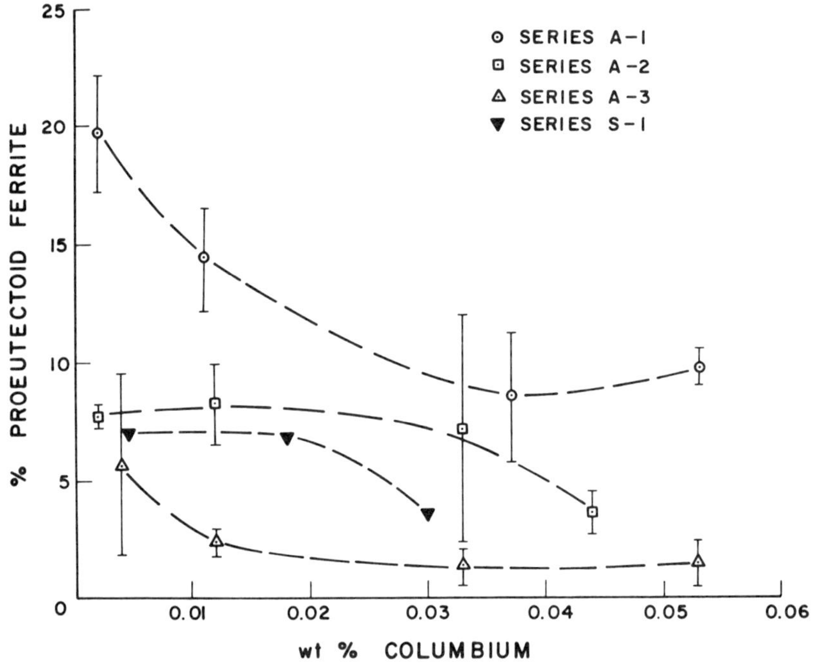

Figure 1 Effect of Cb on the Proeutectoid Ferrite Volume Percent

The % acicular ferrite (AF) is shown as a function of Cb in Fig. 2. Sija et al[2] reported that Cb increases AF. Fig. 2 suggests that in the A series, increasing Cb at first increased AF, but later decreased it. This produced a maximum AF, but the optimum % Cb depended on other alloying elements. Less effect of Cb on AF was found in the S series, which contained more AF than any of the A welds.

Jesseman[10] indicated that as the Cb content was raised, the upper bainite fraction increased. There is some controversy[13] about whether this constituent should be called upper bainite or Widmanstatten side-plates.[12] Both possibilities are included within what we have called % lath structures (LS). No consistent effect of Cb on LS is evident from

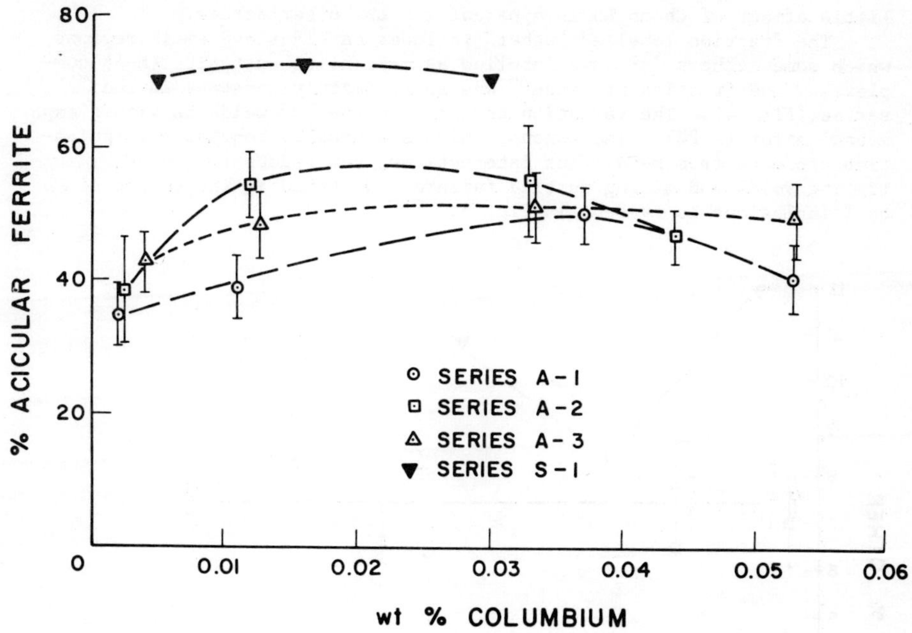

Figure 2 Effect of Cb on the Acicular Ferrite Volume Percent

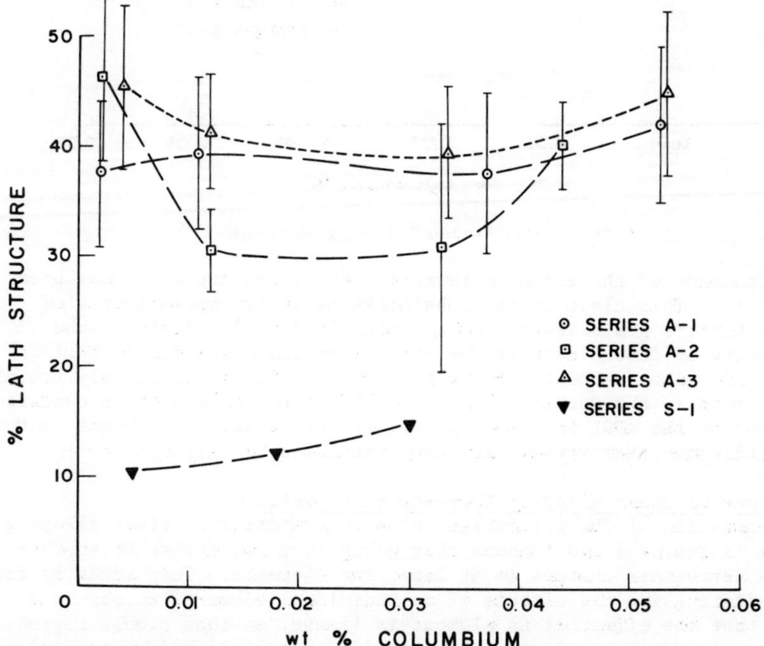

Figure 3 Effect of Cb on the Lath Structure Volume Percent

Fig. 3: although series A-2 showed an apparent minimum in average LS at intermediate % Cb (corresponding to the maximum in AF in Fig. 2), the error bars suggest that this effect is not statistically significant. Little effect of Cb on LS is apparent for the other series.

The fraction labelled "other" includes carbides and small regions which some authors[6,15] have labelled as martensite-austenite (M-A) complexes. The fraction of "other" was approximately constant in the "A" series (Fig. 4). The variation apparent in the "S" welds is within experimental error (± 2%). Inclusions, which are deoxidation and desulphidation products from metal-flux interactions, can reduce toughness by initiating voids and aiding ductile failure. No trend in the effect of Cb on % inclusions is evident in Fig. 5.

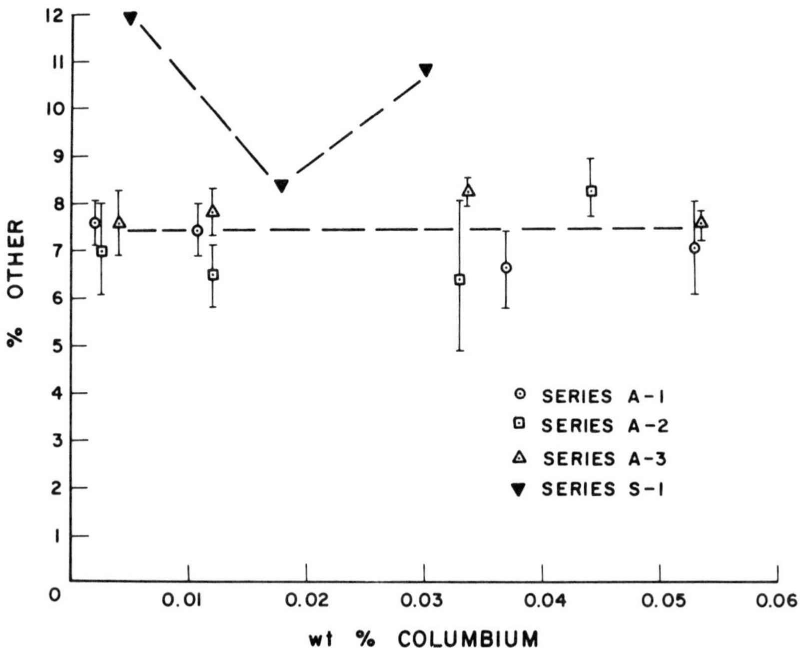

Figure 4 Effect of Cb on the "Other" Volume Percent

Refinement of the acicular ferrite grain sizes due to Cb has been reported[4,6]. This claim is not substantiated by our measurements of acicular ferrite grain length (AFGL) shown in Fig. 6. Instead, the AFGL appears to depend more on the other alloying elements, in that increasing the alloy content reduces the AFGL. Unfortunately, only two of the specimens in the series A-3 were available in the as-welded condition. Comparison of the AFGL in as-welded and stress-relieved specimens of this series indicated that stress-relieving increases the apparent AFGL.

2. Effects of Other Alloying Elements in Cb Welds

Examination of the percentages of more conventional steel alloying elements in Tables 1 and 2 shows that going from one series to another involves percentage changes in at least two elements. This makes it diffcult to distinguish the effects of an individual element, unless it is assumed that the effect of an element is linear, so that simple regression analysis can be employed. Linear effects are assumed by traditional carbon equivalent (CE) equations used to measure hardenability and thus susceptibility to cold cracking. Two such equations are the Dearden and

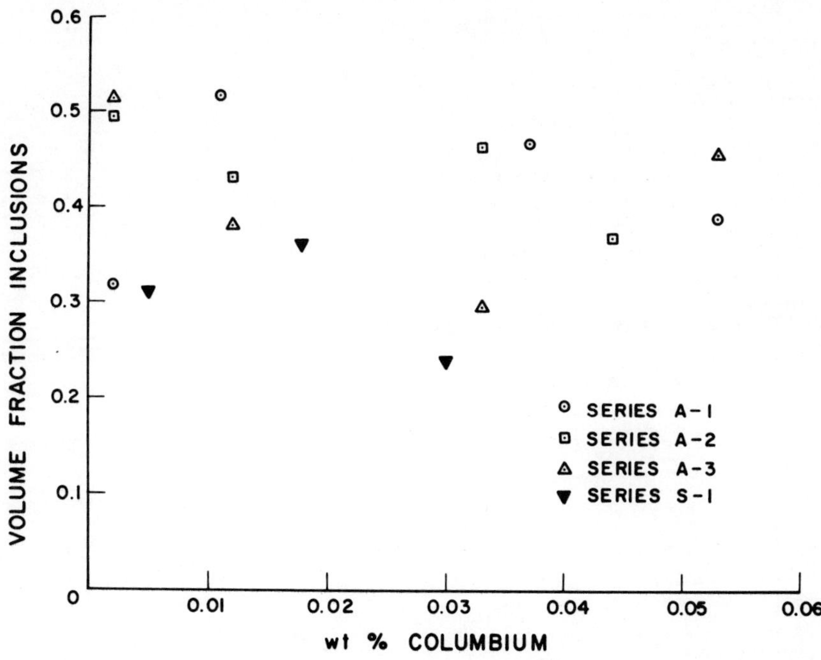

Figure 5 Variations in Inclusion Volume Fractions in Welds Containing Cb

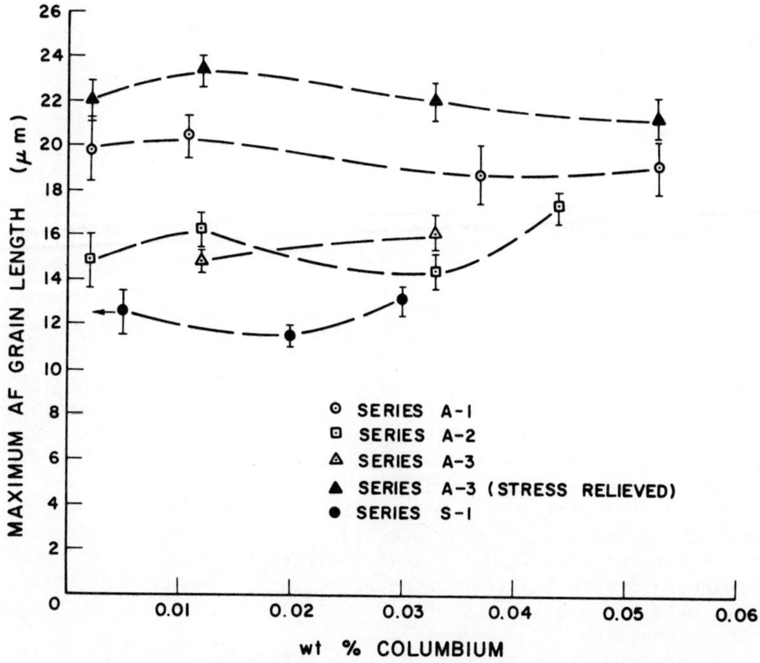

Figure 6 Effect of Cb on the Maximum Acicular Ferrite Grain Length

O'Neill equation:

$$CE = C + \frac{Mn}{6} + \frac{Cu + Ni}{15} + \frac{Cr + Mo + V}{5} \quad (1)$$

and the Ito and Bessyo equation:

$$CE = C + \frac{Mn}{20} + \frac{Si}{30} + \frac{Cu}{10} + \frac{Ni}{60} + \frac{Mo}{15} + \frac{V}{10} \quad (2)$$

Although these expressions are too simple to account for all of the differences between the welds of different series, they permit a rapid measure of the amount of alloying for each series. Both the total CE and CE due to alloying elements (i.e. not due to carbon itself) are shown in Fig. 7 for all of the series. The increased alloy content in going from series A-1 to series A-3 is reflected in increased CE values. Depending on which CE is used, the series S-1 has either a slightly higher or slightly lower CE than series A-3 but it is less alloyed.

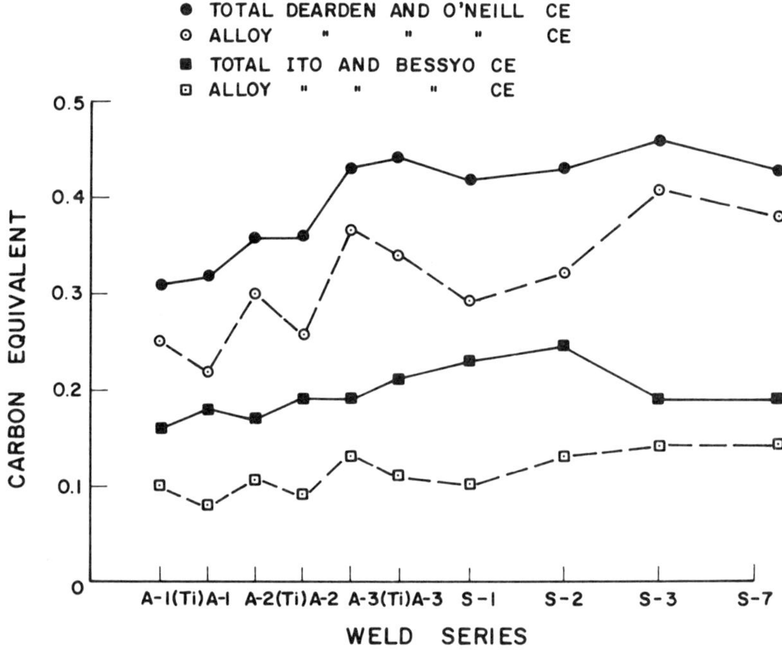

Figure 7 Schematic Representation of Two Carbon Equivalents for the Titanium and Columbium Bearing Welds

Re-examination of Figs. 1-6 for the effects of these alloying elements shows some trends. As the alloy content increased, PF decreased (Fig. 1). This is consistent with increased hardenability, as predicted by the CE equations. AF increased with increasing alloy content, until series A-3 when a decrease was noted (Fig. 2). Series S-1 contained considerably more AF: it contained less Ni, Si and Cr, and more C and Mn than series A-2 and A-3 but was intermediate in Mo content. Ito and Nakanishi (8) have previously noted that intermediate Mo contents give maximum toughness due to high AF contents. Clearly, careful optimization of weld metal chemistry, not predictable by the linear CE equations, is required to maximize AF.

The low fraction of LS evident in all S series welds in Fig. 3 is consistent with their higher AF discussed above. Similarly, although there is considerable scatter, the A series show an apparent decrease, then increase in LS going from series 1 to 3, at intermediate Cb percentages. Both Cr and Mo were increased by approximately 100% in going from series A-2 to A-3. Therefore, the effects of the individual elements cannot be separated.

Little effect of alloying is apparent on the fraction of "other" phases in Fig. 4, or on the volume fraction of inclusions in Fig. 5.

Fig. 6 shows a general decrease in AFGL with increased alloy content, consistent with lower transformation temperatures and high nucleation rates within the acicular ferrite. The minimum values of AFGL were found in the S series welds, despite it having less alloy CE than Series A-3 (Fig. 7). This again illustrates the optimization of alloying elements required for maximizing toughness.

3. Effect of Titanium

The proeutectoid ferrite percentages measured on Ti containing welds are shown in Fig. 8 as a function of Ti content. It has been suggested that increased Ti decreases PF. Little effect of Ti on PF is seen in series S-2 welds. In series S-3 welds PF appears to increase with Ti. In the A series welds the situation is complicated by two levels of Cb. Since increased Cb tends to decrease PF (Fig. 1), higher PF levels, in the absence of Cb, might be expected in Fig. 8 for the A series welds containing <0.01% Ti. Thus the apparent increase in PF with increased Ti in series A-2 and A-3 might simply be an effect of Cb. However, PF decreased in series A-1 (Fig. 8) with increasing Ti, despite the change in Cb level.

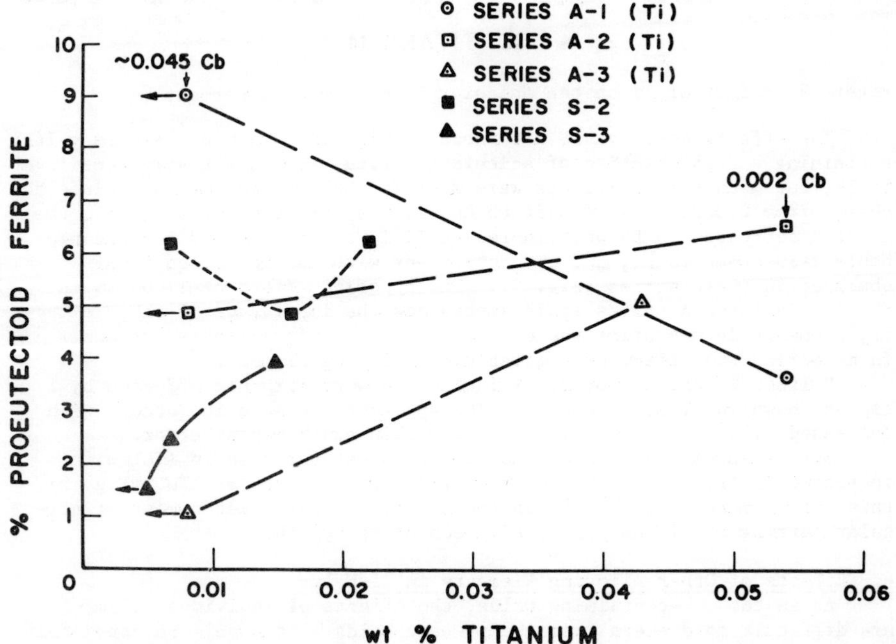

Figure 8 Effect of Ti on the Proeutectoid Ferrite Volume Percent

Fig. 9 shows AF as a function of Ti. Large increases in AF due to increased Ti can be seen in series A-2 and A-3, consistent with the qualitative statements of Jesseman[10]. Since Cb tended to increase AF in other

A series (Fig. 2), even larger increases in AF with Ti might be expected with constant Cb. On the other hand, the S series welds exhibited 80-85% acicular ferrite, independent of Ti content, for Ti levels up to 0.022%.

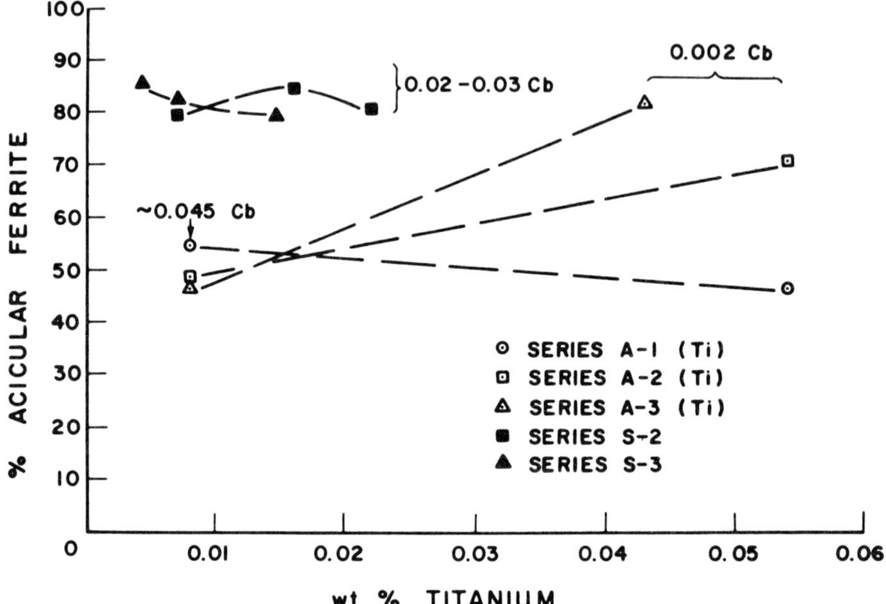

Figure 9 Effect of Ti on the Acicular Ferrite Volume Percent

The effects of Ti on LS are shown in Fig. 10. In the S series welds, containing a high fraction of acicular ferrite, LS was not very sensitive to Ti, but much larger effects were seen in the A series welds. Since the change from 0.002% Cb to 0.045% Cb had little effect of LS (Fig. 3), the drastic decreases in LS with increased Ti in series A-2 and A-3 are probably mainly due to Ti, and are consistent with the increases in AF observed in these series (Fig. 9). The opposite effect of Ti on LS in the less alloyed A series again emphasizes the importance of other alloying elements in measuring the effect of a given microalloying elements in measuring the effect of a given microalloying element.

Neither % "other" nor the % inclusions were strongly affected by % Ti, as shown in Figs. 11 and 12. The apparent increase in "other" with increased Ti in S series (Fig. 11) is within experimental error.

Our measurements of AFGL show a consistent decrease in AFGL with increased Ti (Fig. 13). Since Cb showed little effect of AFGL (Fig. 6), this effect must be due to Ti in the A series welds. Refinement of acicular ferrite by Ti has previously been noted by others.[1,8]

4. Effects of Other Alloying Elements in Ti-Welds

As in the Cb-containing welds, the effects of individual elements are difficult to discern from the present welds. It should be noted that the A weld series containing Ti were slightly more alloyed than their Ti-free counterparts (Fig. 7). Hence, direct comparisons with Ti-free welds are not possible, but certain effects are worth noting.

Unlike the Cb series, where more alloyed welds containined less PF, (Fig. 1), the addition of >0.04 Ti resulted in 5 $\pm$ 1.5% PF in all 3 A series (Fig. 8). At lower Ti contents, more alloyed welds generally

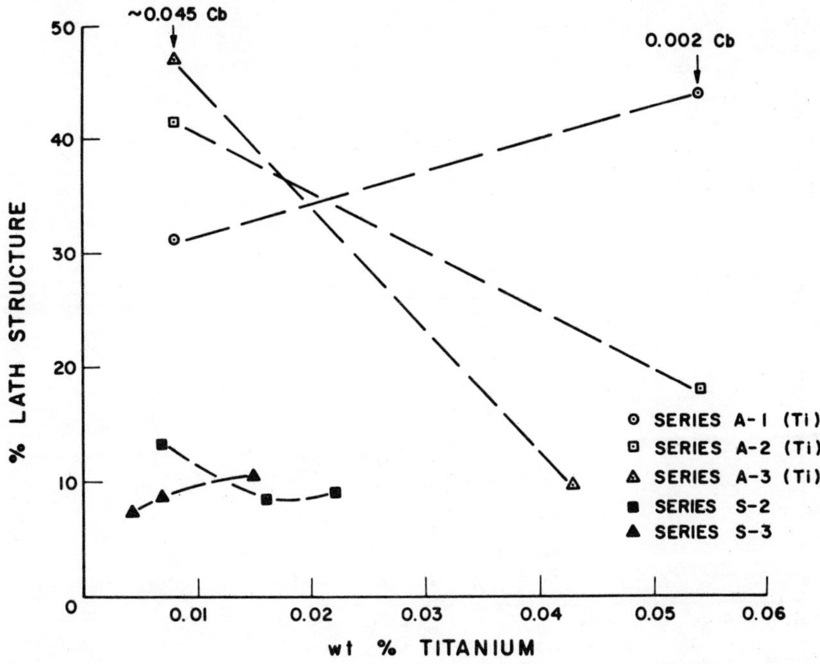

Figure 10 Effect of Ti on the Lath Structure Volume Percent

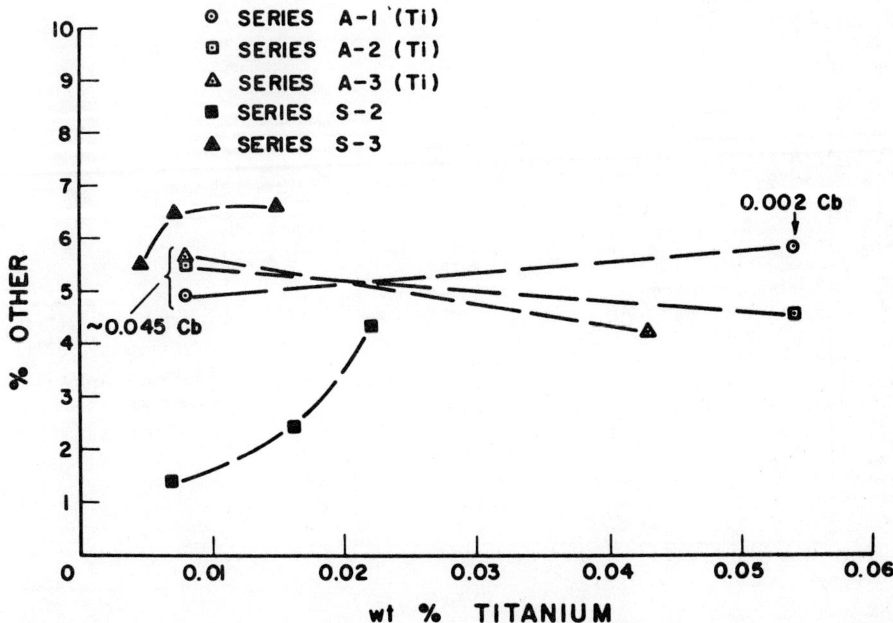

Figure 11 Effect of Ti on the "Other" Volume Percent

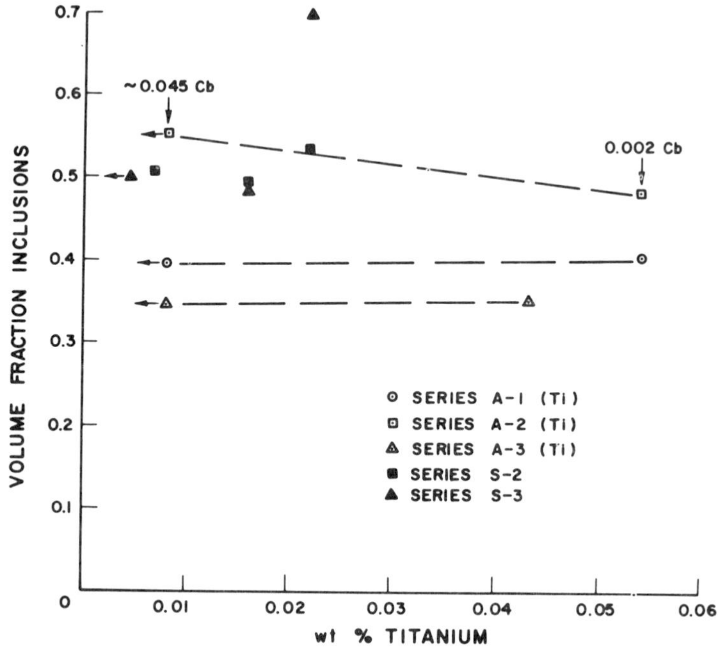

Figure 12 Variations in Inclusion Volume Fractions in Titanium Weld Series

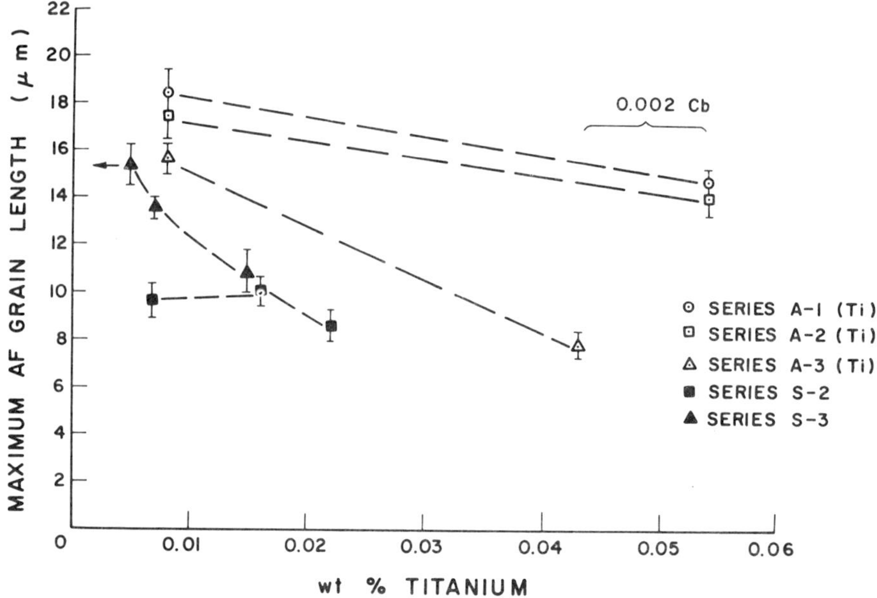

Figure 13 Effect of Ti on the Maximum Acicular Ferrite Grain Length

resulted in less PF (Fig. 8). Hence, the effects of alloying elements on PF expected from hardenability or CE arguments can apparently be modified by additions of Ti.

The importance of balancing the additions of different elements to achieve high AF shown earlier in Fig. 2, is again evident in Fig. 9. In the A series welds, AF > 80% was achieved by the addition of higher Ni, Cr, and Ti contents, but lower Mo content, than in the S series welds. Since only two Ti levels were used in the A series, it is not known whether peaks in AF might occur at intermediate Ti levels, analogous to the effects of Cb. No strong trend is apparent in comparing the two S series, despite doubling the Mo content.

The effects of alloying elements on AF in Fig. 9 are mirrored in Fig. 10: consistently low values of LS in the S series welds with no measurable effect of alloy content, and in the A series welds the lowest LS value corresponds to the weld containing the highest alloy content and a high Ti level. More work to explore intermediate levels of Ti is in progress.

The lower values of "other" in series S-2 compared to series S-3 and the A series (Fig. 11) are at first suprising, since series S-2 has the highest carbon content, which should produce the highest percentage of carbides. Similarly, series S-2 is intermediate in both total and alloy CE (Fig. 7) so that the M-A complex fraction should be intermediate. However, the difference between % "other" in this series and other series is close to the limits of experimental error (± 2% per specimen plus approximately ± .6% within a weld, Table 3) and may not be significant.

As in the Cb-containing welds, a trend to decreased AFGL with increased alloy content is apparent in Fig. 13. Larger AFGL's were found in series S-3 than in series S-2, especially at low Ti contents, despite series S-3 having a large CE according to the Dearden and O'Neill equation (Fig. 7). Fig. 7 shows, however, that Ito and Bessyo would calculate a lower CE for series S-3, comparable to series A-3 (Ti). At the lowest Ti contents AFGL values for series S-3 and series A-3 (Ti) were very similar. This suggests that equation (2) is a better measure of acicular ferrite nucleation than equation (1) but neither formula incorporates Ti.

5. Effects of Vandium

Vanadium is another microalloying element whose effects on microstructure and properties of welds has drawn considerable interest[1-3,7,9,10,13] Jesseman added it to only three A series welds (Table 1): the microstructural effects are given in Table 4.

In series A-1 an increase in V from 0.005 to 0.059%, with a simultaneous decrease in Ti from 0.039% to 0.005% (welds 7 and 8), resulted in a large increase in LS and a large decrease in AF, with little effect on PF. This is opposite to the decrease in LS with increased V noted by Levine and Hill (9,13). The concurrent decrease in Ti might account for the increase in LS, but Fig. 10 shows that for series A-1 (unlike the other A series) LS increased with increasing Ti. Specimens 4 and 8 of the A series have similar Cb (∿0.05) and Ti (<0.01) levels but differ considerably in V (Table 1). Increased V has only caused slight increases in LS and PF and decreases in AF.

In series A-3, comparison of welds 17 and 22 also shows an increase in LS and decrease in AF with increased V content. The Cb levels are also different in these welds (0.033 vs 0.045%), but this difference had little effect on LS or AF in V-free welds of series A-3 (Figs. 2 and 3).

Thus both series A-1 and A-3 showed increases in LS and decreases in AF with increased V, contrary to the observations of others[7,9,13]. It has also been claimed that increasing V decreased PF, similar to the effects of Cb[2]. This statement is not supported by our observations; if anything, V additions tended to increase the PF content.

It should be noted that the V additions in the A series welds were large compared to Nb additions, especially in atomic percent. Interesting effects may occur at intermediate V levels, more consistent with the observations of others. Series S-2 contains ∿0.05 wt % V, but its effects cannot be separated from those of Mo, Ni and C.

6. Regression Analyses

The above presentation of the effects of alloying elements on microstructure is not simple. The balance of different alloying elements to achieve a high fraction of acicular ferrite has been noted. At the same time, interpretation of the effects of individual elements is hampered by simultaneous changes in the percentages of different elements. Regression analysis is sometimes useful in sorting out the effects of individual elements, and was attempted here for the A series welds. Since different heat inputs were used for the S and A series welds, direct comparison is not possible.

Ideally, the effects of individual elements are linear and additive, such as in the carbon equivalent equations employed above as a measure of alloying. All of the A series compositions were used to construct equations relating composition to different microconstituents. The results are given in Table 6 for the elements which were found to have a significant effect at 95% confidence level. The multiple correlation coefficient, R^2, is a measure of the fit of the data to the equation: $R^2 = 1.0$ when the fit is perfect.

TABLE 6

SUMMARY OF REGRESSION ANALYSIS ON ALL "A" SERIES WELDS

$$PF = 0.16 - 0.3\,Mn + 1.09\,Si - 0.43\,Mo - 1.28\,Cb \quad (R^2 = 0.60) \tag{3}$$

$$AF = 0.36 + 0.14\,Ni + 5.28\,Ti \quad (R^2 = 0.39) \tag{4}$$

$$LS = 0.431 - 4.55\,Ti \quad (R^2 = 0.22) \tag{5}$$

$$AFGL = 0.15 - 0.21\,Mn + 0.72\,Si + 0.79\,Cr - 0.53\,Mo - 0.80\,V - 0.64\,Cb - 1.28\,Ti + 1.42\,O_2 \quad (R^2 = 0.78) \tag{6}$$

Some of the trends discussed earlier are present in these equations. For example, equation (3) indicates that increasing Cb decreases PF, as shown in Fig. 1. The increase in AF and decreases in LS and AFGL with increasing Ti are indicated by equations (4) - (6). Effects due to other alloying elements, such as Ni, Mn, Mo etc., are also indicated by some of the equations. This latter information, which was not apparent from the graphs, is a useful guide in planning new experiments, but must be treated with caution. It is apparent for all equations that the statistical equations are not perfect ($R^2 \neq 1$). Recent statistical interpretation of composition-property relationships employs cross product terms of two elements[16]. Unfortunately, the effects of Cb and Ti were not studied in that work. More welds are required in order to make meaningful analysis of possible cross product terms in the present series.

Nevertheless, the apparent increase in PF due to Si, and the indicated increases in AFGL due to Si, Cr and O_2 should all be detrimental to toughness, suggesting that these elements should be limited.

Although equation (3) indicates that increasing Cb decreases PF, Fig. 1 shows that the exact effect of Cb depends on the other alloying elements present. The effect of Cb on PF was examined by regression analysis for each A series. The results are shown in Table 7. As expected from Fig. 1, equations (7) to (9) indicate decreasing influence of Cb on PF with increasing alloy content. Thus equation (3) must be regarded as only generally indicating the effect of Cb on PF.

TABLE 7

SUMMARY OF REGRESSION ANALYSES OF EFFECT OF Cb on PF in "A" SERIES WELDS

Series A-1

$$PF = 20.8 - 643.3\ Cb + 9222.9\ Cb^2 \quad (7)$$

$(R^2 = 0.53)$

Series A-2

$$PF = 8.51 - 125.1\ Cb \quad (8)$$

$(R^2 = 0.55)$

Series A-3

$$PF = 4.25 - 67.8\ Cb \quad (9)$$

$(R^2 = 0.23)$

DISCUSSION

Two statistical effects are worth emphasizing. Even when quantitative microstructural analyses have been carried out, other authors have usually only examined one section of a weld. Within a given microsection the microstructural analysis is only reproducible to approximately $\pm$ 2% for any one phase. Smaller differences are therefore not significant. In addition, variations along a weld can be considerable. (Table 3). The total error shown in Table 3 includes error due to (i) structural variation along a weld in its as-welded condition and (ii) structural changes introduced by stress relieving. Although the microstructure observed by optical microscopy was not expected to change due to stress relieving, consistent decreases in PF and AF and increases in LS and "other" were found. Thus Table 3 overestimates errors due to structural variation along a weld in its as-welded condition. Detailed investigation of this variation was not possible, since only one section of each as-welded specimen was available. Examination of both sides of one specimen approximately 15 mm thick, showed differences of up to 11.5 volume percent AF, illustrating that real structural variations do occur along a given weld. This further increases the required differences of microstructure between different welds in order to be statistically significant.

In several qualitative microstructural studies, general statements about the effect of a given element on a given microconstituent have been made. Our results show that the temptation to extrapolate the results of one alloy series to another should be resisted. For example, Fig. 1 shows a general decrease in PF with increased Cb, but this effect is modified by other alloying elements, as shown by equation (7) - (9) (Table 7). Similarly, some of our results contradict the effects of a given element on a given microconstituent observed elsewhere. An example is the effect of vanadium on LS. An apparent increase in LS with V was found in the present welds, opposite to the effect of V noted by Levine and Hill[7]. In other cases, the addition of both conventional alloying elements and microalloying elements can either have little effect, or can result in a maximum or minimum in a certain phase. For example, Fig. 9 shows little effect due to the doubling of Mo content in series S-2 and S-3, whereas in Fig. 2 the high value of AF in series S-1 may be due to a Mo content intermediate to those of series A-2 and A-3.

The effects of various elements on microstructure are clearly illustrated by continuous cooling-transformation (CCT) diagrams for the different alloys. Levine and Hill[9,13] have used GTA welding to obtain the transformation temperatures of the microconstituents in some submerged arc welds. Unfortunately, sufficient weld material to carry out this type of experiment was only available for the least alloyed S series weld. The GTA remelt weld clearly showed that proeutectoid ferrite appeared first, originating at austenite grain boundaries at approximately 700°C.

Next some regions of proeutectoid ferrite broke down to form parallel side plates or laths (Fig. 14). Finally, the interior of the grain transformed to acicular ferrite, blocking further lath growth. This sequence of PF→LS→AF with decreasing temperature is consistent with the observations of Levine and Hill[13] but is in contrast to others[5,8,15] who have suggested that fine ferrite grains appear at higher temperatures than lath structures. The decrease in PF with increased Cb in series A-1 is consistent with Cb delaying the appearance of PF: i.e. pushing the appearance of PF to longer times. As discussed above, Ti may have a similar effect on PF, contrary to the suggestion of Shija et al[2]. The quantitative effect on PF can be modified by other alloying elements which generally also delay the appearance of PF.

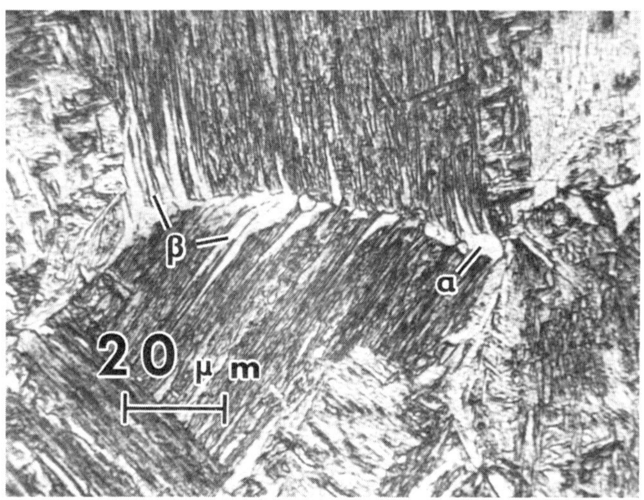

Figure 14 Optical micrograph of S-series G.T.A. remelt specimen showing PF at α, LS at β and low temperature products elsewhere

Further increases of an alloying element often have a consistent effect on a microconstituent, as indicated by conventional CE equations. In some cases, notably Mo, non-linear effects result from the alloying element somehow delaying lower temperature products. Although the interpretation of the effect is still not settled[17] the delay seems related to interation between the alloying element and carbon. All the microalloying elements Cb, Ti and V are strong carbide formers, and might thereby provide similar delays or bays in CCT diagrams. Levine and Hill[13] found that in welds containing V the PF grew very slowly, with little LS formation. This is consistent with a bay in the CCT diagram due to V. In our welds, however, V appeared to favour the lath structure. The possible interaction of an alloying element with carbon at the ferrite-austenite interface is dependent on the segregation of the element there. According to the Fe-V phase diagram[18], V may be less soluble in ferrite than austenite for <0.2% V. Hence, the growth of ferrite would require the rejection of V. If this V reacts with C, then the growth of ferrite might be slowed considerably. Also the solubility of V in ferrite is likely to be changed by the addition of other alloying elements. Only a small change in the phase diagram is required to make vanadium equally soluble in ferrite and austenite. The different effects of V in different weld series may therefore be rationalized.

The Fe-Cb phase diagram[18] shows that Cb is more soluble in ferrite than austenite, unlike V. Hence, welds containing Cb might exhibit a depletion of Cb at an advancing ferrite interface, reducing Cb-C interaction in the austenite and allowing ferrite growth, which is limited only by rapid C diffusion, (i.e. lath formation).

The picture is complicated, however, by possible effects of microalloying elements on the nucleation of acicular ferrite. Ito and Nakanishi[8] have suggested that Ti additions may favour the nucleation of AF, possibly via TiN. This can raise the transformation temperature for AF, limiting the growth of lath structures. Thus, although Levine and Hill[13] observed a temperature range during which little transformation took place, this delay may be modified by nucleation of AF at higher temperatures.

The above concepts can be extended to explain many of the composition-structure results, but the details are too speculative to present here. In research which is in progress, we will measure the transformation ranges of the various microconstituents in different welds in order to better experimentally test these ideas.

SUMMARY AND CONCLUSIONS

Quantitative optical metallography has been used to measure the volume fractions of proeutectoid ferrite (PF), lath structure (LS), acicular ferrite (AF), "other" and inclusions, along with the maximum lengths of the acicular ferrite grains (AFGL), in several weld series with varying Cb, Ti or V levels. Various effects of alloy content on volume fractions of different constituents were observed, but these were not always consistent with previous reports or from series to series. The following conclusions were drawn:

1) Reproducibility of the quantitative microstructural analysis was ± 2 volume percent for a given weld section examined at 352X and 800X.
2) The microstructure can vary considerably along the length of a weld. The observed absolute error in volume percent generally increased as the volume fractions increased to 50%, giving about ± 10% error at 50% volume percent. Part of this error is due to an unexpected increase in LS and "other" with stress relief.
3) Cb produced a decrease in PF content, but this decrease was reduced as the alloy content of the weld increased.
4) Increased alloy content, as measured by conventional carbon equivalent equations, decreased the AFGL but Cb had little effect on AFGL.
5) Ti decreased the AFGL and the Ti effect increased as the alloy content increased.
6) G.T.A. welding was used to obtain the transformation temperatures of the microconstituents in a submerged arc weld. With decreasing temperature, the order of appearance of the transformation products was PF→LS→AF, with PF first appearing at ∿700°C.

ACKNOWLEDGEMENTS

We are grateful to J.E. Hood and D.E. Osborne of Stelco and R.J. Jesseman of Armco Steel for supplying weld specimens. Computer programs for statistical analysis, and useful discussions of statistical interpretation, were provided by G.W. Bennett, G. Hunter, M. Page and J.C. Young. The work was supported financially by grants from Energy Mines and Resources Canada and the National Research Council of Canada.

REFERENCES

1) J.M. Sawhill Jr. and T. Wada: Weld J. 1975, 54, 1S.
2) A. Shija, H. Imura and J. Tsuboi: 11W Doc 1X-1049-77.
3) P.H.M. Hart, R.E. Dolby, N. Bailey and D.J. Widgery: in Microalloying 75 Proceedings, Union Carbide Corp., 1977, p. 540.
4) G. Barnard: ibid p. 552.
5) J.M. Gray: Weld. Res. Counc. Bulletin 213, 1976.
6) J.G. Garland and P.R. Kirkwood: Metal Const. 1975, 7, 272.
7) P.R. Kirkwood and J.G. Garland: Weld. and Metal Fab. 1977, 17.
8) Y. Ito and M. Nakanishi: Sumitomo Search 1975, No. 15, p. 42.
9) E. Levine and D.C. Hill: Metal Const. 1977, 9, 346.

10) R.J. Jesseman: in Microalloying 75 Proceedings, Union Carbide Corp., 1977, p. 578.
11) D. Osborne and J. Hood: Steel Company of Canada, unpublished work.
12) D.J. Widgery: in Microalloying 75 Proceedings, Union Carbide Corp., 1977, p. 566.
13) E. Levine and D.C. Hill: Met. Trans. 1977, 8A, 1453.
14) J.M. Sawhill Jr: Climax Moly Report L-176-115, Oct. 1973.
15) A.G. Glover, J.T. McGrath, M.J. Tinkler, and G.C. Weatherly: Weld J, 1977, 56, 259-S.
16) J.A. Marshall and J. Heuschkel: Weld Res. Coun. Bulletin 228, 1977.
17) R.C. Sharma and G.R. Purdy: Met. Trans. 1973, 4, 2303.
18) M. Hansen and K. Anderko: Constitution of Binary Alloys, McGraw-Hill, New York, 1958.

Microstructure - Property Relationships in High-Strength Steel Welds*

by

J. A. Brooks and D. B. Dawson
Sandia Laboratories
Livermore, CA 94550

Abstract

In a weld, it is difficult to obtain a good structure-property correlation since the microstructure and its associated properties may vary considerably from location to location, depending on the thermal history at each point. To facilitate the correlation of specific weld microstructures with properties and to circumvent possible residual stress problems, resistance heating of specimen blanks has been used to simulate typical weld thermal cycles for high-strength HP9-4-20 (9Ni-4Co-0.20C) steel. The mechanical properties and stress corrosion cracking behavior of microstructures typical of both heat affected zone and fusion zone material were studied after thermal cycles simulating the limited number of weld passes in a thin-section weld. Untempered martensitic microstructures resulting from single-pass welds or thermal cycles above the A_1 temperature exhibit low toughness and low stress corrosion resistance. High-toughness HP9-4-20 welds can be achieved in thin sections if no untempered material exists. Similarly, the stress corrosion resistance of heat affected zone material can be greatly improved by the application of tempering thermal cycles. Resistance to stress corrosion was much lower for fusion zone material than for heat affected zone material. Significant improvement in fusion zone stress corrosion resistance was observed only when a high-temperature austenitizing thermal cycle preceded the tempering thermal cycles. It is shown that weld thermal cycles can be effectively used for tempering of fusion zone and heat affected zone microstructures. Thus multiple fusion passes and tempering passes would be beneficial where post-weld heat treatments are not possible.

*Work supported by U.S. Department of Energy, DOE, under Contract (AT(29-1)-789.

Introduction

The steel alloy HP9-4-20 (9Ni-4Co-0.20C) was designed for applications requiring relatively high yield strength (180 ksi), high toughness (CVN = 55-60 ft-lbs. K_{IC} = 110 ksi-in$^{1/2}$), and good weldability.[1,2] The excellent strength and toughness properties of this alloy after weld fabrication have led to several important aerospace and ordnance applications.[2,3]

Considerable work has been reported which shows that in heavy section weldments (1 to 2 in. thick) using the GTA process, mechanical properties similar to the fully heat treated base metal can be achieved with or without post-weld heat treatment.[1-4] In the absence of post-weld heat treatment, the resulting properties are attributed to tempering of previously deposited weld metal by subsequent weld passes, which in thick-section weldments often consist of fifty or more filler passes.

In HP9-4-20, a microstructure consisting primarily of untempered martensite produced by welding or austenitizing heat treatment normally has high hardness and relatively low yield strength and toughness (Table I). During tempering, both the hardness and ultimate tensile strength decrease, while toughness and yield strength increase. In thick-section GTA weldments, the large number of overlapping filler passes ensures that virtually all of the weld (except for the last few passes) will experience beneficial tempering thermal cycles. In thin-section weldments, however, substantial tempering of the fusion zone (FZ) and heat affected zone (HAZ) is much more difficult to achieve, due to the limited number of weld passes. In practice, this problem is normally dealt with by specifying a post-weld thermal treatment, which has the dual advantage of tempering the weld in addition to reducing residual welding stresses.

TABLE I: HEAT TREATMENT AND MECHANICAL PROPERTIES OF HP9-4-20

Steps	Temp (°F)	0.2 pct YS (ksi)	UTS (ksi)	CVN (ft-lbs)	Hardness (R_C)
Normalized	1650	-	-	-	-
Austenitize & Quench to RT Cool to -100°F	1550	150	220	35	48-50
Temper	1050-4hr	185	205	55	40-44

There are some occasions when post-weld heat treatment is not possible, often due to the physical size of the welded structure or due to the incorporation of associated structures or components which cannot tolerate the high temperatures (900 to 1000°F) specified for post-weld stress relief of HP9-4-20. The present study was designed to address the problem of making a thin-section (0.25 in.) final closure weld in such a heat-critical assembly, where the projected application imposed the additional requirements of high toughness, high strength, and resistance to stress corrosion cracking (SCC). Both GTA and electron beam weld procedures were considered (Figure 1), including the use of tempering passes for GTA welds, and a standing edge for depositing temper passes which can subsequently be machined away.

In evaluating the effectiveness of thin-section weld procedures

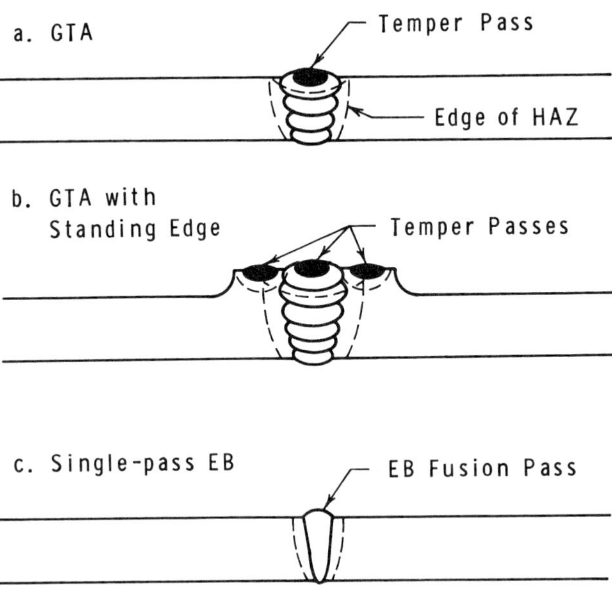

Fig. 1 - Schematic joint designs for HP 9-4-20 thin-section welds: (a) GTA weld with temper pass, (b) GTA weld with three temper passes on standing edge, (c) single-pass electron beam weld.

incorporating tempering passes, it is important to be able to relate microstructures which result from specific sequences of weld thermal cycles to their associated mechanical properties and stress corrosion behavior. In an actual weld, the microstructure (and properties) may vary considerably from location to location within the fusion and heat affected zones, depending on the thermal history experienced at each point. Since the volume of material corresponding to a given thermal history is quite small, mechanical testing of actual weldments must necessarily represent an averaging process over a wide variation in microstructures, which hinders any effort at establishing a specific structure-property correlation. Where post-weld heat treatments are not permitted, as in this case, another residual stresses which can have a major effect on measurements of mechanical properties.

To circumvent the problem of residual welding stresses and to facilitate the correlation of microstructure with properties, resistance heating of specimen blanks was used to simulate typical weld thermal cycles for hardness, Charpy toughness, and stress corrosion testing. This provides a specimen with a large test section of a single microstructure representative of a much smaller region in a weldment. The specific thermal cycles evaluated in this study were based on the HP9-4-20 isothermal transformation diagram of Ault[1] (Figure 2), as well as data from earlier work by Rosenstein et al. for HP9-4-20 simulated HAZ specimens.[4]

Figure 2 shows that partial to complete austenitization will occur for isothermal treatments above about 1180-1200°F (the A_1 transformation temperature). This process can also occur as the result of weld thermal cycles, although peak temperatures for welds must be slightly higher

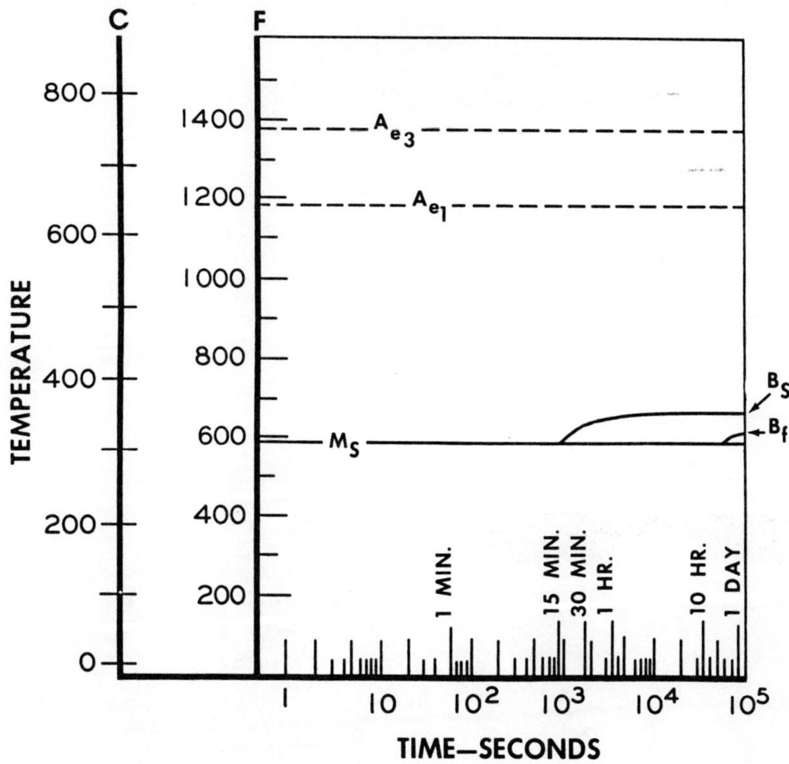

Fig. 2 - HP 9-4-20 isothermal transformation diagram (from Ault, Ref. 1).

because of the short duration of the high-temperature portion of the cycle. Upon cooling, austenite will transform to untempered martensite, with some autotempering being possible due to the high (~ 600°F) M_s temperature. This autotempering process is thought to be partially responsible for the good properties of as-welded HP9-4-20. In a single-pass HP9-4-20 weld (Figure 3), the HAZ will vary from a fairly coarse microstructure near the FZ where peak temperatures are slightly below the melting point, to a very fine or mixed structure near the HAZ/base metal interface where peak temperatures in the intercritical region between A_1 and A_3 (Figure 2) have been reached. The fusion zone typically shows a recast structure, although this can be modified considerably by subsequent weld thermal cycles.

In Rosenstein's study of the mechanical properties of simulated HAZ microstructures[4], it was shown that single thermal cycles above the A_1 temperature increased the hardness and decreased the Charpy toughness of HP9-4-20, with the highest hardness and lowest toughness associated with temperature excursions in the intercritical region (Figure 4). Data showing the effects of multiple thermal cycles are shown in Figure 5. Although some of the thermal cycles are quite complex, several distinct trends are apparent from Rosenstein's data. It is clear that a recovery of

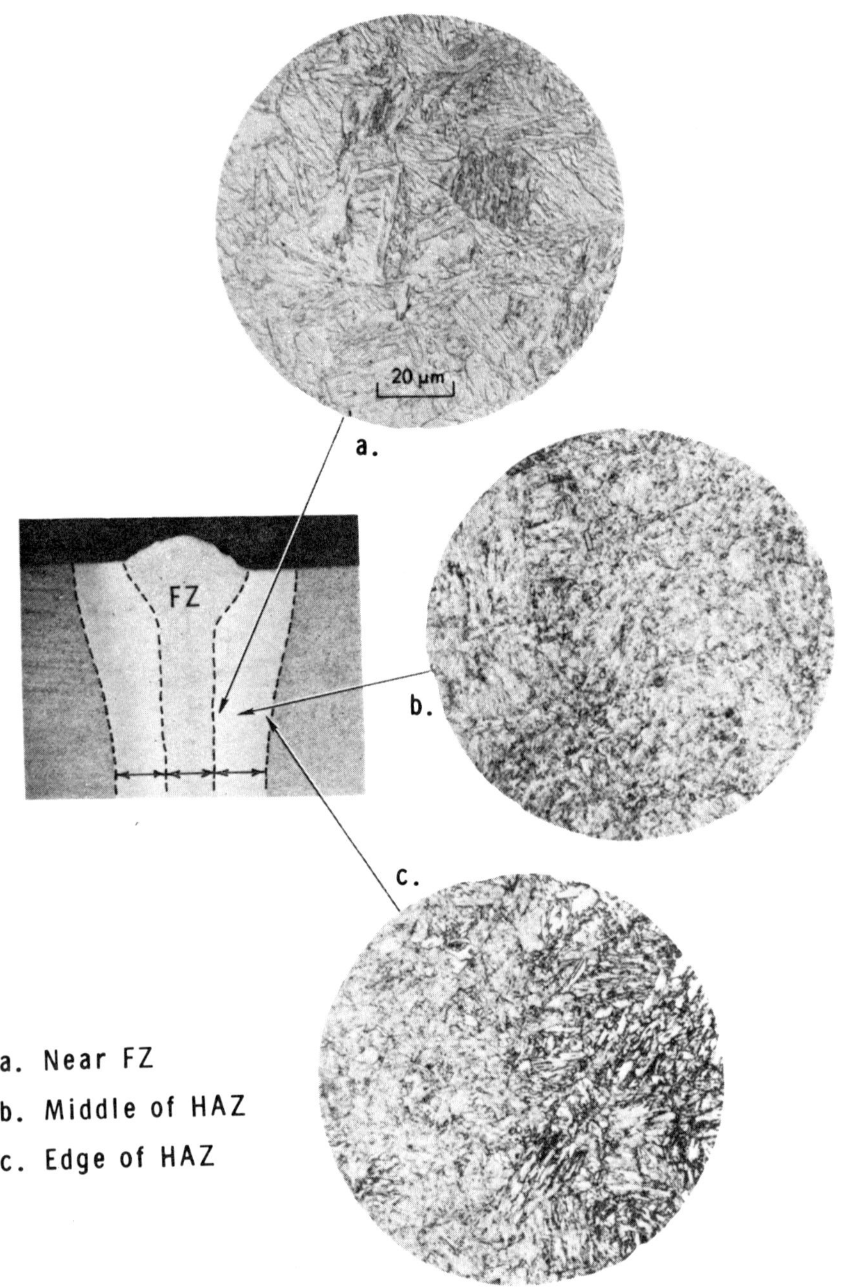

Fig. 3 - Weld HAZ microstructures after 10 inch/min. electron beam pass: (a) coarse-grained structure near fusion zone, (b) fine structure for temperatures just above A_3, (c) interface between fine intercritical structure and base metal at edge of HAZ.

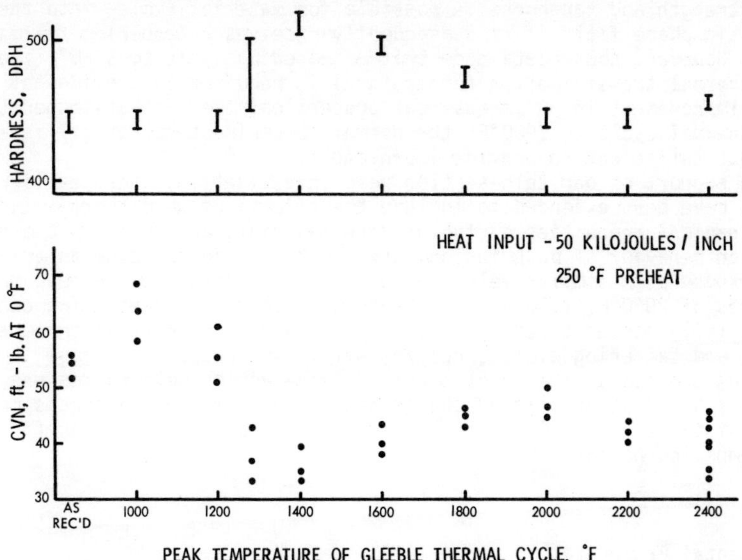

Fig. 4 - Effect of single weld thermal cycles on the hardness and impact toughness of HP 9-4-20 simulated HAZ material (from Rosenstein et al., Ref. 4).

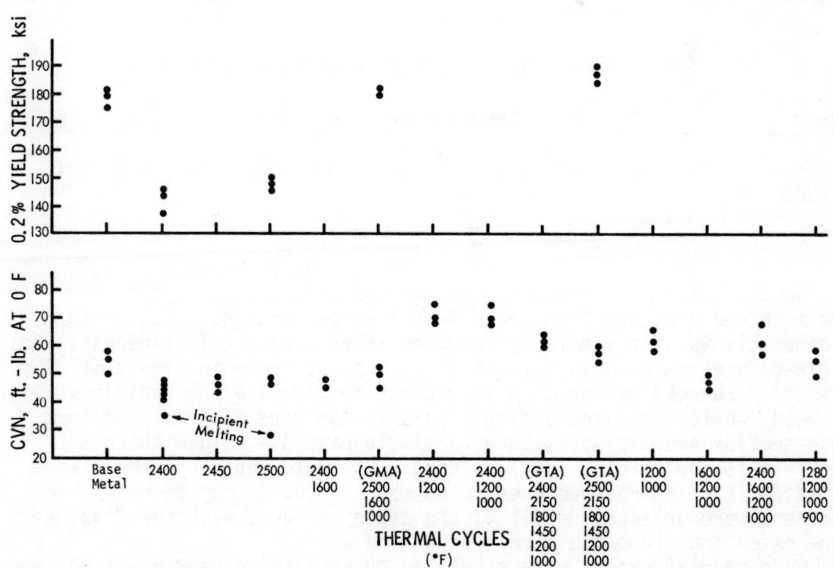

Fig. 5 - Effect of multiple weld thermal cycles on the yield strength and impact toughness of HP 9-4-20 simulated HAZ material (from Rosenstein et al., Ref. 4).

yield strength and toughness is possible for material cycled into the fully austenitic phase field if it subsequently receives a tempering thermal cycle. However, these data show that a tempering cycle to 1200°F (near the A_1 isothermal transformation temperature) is required to provide any significant improvement in HAZ mechanical properties; the time-at-temperature for a thermal cycle to 1000°F (the normal tempering temperature) is apparently not sufficient to provide improvement.

In support of our thin-section weld investigation, these earlier HAZ results have been extended to include the effects of weld thermal cycles on the mechanical properties of fusion zone material, and to study the stress corrosion behavior of both fusion zone and heat affected zone material after exposure to similar weld thermal cycles. Thermal cycle peak temperatures of 2000°F, 1350°F, and 1200°F were chosen, representing coarse-grained fully austenitized cycles, fine-grained intercritical temperature cycles, and tempering cycles, respectively. The results of these studies are being applied to the development of thin-section welding designs and processes aimed at optimizing the mechanical properties and stress corrosion resistance of HP9-4-20 steel through the application of beneficial weld tempering passes.

Experimental Procedure

The composition of the 9Ni-4Co-0.20C steel used in this investigation is shown in Table II. The material was received as 2 in. thick sections cut from 6.25 in. square forged bar. These sections were cut into 0.625 in. thick plates prior to heat treatment and welding. The heat treatment schedule and resulting mechanical properties are shown in Table I. These mechanical properties are typical for material tempered between 1025°F and 1075°F.

TABLE II: COMPOSITION OF HP9-4-20 (WT. PCT.)

Heat	Ni	Co	C	Cr	Mo	V	Mn	Si	P	S	Fe
Republic 3821003	9.3	4.7	.19	0.8	1.0	.08	.35	.04	.008	.007	Bal.

Square 0.5 in. specimen blanks for HAZ weld thermal cycle simulations were machined directly from these heat treated sections. Fusion zone specimen blanks were similarly prepared after making a full-penetration single-pass electron beam weld at 10 in/min. as shown in Figure 6. The slow-speed EB weld was used in an attempt to simulate the heat input of a GTA weld, while providing a single pass fusion zone of fairly uniform cross-section and microstructure (a single-pass GTA weld with this thickness is impossible, and a multipass GTA weld would provide a wide variation in FZ microstructures). Notches in the Charpy toughness and SCC specimens were oriented normal to the plate surface, with the fracture plane extending along the direction of the weld.

Weld thermal cycles were simulated by resistance heating of HAZ and FZ specimen blanks in a Gleeble-type apparatus. The schematic cycle of Figure 7 illustrates the rapid rise to peak temperature and equally rapid decrease typical of a weld thermal cycle. As a result we are seeing the effects of austenitizing and tempering thermal cycles which may be at peak

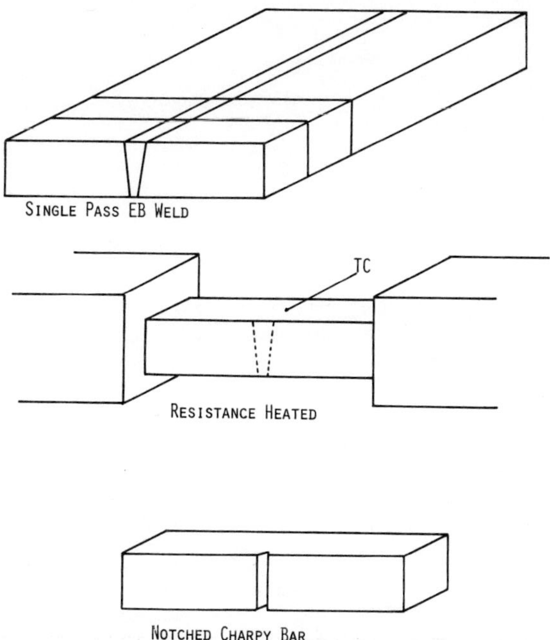

Fig. 6 - Charpy specimen preparation procedure for thermally-cycled fusion zone specimen blanks. Procedures similar for HAZ specimens.

temperature for only seconds, as opposed to isothermal heat treatment cycles lasting hours (Table I).

Standard 0.394 in. square Charpy toughness specimens were machined from thermally-cycled FZ blanks and tested in accordance with ASTM Standard Method E-23 for Notched Bar Impact Testing. Cantilever beam SCC specimens 0.5 in. square were machined from both FZ and HAZ blanks and EDM notched in the orientation shown in Figure 6. The EDM notch was extended to a total depth of 0.150 in. by fatigue cracking, with the maximum stress intensity restricted to 25 ksi-in$^{1/2}$. Stress corrosion specimens were dead-weight loaded in cantilever bending to initial stress intensity levels of 90, 70, 60 or 50 ksi-in$^{1/2}$. Since the number of available specimens was limited, evaluation of SCC performance was based on a comparison of time-to-failure values for given stress intensities, rather than attempting to determine a K_{Iscc} threshold level for each thermal cycle condition. Stress corrosion specimens were tested in 3.5 pct. NaCl solution (salt water), with the solution being changed weekly.

Fracture surfaces of Charpy and stress corrosion specimens were examined in the scanning electron microscope (SEM). Metallographic specimens etched with 3 pct. nital were also prepared for sections normal to the fracture plane, both to delineate the fracture path and to determine the microstructure near the fracture. Hardness measurements were made on thermally-cycled Charpy specimen blanks.

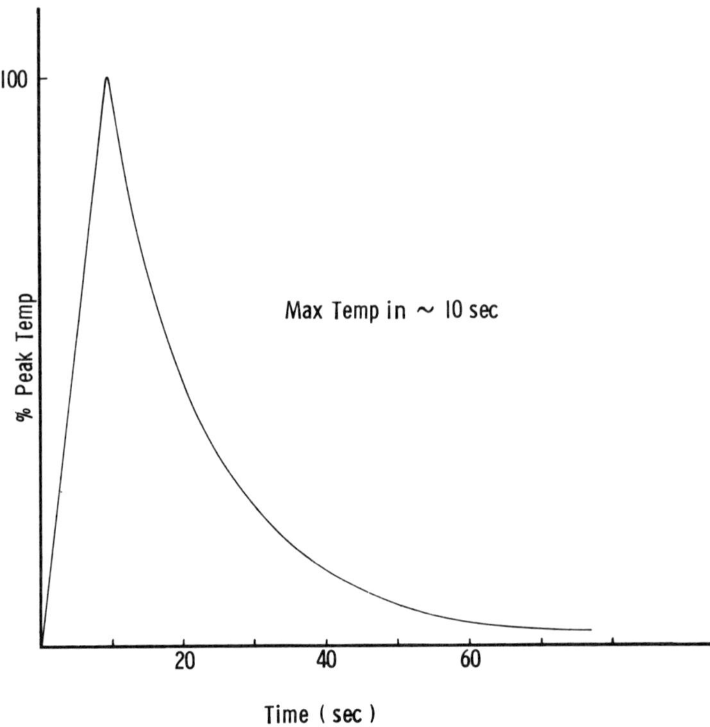

Fig. 7 - Typical resistance heating cycle used in this study to simulate thermal cycles experienced during welding.

Results

Heat Affected Zone Material: SCC Behavior - The data of Rosenstein et al. (Figures 4 and 5, from Reference 4) clearly show that hardness levels of HAZ material can be reduced while strength and toughness can be recovered by the application of a subsequent 1200°F tempering thermal cycle. The data of Table III also support the beneficial effects of a 1200°F thermal cycle on the stress corrosion behavior of HAZ material. Compared to results for base metal, the SCC failure of HAZ specimens cycled to 1350°F or 2000°F is rapid (100 hours or less) at both 70 and 60 ksi-in$^{1/2}$. A K_{Iscc} threshold, if determined, would probably be below 50 ksi-in$^{1/2}$ in either case. Application of a 1200°F tempering thermal cycle to specimens previously cycled to 1350°F or 2000°F results in a significant improvement in SCC resistance, although (unlike Charpy toughness) full base metal properties are not recovered.

SEM examination of fracture surfaces was often restricted due to the buildup of corrosion products, but several distinct types of fracture modes were observed. Ductile dimple rupture (Figure 8) was the normal failure mode for fracture in air in all cases, including fusion zone Charpy specimens. It was also the normal failure mode in the overload zone of SCC specimens tested in salt water. For SCC specimens, the stress

TABLE III: STRESS CORROSION BEHAVIOR OF HP9-4-20

	Stress Intensity, ksi-in$^{1/2}$			
	90	70	60	50
	Time to Failure, hr.			
Base Metal:	324	2003	---	---
Heat Affected Zone:				
1350°F	---	94	48	7.5, 3.4
1350°F + 1200°F	---	747, 645	2743	---
2000°F	---	55	103	---
2000°F + 1200°F	116	740	1017	---
Fusion Zone:				
As Welded	---	6.0	7.5	11.0
1200°F	---	28	855*	147
1350°F	---	---	4.7	50
1350°F + 1200°F	---	5.3	10.8	2577
2000°F + 1200°F	---	1028	80	---

*Crack ran into HAZ

corrosion fracture morphology adjacent to the end of the fatigue pre-crack was a mixture of varying degrees of dimple rupture (probably mechanically-induced) with cleavage-like and/or intergranular fracture (environmentally-induced). Secondary cracking was frequently observed in SCC fractures, particularly at or near the interface with the fatigue precrack. For HAZ specimens which received a single 2000°F cycle, the fracture mode was predominantly intergranular (Figure 9), with the large intergranular facets corresponding to the coarsened microstructure shown earlier for this thermal cycle (Figure 3). For specimens cycled to 1350°F as well as those which received a final 1200°F tempering cycle, the stress corrosion fracture was primarily a mixture of dimple rupture and cleavage-like fracture, with little or no evidence of intergranular fracture (Figure 10).

Fusion Zone Material: Hardness, Toughness, and SCC Behavior - The major emphasis of this program was on the behavior of the weld fusion zone. The simulated GTA fusion zone microstructure produced by the slow EB weld varied from columnar to cellular dendritic, generally being more cellular near the center of the weld (Figure 11). The dendritic microstructure was found to persist after lower temperature (1200°F and 1350°F) thermal cycles, but was essentially eliminated by a 2000°F cycle.

The hardness and Charpy toughness results shown in Figure 12 are similar to the HAZ results shown earlier (Figures 4 and 5).[4] The as-welded fusion zone has high hardness and low toughness. A single 1000°F thermal cycle lowers hardness but has little beneficial effect on toughness. However, a 1200°F tempering cycle returns both hardness and toughness to base metal values. Single 1360°F and 2000°F cycles result in poor fusion zone properties, although toughness values are no worse than for the as-welded fusion zone in spite of the higher hardness associated with the intercritical 1360°F cycle. The addition of a 1200°F cycle to a

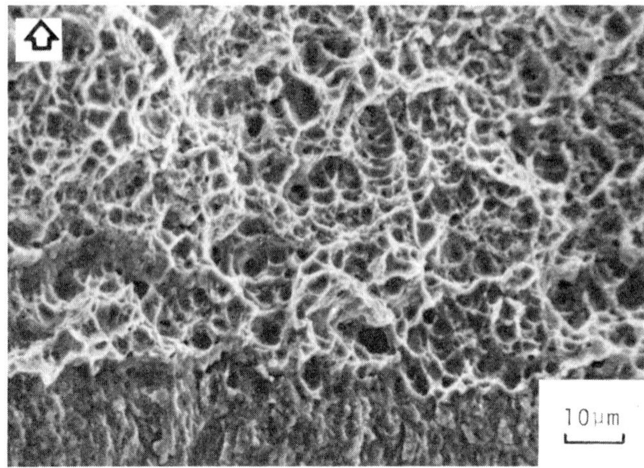

Fig. 8 - Ductile dimple rupture typical of impact fracture in air and overload zone of SCC specimens tested in salt water. Fatigue pre-crack shown at bottom.

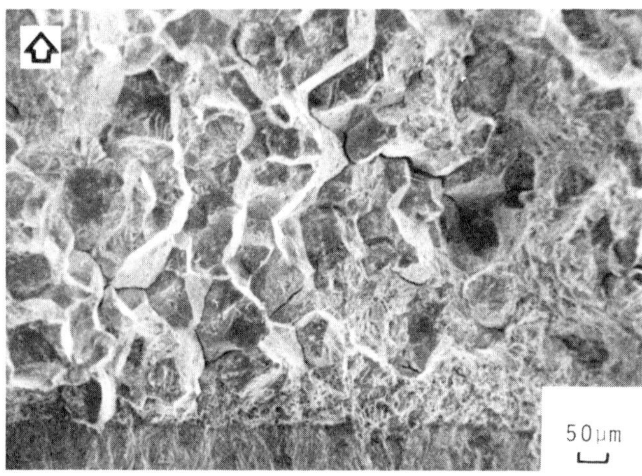

Fig. 9 - Intergranular SCC fracture of HAZ structure resulting from a thermal cycle to 2000°F with no subsequent tempering.

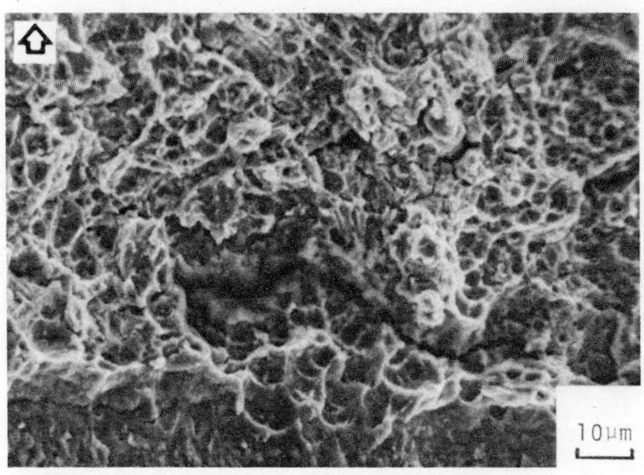

Fig. 10 - Combination of ductile dimple and cleavage-like fracture in stress corrosion zone of HAZ specimen after a thermal cycle to 1350°F.

2000°F cycle once again results in properties similar to the base metal. A third cycle to 800°F does not cause any significant change in toughness. Addition of a tempering cycle to an intercritical temperature cycle was not evaluated, but the beneficial effects would be expected to be similar.

Stress corrosion test results (Table III) show that the as-welded fusion zone is extremely susceptible, with times-to-failure as short as 11 hours even at 50 ksi-in$^{1/2}$. The SCC fracture surfaces for as-welded specimens were almost entirely intergranular (or interdentritic), as shown by the comparison of fracture morphology and microstructure in Figure 13. The SCC overload fracture zone for as-welded specimens was also predominantly interdendritic, rather than dimpled as in all other conditions.

Cycling the fusion zone to 1350°F had little effect on time-to-failure values, although the fracture mode changed substantially from interdentritic (Figure 13) to cleavage plus dimple rupture with secondary cracking (Figure 14), similar to the fracture surface for 1350°F HAZ specimens (Figure 10). Cross-sectioning of the 1350°F FZ fracture surface shows that the fracture path is primarily transgranular with some interdendritic secondary cracks (Figure 15). The dendritic structure is still clearly in evidence.

A single thermal cycle to 1200°F causes only a partial improvement in SCC resistance of the FZ. Time-to-failure increases as a result of this cycle, but the improvement does not approach base metal or HAZ values (the 855-hour data point in Table III for 60 ksi-in$^{1/2}$ is spurious: the pre-crack was mis-located in the HAZ, so the value is more representative of HAZ behavior). Addition of a 1200°F cycle to a 1350°F cycle also increases the time-to-failure, but again the beneficial effect is much

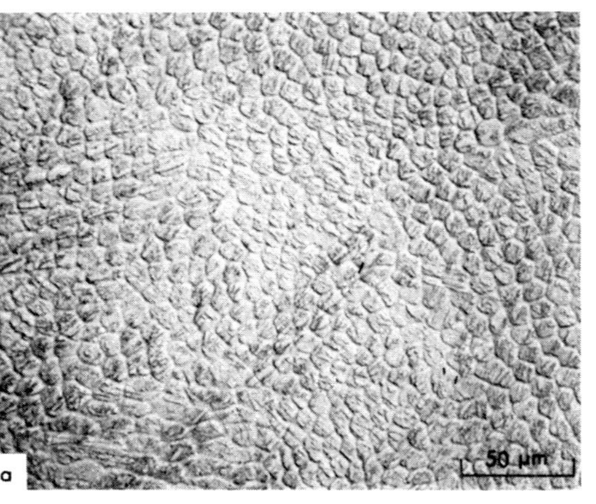

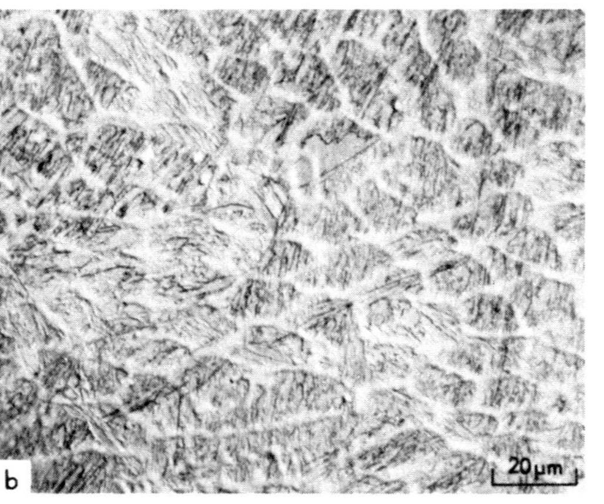

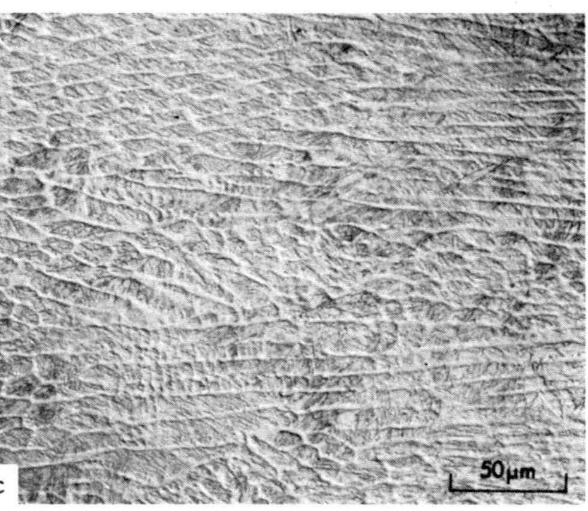

Fig. 11 - As-welded fusion zone structures showing cellular (a,b) and columnar (c) recast structures.

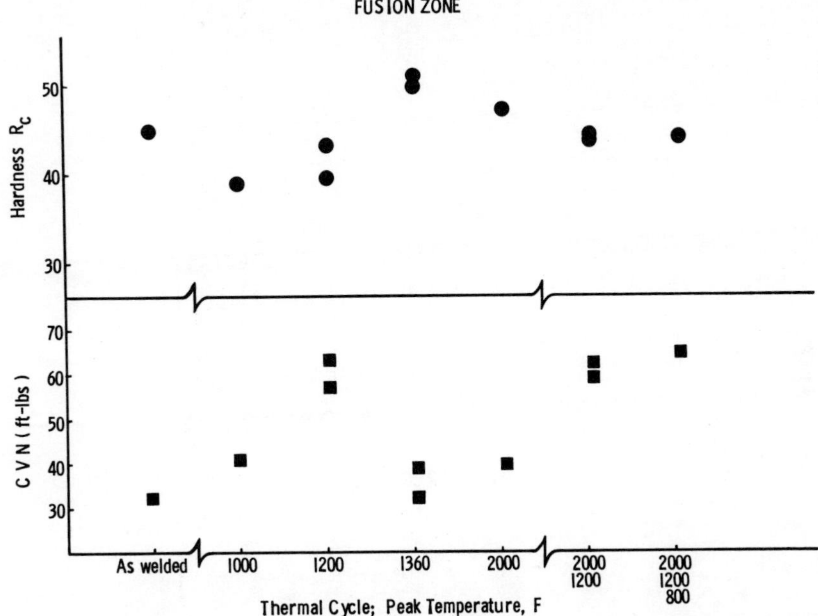

Fig. 12 - Effects of single and multiple weld thermal cycles on the hardness and impact toughness of HP 9-4-20 simulated fusion zone material.

less than that obtained with HAZ specimens. Like the 1350°F cycle, the 1200°F cycle changes the fracture mode from the interdendritic failure seen in as-welded specimens even though the dendritic microstructure is only partially eliminated (Figure 16).

Thermal cycling to 2000°F followed by 1200°F almost completely eliminates the dendritic structure, producing a microstructure similar to the base metal (Figure 17). There also appears to be a more substantial improvement in SCC resistance, although the low value at 60 ksi-in$^{1/2}$ requires further evaluation. Pending further tests, the results for 2000°F + 1200°F are considered promising but incomplete.

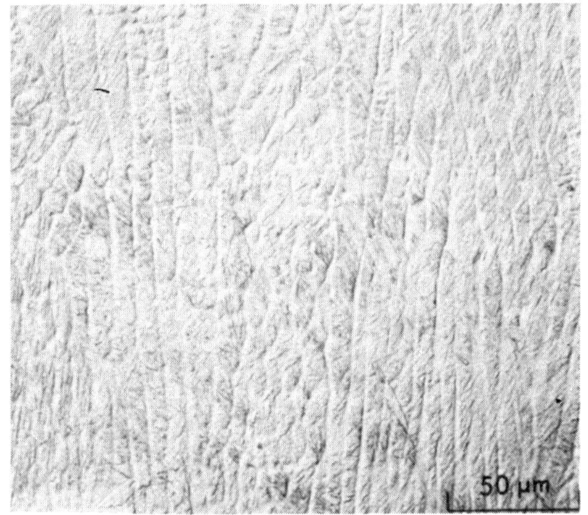

MICROSTRUCTURE

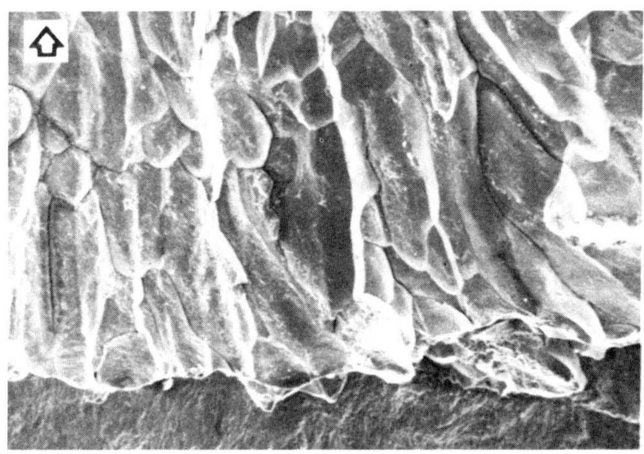

SCC FRACTOGRAPH

Fig. 13 - Intergranular (interdendritic) stress corrosion fracture compared with microstructure for as-welded fusion zone specimen.

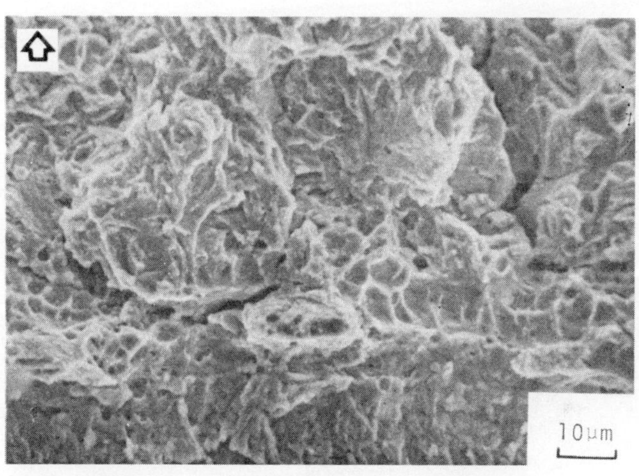

Fig. 14 - Mixed cleavage and dimple stress corrosion fracture for fusion zone specimen thermal cycled to 1350°F.

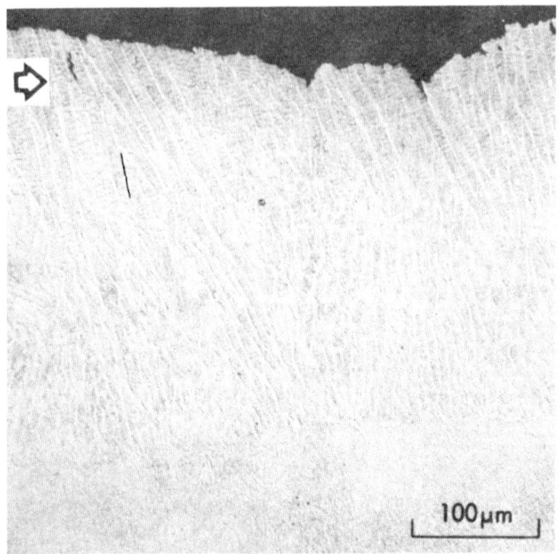

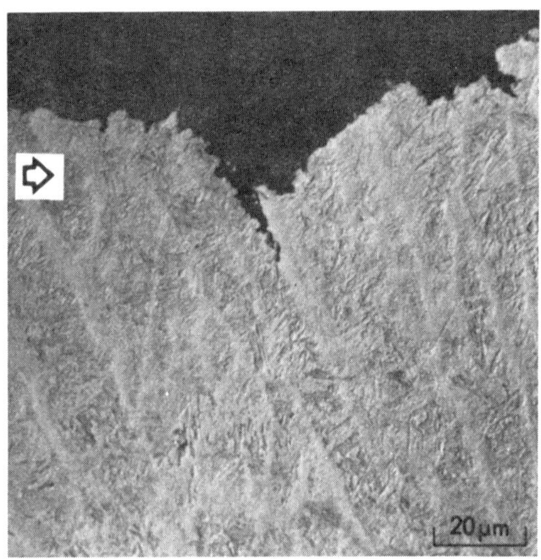

Fig. 15 - Cross-section of SCC fracture surface for fusion zone specimen thermal cycled to 1350°F. Recast structure is still in evidence. Fracture path is primarily across dendritic structure, with some interdendritic secondary cracking.

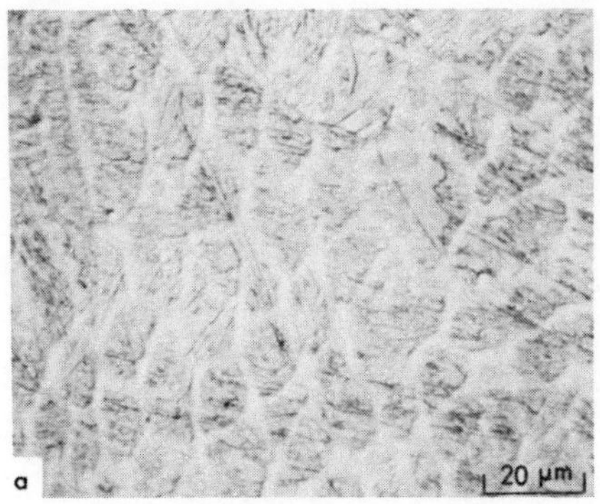

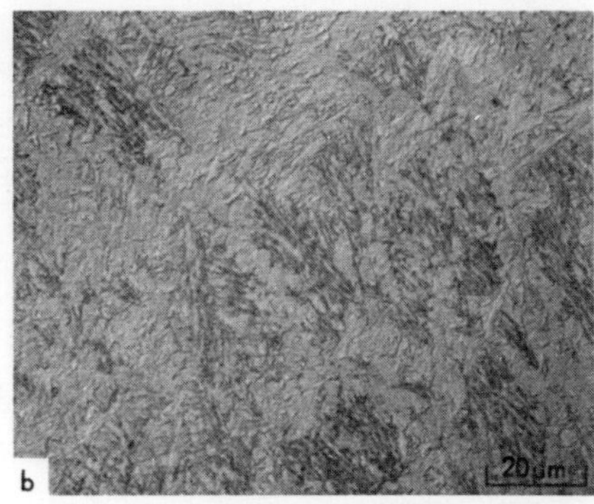

FUSION ZONE: AS-WELDED

FUSION ZONE + 1200°F

MICROSTRUCTURE

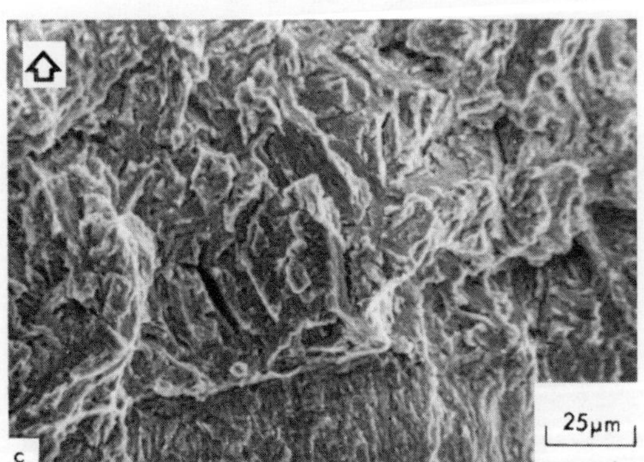

FUSION ZONE + 1200°F

SCC FRACTOGRAPH

Fig. 16 - Comparison of (a) as-welded fusion zone microstructure and (b) fusion zone microstructure after a tempering thermal cycle to 1200°F. SCC fracture surface (c) is mixture of cleavage and dimple fracture.

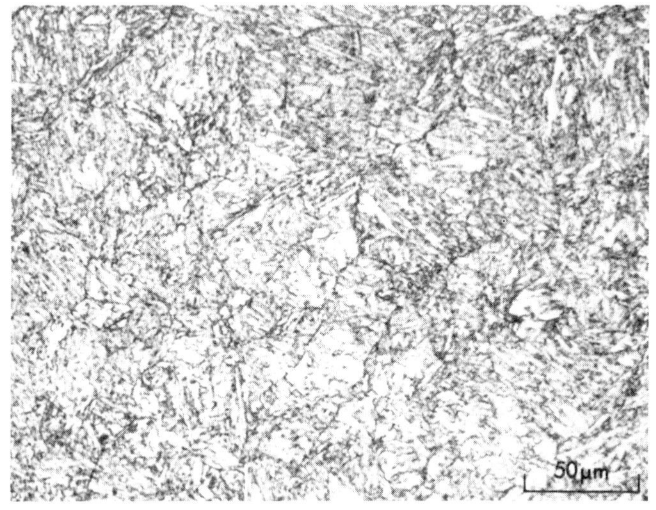

(a) BASE METAL

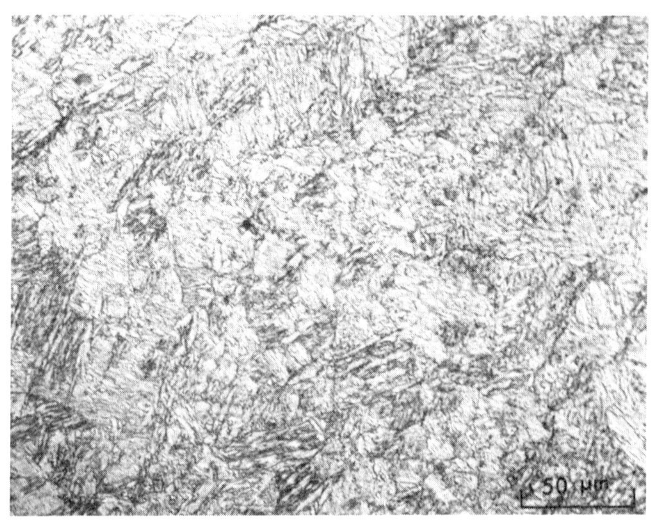

(b) FUSION ZONE + 2000°F + 1200°F

Fig. 17 - Comparison of (a) base metal microstructure and (b) fusion zone after 2000°F + 1200°F thermal cycles.

Discussion

The results of this study in conjunction with the earlier work of Rosenstein[4] show that a significant improvement in HAZ properties can be expected if the material experiences a subsequent thermal cycle at or near 1200°F. This tempering cycle may either be the result of a later filler pass, or some type of tempering pass. The latter is attractive as a means for improving the properties of the material deposited in the last filler passes. The data show that while lower-temperature (e.g., 800 or 1000°F) tempering cycles do not degrade properties, they also do not improve them significantly compared to the benefits of a 1200°F pass. A 1200°F cycle returns the hardness and Charpy toughness properties of HAZ material to that approaching the base material. A 1200°F cycle also improves the SCC resistance of HAZ material, but not to where the resistance is equivalent to that of the base metal. The SCC resistance of HAZ material cycled to either 1350°F or 2000°F is similar, even though the fracture modes are very different. Conversely, the fracture appearance for specimens cycled to 1350°F and for specimens cycled to 1200°F is similar, although the SCC resistance is quite different. Thus there appears to be little correlation between fracture mode and SCC behavior for HAZ specimens. SEM analysis of SCC fracture surfaces was hampered by the amount of adherent corrosion product. Coupling the specimens to a sacrificial zinc anode would have reduced the amount of corrosion, but it was not attempted in this case because of the known detrimental effect of a zinc couple on the SCC performance of HP9-4-20.[5,6]

As shown earlier for HAZ material[4], a 1200°F thermal cycle will also restore the hardness and toughness of fusion zone material to base metal condition. The results for stress corrosion cracking of the fusion zone are less positive, although some benefit is achieved with a 1200°F tempering cycle preceeded by an austenitizing cycle sufficiently high (e.g., 2000°F) to recrystallize the dendritic solidification structure. This less promising recovery in SCC behavior may be related to the very poor SCC resistance of the as-welded fusion zone material itself. An electron microprobe study of the dendritic structure discloses that measurable segregation of Ni, Mo and Mn to the interdendritic regions occurs during solidification (Table IV). While Ni and Mo are not known to have any effect on the SCC behavior of low-alloy steels, manganese in amounts up to 4 wt. pct. has been shown to be detrimental, at least in 4340-type steels.[7] It is not known at this point whether the observed segregation of Mn could have played any role in the interdendritic SCC failure mode. If segregation does play a role, it is unlikely that significant diffusion of alloying elements would occur for thermal cycles to 1200°F or 1350°F, although the microstructure in which the segregation is present would be altered. The recrystallized microstructure which results from a 2000°F thermal cycle is probably more beneficial than any solute diffusion which occurs in such a brief thermal cycle. Microprobe studies of thermally-cycled fusion zone material are continuing in a effort to determine whether any of these thermal cycles cause a significant change in the segregation pattern observed in the as-welded fusion zone. Experiments are also planned for using Auger analysis of clean fracture surfaces to determine whether manganese or any other element is associated with environmental fracture paths in HP9-4-20. It should be noted, however, that changes in fusion zone SCC fracture paths as the result of thermal cycling do not appear to be associated with a substantial increase in SCC resistance, at least for 1200°F and 1350°F thermal cycles.

One factor which governs our inability to draw clear conclusions from the fusion zone SCC studies is the much greater scatter in data when

TABLE IV: ELECTRON MICROPROBE DETERMINATION OF ELEMENTAL SEGREGATION IN AS-WELDED FUSION ZONE

Element	Composition, Wt. Pct.		
	Bulk Alloy*	Grain Bound.	Grain Interior
Iron	Bal.	84	88
Nickel	9.3	9.8	7.7
Cobalt	4.7	5.0	4.2
Molybdenum	1.0	1.8	0.1
Manganese	0.35	0.5	0.2
Carbon	0.19	Not Segregated	

*From Table I

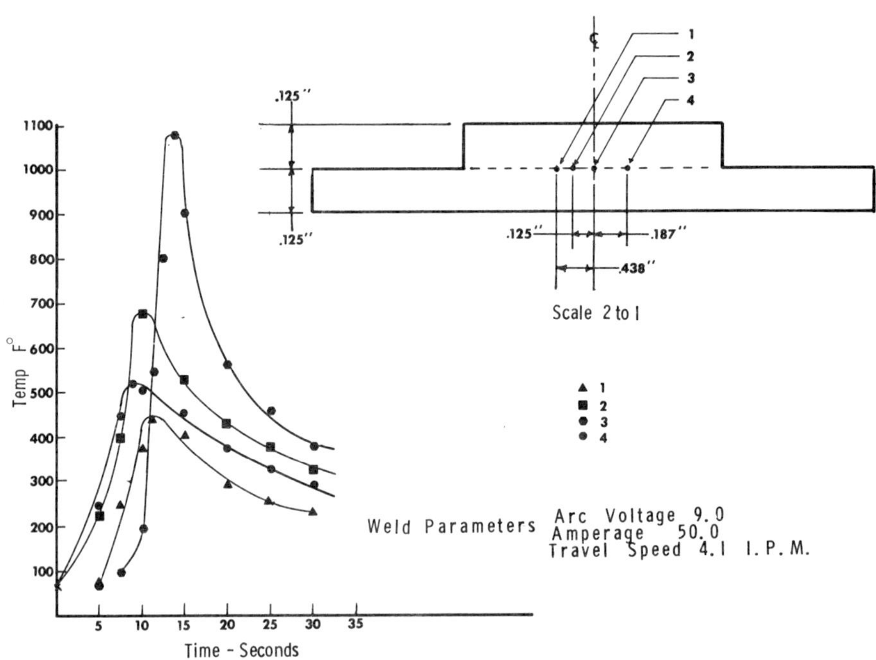

Fig. 18 - Thermal cycles experienced at different thermocouple locations for a temper pass run on the center of a 0.125 in. standing edge on 0.125 in. plate.

compared with HAZ data. Some variability in SCC time-to-failure measurements is always expected (e.g., the duplicate 1350°F + 1200°F HAZ specimens, Table III). For FZ specimens there is also the fact that the fusion zone microstructure itself is not truly uniform, consisting of both columnar and cellular structures. In total, these factors lead us to believe that a larger number of specimens must be tested to be able to draw more definite conclusions about the SCC behavior of thermally-cycled FZ material. Additional tests are in fact now in progress.

The results of these various studies have been incorporated in an experimental thin-section weld joint design similar to the standing-edge configuration shown schematically in Figure 1. In a preliminary study of a 0.125 in. standing edge on 0.125 in. plate (Figure 18), it was found that the peak temperature would reach 1100°F directly below a tempering pass on the center-line of the standing edge, with progressively lower peak temperatures further away. Based on this data, an actual GTA weld was made for 0.250 in. plate with a 0.125 in. standing edge, using three temper passes (Figure 19). In this case almost all the untempered martensite (which etches white) is seen to be in the standing edge, which would be machined off. However, while these tempering results are encouraging, the increased costs and production difficulties associated with the standing edge design have led us to consider other configurations as well. Both weld process design and further efforts aimed at structure-property correlation for HP9-4-20 weld materials are currently in progress.

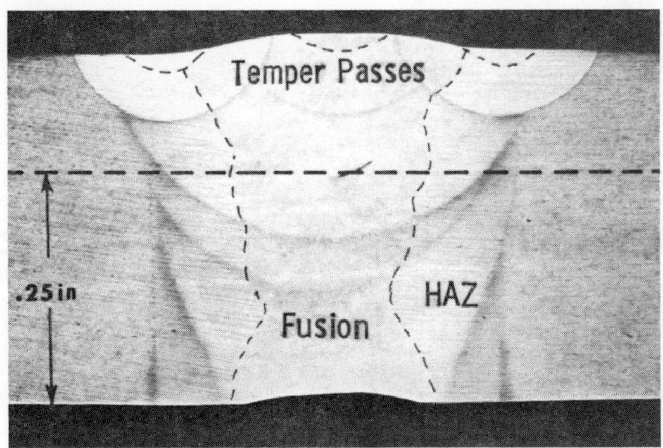

Fig. 19 - Multiple-pass GTA weld in 0.25 in. plate with three temper passes on a 0.125 in. standing edge. Untempered martensite remaining after temper passes etches white, tempered FZ and HAZ material etches darker. In production application the standing edge would be machined away.

Conclusions

1. Untempered HP9-4-20 weld fusion zone and HAZ material exhibits low impact toughness and low stress corrosion resistance.
2. Thermal cycles similar to those experienced during welding can be effectively used for tempering where post-weld heat treatments are not possible.
3. High-toughness HP9-4-20 welds can be achieved in thin sections if no untempered material exists. Fusion zone and HAZ material in the same temper condition exhibited similar toughness.
4. Fusion zone material was more susceptible to stress corrosion cracking than the HAZ.
5. A significant improvement in the stress corrosion resistance of the HAZ is possible after a tempering thermal cycle to 1200°F. Significant improvement in the SCC resistance of the fusion zone is only observed when a 2000°F thermal cycle precedes the tempering thermal cycle.

Acknowledgments

The experimental assistance of D. E. Clark, A. Salmi, T. J. Sage and A. T. Mamaros throughout the program are greatly appreciated. Contributions in the areas of fractographic analysis by C. W. Karfs and electron microprobe studies by W. B. Estill are also acknowledged.

References

1. R. T. Ault, "HP 9-4-20 Steel Data," TR 12018-132, Republic Steel Research Center, March 1968.

2. H. J. Hucek, "A Survey of HP-9Ni-4Co-0.20C Steel Forging," Battelle Columbus Laboratories, October 1975.

3. J. A. Brooks, "Welding Studies for the B-77 Closure Weld," Report SAND77-8225, Sandia Laboratories, May 1977.

4. A. H. Rosenstein, M. R. Gross, W. G. Schreitz, and G. A. Wacker, "Metallurgical Investigations of 9Ni-4Co-0.20C Steel," Report 2678, Naval Ship Research and Development Center, July 1968.

5. G. Sandoz, Met. Trans., Vol. 2, p. 1055 (1971).

6. J. W. Munford, Sandia Laboratories, Albuquerque, N.M., private communication.

7. G. Sandoz, in "Stress-Corrosion Cracking in High Strength Steels and in Titanium and Aluminum Alloys," B. F. Brown, ed., Naval Research Laboratory, Washington, D.C., 1972.

DISCUSSION

J. S. Ahearn (Martin Marietta Labs.): Could you comment on the use of a defocused electron beam for tempering the weldment?

Author: We at one time considered the possibility of tempering with a defocused electron beam, but there are a number of difficulties with this technique when applied to our particular situation. The most important problem is the narrow temperature range in which we see significant beneficial effects of tempering thermal cycles. Both our study and that of Rosenstein et al[4] show that relatively high (1100 to 1200°F) temperature cycles are required for tempering, yet temperatures much above 1200°F are very damaging. Thus, using a surface heating technique to temper the lower region of an electron beam weld (Figure 3) without overheating the outer surface is difficult, and perhaps more so if the welded structure contains temperature-sensitive components. The difficulty in obtaining reproducible heating with an electron beam further complicates this technique. A better solution to the problem of tempering an electron beam weld may be localized induction heating.

WELD POOL MOTION IN GTA WELDING: IMPORTANCE AND DESCRIPTION

G. Scott Mills

Rockwell International
Atomics International Division

Studies of material and arc effects on penetration in GTA welding are summarized and the resulting conclusions given. Metal flow in the weld pool, the apparent control mechanism, is described as having two basically different patterns which lead to the observed fusion zone shapes. Experimental evidence for these patterns and their likely mode of interaction with the arc are also presented.

INTRODUCTION

In seeking the cause of problems encountered in making deep penetration GTA welds in high manganese stainless steel (21Cr-6Ni-9Mn), temperature profiles in the arc plasma immediately above the weld pool (Manganese vapor dominated) for identical arc parameters and different heats of steel were measured. These measured profiles were then simulated by a computer solution to the radiation equations using assumed radial temperature and vapor density distributions. The results of these measurements and calculations showed that the temperature distributions (current density distributions) were essentially the same even though the depth-to-width ratios (D/W) of the fusion zones in the different heats of steel varied by a factor of 1.8. There were no definitive variations in heat flow within the arc which related consistently to the changes in D/W.

Looking at other influences on D/W, the effect of helium on welding arcs was considered. Typically, helium is added to argon shielding gas to obtain an increase in weld penetration without the increase in weld width usually obtained with an increase in weld current. This suggests some subtle effect of increased energy or a sharpening of the energy (current) density in the arc. Therefore, current density distribution measurements were made in arcs on a chilled copper block (no molten pool) and the results indicated that the current density distribution is slightly broader (less heat concentration) in a 50%He-50%Ar arc than in a 100% Ar arc. This is contradictory to the weld behavior and suggests that heat flow distribution in the arc does not have primary influence on D/W.

To get a broader base of information on GTA welding behavior, an extended series of welds was made on rotating stainless steel bars using a representative range of arc length, current and gas mixture values. From these welds, values of fusion zone width and area were determined for each set of parameters. Then, a computer was used to calculate the heat input and "arc width" required to duplicate the experimental fusion zone area and width. Compilation of these data revealed that the heat input widths determined from the calculations did not correlate with experimental values of current density full-width half-maximum (FWHM). More significantly, the larger values of D/W could not be accounted for by diffusive heat conduction alone. Further details and results of these investigations may be found elsewhere[1,2,3].

The conclusions drawn from all of the work discussed so far are that weld fusion zone shape (D/W) does not correlate with the energy distribution in the arc, the differences in heat flow which yield various D/W's must occur in the weld pool, and the arc has a secondary effect on D/W.

PENETRATION CONTROL MECHANISM

Having drawn the conclusions just mentioned, the following question became most pertinent: How does the arc (and other factors such as material properties) influence the flow of heat in the weld pool and what are the fundamental differences that result in high or low penetration welds? The mechanism which fit all of the welding results obtained to date was "inverted convection" driven by the flow of weld current through the liquid metal. Current flowing into the weld pool exerts a (Lorentz) force distribution on the liquid metal which is the analog of the force exerted by the acceleration of gravity acting on a thermally induced density gradient in a fluid (normal convection). Under proper conditions, this force on the weld pool will induce a circulation of metal downward at the center and radially inward at the surface with outward flow at the bottom completing a stable convection cell (Figure 1). The flow of metal enhances the flow of heat to the perimeter of the pool and produces a relatively broad, shallow weld.

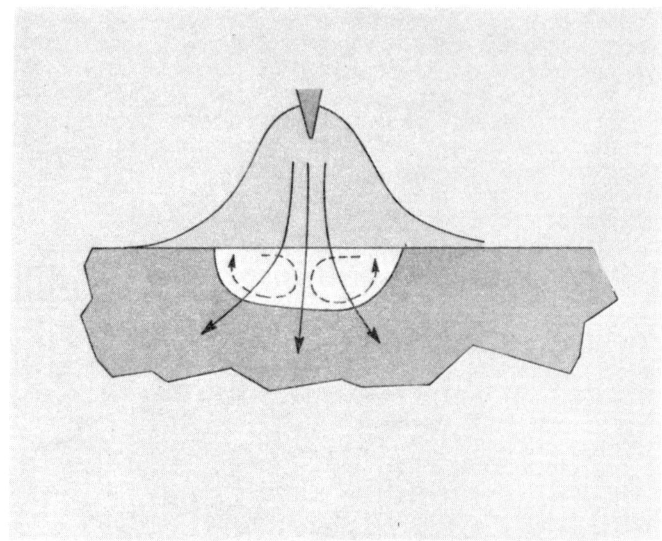

Figure 1 - Schematic of the current flow (solid lines) and liquid metal flow (dashed lines) as seen in a transverse (perpendicular to the welding direction) section of a convecting weld pool.

With such a convection analogy in mind, the effects of the arc current density distribution could be understood. If the distribution was narrow (argon arc) rather than broad (helium-argon arc), convection was more likely; just as in heating a beaker of water with a constricted (Figure 2) rather than broad (Figure 3) heat source. In the case of a welding arc, another phenomenon enters which explains the exceptionally deep, narrow fusion zones. Instead of a strong transverse flow developing as above, a predominantly front-to-back flow can occur in which heat is carried down by the metal flow at the arc center but is held near the weld centerline as the metal flows backward and then to the surface at the rear of the weld pool (Figure 4). This flow develops if the convection flow is suppressed by a broad current density distribution or surface forces which restrain spreading of the liquid metal.

EXPERIMENTAL CORROBORATION

Flow patterns such as these have been suggested, simulated and (in the second case) observed by others[4-8], but a clear direct observation of the vertical convection cell in a weld pool seems usually to be masked by single or double swirling flows in the horizontal plane. During the course of this present research, indirect evidence of vertical convection flow, possibly superimposed on the swirling flow, and the front-to-back flow was obtained in the case of the high manganese stainless steels dealt with earlier. The surfaces of weld pools were observed with a video camera (and recorder) through a narrow band optical filter which singled out manganese characteristic emission

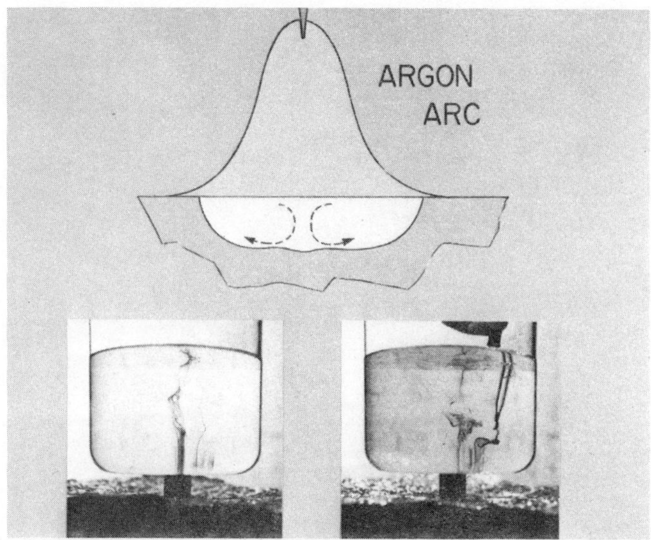

Figure 2 - Illustration of the convection flow which results from a constricted heat input to a beaker of water and the analogous welding situation. Note that the ink rises above the heat source in a well defined column and sinks outside the diameter of the copper heat conductor.

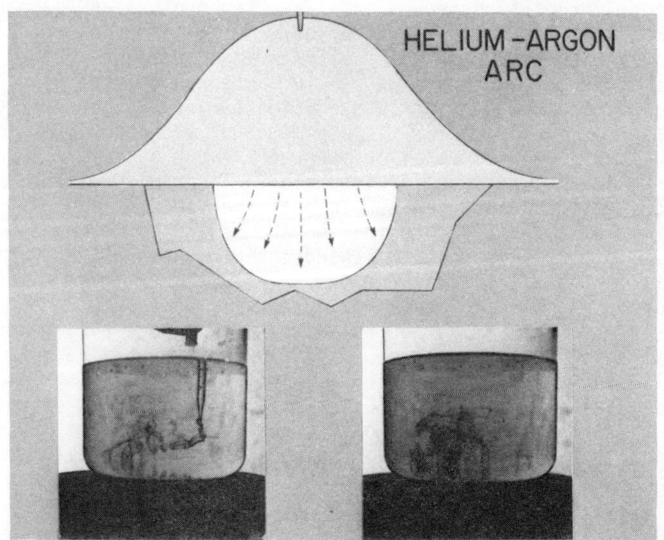

Figure 3 - Illustration of the absence of convection in the case of a broad heat source and the analogous welding situation. Note that there is only random stirring of the ink and no overall circulation.

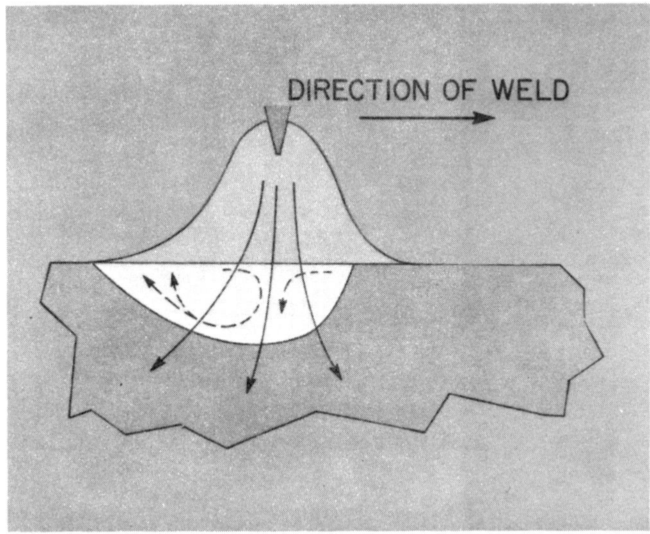

Figure 4 - Schematic of the current flow (solid lines) and liquid metal flow (dashed lines) as seen in a longitudinal section of a weld pool exhibiting front-to-back flow. Enhances weld penetration.

light whose intensity depended on plasma temperature and manganese vapor density. The physical set up of camera, torch and rotating bar is shown in Figure 5. For welds of intermediate to poor D/W, a ring of light concentric with the arc center appeared (Figure 6). For welds of high D/W, a plume of light was emitted behind the arc center and there was no ring (Figure 7). These areas of enhanced light emission resulted from local increases in manganese vapor density [1 2] caused by the surfacing of manganese-rich metal. The ring of light suggested strong convective flow which spread heat transversely and gave a weld of low D/W; the plume behind the arc center corresponded to the front-to-back flow that increased the D/W over that expected on the basis of thermal diffusion alone. While these varied distributions of manganese vapor immediately above the weld pool gave an indication of increased surface temperatures and rising manganese-rich liquid metal, in agreement with the proposed flow patterns, direct indication of the liquid metal flow would be more satisfying.

Conceptually, the flow could be revealed by a series of welds which were terminated at varying intervals after a dopant had been introduced to the weld pool. Various dopants, such as gold and tungsten powder which could be located by radiography plus a palladium-colbalt alloy which could be located by sectioning and chemical etching, were tried. The latter proved to be most satisfactory but a basic fault in the scheme developed as shown in Figure 8. The dark streakes in this longitudinal section of a 150 amp weld in 304 stainless steel depict the progress of the dopant during the approximately 0.2 second interval between introduction of the Pd/Co dopant and freezing of the weld pool. As the dopant prevaded almost the entire weld pool during the shortest interval possible, the technique was clearly not satisfactory. The rapid mixing was evidently due to the swirling motion mentioned earlier and clearly shown in Figure 9.

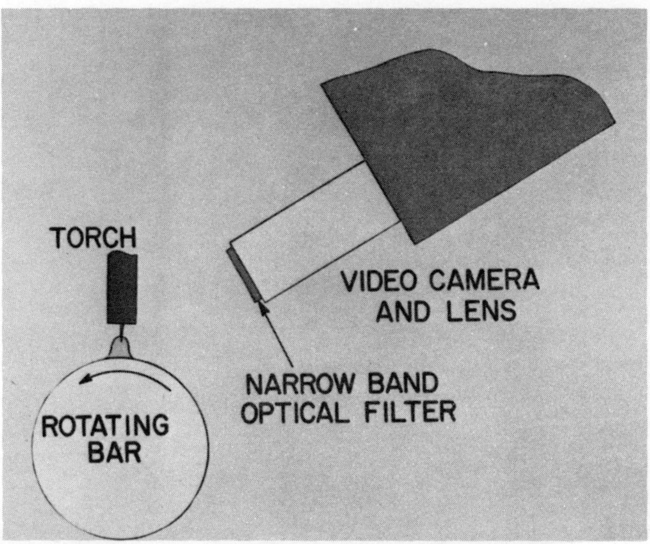

Figure 5 - Schematic of the video camera and weld setup used to make the manganese characteristic emission light images shown in Figures 8 and 9.

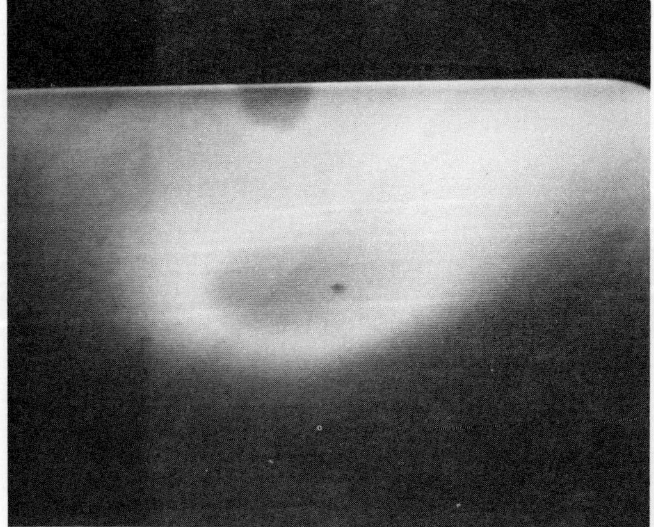

Figure 6 - Photograph of a single video frame from a recording of manganese characteristic light emission above the weld pool of a 100 amp arc on low D/W stainless steel. The electrode tip is visible at the top and the leading edge of the weld pool is visible at the bottom. The black dot is a reference mark on the vidicon face used in obtaining quantitative data.

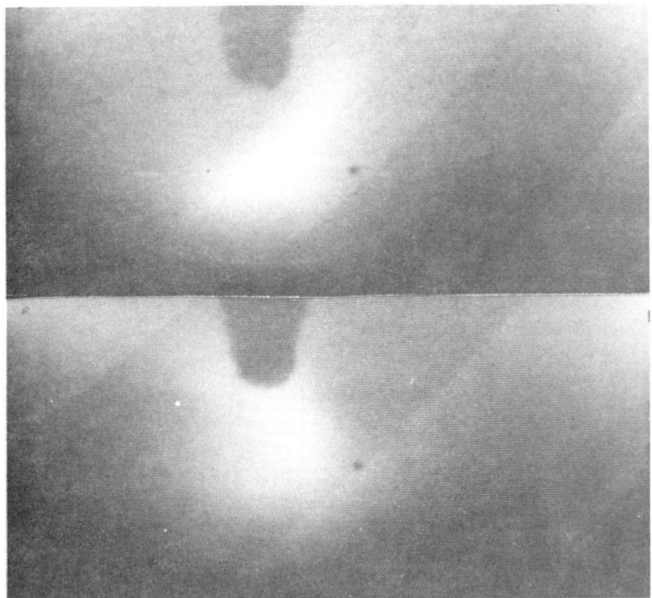

Figure 7 - Photographs of two video frames from a recording of manganese characteristic light emission above the weld pool of a 100 amp arc on high D/W stainless steel. The two views indicate the extent of random osciallation from side to side of the plume as it trails out horizontally over the back of the weld pool.

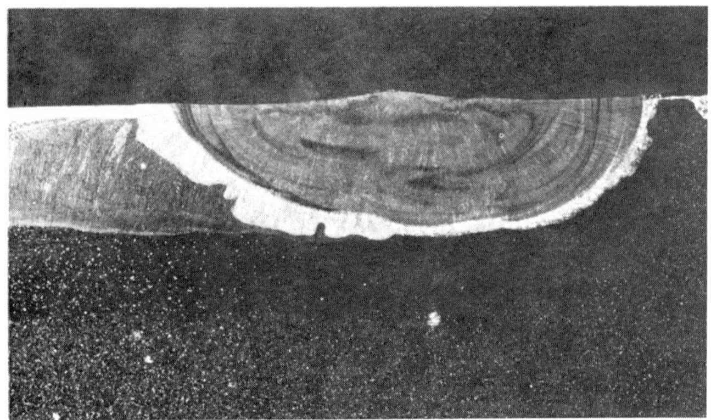

Figure 8 - Transverse section of a 150 ampere weld in 304 stainless steel. Dark streaks in the weld pool show the extent of mixing of the Pd/Co dopant in approximately 0.2 second. Magnification is 10x.

Figure 9 - Planar section of a 150 ampere weld in 304 stainless steel. Chemically stained Pd/Co dopant clearly reveals rotational motion in the horizontal plane. Magnification is 10x.

The speed of mixing in the weld pool was shown even more emphatically by a real-time video recording of the mixing of gold dopant in a similar weld. An x-ray image intensifier system, made available by the Los Alamos Scientific Laboratory, was used to look through the weld pool as the gold was introduced. Counting video frames of the recording made indicated that the mixing took place approximately 0.1 second.

CONCLUSIONS

The seemingly anomalous variations in fusion zone shape (D/W) relative to the current density distributions in welding arcs are explained by the liquid metal flow patterns in the weld pools. Two distinct patterns proposed account for the extremes in D/W observed and qualitative experimental data support the existence of these two flow patterns in welds having the corresponding D/W's. A mode of interaction analogous to thermal convection explains how the welding arc influences metal flow and therefore has a second-order effect on D/W. Better knowledge of liquid metal fluid properties and flow patterns would help to explain the effects of subtle material property changes on weld penetration and is likely necessary for understanding the details of solidification phenomena.

REFERENCES

1. Mills, G. S., "Analysis of a High Manganese Stainless Steel Weldability Problem", Welding Journal, 56(6), June 1977, Res. Suppl., pp 186s-188s.

2. Mills, G. S., "Use of Emission Spectroscopy for Welding Arc Analysis", Welding Journal, 56(3), March 1977, Res. Suppl., pp93s-96s.
3. Mills, G. S., "Fundamental Mechanisms of Penetration in GTA Welding", RFP-2681, Rockwell International, Golden, CO., Nov., 1977.
4. Bradstreet, B. J., "Effect of Surface Tension and Metal Flow on Weld Bead Formation", Welding Journal, 47(7), 1968, Res. Suppl., pp314s-322s.
5. Woods, R. A., and Milner, D. R., "Motion in the Weld Pool in Arc Welding", Welding Journal, 50(4), 1971, Res. Suppl., pp163s-173s.
6. Demyantsevich, V. P. and Matyukhin, V. I., "Characteristics of the Movement of Molten Metal in the Weld Pool During Welding with a Non-consumable Electrode", Welding Prod., 19(10), 1972, pp1-3.
7. Lawson, W. H. S. and Kerr, H. W., "Fluid Motion in GTA Weld Pools Part I: Flow Patterns and Weld Pool Homogeneity", Welding Res. Int., 6(5), 1976, pp63-77.
8. Lawson, W. H. S. and Kerr, H. W., "Fluid Motion in GTA Weld Pools Part II: Weld Pool Shapes", Welding Res. Inst., 6(6), 1976, pp1-7.

Prepared under contract EY-76-C-04-3533 for the Albuquerque Operations Office, U.S. Department of Energy.

DISCUSSION

T. W. Eager (M. I. T.) Questioner

Author: This presentation does not deal directly with the effects of material differences on pool flow. However, work by others at Rocky Flats have suggested that the aluminum affects the surface tensions and the "wetting" of the liquid metal on the parent material.

G. A. Knorovsky (Sandia Labs) Questioner

Author: I really don't have any idea what a chisel-shaped electrode might do but in any case, electrode effects tend to be rather subtle, i.e., easily overcome by other effects. Grounding position is very important because the asymmetry in current flow is what drives the azimuthal rotation.

ARC-WELD POOL INTERACTIONS

Stanley S. Glickstein

Westinghouse-Bettis Atomic Power Laboratory
P. O. Box 79
West Mifflin, Pennsylvania 15122

Abstract

The mechanisms involved in arc-weld pool interactions are extremely complex and no complete theory is presently available to describe much of the phenomena observed during welding. For the past several years, experimental and analytical studies have been undertaken at the Bettis Atomic Power Laboratory to increase our basic understanding of the gas tungsten arc welding process. These studies have included experimental spectral analysis of the arc in order to determine arc temperatures, and analytical modeling of the arc and weld puddle. The investigations have been directed toward determining the cause and effects of variations in the energy distribution incident upon the weldment. In addition, the effect of weld puddle distortion on weld penetration was investigated, and experimental and analytical studies of weld process variables have been undertaken to determine the effects of the variables upon weld penetration and configuration. A review of the results and analysis of these studies are presented.

I. Introduction

Gas Tungsten Arc Welding (GTAW) is a process wherein coalescence of metal is produced by heating the metal with an arc discharge formed between a tungsten electrode and the workpiece. The resultant weld is a product of complex interactions that occur between the electrode, the arc, and the weld pool. A review of the GTAW process was made several years ago by this writer and it was concluded that efforts devoted to improving our understanding of these complex arc-weld pool interactions have found little support in the research and development laboratories in the United States. As a result, a program at the Westinghouse-Bettis Atomic Power Laboratory devoted to fundamental studies of the GTA welding process was initiated. These studies have focused on three major areas of investigation 1) arc physics, 2) heat transfer within the weldment and 3) weld pool distortion.

While we have tried to analyze each of these areas of concern separately, the interaction between the arc, the weld pool, and the heat transfer mechanisms play a dominant role in the weld process. For example, distortion of the weld pool can affect both the size and shape of the weld. The distortion is a result of arc pressure exerted on the surface of the weldment. This pressure results from a changing arc configuration and is influenced by the thermal and electrical properties of the gas. The properties of the gas are affected by vapor emission from the surface of the weldment. The type, and rate of vapor emission is determined by the temperature of the weld pool which in turn is established by the heat transfer processes within the weldment. In addition, the shape of the electrode has a strong influence on the configuration of the arc discharge. Vapor emission can erode the electrode and thus perturb the shape of the arc.

These feedback mechanisms between the electrode, the arc, and the weldment are extremely complex and have yet to be incorporated into a complete analysis. What has been done at this stage of development is to study the arc, weld pool distortions, and the heat transfer process separately. The objective in each of these studies was to establish the important factors that influence the welding process.

The following sections of this paper briefly review and summarize the results of our investigations in these three areas of concern. Details of the studies can be found in References 1-5.

II. Arc Studies

The size and shape (width/depth) of the weld is a strongly dependent upon the magnitude and distribution of the input energy[3]. Because the welding arc is a principle factor in determining the input energy to the weldment, calculational[4] and experimental[1-3] studies have been initiated to investigate the properties of a typical welding arc.

Calculationally, a one dimensional axisymmetric model of the arc as expressed by the energy conservation equation

$$\sigma E^2 - U = \frac{1}{r} \frac{d}{dr} \left(r K \frac{dT}{dr} \right) \tag{1}$$

where

σ = electrical conductivity

K = thermal conductivity

E = electric field strength

U = radiated power per unit volume

r = distance from the axis of the arc

T = temperature

was used to study the radial temperature distribution as it is effected by the temperature dependent thermal and electrical properties of the gas. In particular, helium, argon, and various mixtures of these gases were investigated. The importance of the temperature distribution resides in the fact that once the temperature distribution is known, the current density j can readily be inferred assuming knowledge of the electric field intensity E. This can be obtained from the following equation:

$$j = \sigma(T)E \tag{2}$$

From j, the energy distribution $\mathcal{E}(r)$ can be obtained from

$$\mathcal{E}(r) = j\ (r) \left[V_a + kT + W \right] \tag{3}$$

where

V_a = anode potential drop

kT = thermal energy in plasma

W = anode work function

It should be stressed that a one-dimensional analysis represents the effect in the column region of a long arc. The actual input energy at the arc-weldment boundary is complicated by convection and electromagnetic interactions occurring within the arc. It is these interactions at the weldment boundary that have made the development of the welding arc model extremely complex, difficult, and incomplete. Typical temperature and current density distributions for helium, argon, and argon-helium gas mixtures are shown in Figure 1. Unfortunately, under practical welding conditions the arc contains metal vapors emitted from the weldment during the melting of the workpiece. These vapors interact with the "clean" gas environment. The arc analysis has been extended to include the effects of minor addition of aluminum vapor entering an argon-helium discharge. The results shown in Figure 2 reveal a significant perturbation of the radial temperature and current density distributions of a helium discharge when a small addition of aluminum vapor is present. These results are a typical representation of any low range ionization potential element such as Al, Fe, Ni, or Cr, etc. in helium, whose ionization potential is 24.46 eV. For argon whose ionization potential is 15.08 eV, the effect of impurity contamination due to these lower ionization potential elements is less significant.

Also studied experimentally and analytically were shielding gases containing mixtures of argon and helium. The results show that less than 1% addition of argon gas significantly perturbs a pure helium discharge, while small additions of helium to an argon discharge have little effect on the arc configuration. Measurements of the electric field and analysis of the potential drop associated with each gas indicate that the greater energy of a helium welding arc compared to argon can be attributed to a larger anode drop potential in helium relative to argon. This accounts for the so called "hotter" arc of helium and is not due to the higher arc temperature of the helium discharge as is commonly believed.

The electric field and configuration of the arc are also influenced by the shape of the electrode tip in the GTA welding process. Vapor emission from the weldment can interact with the electrode causing erosion of the tip[6], thus perturbing the arc discharge.

Although these arc studies have emphasized the effects of minor contaminants in the welding gas and their effect on arc configuration, minor elements in the weldment can significantly affect weld penetration through other type interactions that have been discussed elsewhere[6].

While radiation from the welding arc contributes a negligible amount of energy to the weldment, it is an important factor in the energy conservation equation for an argon arc with temperatures greater than 10 000 K. The analytical sutdies clearly revealed that radiation losses reduce the peak temperature along the arc axis and broaden the radial temperature distribution, thus influencing the energy distribution.

An important factor controlling the rate of vapor emission into the arc from the surface of the weldment is the heat transfer mechanism within the weldment. In order to understand the important factors that influence heat transfer, calculational and experimental studies were initiated. The results of these investigations are discussed in the next section.

III. Heat Transfer

Variations in the input energy distribution at the surface of the weldment can be attributed to the arc-weld pool interaction such as explained previously in the arc analysis. The consequences of such variations have been studied analytically[3] using a heat transfer model and the finite element method for calculating the transient temperature, stress, and distortion resulting from the welding process.

Briefly, the energy input from the welding arc is assumed deposited on the surface of the weldment as a radially symmetric normal distribution function. Heat flow is assumed axisymmetric about the center of the heat source. A unique feature of the model is the inclusion of phase change effects such that latent heat is absorbed during melting and liberated during solidification of the weld metal.

Temperature-dependent specific heat and thermal conductivity are employed, and surface heat losses due to convection and radiation are included in the model. Temperature transients were calculated and compared to experimentally measured values at discrete points on the weldment. Typical results are shown in Figure 3. Such comparisons permit an evaluation of the strength of the heat source Q_o, and an esimation of the weld process efficiency η given by

$$\eta = \frac{Q_o}{VI}$$

where

Q_o = strength of heat source
V = welding arc voltage
I = welding current

The analytical and experimental investigation of the welding thermal cycle for stationary gas-tungsten arc welds led to the following conclusions:

1. The welding thermal conduction analysis produces a thermal cycle that is in very good agreement with thermocouple data at locations outside the weld metal and heat-affected zones.

2. Though calculated depth of penetration of the weld was in line with depths measured from metallographic sections, the humped, shallow weld metal configuration observed for stationary test welds could not be reproduced analytically. Introduction of anisotropy of the effective weld pool conductivity, though producing a more shallow weld bead, resulted in bead dimensions that are smaller than the measured values.

3. The nature of the weldment thermal response and the resulting weld bead penetration and shape characteristics were very much dependent on the thickness of the weldment relative to the heat input from the arc. The differentiation between "thin" and "thick" weldment types of thermal response is crucial for proper interpretation of any experimental or analytical welding analysis. When attempting to identify certain characteristics of a welded joint by performing bead-on-plate welds, the thickness of the test weld must be compatible with the actual thickness of the weldment for which the test results are to be applied.

4. The analytical model demonstrated that, for a weld whose depth of penetration is as low as 60% of the thickness, a small increase in current can produce a full-penetration weld.

5. In addition to the magnitude of the heat input from the welding arc, the area over which that heat is distributed on the weldment surface is an additional parameter that strongly influences penetration and weld bead shape characteristics.

6. The finite element welding thermal analysis method is very well suited for determining the interrelationship of heat input magnitude, heat input distribution, duration of the heat input, and weldment geometry on the thermal response. In particular, it can be used for calculating that combination of thermal and geometric parameters that will produce an optimum weld configuration for a given joint.

7. Though the means have been established to specify magnitude and distribution of the heat input from the arc so that an optimum weld metal configuration and weldment thermal response will be obtained, the problem remains as to how these parameters may be controlled by varying specific welding parameters such as current, voltage, electrode shape, arc gap, shielding gas, and insert additives. The previous discussion of the arc studies attempted to direct itself to the effects of various shielding gases and impurities in the arc on just this problem.

As a result of this study on heat transfer within the weldment, it was concluded that heat transfer exclusively by conduction is inadequate to completely characterize thermal conditions in the weld puddle. To accurately predict both depth of penetration and width of the weld metal cross-section, work on weld pool heat transfer, as well as effects of weld pool distortion need to be considered. Distortion of the weld pool is affected by the configuration of the arc. The results of an analytical study which investigated the effects of weld pool distortion on weld penetration is discussed in the next section.

IV. Weld Pool Distortion

The change in the radial current density distribution along the axis of the welding arc coupled with the electromagnetic interaction induces the so called "arc jet". This phenomenon had been discussed and explained in detail by Finkelnburg and Maecker[7] many years ago. The effect of the arc jet is to distort the surface of the weld puddle as shown for example in Figure 4.

An analytical model has been developed by Friedman[5] using the finite element method of analysis, and considers both heat flow in the weldment and changes in pool shape due to both the arc jet and gravitational forces.

The heat supplied by the arc is conducted through the weldment in its distorted configuration. Under the action of the arc pressure and gravitational forces, the heated surface of the weld puddle in the vicinity of the weld centerline is depressed below its original elevation as shown by the distorted finite element grid in Figure 4. It is this depression that significantly affects the penetration characteristics of the weld. Calculations were carried out in increments of time, so that weld puddle distortion progressed in discrete steps, rather than continuously. The distortion of the molten surface is inhibited by the surface tension existing at the interface between the puddle and atmosphere. In the analytical model, it is simulated by a membrane with uniform isotropic tension. The pressure supplied by the arc to the surface of the weldment is taken to be distributed as a radially symmetric normal distribution function.

The effects of arc pressure and gravity forces on penetration and weld bead dimensions are illustrated by determining the growth of the weld pool with time neglecting pool distortion (i.e., neglecting arc pressure and gravity forces or assuming very large surface tension), and then comparing the results with those obtained using representative pool distortion parameters.

The results plotted in Figure 5 show that the distortion of the pool contributed significantly to the penetration characteristics of the weld - especially when the depth of penetration approaches the weldment thickness. When including the effects of puddle distortion due to arc pressure and gravity forces, full penetration occurs after 32 s of heating; neglecting pool distortion, full penetration is not yet attained at 40 s.

The maximum arc pressure resulting from electromagnetic forces is proportional to the square of the current. It is thus of interest to determine the effects of arc pressure on the growth and penetration of the weld puddle. (Although an increase in current produces increases in both heat input and arc pressure, only the latter is considered here).

The results are presented in Figure 6, in which the growth of the pool dimensions with time is plotted for a number of values of the maximum arc pressure. Figure 6 also demonstrates that arc pressure significantly influences weld penetration when the puddle penetrates to approximately 60 percent of the weldment thickness. For these penetrations, the depression of the heated surface of the pool is of sufficient magnitude that the rate of further penetration depends strongly on the degree to which the surface is depressed. The effect on bead width (at the heated (top) surface) is less pronounced, since the influence of pool depression on the radial transfer of heat in the weldment is less than on the heat transfer through the thickness.

From this study, the following was concluded:

1. Distortion of the weld pool contributes significantly to the penetration characteristics of the weld, and the propensity of a weld bead to fully penetrate the thickness of the weldment is enhanced by weld pool depression.

2. Arc current influences weld penetration from two standpoints: (a) increased current results in more heat input from the arc to the workpiece, causing more material to be melted; and (b) arc pressure increases with the square of the arc current, producing a greater degree of pool depression. Application of heat at surfaces beneath the undistorted surface of the weld further enhances penetration. Penetration becomes sensitive to small changes in current when the depth of the pool exceeds 60 percent of the weldment thickness.

3. For partial penetration welds, depression of the top surface of the weld pool varies linearly with arc pressure. When full penetration is achieved, weld bead shape changes are significantly more sensitive to changes in pressure.

4. Both pool depression and weld penetration are strongly dependent on surface tension at the puddle surfaces. The understanding and controlling of surface tension is essential to controlling weld pool shape change, as well as enhancing penetration. But surface tension can be affected by minor elements in the weldment as well as the temperature of the weld pool. Thus, the need for an accurate heat transfer model is again emphasized.

Summary and Conclusions

In order to improve our current understanding of the arc-weld pool interactions that occur during the GTAW process, experimental and analytical studies of the welding arc and heat transfer process have been performed. Temperature in arcs and metal vapor emission from the weld pool produced under typical welding conditions were investigated spectroscopically. This information has provided a basis for comparing experimental results with those of a simplified analysis of the arc discharge. The present analysis employs a one-dimensional model of the welding arc that considered heat generation by the Joule effect and heat losses by radiation and conduction. It has been used to study the effects of various gases and gas mixtures currently employed for welding applications. Minor additions of low ionization potential impurities to these gases have been shown to significantly perturb the electrical properties of the parent gas causing gross changes in the radial temperature distribution of the arc discharge. Such changes are reflected in the current density distribution, input energy distribution of the weldment, and ultimately in weld penetration and configuration.

One of the more important arc-weld pool interactions is the distortions of the weld puddle due to pressure from the weld arc, and inherent changes in weldment heat transfer that accompany such distortion. An analysis of weld pool distortion which treats the effect of arc pressure and surface tension on the pool configuration has recently been completed. The results of this study has significantly contributed to our understanding of how welding parameters influence weld penetration and configuration.

Variations in input energy distribution to the weldment that may occur due to arc perturbations have been input to an analytical model developed for calculating transient temperatures, stresses, and distortions resulting from the welding process. Verification of the thermomechanical analysis has been achieved through experimental studies. An in-depth analytical investigation has also been performed to study the effects of varying a number of welding parameters on the thermal response of the weldment.

While each of these studies have been treated separately, it has been stressed repeatedly that the interaction between the arc, the surface of the pool, and the heat transfer mechanisms are intertwined such that each study is strongly influenced by the other. In additon, the important role of weld pool motion has been omitted at this stage of development. It too must be considered as a major parameter in establishing the solidification pattern of the weld pool. In the future, more effort must also be directed toward incorporating into the analysis the complex feedback mechanisms that have been discussed in this paper.

References

1. S. S. Glickstein, E. Friedman and W. Yeniscavich, "Investigation of Alloy 600 Welding Parameters," Weld. J., 54, Res. Suppl. 113-s, April 1975
2. S. S. Glickstein, "Temperature Measurements in a Free Burning Arc," Weld. J., 55, Res. Suppl., 222-s, August 1976
3. E. Friedman, and S. S. Glickstein, "An Investigation of the Thermal Response of Stationary Gas Tungsten- Arc Welds, "Weld. J., 55, Res. Suppl. 408-s, December 1976
4. S. S. Glickstein, "Arc Modeling For Welding Analysis," WAPD-TM-1382, April 1978
5. E. Friedman, "Analysis of Weld Puddle Distortion and Its Effect on Penetration," Weld. J., 57, Res. Suppl. 161-s, June 1978
6. S. S. Glickstein and W. Yeniscavich, "A Review of Minor Element Effects on the Welding Arc and Weld Penetration," Weld. Res. Council, Bulletin No. 226, May 1977
7. W. Finkelnberg and H. Maecker, "Elektrische Bögen und Thermisches Plasma, Handbook der Physik, Bd XXII (Springer-Verlag, Berlin 1956), 254, English Translation: ARL 62-302 (1962)

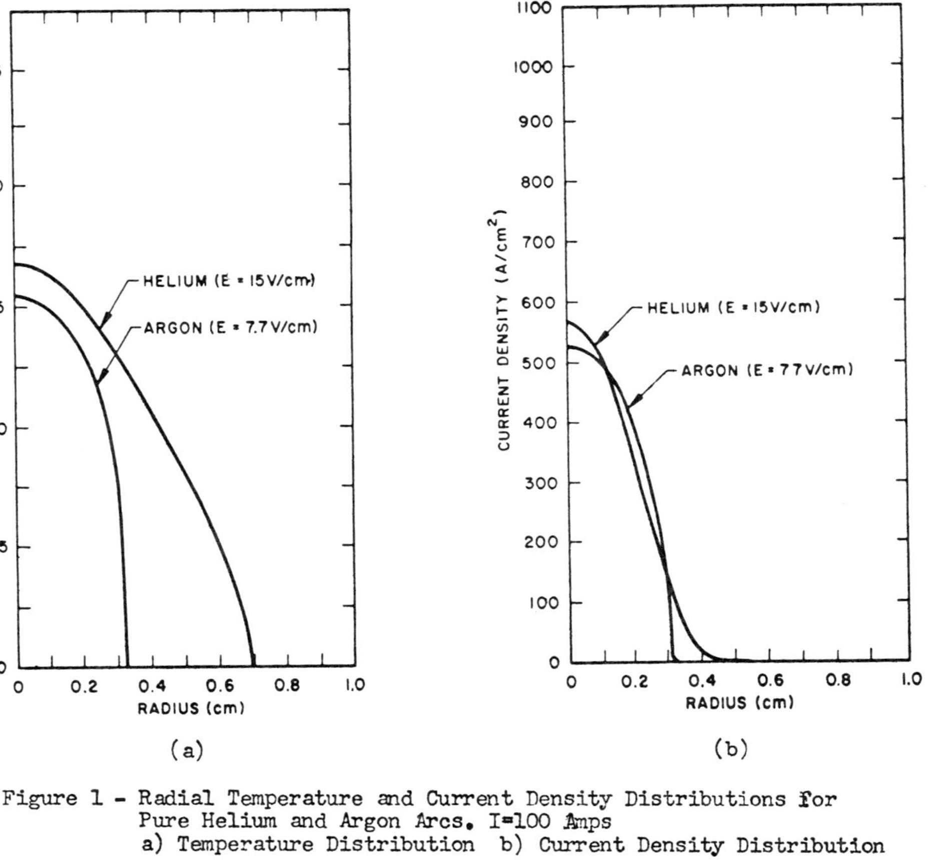

Figure 1 - Radial Temperature and Current Density Distributions for Pure Helium and Argon Arcs. I=100 Amps
a) Temperature Distribution b) Current Density Distribution

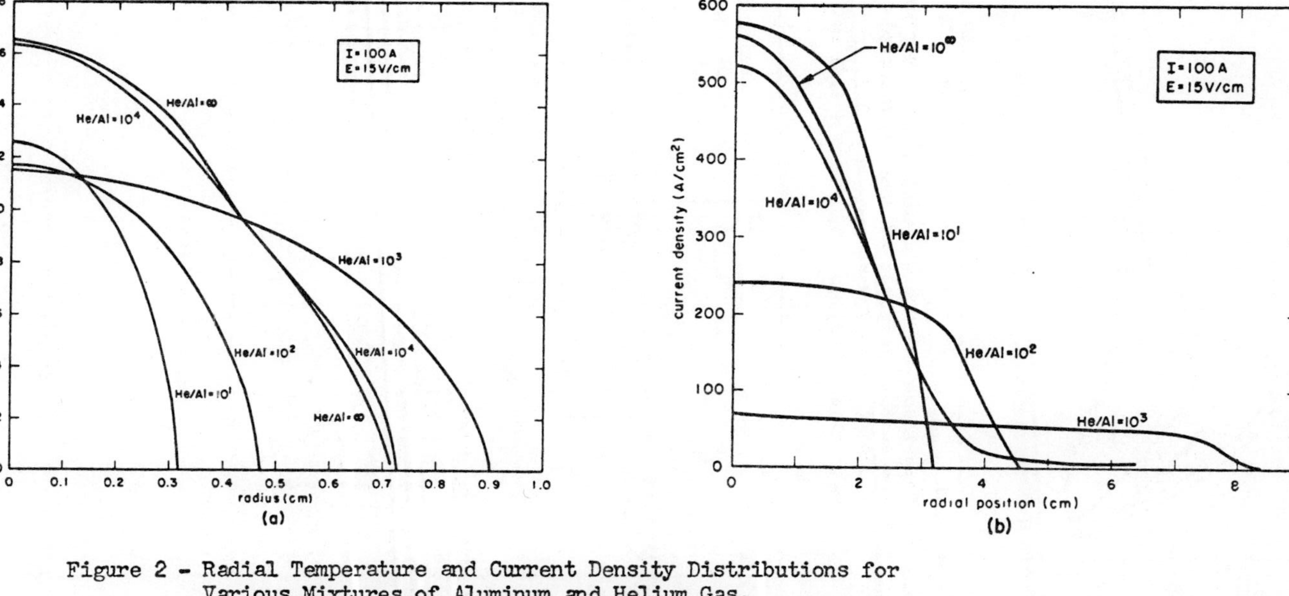

Figure 2 - Radial Temperature and Current Density Distributions for Various Mixtures of Aluminum and Helium Gas.
a) Temperature Distribution b) Current Density Distribution

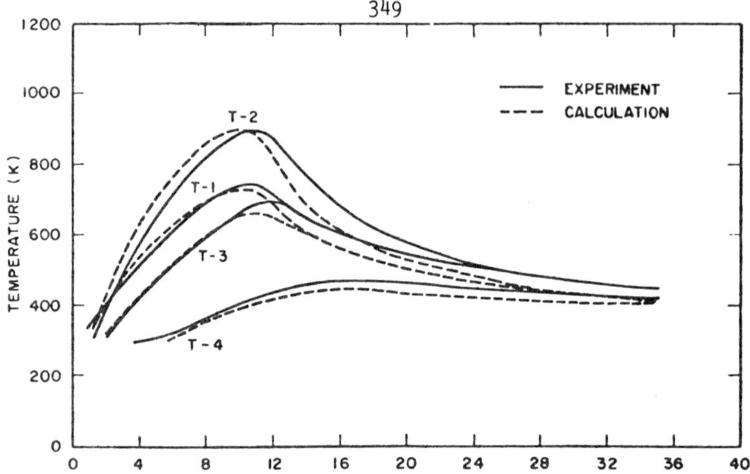

Figure 3 - Comparison Between Calculated and Experimental Transient Temperature Response at Four Thermocouple Positions. I = 91 A, V_a = 7.7 V, Arc Gap = 0.76 mm Weld Time = 10s, Q_o = 530 W, r = 3.81 mm.

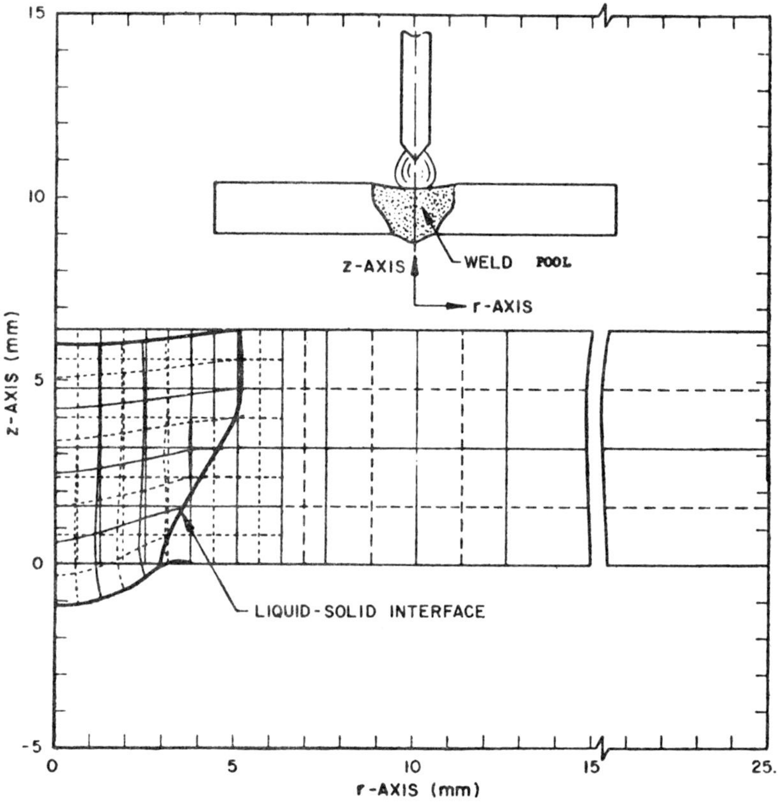

Figure 4 - Weld Bead Distortion As Depicted By Deformed Finite Element Mesh

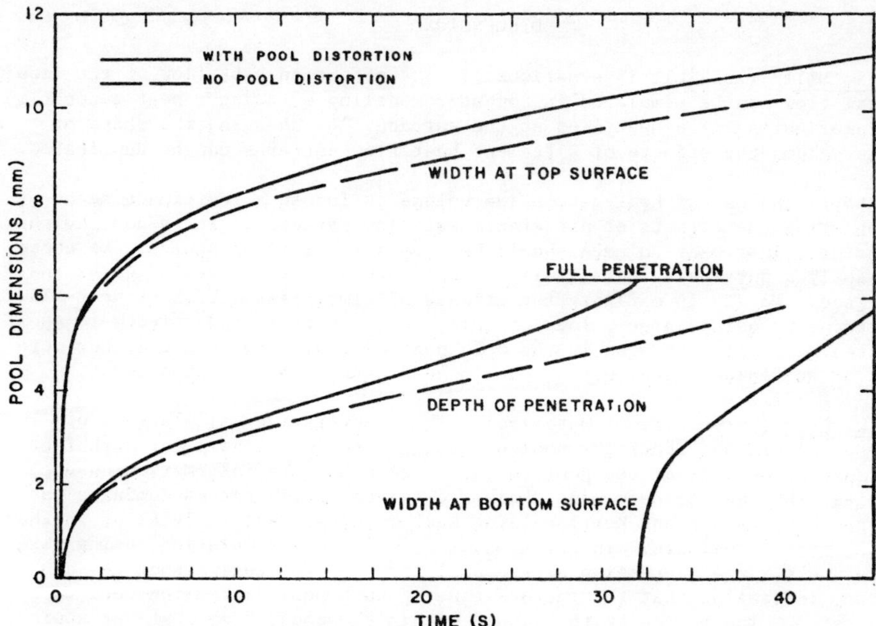

Figure 5 - Effect of Pool Distortion on Pool Dimensions. Pressure = 1000 Pa.

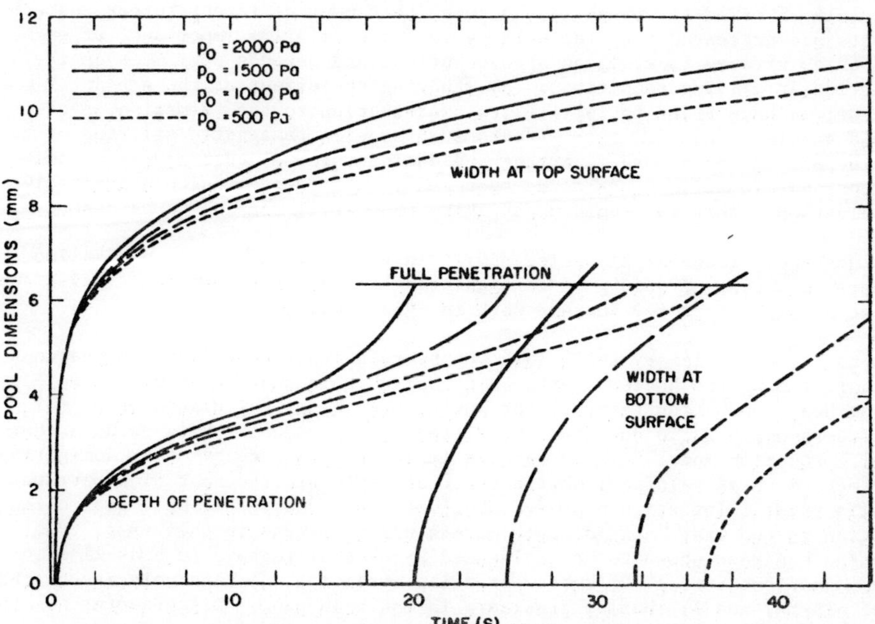

Figure 6 - Effect of Arc Pressure on Pool Dimensions

DISCUSSION

G. S. Mills (Rockwell International): The effect on heat flow of the liquid metal flow can be simulated in computer modeling by using a heat generation volume instead of a heat flux at the surface. By changing the shape of this volume the effects of different heat flow patterns can be duplicated.

Author: The use of heat generation volume is indeed a legitimate means of simulating the effects of different heat flow patterns. It is not obvious a priori, just what volumes should be used as an initial source. We chose to apply a physically more realistic approach by using a heat flux at the surface. We did investigate the effects of simulating different heat flow patterns by using large values of anisotropic thermal conductivity in the molten material. Changes in the weld pool dimensions were noted, but were not of sufficient magnitude to obtain good agreement with experiment.

H. W. Kerr (University of Waterloo): The quantitative calculations of forces within and above the pool depend on assuming numbers for such factors as current density at the pool surface. We need more information on such topics, in order to determine whether electromagnetic forces dominate (as proposed by Lawson and Kerr, Welding Res. International in 1976) or whether other forces dominate. In our studies of weld pool simulation, using mercury pools, some depression of the pool surface was seen without any arc force, indicating that the factors causing such pool depression need to be separated. The motion of the liquid within the pool, from whatever source, will affect the pool surface.

Author: I am in total agreement with Prof. Kerr that we are in need of more information on the phenomena occurring at the surface of the weld. During my talk, I tried to emphasize the point that many different forces and effects are present during the weld cycle. The relative importance of each of these effects is still an open question, and depends very much on the particular welding conditions. By studying the effect of the arc jet force alone, we have tried to isolate its contribution to the formation of the weld puddle. Indeed, other effects such as electromagnetic stirring of the weld pool may be a more important and dominant force under certain conditions, but it still needs to be factored into a comprehensive quantitative evaluation. This is a bold undertaking that still remains to be done.

P. Dembowski (General Electric, MAO): In your presentation, you implied that simulation of weld pool configuration is not reproducible. Please comment on directions for further work in this area.

Author: In stationary arc welds, we observed experimentally a humped configuration at the bottom of the weld pool which cannot be predicted solely from heat conduction theory. For moving arc welds, calculated weld pool width-to-depth ratio could not be fitted to experimental data using a thermal conduction model. These results imply the presence of other dominating effects such as weld pool motion that can influence the heat flow pattern. While several investigators have observed and investigated weld pool circulation in the past, considerably more study is needed in this area. Pool motion has been shown to be influenced by several factors such as 1) electromagnetic stirring, 2) surface tension variation, 3) high velocity of the arc plasma, and 4) thermal gradients in the weld pool. Disagreement exists as to the absolute relative effect of each of these factors and additional detailed qualitative analyses of these effects are needed. These are difficult problems that remain to be evaluated.

Effect of Intermetallic Phase Formation on Electrical and Mechanical Properties of Flash-Welded Al-Cu Couples

J. A. Rayne and C. L. Bauer

Center for the Joining of Materials
Carnegie-Mellon University
Pittsburgh, Pa. 15213

Abstract

Interfacial reactions and resulting degradation of electrical and mechanical properties have been investigated in flash-welded couples of aluminum and copper to correlate microstructure and chemical composition with important engineering properties. In addition to its special practical importance, this system affords opportunity to study nucleation and growth of as many as five intermetallic phases involving six separate interfaces. These phases are characterized by properties which differ greatly from those of their parent elements.

Although all five intermetallic phases are discernible following most degradation processes, total measured phase width w provides a useful overall measure of the reaction products at the original interface. In turn, many important properties are a sensitive function of w. For example, ultimate tensile strength varies from that of the weaker material (aluminum) for $w < 2$ μm to nearly zero for $w > 100$ μm, whereas electrical resistivity varies in proportion to widths of individual phases which define w. General results of this investigation, such as effect of intermetallic phase growth on the transition from ductile to brittle failure and preferred sites for crack nucleation, are presented and possible failure mechanisms suggested.

1. Introduction

One of the most insidious of all natural paradoxes is that joining of two of the best electrical conductors (aluminum and copper) often results in formation of a near electrical insulator at the original interface. The key element of this paradox, of course, is revealed by inspection of the aluminum-copper phase diagram, where it is noted that, below 548°C, as many as five intermetallic phases may coexist. Moreover, these phases nearly represent chemical compounds and, thus, are characterized by properties more closely representing non-metals, such as low resistance to fracture, electrical conductivity, and thermal expansion coefficient. The end result is that utility of both copper and (especially) aluminum, as well as bimetallic connections, is greatly restricted.

Due to formation of several intermetallic phases, usual joining operations must minimize opportunity for phases to form and grow by either mechanical joining at room temperature (cladding, crimping, or bolting) or expulsion of reaction products while still in the liquid phase (flash welding). The principal objective of this research is to relate selected electrical (resistance) and mechanical (ultimate tensile strength and elongation) properties to characteristics of the intermetallic phases in the aluminum-copper system, and thereby deduce possible failure mechanisms. The remainder of this article is divided into several sections: First, relevant background material is reviewed in Sec. 2. Thereafter, experimental procedures are described in Sec. 3, preliminary experimental results are presented in Sec. 4, and significance of these results is discussed in Sec. 5. Lastly, important conclusions stemming from this research are summarized in Sec. 6.

2. Background

The process of flash welding involves establishment of an arc between members to be joined (flashing) followed by an upset forging operation, which produces continuous contact at the interface while extruding most of the molten material. In this manner, reliable welds which exhibit satisfactory mechanical properties can be produced at moderate operating temperatures. Extensive tests by Dixon and Nelson[1] have demonstrated that properties of flash-welded aluminum-copper couples, such as tensile strength, ductility, impact resistance, and electrical resistance do not noticeably degrade at combinations of time and temperature ranging from 2 yr at 149°C to 5 min at 371°C. These combinations may be expressed in terms of an Arrhenius relationship between time and temperature with an activation energy of about 34 kCal/mol. Presence of intermetallic phases at the interface could not be definitively detected for any combination of the above conditions.

A more recent investigation of mechanical properties of aluminum-copper couples, produced by roll bonding, has been conducted by Wallach and Davies.[2] In this case, the couples were subjected to various thermal treatments in order to investigate effect of total intermetallic phase width w on tensile, shear, and impact strength. Such mechanical properties are reduced strikingly as w increases; for example, impact strength is reduced to near zero for values of w of 2 μm. Moreover, fracture occurred by propagaton of a crack between the brittle phases $CuAl_2$ and CuAl for $w < 25$ μm and through the latter phase for $w > 25$ μm. Therefore, important engineering properties of aluminum-copper couples are affected, if not controlled, by presence of detrimental intermetallic phases, which may form during the joining operation and/or during subsequent service.

Important characteristics of the intermetallic phases of aluminum and copper (below 548°C) as well as those of the base metals are reported in Table 1. Growth kinetics of all five phases can be described by a parabolic growth law given by the expression[3]

$$w = kt^{\frac{1}{2}}, \tag{1}$$

where w denotes phase width, t denotes time, and k denotes a time-independent but temperature-dependent rate constant. This constant, in turn, is given by the expression

$$k^2 = k_o^2 \exp(-Q/RT), \tag{2}$$

where k_o denotes a temperature-independent rate constant, Q denotes the apparent activation energy, R denotes the gas constant and T denotes absolute temperature. Measured values of k_o^2 and Q are reported in Table 1, from which it is evident that growth kinetics of the phases also differ greatly as a function of temperature. Many other details of the aluminum-copper system may be found in the extensive review by Vol.[4]

Table 1. Important Characteristics of Low-Temperature Phases in the Al-Cu System: phase designation, nominal composition, crystal structure, square of temperature-independent rate constant k_o^2, activation energy Q, and electrical resistivity ρ at 20°C.[5]

Phase Designation	Nominal Composition	Crystal Structure	k_o^2(cm/sec)	Q(kCal/mol)	ρ(μΩcm)
α	Al	fc cubic	-	-	2.74
θ	$CuAl_2$	bc tetragonal	9.1×10^{-3}	29.3	7
η_2	CuAl	bc orthohombic	1.7×10^{-6}	19.6	9
ζ_2	Cu_4Al_3	monoclinic	2.7×10^{6}	61.2	11
δ	Cu_3Al_2	bc cubic	2.6×10^{-1}	33.5	28
γ_2	Cu_9Al_4	bc cubic	3.2×10^{-2}	31.6	21
β	Cu	fc cubic	-	-	1.70

By inspection of Table 1 it is evident that the value of Q extracted from the data of Dixon and Nelson corresponds to some weighted value of Q for individual phases. Moreover, substituting values of k_o^2 and Q into Eq. 1, yields values always less than 3 μm for w, which empirically corresponds to the onset of degradation of mechanical properties. Since these properties are such a sensitive function of w, it is extremely important to appreciate conditions which affect nucleation and growth of the corresponding intermetallic phases, to interrelate such properties to structure and morphology of these phases, and to develop methods for monitoring concomitant reactions, especially on a submicron scale. A preliminary attempt to accomplish these objectives is described in the following sections.

3. Experimental Procedure

Plates of EC aluminum and OFHC copper were flash welded at the Alcoa Technical Center, New Kensington, Pa., through the courtesy of F. R. Hoch. Specimens were prepared from these plates for standard ASTM (square cross-section) tensile tests, measurement of electrical resistance, and metallographic examination, and then subjected to various annealing cycles under a controlled atmosphere of argon in order to promote formation and growth of anticipated intermetallic phases. Temperatures and times of these anneals ranged from 337 to 516°C and from 3.5 to 80 hrs, respectively. Electrical resistance was measured between 4.2 and 300°K on specimens measuring approximately 2 x 2 x 15 mm on an edge (with the original interface perpendicular to the longitudinal axis) by a standard four-point probe technique, involving a semi-automatic apparatus interfaced to a PDP-11 minicomputer. Tensile tests were conducted on an Instron machine at a strain rate of 0.1 in/min. Interfacial regions were examined metallographically by polishing with, successively, 6 and 1 μm diamond paste and 0.05 μm Al_2O_3, etching in a solution consisting of 10 ml H_3PO_4 diluted with 90 ml H_2O, and then photographically recording selected cross-sections with a standard metallograph at magnifications ranging from 100 to 800X.

4. Experimental Results

Several specimens were annealed under various conditions and then examined metallographically in order to determine widths of intermetallic phases. Typical results are displayed in Fig. 1, wherein all five intermetallic phases are present in a couple subjected to an anneal at 512°C for 10 hr. Individual phases are generally delineated by distinct boundaries

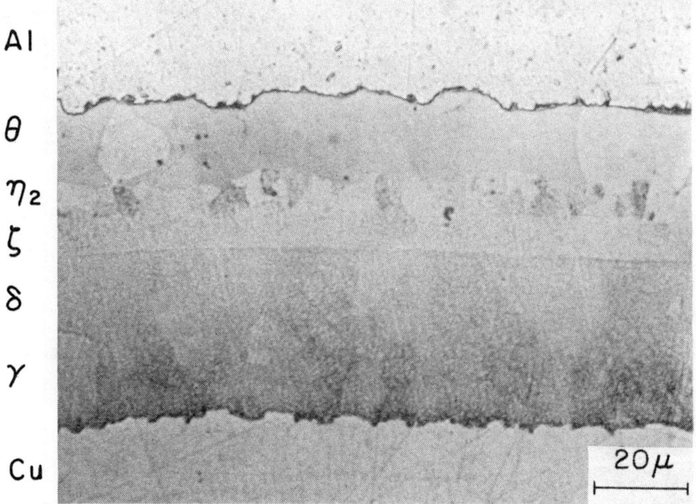

Fig. 1. Interfacial region of a flash-welded aluminum-copper couple following an anneal at 512°C for 10 hr. Five intermetallic phases are discernible.

and are primarily identified by comparison of widths with values predicted from Eq. 1, since neither chemical composition nor crystal structure was determined. Small dimensions of the η_2 and ζ phases and identical crystallographic structure of the δ and γ_2 phases (cf. Table 1) explain why all five intermetallic phases are not always observed. In fact, existence of separate δ and γ phases is not unanimously recognized.[2] In this particular investigation the δ and γ_2 phases appeared to be distinct in some cases and indistinguishable on others (cf. Fig. 1). Intermetallic phases were not discernible in the as-received, flash-welded specimens by optical microscopy, implying $w < 2$ μm.

Although five phases were discernible in most cases, the total measured phase width (thickness) w provides a composite measure of the degradation process. Typical results for 337, 400, and 505°C are presented in Fig. 2 along with values of w calculated from Table 1 (solid lines). Agreement is considered to be satisfactory in light of experimental errors and the difference in method of joining (diffusion bonding vs. flash welding).

Typical results for normalized resistance are presented in Fig. 3 for an as-received specimen and one annealed for 5 hr at 500°C. It is evident that both absolute values of resistance R and resistivity ratio $R_{300}/R_{4.2}$ are affected by the annealing treatment. The values of $R_{300}/R_{4.2}$ are 60 ± 4 and 15 ± 1 for the as-received and annealed specimens, respectively.

Effect of intermetallic phase formation and growth on mechanical properties was determined by measuring maximum tensile stress σ_m and elongation $(\Delta\ell/\ell)_m$ of the standard ASTM specimens. Results are plotted as a function of w in Fig. 4, which shows that σ_m and $(\Delta\ell/\ell)_m$ decrease rapidly when w exceeds a few micrometers. Failure occurred during cooling or handling for estimated values of w greater than about 100 μm. Examination of such failures (or near failures) indicates that cracks form and propagate in the θ phase, whereas, in other cases, cracks form in the θ phase during mechanical grinding and polishing. A typical result is presented in Fig. 5, wherein a crack is clearly visible in the θ phase and/or at a neighboring interface. Significance of these results is discussed in the following section.

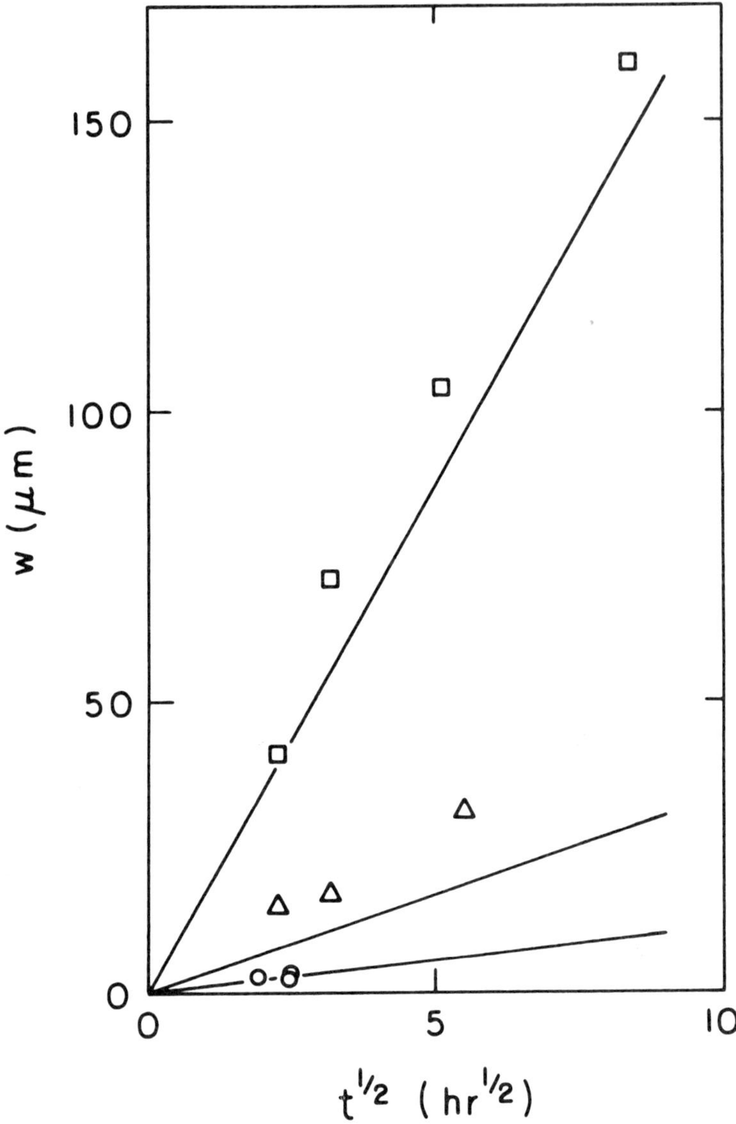

Fig. 2. Variation of total phase width (thickness) w for flash-welded aluminum-copper couples annealed at 337 (○), 400 (△), and 505°C (□) with square root of annealing time $t^{\frac{1}{2}}$. Solid lines correspond to values calculated from Eq. 1 and data of Funamizu and Watanabe[3] at equivalent temperatures.

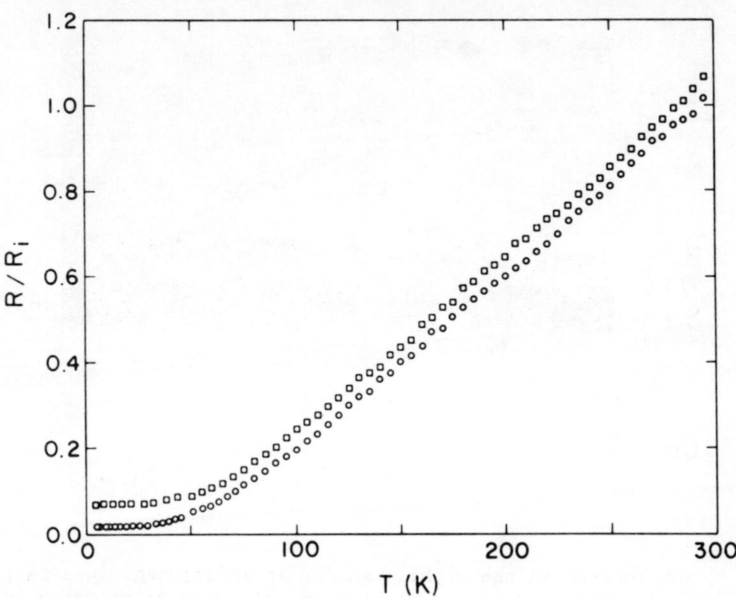

Fig. 3. Variation of normalized electrical resistance R/R_i with temperature T for as-received (o) and annealed at 500°C for 5 hr (□) flash-welded aluminum-copper couples. Data are normalized by the ideal resistance $R_i = R_{295} - R_{4.2}$.

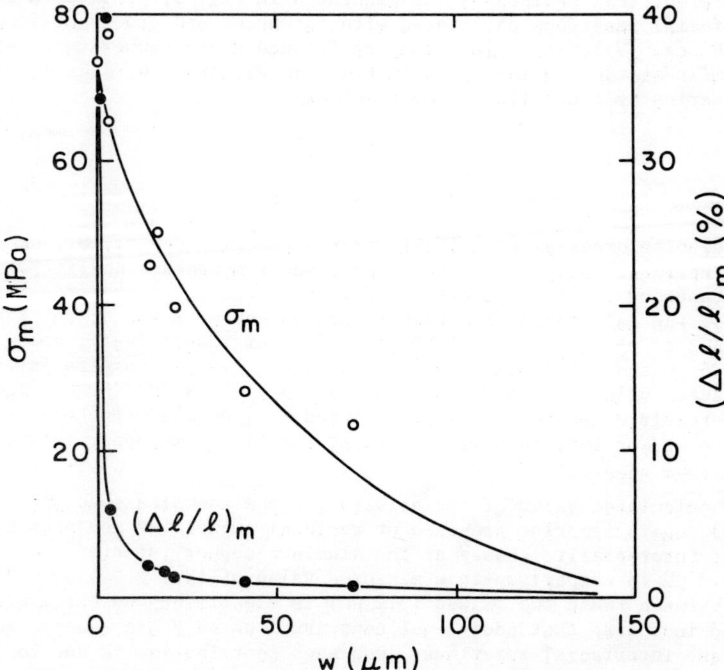

Fig. 4. Variation of ultimate tensile stress σ_m (o) and maximum elongation $(\Delta\ell/\ell)_m$ (●) with total intermetallic phase width (thickness) w.

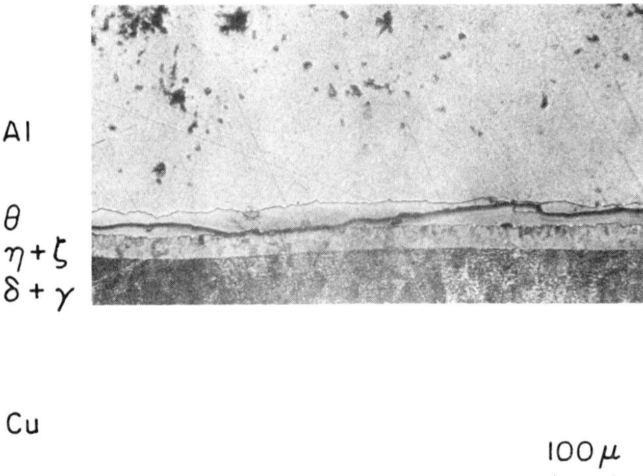

Fig. 5. Crack located in the θ phase and/or at an adjacent interface. These cracks were produced during cooling, handling, or metallographic preparation and, presumably, identify the most brittle phase in the aluminum-copper system.

5. Discussion of Results

The electrical resistance, as presented in Fig. 3, provides a measure of interfacial reactions associated with formation and growth of intermetallic phases. Values of electrical resistance R and temperature coefficient of resistance α may be estimated from weighted averages of resistors in series by the following expressions

$$R = (1/A)\ \Sigma\ \rho_i w_i \tag{3}$$

$$\alpha = \Sigma\ (d\rho/dT)_i\ \ell_i/\Sigma\ \rho_i w_i, \tag{4}$$

where A denotes cross-sectional area of the specimen and ρ_i, w_i, and $(d\rho/dT)_i$ denote electrical resistivity, thickness, and temperature coefficient of resistance of the ith component. For the as-received specimen reported in Fig. 3, the calculated room-temperature values of R and α, using handbook values of ρ_{Cu} and ρ_{Al} of 1.70 and 2.74 μΩcm respectively, are 28.2 ± 0.5 μΩ and 4.0 ± 0.1 x 10^{-3} $^oC^{-1}$. These figures compare favorably with measured values of 28.8 ± 0.3 μΩ and 3.9 ± 0.1 x 10^{-3} $^oC^{-1}$. Therefore, as-received specimens may be analyzed by ignoring effects of residual impurities and/or intermetallic phases at the aluminum-copper interface, at least at room temperature.

The calculated value of resistivity for the annealed specimen reported in Fig. 3, again ignoring presence of residual impurities and possible formation of intermetallic phases at the aluminum-copper interface, is 15.5 ± 0.3 μΩ in comparison to a measured value of 18.7 ± 0.4 μΩ. The difference between these two values (3.2 μΩ) is clearly beyond experimental error and indicates that additional contributions to R are associated with concomitant interfacial reactions. One such contribution is due to increased resistivity of the intermetallic phases (cf. Table 1). The corresponding increment of resistance ΔR may be estimated from the measured thickness and increment of electrical resistivity $\Delta\rho_i = \rho_i - \frac{1}{2}(\rho_{Al} + \rho_{Cu})$

of each phase by a modified version of Eq. 3. Using values for ρ_i reported by either Bauer and Lessmann[5] or Rabkin et al.,[6] ΔR is computed to be about 1.5 μΩ for a total thickness of intermetallic phase (Σw_i) of 41.3 μm. Solid solution alloying can also increase electrical resistivity; for example, 4 w/o of aluminum in copper increases room-temperature resistivity by a factor of 5.[4] The contribution of this increase to ΔR may be estimated if an average diffusion length ℓ in the solid solution regions is known. Assuming a value for the average interdiffusion coefficient of 2×10^{-10} cm^2/sec, ℓ is computed to be aobut 19 μm, which is equivalent to an increment of resistance of about 0.3 μΩ in each terminal solid solution. Accordingly, the total calculated increment of resistance (2.1 μΩ) is in reasonable agreement with the measured value (3.0 μΩ), especially considering uncertainty of reported values of ρ_i and non-planarity of interfacial boundaries. It is not surprising that Dixon and Nelson did not note a tendency for R to increase with increasing annealing time and temperature, since the estimated maximum value of w was always less than 3 μm.

Although resolution of the resistance technique is presently limited to a few micrometers, improved resolution can be achieved by extending analysis to low temperatures and increasing sensitivity of the amplification system. Of course such indirect measurements must be supplemented by direct examination of interfacial morphologies because it is unlikely that each intermetallic phase nucleates and grows in a purely planar configuration. Nevertheless, this particular technique provides opportunity to monitor continuously early stages of phase formation and growth under a variety of experimental conditions.

Mechanical properties provide an even more sensitive indication of intermetallic phase formation and growth. It is evident that maximum elongation (2 in. gauge length) is more sensitive to w than maximum tensile stress, since the value of w necessary to reduce $(\Delta\ell/\ell)_m$ to one-half of its original value is about 5 μm compared to a value of about 30 μm for σ_m, in good agreement with results of Wallach and Davies. Since specimens with a predicted value of w greater than 100 μm were so brittle that they could not be removed from the furnace and mounted for mechanical testing without premature failure, it is believed that differential thermal expansion plays an important role in nucleation of cracks at various interfaces. Although insufficient information is available to accurately estimate thermal stresses generated during an annealing cycle, it is clear that an additional increment should be added to the ordinate of Fig. 3. Similar conclusions were reached by Wallach and Davies.

Although attempts to identify the path of crack propagation by fractography were inconclusive, some indication of the most likely origin of crack nucleation was obtained by noting crack formation and growth which occurred in several metallographic specimens during mechanical grinding and polishing (cf. Fig. 5). In all cases, these cracks were located within the θ phase or at an adjacent interface. The largest difference between thermal expansion coefficients of adjacent intermetallic phases occurs between the θ and η_2 phases (5×10^{-6} $^oC^{-1}$). This difference, coupled with the fact that the θ phase exhibits the lowest tensile strength (8.8 MPa), could explain why failure seems to occur in or near this phase. These observations differ slightly from those of Wallach and Davies, who observed transgranular fracture through the η phase for values of w greater than 25 μm. In this case, however, only three (θ, ζ_2, and δ) of the five intermetallic phases were detected.

Other curious effects observed during growth of intermetallic phases are depicted in Fig. 6. The blocking of phase formation near an inert particle located at the original interface is illustrated in Fig. 6a. Likewise, suppression of phase growth in certain regions is illustrated in Fig. 6b. Finally, a curious eutectic-like region at the interface is illustrated in Fig. 6c. It is likely that these curiosities are related to

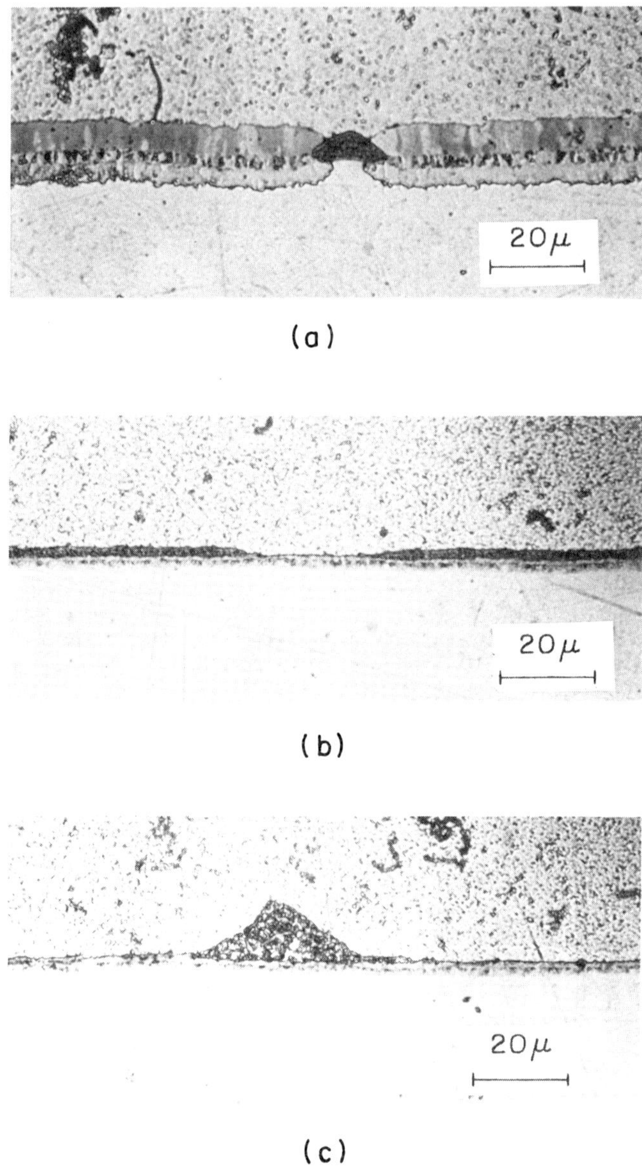

Fig. 6. Selected characteristics of intermetallic phase morphology: (a) Blocking of intermetallic phase formation at an inert particle situated at the original interface, (b) suppression of intermetallic phase growth at certain localized regions, and (c) eutectic-like structure formed by localized melting and solidification during the flash welding operation.

surface preparation and the subsequent flash-welding operation. Indeed, observation of the eutectic-like morphology was observed on numerous occasions and is attributed to imperfect extrusion of pockets of liquid during the upset forging phase of the flash-welding process and subsequent solidification in a two-phase mixture. Some coarsening of this mixture ($\alpha + \beta$) undoubtedly occurs during subsequent annealing processes. Such differences may be at the root of variations in experimental observations on aluminum-copper couples produced by roll bonding (Wallach and Davies), diffusion bonding (Funamizu and Watanabe), and flash welding (Dixon and Nelson, and Rayne and Bauer). Further investigations are in progress to resolve some of these questions.

6. Summary

Interfacial reactions and concomitant degradation of electrical and mechanical properties have been investigated in flash-welded couples of aluminum and copper in order to correlate microstructure and chemical composition with important engineering properties. Although all five intermetallic phases are discernible following most degradation processes, total measured phase width w provides a useful overall measure of the reaction products at the original interface. In turn, many important properties are a sensitive function of w; for example, ultimate tensile strength varies from that of the weaker material (aluminum) for $w < 2$ μm to nearly zero for $w > 100$ μ, whereas electrical resistivity varies in proportion to widths of individual phases which define w. It is concluded that observation of intermetallic phase formation and growth through measurement of electrical resistance coupled with selected mechanical tests can provide valuable information concerning eventual failure of dissimilar metal couples.

Acknowledgements

The authors express appreciation to I. Dustin and P. M. Martin for valuable experimental assistance, to F. R. Hoch and the Alcoa Technical Center for preparation of the flash-welded members, and to the Materials Research Laboratory Section, National Science Foundation for financial support of this research under grant DMR76-81561.

References

1. C. R. Dixon and F. G. Nelson, Communication and Electronics, 1, Jan. 1960.
2. E. R. Wallach and G. J. Davies, Metals Tech. 183, Apr. 1977.
3. Y. Funamizu and K. Watanabe, Trans. JIM 2, (1971).
4. A. E. Vol, "Handbook of Binary Metallic Systems--Structure and Properties" (translated by Israel Translation Program for Scientific Translation Service, Jerusalem) 1, 81 (1966).
5. C. L. Bauer and G. G. Lessmann, Ann. Rev. Mat. Sci. 6, 367 (1976).
6. D. M. Rabkin, V. R. Ryabov, A. V. Lozovskaya and V. A. Dovzhenko, Soviet Powder Metallurgy and Ceramics 92, 182 (1970).

DISCUSSION

G. A. Knorovsky (Sandia Labs): You mentioned that some of your couples fell apart in the annealing oven; did they do this at temperature? or during cooling? If at temperature, perhaps densities would be more important than thermal expansion coefficients. (similar to the Pilling-Bedworth ratio for after oxide coatings on metals).

Author: It was merely noted that specimens, characerized by a total thickness of intermetallic phase exceeding about 50 μm, could not be handled after cooling to room temperature without breaking. Whether the cause of failure is due to disparities in specific volume, thermal expansion coefficients, or to some other phenomenon could not be ascertained. This question, however is certainly important and should be addressed in future investigations.

Fusion Welding of Irradiated AISI 304L Stainless Steel Tubing*

M. M. Hall, Jr., A. G. Hins, J. R. Summers, and D. E. Walker

Materials Science Division
ARGONNE NATIONAL LABORATORY
Argonne, Illinois 60439

ABSTRACT

Fast reactor irradiated AISI 304L stainless steel tubing was fusion-welded using conventional inert gas-tungsten arc welding (GTAW) procedures which were adapted for remote operation. Metallographic examination of weld joints sometimes revealed porosity in the weld zone and cracks at the tube inner wall within the heat-affected zone. A stress analysis was performed to evaluate these defects as sites for weld failure and to establish procedures for the detection and rejection of weld joints likely to fail in service.

INTRODUCTION

An in-reactor creep experiment that is currently being conducted by Argonne National Laboratory in Experimental Breeder Reactor (EBR)-II uses gas-pressurized creep capsules that were fabricated from sections of previously irradiated stainless steel tubing. Since there appeared to be no published accounts of applicable past welding experience, we decided to adapt conventional welding procedures to weld end fittings to the irradiated specimen tubes. One purpose of this paper is to describe the weld-related defects discovered during metallographic examination of qualification-weld samples. Since these weld defects occurred infrequently, no attempt was made to determine why they occurred or to discover ways to control or eliminate them. We relied instead on nondestructive techniques to detect the occurrence of defects, and on proof-tests to reject creep capsules susceptible to weld-related failures. A further purpose of this paper is to present the results of the stress analyses that were performed to evaluate the observed defects as sources of creep-capsule failure. The

---

*Work supported by the U.S. Department of Energy.

results of this evaluation were used to set limits on the size of acceptable defects and to determine the details of the proof-testing procedures. Finally, we speculate on the causes of the observed defects and discuss modifications of the weld-joint design and welding procedures that may improve the reliability of fusion welds of irradiated materials.

MATERIALS AND PROCEDURES

Materials

The AISI 304L solution-annealed stainless steel tubing used in this work had previously been used to encapsulate EBR-II reactor-fuel pins as part of a driver-fuel surveillance program. The chemical composition of the as-fabricated tubing material was analyzed by wet-chemistry techniques, and the results are given in Table 1. Tube sections with a length of 1.375 in. (34.93 mm), a nominal outer diameter of 0.290 in. (7.37 mm), and a wall thickness of 0.020 in. (0.51 mm) were taken from capsule tubes that had previously been irradiated at temperatures between 849°F (454°C) and 908°F (487°C) to integrated fast fluxes between 1.38 x 10^{22} n/cm^2 and 7.51 x 10^{22} n/cm^2 (E > 0.1 Mev). The tubing end plugs were fabricated from AISI 304 stainless steel with the chemical composition given in Table 1.

Table 1. Chemical Composition of Specimen Tubes and End Fittings

	Element (wt%)										
	C	Mn	P	S	Si	Ni	Cr	Ti	Cu	Mo	Co
Tube (AISI 304L)	0.03	1.37	0.01	0.007	0.62	9.26	18.3	0.02	0.074	0.02	0.05
End Fitting (AISI 304)	0.074	1.73	0.034	0.009	0.53	9.35	18.35	-	0.38	0.43	0.15

Weld Procedure

Conventional inert gas-tungsten arc welding (GTAW) procedures were adapted for remote operation. The tubing, end-fitting, and tungsten-electrode configurations are shown in Fig. 1. Welding is accomplished remotely in a shielded cell by fusing the tube and end fitting (without filler metal) while rotating them horizontally under a He-30% Ar cover gas. The welding current and the rotation speed of the work piece were preselected and automatically controlled. Weld parameters are summarized in Table 2. Split copper chills were placed 0.200 in. (5.08 mm) from the tube-end fitting interface to prevent overheating of the tubing. The weld current used for the irradiated tubing (Table 2) was found to be one to two amperes less than that required to produce similar welds with unirradiated tubing.

Inspection Techniques

Metallographic examination of weld joints was accomplished with a remotely operated optical microscope. Nondestructive examination was accomplished by means of x-ray radiography. Unwanted film exposure, caused by gamma radiation emitted from the irradiated tubing, was minimized by using low-speed film and short contact time of the part with the film.

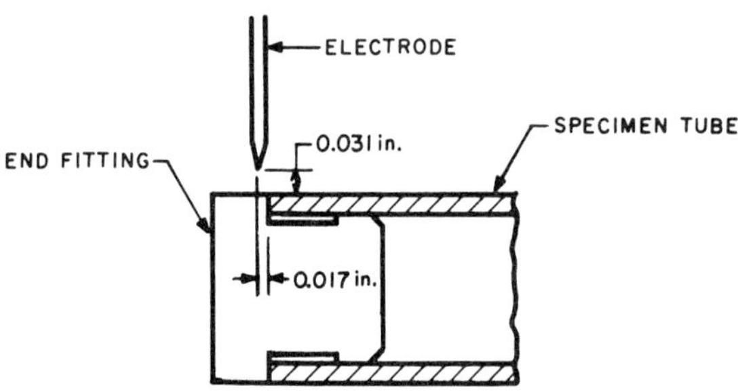

Fig. 1 Tungsten Electrode Placement Relative to Specimen Tube and End Fitting.

Table 2. Weld Parameters

Weld Current	21 A
Final Current	Slope decay to 9 A
Rotation Speed	11 r/min
Rotation Delay	0.1 S
Weld Overlap	40° minimum
Shield Gas	70% He/30% Ar mixture
Gas Flow Rate	20 ft^3/h (9.4 l/s)
Electrode	0.040 in. (1.0 mm) EWTH-2, 25° included angle, 0.010 in. (0.3 mm) blunt end
Weld Chills	Two-piece split copper rings 0.200 in. (5.08 mm) from tubing-end fitting interface

Pressure Prooftest

All creep capsules were prooftested at the design operating pressure and temperature. Each helium-pressurized creep capsule was placed in a vacuum furnace at room temperature, heated to 932°F (500°C), and held for 2 h; during this time, a helium mass spectrometer was used to detect specimen leaks.

RESULTS OF WELD TESTS AND INSPECTIONS

Two weld-related defects were observed during metallographic examination of developmental welds. First, spherical cavities or pores were found in the weld zone; these are shown schematically in Fig. 2, which illustrates typical weld-joint geometry. Pores were, in most cases, <0.001 in. (0.025 mm) in diameter and were found most often near the end fitting-weld zone interface. However, pores were occasionally found near the specimen tube-weld zone interface and could be as large as 0.010 in. (0.25 mm) in diameter. Pores shown in Fig. 3 are typical of those observed most often. The sensitivity of our radiographic inspection technique was found to be sufficient for the detection of pores with diameters

≳0.004 in. (0.10 mm). Of 56 welds inspected by radiography, only one was rejected because of detectable pores too near the tube-weld zone interface.

A second type of weld-related defect consists of cracks, which appear to be grain-boundary separations, along the specimen tube's inner diameter within the heat-affected zone, i.e., within approximately 0.150 in. (3.81 mm) of the tube-weld zone interface. The cracks shown schematically in Fig. 2 and visible in Fig. 4 are typical. These cracks were observed in only 2 of 11 developmental welds examined. Since, however, these defects are very likely sites for creep-capsule failure and are not easily detected by nondestructive examination, all creep capsules were pressure prooftested by means of the test described above. Of 29 creep capsules so tested, only two failures were noted. One failure occurred as a minute pinhole leak near the tube-weld zone interface. The other failure occurred as a catastrophic rupture that included a circumferential separation of the tube and end fitting, also very near the tube-weld interface, as shown in Fig. 5. Both failures appear to be associated with the heat-affected zone. The difference between the pinhole failure and the rupture is probably due to the somewhat higher initial pressure in the latter case.

STRESS ANALYSIS

Porosity

Porosity was observed to occur most often near the end fitting-weld zone interface. It is clear that porosity in this region (Fig. 2, region A) should have little effect on weld integrity. However, the porosity that sometimes occurs near the specimen tube-weld zone interface (region B) reduces the effective weld-tube wall thickness so that abnormally high stresses are expected in this region when the tube is internally pressurized. We could find no solution to this stress problem in the literature. However, a solution to the problem of a strip perforated with a cylindrical hole and subjected to tension, bending, and shear is available.[1,2] We are, therefore, able to obtain an approximate (and, we think, conservative) idea of the actual stresses if we apply the perforated-strip solutions to the longitudinal tube-wall element shown in Fig. 6. This element is taken through the weld zone so as to include a cross section of a spherical pore. That the stresses obtained in this way are conservative can be shown by comparing the solution for a cylindrical hole in a strip of infinite width under uniaxial tension T with that of a spherical hole in a tensile bar of infinite diameter, also under uniaxial tension T. The maximum tensile stress at the edge of the cylindrical hole is 3T, while the maximum tensile stress at the surface of the sphere[3] is 2T. We therefore expect that our approximate solution will be conservative by the ratio of these solutions.

We consider the pore to be located as shown in Fig. 2 and look to Ref. 2 for the bending stress σ_{xx} and the shear stress σ_{xy} across the section mn, caused by the bending moment M and the shearing force Q, respectively. In addition, Ref. 2 gives us the tensile stress $\sigma_{\phi\phi}$ at the edge of the hole as a result of M and Q. To these stresses we must add $pr/2t$, which is the uniform axial membrane stress for a thin-walled tube of radius r and wall thickness t under internal pressure p. Figure 7 shows the resulting axial stress distribution through section mn (see Fig. 2), from the tube midwall to the tube inner diameter, for the case of no pore and for pores having diameter to wall-thickness ratios $R = 0.2$, 0.3, 0.4, and 0.5. Note that for $R > 0.2$, the axial stress at the pore surface becomes greater than the maximum stress that occurs at the tube

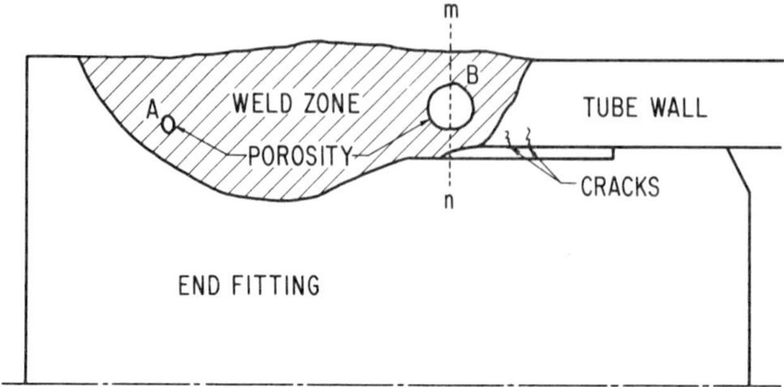

Fig. 2 Typical Weld-joint Geometry Showing Location of Weld-related Defects.

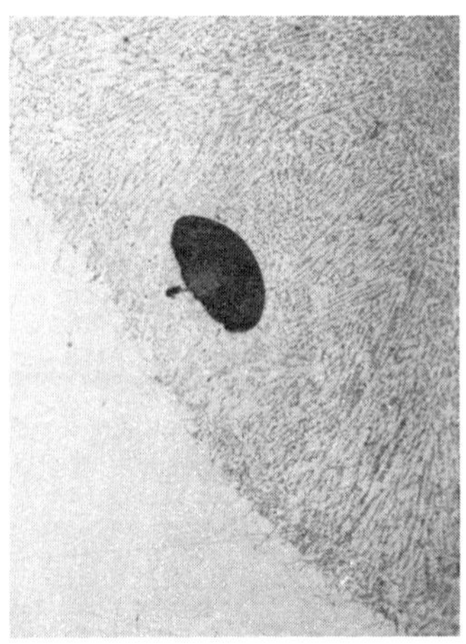

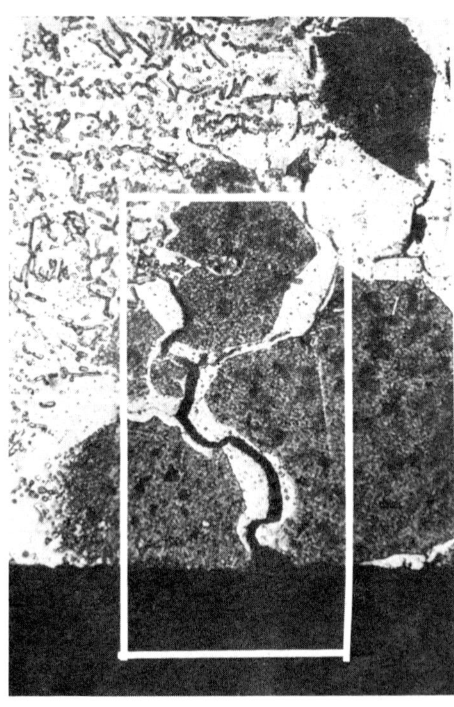

Fig. 3 Porosity in GTAW Fusion Weld of Irradiated AISI 304L Tube. 200X

Fig. 4 Grain-boundary Cracks Appearing in Heat-affected Zone during GTAW Fusion Welding of Irradiated AISI 304L Tube. 500X

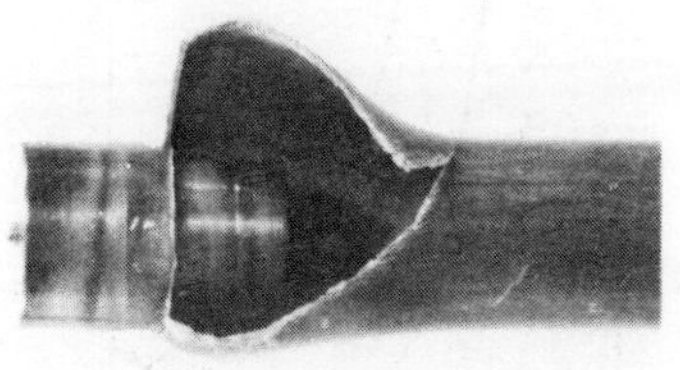

Fig. 5 Creep Capsule Failure that Appears to Have Been Initiated in Heat-affected Zone.

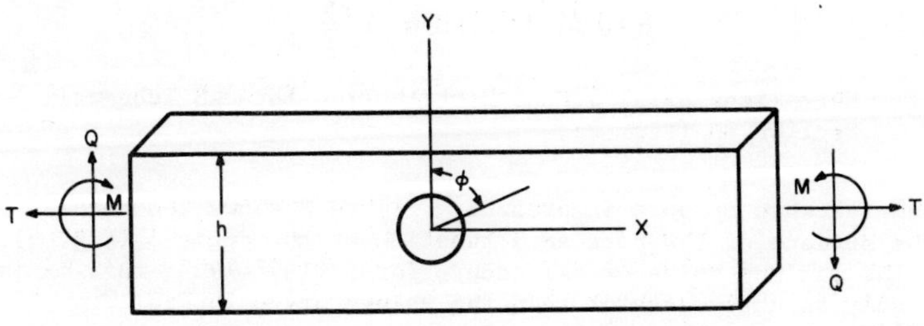

Fig. 6 Longitudinal Section of Specimen Tube Wall, Taken to Include a Cross Section of a Weld Pore.

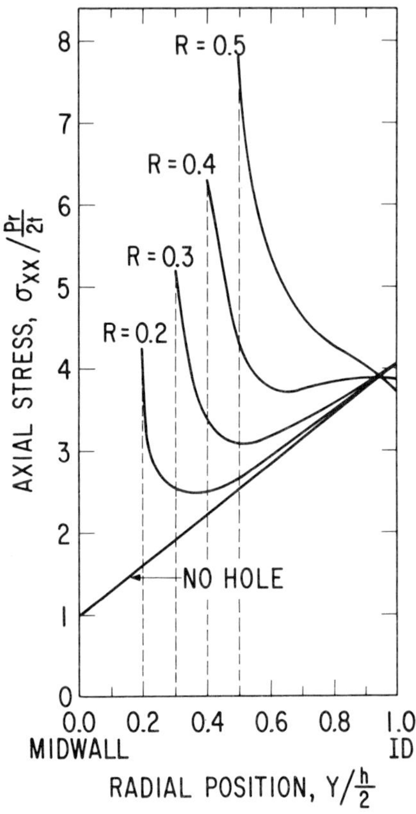

Fig. 7 Normalized Axial Stress Distribution Through Tube-wall Section mn (Fig. 2).

inner diameter when no pore is present. Figure 8 shows the tensile stress $\sigma_{\phi\phi}$ at the surface of the pore as a function of the angle ϕ (Fig. 6). We see that the maximum value of $\sigma_{\phi\phi}$ occurs for $\phi \approx 189°-190°$; this maximum value is only slightly greater than the values given in Fig. 7 [$\sigma_{\phi\phi}$ (180°) = σ_{xx} of Fig. 7].

Fig. 9 shows the distribution of the shear stress σ_{xy} through the section mn (see Fig. 2), from the tube midwall to the tube inner diameter, for the case of no pore and for pores with R = 0.2, 0.3, 0.4, and 0.5. We see that even when R is as large as 0.5, the maximum shear stress is only 0.72 of the pressure stress pr/t, which is the hoop membrane stress, σ_θ, for a pressurized thin-walled tube. Since generalized or uniform yielding occurs for a capped-end tube only when $\sigma_\theta > 1.15\ \sigma_y$ (where σ_y is the yield stress), yielding in the specimen gauge section will occur before σ_{xy} can cause yielding in section mn, even when pores with R as large as 0.5 are present.

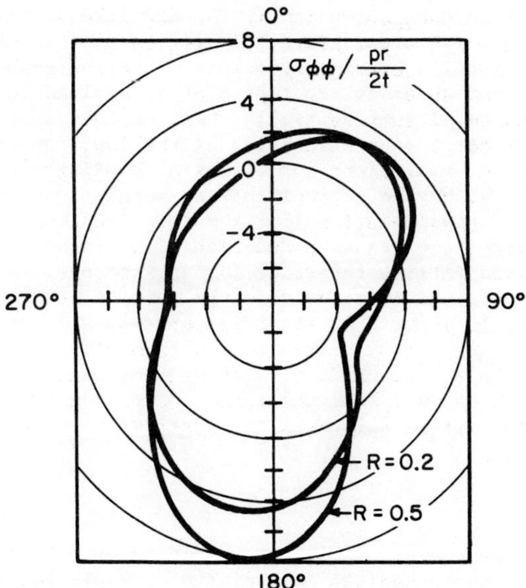

Fig. 8 Normalized Tangential Stress Distribution at a Pore Surface.

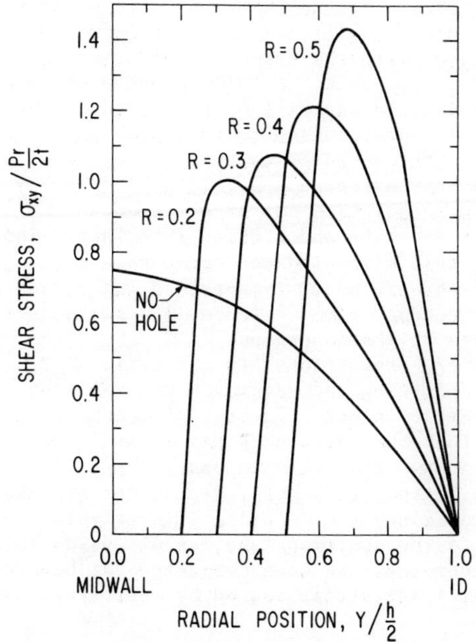

Fig. 9 Normalized Shear-stress Distribution Through Tube-wall Section mn (Fig. 2).

Cracks

Cracks that occur near the weld joint on the inner diameter of a pressurized tube, like ones shown in Fig. 4, are likely sites for initiation of creep-capsule failures, since this region of the tube wall is subjected to large tensile bending stresses. These stresses are caused by the bending moment and shear forces that must be applied to the tube end (which tends to expand circumferentially when the tube is pressurized) in order to prevent diameter discontinuities at the junction of the tube with the rigid end fitting. Solutions for the axial bending stresses, σ_x, and the hoop stresses, σ_θ, may be accomplished by means of cylindrical shell theory.[4] Figure 10 shows solutions for the axial and hoop stresses at the inner and outer tube diameters as a function of a dimensionless distance, βx, from the tube-end fitting interface for an internally pressurized capped-end tube.* For the tubes used in the current study, $\beta = 23.6$ in.$^{-1}$ (0.93 mm^{-1}). Figure 10 shows that the axial stress, σ_{xi}, at the inner diameter can be as large as 3.5 times the uniform midwall stress, $\sigma_x = pr/2t$, and falls to the uniform value at a distance $\beta x = 0.6$, that is, x = 0.025 in. (0.64 mm), from the tube-end fitting interface. Therefore, cracks located within this area are particularly likely sites for failure initiation.

Restrained Swelling

An additional source of stress that could in principle cause creep-capsule failure is the mechanical restraint of irradiation-induced swelling.** The restraint comes about when the tubing, which was previously irradiated well beyond the incubation period for the onset of significant swelling, is welded to previously unirradiated, and therefore low-swelling, rigid end fittings. The difference between the swelling rates of the end fittings and that of the irradiated tubing leads to large discontinuity stresses at the tube-end fitting interface. The tube creeps in response to these swelling-induced stresses and a saturation stress level is attained, with the stress magnitude proportional to the ratio of swelling and creep rates. A full solution of this stress problem is presently being considered and will be submitted elsewhere.[5]

Briefly, a conservative solution to this problem, which ignores local reduction of swelling rate caused by thermal annealing of irradiation microstructure in the heat-affected zone, is arrived at by analogy with the linear thermo-elastic solution for stresses caused by a step increase in tube temperature at the tube-end fitting interface. This analogy[6] is possible because the irradiation-induced creep rate is linearly proportional to the applied stress, under temperature and stress conditions relevant to the present discussion. Figure 11 shows solutions for the axial and hoop stresses at the inner and outer tube diameters as a function of βx. Stresses are proportional to the ratio $\Delta\dot{S}/B$, where $\Delta\dot{S}$ is the difference between end-fitting and specimen-tube swelling rates and B is the irradiation-creep coefficient. In-reactor measurements of these quantities[7] indicate that the ratio $\Delta\dot{S}/B$ may be as large as 45,400 psi (313 MPa). Figure 11 shows that in this case, the axial bending stress may be as large as 30,264 psi (209 MPa). We do not believe that stresses this high are attained, since the actual axial variation in swelling rate from the end fitting to the specimen tube, as discussed below, is more gradual than a step increase. We have demonstrated, however, that there is potential for significant stress caused by restrained swelling.

---

*See Ref. 4 for a definition of β.

**Decreases in density accompanied by the formation of voids.

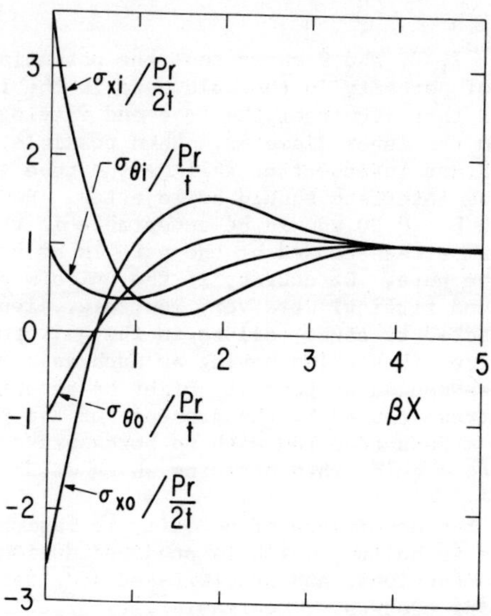

Fig. 10 Axial and Circumferential Discontinuity Pressure-stress Distributions Near the Tube-Weld Zone Interface.

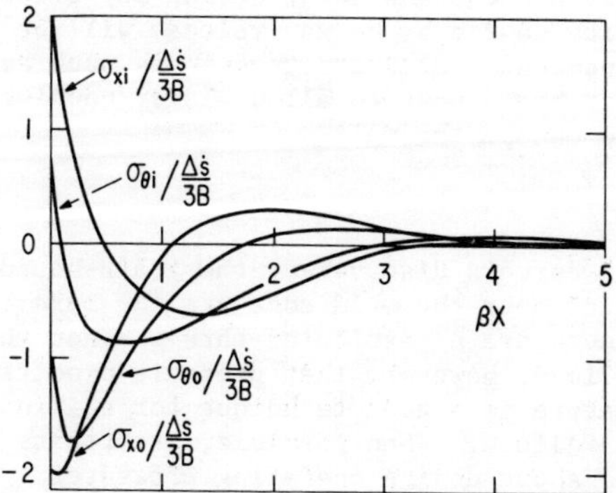

Fig. 11 Axial and Circumferential Swelling-restraint Stress Distributions Near the Tube-Weld Zone Interface.

DISCUSSION

Porosity

Comparison of Figs. 7, 8, and 9 shows that the potentially most damaging stress effect of porosity in the weld zone is the increase in the axial tensile stress that occurs at the tube-end fitting interface from the tube midwall to the inner diameter. When possible, all welds that have pores which either intersect or fall on the tube side of the original tube-end fitting interface should be rejected. However, Fig. 7 suggests that pores with $R < 0.20$ should be acceptable in this region. For $R < 0.20$, the maximum stress caused by the pore is no greater than that expected without the pore. Of course, if the uniform pressure stresses (far from the end fitting) were very low, e.g., less than one-tenth of the stress expected to cause failure in the uniform tube section during the design lifetime, then welds having as much as a 50% reduction in minimum wall thickness caused by porosity might be acceptable. In this case the maximum stress caused by the presence of the pore would be less than ten times the stress expected with no pores present. In any case, Fig. 6 can serve as a guide when deciding which welds to accept or reject.

We conjecture that the occurrence of porosity in fusion welds of our irradiated tubing is due to helium, which is produced during irradiation by (n, α) transmutation reactions, and precipitated and trapped as gas bubbles during the welding process. The feasibility of this explanation is supported by our estimate that, since the He production rate for AISI 304L stainless steel in EBR-II is approximately 6.3×10^{-7} appm/sec, enough He is available within the weld zone to form up to 2000 equilibrium gas bubbles as large as 0.010 in. (0.25 mm) in diameter during the welding of tubing having a fast-neutron exposure of 6×10^{22} n/cm^2 ($E > 0.1$ MeV). However, we do not observe this many bubbles, probably because much of the gas escapes the melt. Other explanations, such as cover-gas entrapment or vaporization of impurities, seem improbable since we made many welds of the same lot of tubing in the unirradiated condition by means of the same weld procedure without seeing any evidence of porosity. Control of porosity caused by He gas release will be difficult, but could perhaps be enhanced by standard techniques such as slower rotational speeds and greater heat input to allow slower cooling and therefore more time for the gas to escape the melt.

Cracks

Of the two types of defects discovered, the grain-boundary separations at the tube inner diameter near the weld zone are the more likely sites for failure. These defects are a particular threat since they are not easily detected. We believe, however, that pressure prooftesting at the design operating temperature is a good technique for elimination of welds that are susceptible to failure. When possible, prooftests should be carried out at pressures above design operating pressures. Although the tubing near the weld zone is softened by the heat of the weld, it is most likely also embrittled by extensive grain-boundary precipitation of He gas.[8] Prooftesting at high pressures should then propagate cracks and eliminate those creep capsules likely to fail in service. Helium embrittlement by precipitation to grain boundaries, coupled with thermal shrinkage of the tube following weld solidification, is in fact the probable cause of the cracks.

If the prooftest temperature is high enough for thermal creep or plastic flow, an additional benefit of the pressure prooftest is that discontinuity stresses may be relieved before insertion of the pressurized creep capsule into the reactor. This may be important since, although a creep capsule may survive the out-of-reactor pressure prooftest at the operating temperature, it may fail in-reactor when swelling-restraint stresses are added to the discontinuity stresses caused by pressure loading alone.

Swelling-restraint stresses, unlike the pressure-induced discontinuity stresses, are not eliminated by creep strain but instead quickly reach a saturation value proportional to the ratio of swelling and creep rates. However, the magnitude of these stresses will not be as large as predicted for a step increase in swelling rate at the tube-end fitting interface, since the actual swelling-rate increase is more gradual. Temperatures in excess of approximately 900°F (482°C) destroy the microstructure responsible for high swelling rate, and cause the full swelling-rate transition to take place over an axial distance within the heat-affected zone. Since the magnitude of swelling-restraint stresses also depends upon the rate of transition along the tube length from the low swelling rates of the unirradiated end fittings to the high swelling rates of the irradiated tubing, swelling-restraint stresses should be reduced by applying procedures which insure a large heat-affected zone. Two procedures that may help are (1) use of no weld chills during the welding process, and (2) application of an annealing heat treatment to the end of the welded tube. As mentioned above, the details of this problem are being considered and are to be reported.

Circumferential collars placed over the heat-affected zone would appear to reduce the probability of capsule failure by lowering the pressure stresses in the area most susceptible to failure by crack initiation and growth. However, these collars would effectively create the step increase in circumferential strain rate discussed above and would therefore insure the maximum possible swelling-restraint stresses, although these would of course occur beyond the heat-affected zone.

Only experience with these various weld designs and procedures will determine which produces the most reliable weld joints. At the time of this writing our creep capsules, which were fabricated by means of the weld procedures described above, have survived 2200 hours in-reactor, without any evidence of failure.

CONCLUSIONS

1. Although the irradiation-induced microstructure of the irradiated tubing will sometimes cause porosity to occur in the weld zone and tube cracks to occur in the heat-affected zone, the nondestructive examination and prooftesting procedures described above can be used to identify and reject those welds likely to fail in service.

2. AISI 304L stainless steel tubing having fast-flux exposures as high as 7.51×10^{22} n/cm^2 (E > 0.1 MeV) may be successfully welded using conventional GTAW fusion welding procedures.

ACKNOWLEDGMENTS

The authors would like to thank F. E. Savoie, who carried out the welding program.

REFERENCES

1. R. C. J. Howland, Proc. R. Soc. London, Ser. A 229, 49 (1930).

2. R. C. J. Howland and A. C. Stevenson, Proc. R. Soc. London, Ser. A 232, 155 (1933).

3. S. P. Timoshenko and J. N. Goodier, Theory of Elasticity, 3rd ed. (McGraw-Hill, New York, 1970), pp. 396-398.

4. S. P. Timoshenko and S. Woinowsky-Krieger, Theory of Plates and Shells, 2nd ed. (McGraw-Hill, New York, 1959), pp. 466-501.

5. M. M. Hall, Jr., "Irradiation Swelling-Induced Stresses in a Circumferentially Restrained Circular Cylinder," to be submitted to Nuclear Technology.

6. G. L. Wire and J. L. Straalsund, "A Simple Method for Calculations of Steady-State Creep Rates in Nonconservative Plastic Deformation," Nucl. Technol. 30, 71-76 (1976).

7. L. C. Walters, G. L. McVay, and G. D. Hudman, "Irradiation-Induced Creep in 316 and 304L Stainless Steels," ANS-AIME Intl. Conf. on Radiation Effects in Breeder Reactor Structural Materials, ed. M. L. Bleiberg and J. W. Bennett (1977), 191-207.

8. M. L. Grossbeck, J. O. Stiegler, and J. J. Holmes, "Effects of Irradiation on the Fracture Behavior of Austenitic Stainless Steel," ANS-AIME Intl. Conf. on Radiation Effects in Breeder Reactor Structural Materials, ed. M. L. Bleiberg and J. W. Bennett (1977), 95-116.

DISCUSSION

D. J. Gooch (CEGB): Do you consider it necessary to perform elevated temperature crack growth tests on embrittled material to determine the maximum defect size which can be tolerated under extended service conditions?

Author: Such tests could possibly have improved our confidence in the ability of our welded tubes to survive the duration of the in-reactor tests. The elevated temperature exposure given each of our test specimens was in fact designed to detect those specimens likely to fail due to growth of weld-related cracks in the heat-affected zone. We have reason to believe, however, that the rate of crack growth in the irradiation environment may be somewhat lower than that in a thermal environment alone. Thus, a limit placed on defect size based on thermal crack growth tests may be too restrictive.

A. Glover (Canadian Welding Institute): Does your stress analysis on porosity take into account arrays of small diameter pores and what is the resolution for your N.D.E?

Author: Our stress analysis does not take into account arrays of pores. However, our analysis can be used to evaluate a situation of multiple pores so long as pores are separated greater than approximately two diameters between centers, since their cumulative effect is then simply additive. The resolution of our x-ray technique sufficient for detecting pores having diameters greater than approximately 0.004 in. (0.010 cm).

R. Carter (General Electric Company): Was helium detected in the pores shown in the weld zone? Could these pores have resulted from coalescence of irradiation generated vacancies in the stainless steel?

<u>Author</u>: We did not test for helium gas in weld zone pores. We cannot, therefore, give direct evidence to support our conjecture that these pores occur when transmutation-product atomic helium is precipitated from the molten metal and trapped during weld solidification. It is highly unlikely, however, that these pores are due to the coalescence of irradiation generated lattice vacancies or voids (conglomerations of many vacancies). The equilibrium thermal vacancy concentration at the melt temperature is three to four orders of magnitude greater than the irradiation generated excess vacancy concentration at the irradiation temperature. The driving force for agglomeration of this small excess of vacancies at the melt temperature is therefore very small. Similarly, the rate of thermal emission of vacancies from voids is quite high at the melt temperature due to the high equilibrium concentration of vacancies, causing voids to anneal away, not coalesce.

Crack Growth During Elevated Temperature Embrittlement of a Gamma Prime Hardened Nickel-Base Alloy

Martin Prager
Consultant, N.Y.

and

George Sines
Professor, University of California at Los Angeles

Abstract

Certain gamma prime hardened nickel-base alloys may experience extensive crack growth during postwelding heat treatment. Preliminary examination of the conditions necessary for cracking indicated that propagation proceeds at relatively low stress intensities and might be due to embrittlement.

To study this behavior, precracked specimens were tested in air at temperatures of 800 to 855 C (1475 to 1575 F) and crack growth was monitored using the voltage drop method. The material selected was a commercial Ni-Cr-Co-Mo alloy containing 1.5% Al and 3.1% Ti. The applied stresses were substantially below the yield strength of the alloy and comparable to residual stresses in welds. Crack growth rates observed were in the range of 1.5×10^{-3} to 5×10^{-1} cm/min (6×10^{-4} to 2×10^{-1} in/min).

The fracture surfaces formed during stable crack propagation were intergranular and similar to those found following postwelding heat treatment. However, all tests terminated with fast transgranular rupture.

It was found that intergranular crack propagation was extremely sensitive to stress intensity at all test temperatures. Specifically, crack growth rates typically appeared to increase with at least the 12th power of stress intensity. In addition, crack growth appeared to be a continuous process. Study of the intergranular portions of the fracture surfaces did not disclose features characteristic of discontinuous growth or plastic deformation.

It is suggested that the high exponential dependence of crack growth rate on stress intensity is indicative of an embrittling reaction between the alloy and oxygen in the test environment. Further, it is shown that the observed behavior parallels embrittlement displayed by the same alloy during tensile tests. Sensitivity to oxygen appears to be influenced by heat treatments which alter the size and distribution of gamma prime. The microstructure expected to be most sensitive to embrittlement would be that found in "as-welded" structures, that is, prior to final heat treatment. However, sensitivity to embrittlement might persist following heat treatment and influence performance in service.

Introduction

Successful fabrication of precipitation-hardenable nickel-base alloys requires more than just depositing defect free welds. The greatest problems are often encountered during the crucial and essential step of postwelding heat treatment. Cracking which can occur at that time may extend far beyond the weld heat affected zone and may even prove to be impossible to repair. In some cases cracking is so severe that length is measured in feet rather than inches. This is somewhat remarkable in light of the fact that it is primarily residual stresses which are responsible for initiation and propagation of these cracks. The transient thermal stresses of heat treating are not an important factor. It is generally accepted that slow heating only aggravates the cracking, but that very rapid heating even with commensurately large thermal stresses, is an effective means of preventing cracking.

Other methods of reducing or eliminating cracking are minimizing weld energy input, using low strength filler alloys, overaging prior to welding, and heat treating in an inert atmosphere. Each of these methods has been found to be beneficial under specific circumstances depending on, among other things, the particular alloy involved.

One of the gamma prime hardened alloys which has been found to be strongly influenced by the heat treating atmosphere is René 41, a commercial Ni-Cr-Co-Mo alloy containing 1.5% aluminum and 3.1% titanium. Elimination of oxygen and water vapor from furnace environments prevented cracking of severely restrained, welded test specimens and was a successful remedy on heavy section production hardware which had cracked consistently when heat treated under an oxidizing (exothermic) atmosphere.[1]

In a parallel study it was found that the ductility of the alloy varied dramatically with the environment and, to a lesser extent, heat-to-heat differences in composition and grain size.[2,3,4] Furthermore, it was found that there was a correlation between ductility and cracking both during restrained patch tests and in manufacturing operations.

The ductility was measured by short-time elevated temperature tensile tests of material in the unaged or solution annealed condition, the condition in which the alloy is usually welded. Thus, the tests were performed on material in a metallurgically unstable condition with nucleation and growth of precipitates continuing during the elevated temperature exposure. In order to obtain meaningful and reproducible data, heating rates, hold time and strain rate all had to be strictly controlled. Details and development of the procedure have been reported elsewhere.[2] It was concluded that for René 41, oxygen promoted the initiation of intergranular cracks and thereby reduced elongation to values of only 1 to 4%, while in inert atmospheres elongations of 10 to 15% were typical. Sensitivity to environmental embrittlement increased with grain size and decreased with gamma prime particle size. The influence of grain size followed the Petch-Stroh model[5]. The gamma prime particle size is important because it determines the tendency for planar slip. Coarse gamma prime, 1000 angstroms or larger favors relatively homogeneous deformation and prevents the stress concentrating pile-ups of dislocations.

While the mechanism proposed adequately accounted for crack initiation, it did not deal directly with crack propagation. However, it was apparent from fractured hardware that crack propagation was intergranular and occurred at relatively low values of stress intensity and was probably rapid since heat treating took only a short time. In a more recent study there has been further confirmation of an environmental effect on the crack growth of this alloy. D'Annessa and Owens[6] used acoustic emission to demonstrate differences in crack growth in air and inert atmospheres. For example, the temperature range over which cracking occurred was reportedly shorter in vacuum and this was accompanied by what was termed "increased resistance to unstable crack propagation." Emissions in hydrogen, helium and vacuum were of lower intensity than those reported in air. It was observed that "the mode of cracking shifts from predominantly rapid (unstable) crack growth in air to less rapid (stable) crack extension in vacuum.../and/... microplasticity features were more prevalent in the fractures of specimens heat treated in vacuum...."

It has been recognized for many years that environmental interactions can accelerate crack growth over particular ranges of stress intensity; however, in most cases the mechanism is poorly understood. Most recently Sadananda and Shahinian[7] have reported that for one of these precipitation hardenable nickel-base alloys "experiments in vacuum show that environment has a significant effect on crack growth behavior...the crack grows rapidly across the whole thickness in a specimen tested in air...but in a vacuum the crack grows less rapidly and it grows mostly in the mid-thickness region..." They concluded that in Alloy 718 crack growth "results from the interaction of two processes; grain boundary diffusion process along with environmental interactions at the crack tip." However, they offer no more specific details of the mechanism of crack growth. It is noteworthy that, like René 41, Alloy 718 has been identified as a material which exhibits embrittlement during specific conditions of short time tensile tests provided the test specimens are previously heat treated in a particular fashion.[3]

A strong environmental effect on elevated temperature ductility of Alloy 718 was first reported by Carlton and Prager.[3] They used specimens which had been solution annealed at 1040C (1900 F) and air cooled prior to aging. At 815C (1500 F) elongation at fracture was observed to be about 15% in pure argon but only about 3% in air. Keiser[8] subsequently performed a broader study which established that for Alloy 718 susceptibility to heat treatment cracking increased when annealing temperatures of 1040C (1900 F) and above were used and when subsequent cooling rates were rapid. Slow cooling from 950C (1750 F) provided a degree of immunity. Keiser found that heat treating in argon assisted in preventing postwelding heat treatment cracking "if occluded oxygen in the 718 is not too high."

Chang[9] identified a number of conditions necessary for what he referred to as "dynamic" embrittlement of gamma prime hardened nickel-base alloys. He reported that this embrittlement occurred "instantaneously during testing in air when the gamma prime network is exposed by surface cracking. The dissolution and/or adsorption of the interstitials is believed to lower the surface energy and fracture strength of the gamma prime phase." Embrittlement was found to be most severe at 870C (1600 F) and at strain rates of 0.05 to 0.5/min. He attributed the strain rate effect on ductility to its influence on the location of crack initiation and propagation. At high strain rates cracks appeared in the grain boundary where the gamma prime network was "susceptible to instantaneous reactions with interstitials." He postulated that the interstitials detrimental to the alloys studied could be nitrogen or oxygen. In general agreement with the earlier work on René 41,[5] Chang concluded the "gross mechanism of embrittlement appears to lie in reduction in surface energy and hence fracture strength of the gamma prime....The surface energy is reduced either by dissolution of the interstitials during air exposure or by their instantaneous adsorption upon testing in air."

While Chang, Keiser, and Prager et al dealt with embrittlement in very short time tests, detrimental environmental interactions are also to be found in tests lasting many hours. However, the mechanisms proposed have not been very different. Chaku and McMahon[10] called attention to the similarity between stress-rupture tests of Udimet 700 in air and stress corrosion tests. They observed that there were 5 times more intergranular surface cracks in air than in vacuum and ductility in air was only about one-half as great as in vacuum. From these tests which lasted on the order of 100 to 1000 hours they concluded that the oxygen effect "comes from internal oxidation along grain boundaries, rather than adsorption of a free surface."

In a later work McMahon and Coffin[11] examined surface cracking and crack propagation in low-cycle fatigue specimens of Udimet 500 (3% Al and 3% Ti). They were unable to distinguish between mechanisms involving oxidation of gamma prime and those where cracks ahead of the main one acted as oxygen sinks. However, they did note that the "present work suggests that the atmosphere plays an important role in the crack propagation process."

Organ and Gell[12] were more specific in proposing a mechanism to explain intergranular cracking during elevated-temperature, high stress fatigue tests of Udimet 700 (4.25% Al, 3% Ti). They suggested that "intergranular oxidation occurs because of the presence of $Cr_{23}C_6$ and other brittle phases having poor oxidation resistance...and accelerated oxygen diffusion along the boundaries. The oxide in the boundary is poorly bonded to the matrix and serves as an easy site for cracking."

Environmental effects on these nickel-base alloys seem to be most pronounced under conditions of high stress and correspondingly relatively short exposure times. This was the conclusion of Gell and Duquette[13] after an extensive survey of the performance of superalloys. They observed that at high stresses fatigue lives in vacuum tend to be superior to those in air, while at low stresses this effect disappears. They identified two possible mechanisms to account for this behavior. In the first "oxygen adsorption at the crack tip accelerates the rate of crack growth in air." Alternatively, "oxidation of grain boundary phases accelerates the rates of intergranular crack initiation and propagation."

In light of the abundant evidence of environmental effects on crack growth it is surprising that environmental effects are usually ignored when properties are determined at elevated temperatures. The dramatic effects of environment on tensile ductility of René 41 and its susceptibility to cracking during postwelding heat treatment, indicated that a study should be undertaken to determine if there is a similar embrittling mechanism in the crack propagation in air.

Test Specimen

To study crack growth in René 41, precracked specimens were tested to failure under constant loads in air at temperatures of 800 to 855 C (1475 to 1575 F). The temperature range chosen was to be that of greatest embrittlement in short time tensile tests. The specimens used were flat tensile bars, 0.150-in. thick by 0.5-in. wide. The material was solution annealed at 1075 C (1975 F) for one hour in argon and water quenched after machining. The heat treatment was the same as that normally used before welding.

A precrack was introduced in each specimen by cantilever bending on a Krouse machine. A small notch was ground on one corner close to the clamped end of the specimen in order to ensure the position of the precrack. Deflections were adjusted to give a maximum surface stress approximately equal to the room temperature yield strength of annealed material. Typically, cracks initiated in less than 100,000 cycles and grew at the rate of 25 to 75 nanometers (1 to 3 microinches)/cycle. Growth was monitored with a 40x hand microscope so cracks could be grown to any desired size.

The crack profile emanating from the corner notch was essentially a quarter of an ellipse with about a 2:1 ratio of major to minor axes. Crack length was defined as the distance on the face of the specimen from the edge of the specimen to the tip of the crack.

Crack growth was monitored by continuous recording of the potential drop across the crack at constant current. Typically, at one ampere the initial potential drop was of the order of 50 microvolts. Sensitivity of the measuring potentiometer was better than 0.25 microvolts.

The potential leads were spot welded above and below the notch (approximately 0.75mm (0.030-in.)) apart. Nichrome was used to reduce thermoelectric effects at the junctions of the leads and the specimen. Stainless steel and other iron-base wires were unsatisfactory because minor temperature differences between the junctions generated an EMF of the order of 1 microvolt.

Potential Drop Versus Crack Length Calibration

The potential drop across a crack is a complex function of the material, temperature, specimen configuration, crack shape and crack size. However, empirical and analytical studies have indicated that for small increases in crack size, the potential drop increases in an approximately linear fashion[14,15], thereafter, behavior is more complex; our study verified this.

Crack size calibration was made in two ways. First, the potential drop at constant current was measured for specimens with initial crack sizes from 1.25 to 2.7mm. (0.051- to 0.107-in.). A second set of data was obtained by interrupting crack growth on several occasions during a test by reducing the load. Various oxide colors then were observed on the final fracture surface. By comparing the resistivity data to the positions of the color changes, a calibration curve for tensile crack size was obtained.

Tensile and bending fatigue resulted in different growth profiles. When the fatigue cracks grew, they tended to maintain their elliptical shape. When the tensile cracks grew, they tended to become more circular. Their different profiles introduces an uncertainty into the calibration.

However, expecting an essentially linear change in resistivity for small increases in crack size, one may use the fatigue data to approximately determine the constant relating the change in resistance to the change in crack length. This is the calibration factor. It was found that large changes in the calibration factor were not critical in determining the crack growth rate and trends. For large increases in crack size, the calibration curve based on the oxidation pattern was used.

Test Method

Crack growth rates were studied by heating the specimen to the test temperature in a clam shell furnace at a controlled rate, rapidly loading to a preselected stress, and then maintaining the load and temperature constant until failure.

The temperature was monitored with a chromel-alumel thermocouple spot welded in line with the crack. Enclosing the specimen in the furnace, the power supply, potentiometer, thermocouple, recorder and tensile machine accessories were permitted to stabilize for at least one hour. Electrical continuity through the tensile machine was interrupted by insulating the load cell. This was necessary so that the potential drop measured would be through the test specimen and not through the machine. The heating rate to the test temperature was held to 14 to 17 C (25 to 30 F)/min through the critical portion of the aging range.

Due to the metallurgical changes which occurred during heating, large changes in resistivity were noted; however, by the time 830 and 858 C (1525 and 1575 F) were reached it was found that resistivity was stable with time. The slower kinetics of the metallurgical changes at 800 C (1475 F) introduced uncertainty in the initial period of tests at that temperature.

Stresses applied to the specimens were approximately 50% of the yield strength at temperature. They were equal to or even less than those which might be expected in the heat affected zone of a weld.

For this study the stress intensity was defined as the product of the stress and the square root of the crack length. Further refinement was considered inappropriate considering all the uncertainties in the measurements and the crack profile. However, since usually only relatively small changes in crack length occurred in the course of most of the data gathering, it is felt that stress intensity as defined above was adequate for the purposes intended herein.

Crack Growth Rates

Initial stress intensities versus failure times are plotted in Figure 1 for the three test temperatures. Times to complete fracture varied from only a few seconds to about 1,000 seconds. The rupture times are corrected for small differences in initial crack sizes as described in Reference 16. From Figure 1, it appears that time to failure decreased with temperature.

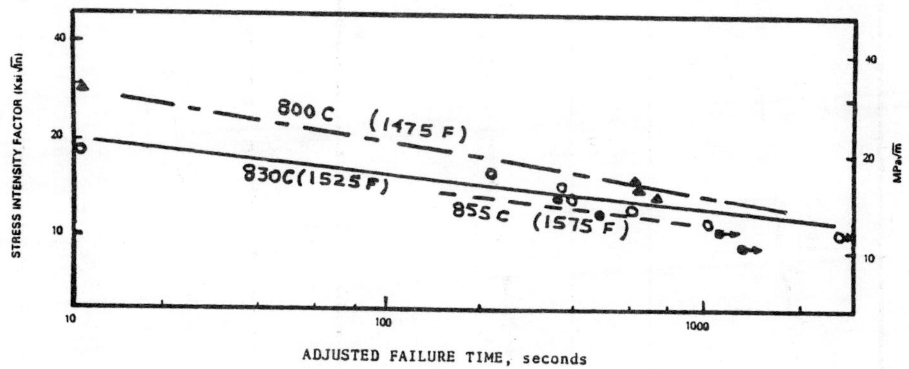

FIGURE 1. Failure time versus initial stress intensity at three test temperatures.

Crack growth rates were determined from the potential drop data and for each crack length the instantaneous value of the stress intensity was also calculated. A least squares curve fitting technique was then used to determine the exponential dependence of crack growth rate on stress intensity. These operations were carried out by computer for several values of the calibration constant. Crack growth rates are plotted versus stress intensity in Figure 2 for an intermediate value of the calibration factor. Crack growth rates observed were in the range of 1.5×10^{-3} to 5×10^{-1} cm/min (6×10^{-4} to 2×10^{-1} in/min). The slope of the line, the crack growth exponent, was high and appeared to increase with increasing temperature. It is noteworthy that all the data fell within narrow scatterbands. Since the stresses were different it appears that growth rate was actually dependent on the stress intensity.

Crack growth, once begun, was a continuous process; observations of the fracture surfaces at up to 1000X failed to reveal any bands, striations, or other markings indicative of discontinuous crack growth. In all respects, the fractures obtained by slow crack growth were identical to those obtained by tensile test and in each case, the tests terminated in fast transgranular shear fractures. Under the microscope the intergranular portions were free from indications of deformation.

During the initial period of crack growth there did appear to be a slight departure from the general behavior. After loading, there was a short period during which crack growth rate decreased. The duration of this period decreased

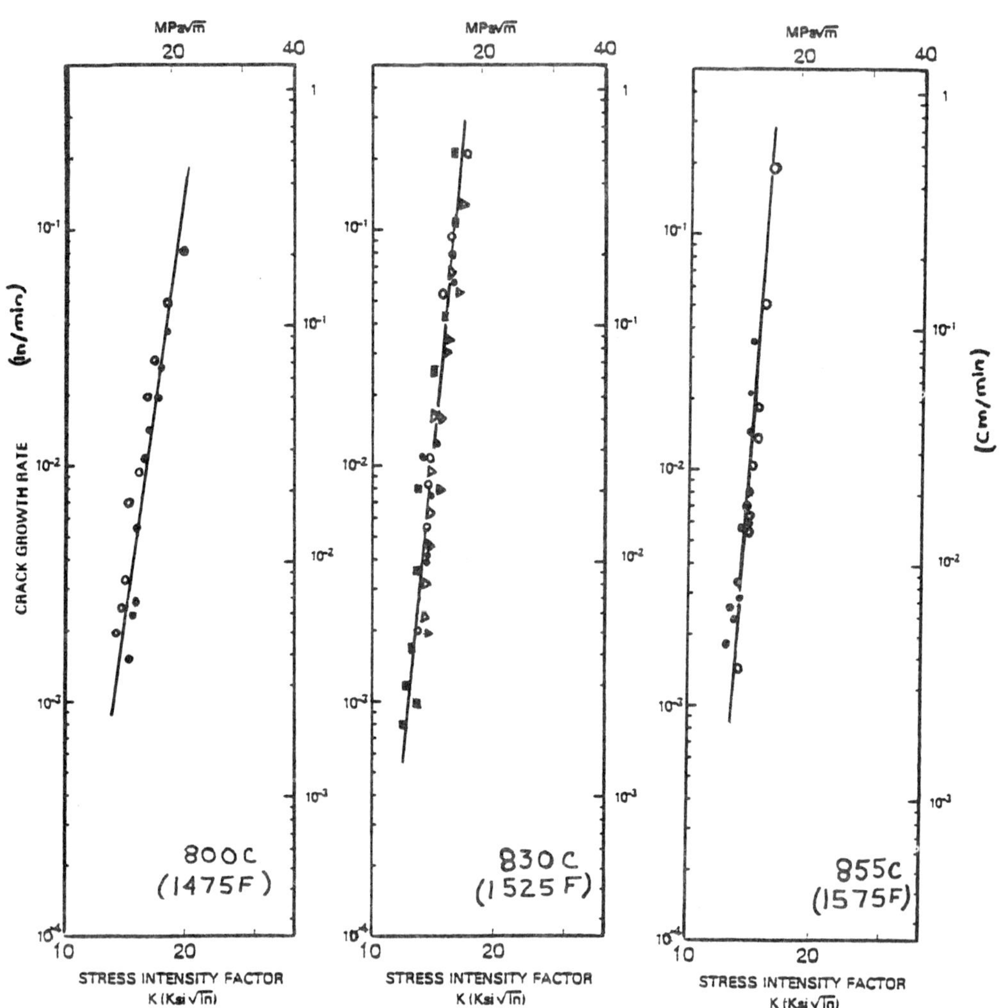

FIGURE 2. Crack growth rate versus stress intensity.

with increasing stress intensity, while the growth rate increased with increasing stress intensity. After a critical crack size was reached, it appeared that the crack accelerated in accord with the high exponential dependence noted above. It was surmised that this behavior was associated with the passage of the crack through the cold worked zone at the tip of the precrack. However, it was concluded that there was probably no incubation period.

Effect of Exposure at Elevated Temperatures

In one series of tests an evaluation was made of the influence of exposure at elevated temperature. The crack growth rate of a specimen preexposed 100 minutes at 830 C (1525 F) was much lower than that for a specimen with only nine minutes exposure. A similar effect was noted at 800 C (1475 F) on a specimen aged forty-five minutes at that temperature prior to the application of load. It lasted 50% longer than an unaged specimen.

Discussion

The crack growth reported displayed four characteristics. It was intergranular, progressed at low values of stress intensity (considering the strength of the alloy involved), was extremely sensitive to the stress intensity and was rapid.

The observation that crack growth exponents were high, at least 12 and possibly higher deserves special comment. Any attempt to refine the estimate of the calibration factor by compensating for some of the probable sources of error would have resulted in still higher values for the exponents. The relatively small scatter in the crack growth rate versus stress intensity data shows that the calibration procedures were adequate.

A survey of crack growth behavior of other materials for comparable exponents revealed some parallels. Generally, very high exponential dependence is associated with adsorption controlled processes such as liquid metal or complex-ion embrittlement or static fatigue of glass. Stress corrosion of aluminum alloys, season cracking of brass or strain aging of steels, processes controlled by diffusion or film formation are typically characterized by exponents which are smaller, usually less than 4. Similarly, diffusion controlled processes typically yield slower crack growth rates than were observed in this study.

High exponents and rapid crack growth at low stress intensities have been reported by Floreen and by Sadananda and Shahinian in studies of Alloy 718. Floreen[14] largely ignored the possibility of an environmental effect, but Sadananda and Shahinian[7,17] noted that their tests in vacuum showed that the high propagation rates observed were attributable to an environmental effect rather than simply creep processes at the crack tip.

Examination of a number of models for interactions at the crack tip shows that the growth rate could be consistent with the kinetics of chemisorption at the tip of a narrow crack filled with two gases, one reactive (oxygen) and one inert (nitrogen). Such a model requires that the surface energy be significantly reduced when the most rapidly reacting adsorption sites are occupied. Transport of oxygen to the crack tip then may be shown to be capable of limiting crack growth over the full range of rates observed.[16] This may be done using diffusion theory and considering the length and height of the crack, the temperature and molecular weight of the gases. Then from the flux toward the crack tip a significant fraction must be subtracted because the oxygen will react with material which has been exposed on the walls of the crack. As a result the partial pressure of oxygen at the tip of the crack is considerably less than the value in the surrounding air.

When these calculations are performed it is found that the flux of oxygen atoms to the crack tip could sustain the observed crack growth rates.[16] At the slowest rates there is sufficient time for the tip of the crack to be saturated with chemisorbed oxygen atoms. At the more rapid crack growth rates a full monolayer of oxygen can not form as the crack grows. However, saturation of the surface is not necessary since the model calls for the greatest reductions in surface energy, and hence strength, to be associated with occupation of the preferred adsorption sites at the tip. This reasoning leads to the conclusion that the crack growth rate can appear to be sensitive to the stress intensity factor in the manner observed i.e. a high exponential dependence.[16]

The foregoing assumes that processes leading to crack tip blunting are not important. As shown by Sadananda and Shahinian[17], blunting can result in decreasing growth rate with increasing temperature. The reduction in crack growth rate with increasing time at temperature observed in this study may well be associated with blunting and more homogeneous deformation expected at the crack tip when the gamma prime is coarse. The well known success of overaging and related preweld heat treatments in preventing cracking during postwelding heat treatment suggests that this is true. Cracks formed in the heat affected zone of overaged material are not likely to propagate into parent metal. The results of this study point toward the conclusion that the threshold stress for crack propagation in overaged material is higher than in material which has been quenched during or prior to welding and, as a result, generates fine gamma prime in the first stages of postwelding heat treatment.

In earlier work of the authors[4,5] it was shown that the extremely fine gamma prime which precipitates in the heat affected zone as the weld cools and then throughout the parent metal on heating during postwelding heat treatment is most susceptible to crack initiation. This study indicates that the same microstructural condition may also be the least resistant to crack growth.

It appears that too little attention has been paid to environmental effects on precipitation-hardenable nickel-base alloys. It is often assumed that because they are oxidation resistant and can be used under oxidizing conditions, precautions need not be taken during heat treatment to prevent exposure to oxygen or other interstitial contaminants. However, where stress or temperature is high, even when exposure is short, detrimental environmental interactions appear possible. The work of Chang[9] and similar studies of one of the authors with René 41 indicates that pick-up of interstitials during exposure in air at temperatures of 900 C (1650 F) and above can degrade performance at lower, but nevertheless elevated, temperatures. The coarsening of gamma prime which occurs during the normal aging operations for many of the alloys tends to alleviate sensitivity to the environment, but the studies cited demonstrate that the higher the stress the more significant will be the environmental effect. This is true even when exposure is of short duration.

The fact that design stresses normally are low tends to keep susceptibility to embrittlement from being a major problem in service. However, the same is not true during fabrication when residual stresses may be high. Once cracks are formed they can propagate readily. The threshold stress for crack growth is so low that defects on the order of grain dimensions can start propagation. Since these gamma prime hardened alloys contain elements which are strong oxide formers, prevention of environmental interactions requires high purity environments especially if exposure under stress is of long duration as during heat treatment.

References

[1]Thompson, E.G., Nunez, S., and Prager, M., "Practical Solutions to Strain-Age Cracking of René 41," Welding Journal, 47 (7), 2995-3125 (1968).

[2]Fawley, R.W., and Prager, M., "Evaluating the Resistance of René 41 to Strain-Age Cracking," Welding Research Council, Bulletin No. 150, 1-12 (1970).

[3]Carlton, J.B., and Prager, M., "Variables Influencing the Strain-Age Cracking and Mechanical Properties of René 41 and Related Alloys," Welding Research Council, Bulletin No. 150, 13-23 (1970).

[4]Prager, M., and Sines, G., "A Mechanism for Cracking During Postwelding Heat Treatment of Nickel-Base Alloys," Welding Research Council, Bulletin No. 150, 24-32 (1970).

[5]Prager, M., and Sines, G., "Embrittlement of Precipitation Hardenable Nickel-Base Alloys By Oxygen," Trans. ASME, Journal of Basic Engineering, Vol. 93, Series D, (2), pp.225-230, (1971).

[6]D'Annessa, A.T., and Owens J.S., "Effects of Furnace Atmosphere on Heat Treat Cracking of Rene 41," Welding Journal, 52 (12), 568s-575s (1973).

[7]Sadananda, K., and Shahinian, P., "Effect of Specimen Thickness On Crack Growth Behavior in Alloy 718 Under Creep and Fatigue Conditions," Characterization of Materials for Service at Elevated Temperatures, G.V. Smith Ed., MPC-7, a publication of the Metals Properties Council, distributed by ASME, pp. 107-119, (1978).

[8]Keiser D.D., Heat Treat Cracking In Inconel 718: Its Cause and Prevention, Aerojet Nuclear Co., TID-26773, 224 pages, (1975).

[9]Chang, W.H., "Tensile Embrittlement of Turbine Blade Alloys After High-Temperature Exposure," Proceedings of an International Symposium on Superalloys, Seven Springs, Pa., Sponsored by High Temperature Alloys Committee AIME pp.v-1 to v-41, (1972).

[10]Chaku, P.N., and McMahon C.J., "The Effect of an Air Environment on the Creep and Rupture Behavior of a Nickel-Base High Temperature Alloy," Met. Trans. 5(2), 441-450 (1974).

[11]McMahon C.J., and Coffin, L.F., "Mechanisms of Damage and Fracture in High-Temperature, Low-Cycle Fatigue of a Cast Nickel-Based Superalloy," Met. Trans. 1(12), 3443-3450 (1970).

[12]Organ, F.E., and Gell, M., "The Effect of Frequency on the Elevated Temperature Fatigue of a Nickel-Base Superalloy," Met. Trans. 2(4), 943-952 (1971).

[13]Gell, M., and Duquette, "The Effects of Oxygen on Fatigue Fracture of Engineering Alloys," Corrosion Fatigue, Proceedings of a NACE Conference, University of Conn., pp.366-378, (1971).

[14]Floreen, N., "The Creep Fracture of Wrought Nickel-Base Alloys by a Fracture Mechanics Approach," Met. Trans. A, 6A (9) 1741-1749, (1975).

[15]Anctil, A.A., Kula, E.B., and DiCesare E., "Electric Potential Technique for Determining Slow Crack Growth," Proc. ASTM 63, 799-808 (1963).

[16]Prager, M., Mechanisms of Environmental Embrittlement, The Influence of Oxygen on Nickel-Base Alloys, Ph.D. dissertation, UCLA, 1969.

[17]Sadananda, K., and Shahinian, P., "Creep Crack Growth in Alloy 718," Met. Trans. A, 8A(3), 439-449 (1977).

DISCUSSION

J. S. Haydon (General Electric Co., GTEMD): Did you do crack growth studies in vacuum?

Author: The crack growth studies in vacuum noted were reported in References 1 and 6. It is bleieved that more quantitative tests should be conducted in vacuum and in environments of controlled oxygen potential. Such tests could aid in clarifying the mechanism of embrittlement. For the present, it is felt that there is abundant evidence in the literature to prove that oxygen plays a significant role in crack propagation at the identified temperatures and stresses.

L. W. Sandor (Sun Shipbuilding and Dry Dock): Have you given thought as to how to keep in a practical sense the oxygen, as the culprit, from reacting with a given hardware to minimize crack initiation (nucleation) and subsequent propagation?

Author: Dr. Sandor has touched on two important practical aspects of oxygen embrittlement. The use of inert atmospheres for postwelding heat treatment has been found to be beneficial. However, the requirements on atmosphere quality are stringent since exposure during heat treatment is relatively long and all of the gamma prime hardened alloys contain strong oxide formers. A practical alternative is the use of preweld heat treatments (e.g. overaging) which reduce susceptibility to cracking even in the presence of oxygen. The same philosophy has been applied when selecting heat treatments for service. It is not reasonable to contemplate altering the service environment, therefore, other factors contributing to embrittlement are controlled as far as possible. Experience has shown that, overaged microstructures containing coarse gamma prime may provide for enhanced service life. Low design stresses are naturally beneficial, but residual welding or assembly stresses can initiate the intergranular cracking. Grain-size is another important variable which is often controlled. Coatings are used, but they should not be the last line of defense if they may crack and expose the substrate.

L. W. Sandor (Sun Shipbuilding and Dry Dock): I was not so much thinking about keeping the oxygen away during heat treatment of these materials - it is well established - as I was to aim at in-service situations. These materials are used in high temperature, oxidizing environments such as jet engines.

HIGH TEMPERATURE CRACK PROPAGATION IN $2\frac{1}{4}$Cr-1Mo MANUAL METAL ARC WELD METALS

D.J. Gooch and B.L. King

Central Electricity Research Laboratories
Kelvin Avenue
Leatherhead, Surrey
U.K.

ABSTRACT

High temperature crack propagation tests on $2\frac{1}{4}$CrlMo MMA weld metals showed that overall growth rates were primarily dependent upon creep strength and not on residual element level, the fastest rates and the lowest accompanying displacements being found for weld metals with the highest creep strength (which was not dependent upon carbon level). The material with the highest carbon, phosphorus and arsenic levels had the slowest crack growth rates and largest displacements in the tempered condition; this was consistent with its low creep strength and high creep ductility

INTRODUCTION

Transverse cracks in the $2\frac{1}{4}$CrMo weld metal used to joint components in CrMoV steels are the most common cause of repairs to welded joints in HP and IP steam pipework systems of C.E.G.B. plant. It is known that the cracks are intergranular and form preferentially in the columnar regions of the weld metal, often close to the fusion boundary, and are the result of the nucleation, growth and linkage of creep cavities.

It has been shown (other factors being equal) that weld metals with high manganese and silicon levels have increased susceptibility to stress relief cracking[1], but there is no clear evidence on the role of carbon or trace impurities such as phosphorus, arsenic, tin and antimony. The latter elements are known to promote temper embrittlement in $2\frac{1}{4}$CrMo and creep embrittlement in $\frac{1}{2}$Cr$\frac{1}{2}$Mo$\frac{1}{4}$V [2,3] and this has led to the suggestion that they will also enhance creep embrittlement in $2\frac{1}{4}$CrMo [2]. It is clear, however, that in short term studies [4] $2\frac{1}{4}$CrMo is more tolerant of trace impurities than $\frac{1}{2}$Cr$\frac{1}{2}$Mo$\frac{1}{4}$V and that this may be due to the lower creep resistance and more rapid tempering response of $2\frac{1}{4}$CrMo.

This paper presents the results of creep crack propagation studies at 565°C (1049°F) on four weld metals of varying carbon and trace element levels produced from proprietary flux coated manual metal arc electrodes.

EXPERIMENTAL PROCEDURES

Pads of $2\frac{1}{4}$CrlMo weld metal, 300mm x 180mm x 30mm (11.81 in x 7.09 in x 1.18 in), were deposited onto mild steel substrates using four different commercial basic coated electrodes. The chemical compositions of the weld deposits are given in Table 1.

Single edge notched tension (SENT) specimens (W, 20mm (0.787 in) B, 12mm (0.472 in)) were machined with their axes parallel to the welding direction and with the narrow faces of the gauge length parallel to the plane of deposition to ensure that the crack plane corresponded to that for transverse cracking. After final heat treatment the uppermost of these faces was notched to an initial a/W of 0.2 by spark machining giving a notch root radius of ∿0.15mm (0.006 in). Most specimens were side-grooved to provide increased constraint and thereby to reduce crack front bowing. In these cases each groove was 0.6mm (0.024 in) deep with a 60° included angle.

The applied load was taken as that required to give an initial nominal linear elastic stress intensity of 14 MN $m^{-3/2}$ (12.7 ksi$\sqrt{in}$). However, it is emphasized that this should not be taken to imply an anticipated correlation of crack growth rates with stress intensity. When sidegrooving was employed the reduced load was calculated on the basis of the stress intensity modification suggested by Freed and Krafft [5].

The loads applied to the plain-sided specimens corresponded to an engineering reference stress[6] of ∿120 MN m^{-2} (17.4 ksi) (plane strain) or ∿140 MN m^{-2} (20.3 ksi)(plane stress).

The various heat-treatment conditions used to investigate the effects of tempering, pre-strain and grain refinement are shown in Table 2.

Concurrent tempering and pre-strain was carried out under stress relaxation conditions at $650^{\circ}C$ ($1202^{\circ}F$) and $700^{\circ}C$ ($1292^{\circ}F$). Specimens were loaded to the 0.25% proof stress at a nominal strain rate of 2.4×10^{-5} s^{-1}, at which point the testing machine cross head was locked to allow stress relaxation for a period of 3h. The tensile strains accumulated during stress relaxation were largely due to the conversion of elastic displacements in the frame of the testing machine to plastic strain in the specimen gauge length.

Structures of substantially uniform equiaxed prior austenite grains were obtained by solution treatment at either $1300^{\circ}C$ ($2372^{\circ}F$) or $950^{\circ}C$ ($1742^{\circ}F$) followed by oil-quenching with or without subsequent tempering for 3h at $700^{\circ}C$ ($1292^{\circ}F$). Solution treatment for 5 minutes at $1300^{\circ}C$ ($2372^{\circ}F$) produced a grain size of 80-100μm (ASTM 4) in each of the four weld deposits. Deposits W2, W3 and W5 had a grain size of ∿30→40μm (ASTM 6-7) after 30 minutes at $950^{\circ}C$ ($1742^{\circ}F$). A specimen of deposit W1 was solution treated for 10s at $1300^{\circ}C$ ($2372^{\circ}F$) and this produced a grain size of ∿50μm (ASTM 5-6).

Crack lengths were monitored using the direct current potential drop technique. Where bowed crack fronts were obtained the values quoted in this paper refer to the centre of the front.

Beam displacements were measured using dial gauges. In addition the distance between pyramidal hardness indentations placed ∿1mm on either side of the notch on the notched face was measured before and after testing. This enabled the plastic component of the edge opening displacement (EOD), Δd, at the conclusion of each test to be determined. EODs during the course of testing were estimated by linearly relating beam displacements to EODs using the final EOD as a reference.

Uniaxial creep tests were performed on as welded samples tempered for 3h at $650^{\circ}C$ ($1202^{\circ}F$) or $700^{\circ}C$ ($1292^{\circ}F$). Specimens were of round section with diameter of 6.5mm (0.256 in) and gauge length of 20mm (0.787 in).

CHEMICAL COMPOSITION AND TEMPERING RESPONSE

The phosphorus and arsenic levels of W1 were significantly higher than W2, W3 and W5. These were lowest in W2 which was supplied as a high purity, low carbon electrode. W3 was the standard purity electrode from the same manufacturer as W2 and also had a higher carbon level. W5 was

Element	wt.% W1	W2	W3	W5
Aluminium	0.012	< 0.01	< 0.01	< 0.01
Antimony	< 0.002	< 0.002	< 0.002	< 0.002
Arsenic	0.050	0.019	0.030	0.031
Boron	< 0.001	< 0.001	0.002	< 0.001
Carbon	0.06	0.02	0.05	0.04
Chromium	2.46	2.64	2.72	2.26
Cobalt	0.027	0.011	0.015	0.01
Copper	0.030	0.02	0.05	0.04
Magnesium	< 0.01	< 0.01	< 0.01	< 0.01
Manganese	1.03	0.78	1.02	0.96
Molybdenum	1.00	1.08	1.00	1.20
Nickel	0.05	0.04	0.06	0.03
Niobium	0.02	0.02	0.03	0.01
Phosphorus	0.031	0.013	0.017	0.012
Silicon	0.44	0.33	0.37	0.22
Sulphur	0.025	0.009	0.026	0.014
Tin	< 0.01	< 0.01	< 0.01	< 0.01
Titanium	0.03	0.02	0.02	0.01
Tungsten	0.03	0.03	0.03	0.02
Vanadium	0.03	0.03	0.03	0.01
Total Nitrogen	0.018	0.008	0.012	0.010
Total Oxygen	0.045	0.057	0.043	0.071

Table 1 (left)
Chemical Compositions

Table 2 (below)
Summary of test conditions and data

Heat Treatment	Specimen	H_v20 Pre-Test	H_v20 Post-Test	Final Crack Extension Δa mm	Final Aspect Ratio $(\Delta a+a_o)/\Delta d$	Failure Time h
As Welded	W1/5	278	224	3.4	20.0	650
Tempered 3 h 650°C	W1/9	260	225	4.4	5.97	1600
Tempered 3 h 650°C	W1/11(S)	245	220	3.3	8.11	3000
Tempered 3 h 700°C	W1/2	230	230	1.4	8.16	~1500
Tempered 3 h 700°C	W1/12(S)	235	225	3.2	11.3	3500
Pre-Strained 3 h 650°C	W1/10	220	210	6.9	17.5	1250
Pre-Strained 3 h 700°C	W1/7	187	186	1.8	3.24	~ 600
10s 1300°C F.O.Q.	W1/4	334	238	3.3	9.82	1100
5 min 1300°C F.O.Q.	W1/8(S)	330	215	2.7	6.69	1150
As Welded	W2/5	235	205	5.9	21.9	1500
As Welded	W2/1(S)	250	200	1.5	61.8	~1100
Tempered 3 h 650°C	W2/2(S)	210	200	4.3	27.0	1050
Tempered 3 h 700°C	W2/7	210	205	4.0	25.1	1000
Tempered 3 h 700°C	W2/3(S)	200	200	3.8	28.9	1100
Pre-Strained 3 h 650°C	W2/6(S)	185	180	5.0	54.1	700
Pre-Strained 3 h 700°C	W2/11(S)	180	180	3.7	38.7	575
30 min 950°C F.O.Q.	W2/8(S)	270	215	3.8	74.8	400
5 min 1300°C F.O.Q.	W2/4(S)	260	210	3.7	52.9	600
30 min 950°C F.O.Q. + Tempered 3 h 700°C	W2/10(S)	200	195	3.4	70.9	575
5 min 1300°C F.O.Q. + Tempered 3 h 700°C	W2/9(S)	240	210	3.0	33.5	1750
As Welded	W3/12	305	235	4.0	29.6	850
As Welded	W3/1(S)	310	220	4.3	32.7	1000
Tempered 3 h 650°C	W3/6	245	220	4.2	30.3	1700
Tempered 3 h 650°C	W3/2(S)	250	230	5.3	22.9	1050
Tempered 3 h 700°C	W3/7(S)	225	205	2.5	21.0	950
Pre-strained 3 h 650°C	W3/3(S)	225	210	4.1	43.4	900
Pre-strained 3 h 700°C	W3/11(S)	195	190	3.9	35.1	925
30 mins 950°C F.O.Q.	W3/8(S)	340	250	4.2	63.7	300
5 min 1300°C F.O.Q.	W3/4(S)	335	225	3.8	53.2	525
30 min 950°C F.O.Q. + Tempered 3 h 700°C	W3/10(S)	210	190	2.4	16.2	900
5 min 1300°C F.O.Q. + Tempered 3 h 700°C	W3/9(S)	220	210	4.9	16.1	1650
As Welded	W5/1(S)	275	195	3.4	15.5	1400
Tempered 3 h 700°C	W5/2(S)	190	180	3.8	6.90	2300
30mins 950°C F.O.Q. + Tempered 3 h 700°C	W5/3(S)	190	170	4.1	16.7	1950
5 mins 1300°C F.O.Q + Tempered 3 h 700°C	W5/4(S)	185	180	3.3	10.6	2350
Pre-strained 3 h 700°C	W5/5(S)	165	160	3.4	8.45	3500
Pre-strained 3 h 650°C	W5/6(S)	180	175	4.4	27.2	1000

Notes. F.O.Q. Fast Oil Quench. i.e. capsule broken under oil

(S) Sidegrooved Specimens

slightly lower in chromium and silicon than the other electrodes and the carbon level was intermediate (Table 1).

The average tempering responses of as-deposited weld metal are shown in Fig. 1. The softening rate of W5 is significantly faster than for the other deposits, and it is thought that this may be a major contributory factor to the wide as-welded hardness range. W5 is lower in the carbide stabilising elements vanadium and niobium than the other deposits and this may be partly responsible for the rapid softening.

CRACK GROWTH TESTS

Crack extension rates for W2 and W3 in the 'welded and tempered' conditions were very similar and substantially faster than for either W1 or W5 (Fig. 2). In general, initiation was followed by continually accelerating crack growth. However, periods of apparently constant growth rate were observed during testing of tempered specimens of W1 and W5 and also the as-welded sample of W5. Similar behaviour has previously been observed during tests on refined structures of CrMoV steels[3]. In the as-welded condition, W1 showed a region of apparently decreasing growth rate after loading. This is attributed to the formation of cavitational damage and micro-cracks on columnar grain boundaries normal to the stress axis ahead of the crack tip soon after loading. The decreasing growth rate then corresponds to the growth of the crack through this region and into the next band of more crack resistant refined material.

In all cases there was negligible difference in growth rates between material tempered at 650°C (1202°F) and 700°C (1292°F). This is consistent with the creep data for W1 (Fig. 7) but not for W2 or W3 where tempering at 700°C (1292°F) resulted in significantly faster secondary creep rates.

The failure times for as-welded specimens were approximately 0.4 and 0.6 of those for tempered specimens in the case of W1 and W5 respectively but for W2 and W3 there was no significant effect of tempering (Table 2).

The edge opening displacements (Δd) during crack growth are shown in Fig. 3 and the aspect ratios, $(a_o+\Delta a)/\Delta d$, at the termination of each test are shown in Table 2. It is apparent that the faster crack growth rates of W2 and W3 are accompanied by displacements approximately one third of those for the other deposits. With the exception of specimen W2/1 the final aspect ratios for W2 and W3 in both tempered and as-welded conditions were very similar. Once again this is in contrast to W1 and W5 where the aspect ratios for as-welded structures were significantly greater than for tempered material.

It is important to note that, although the failure times for W1 in the tempered conditions were superior to the other three welds the life of the as-welded specimen was the shortest. It is thought that this is largely due to the high arsenic and phosphorus levels in W1 which promote extensive columnar boundary cavitational damage before the matrix softens during testing. The superior failure time of W5 in the as-welded condition is due to the lower residual element levels accompanied by the rapid matrix softening.

Grain refinement of W1 by solution treatment at 1300°C (2372°F) resulted in increases in (a) specimen life and (b) displacement by a factor of approximately 2 relative to the as-welded condition (Table 2).

In contrast to this behaviour samples of W2 and W3, retransformed at 950°C (1742°F) or 1300°C (2372°F) and tested in the untempered condition, failed with displacements similar to or lower than the as-welded material and in times approximately one half those of the latter. Tempering for 3 hours at 700°C (1292°F) increased failure times and displacements but for the 40μm grain size material retransformed at 950°C (1742°F) lives were

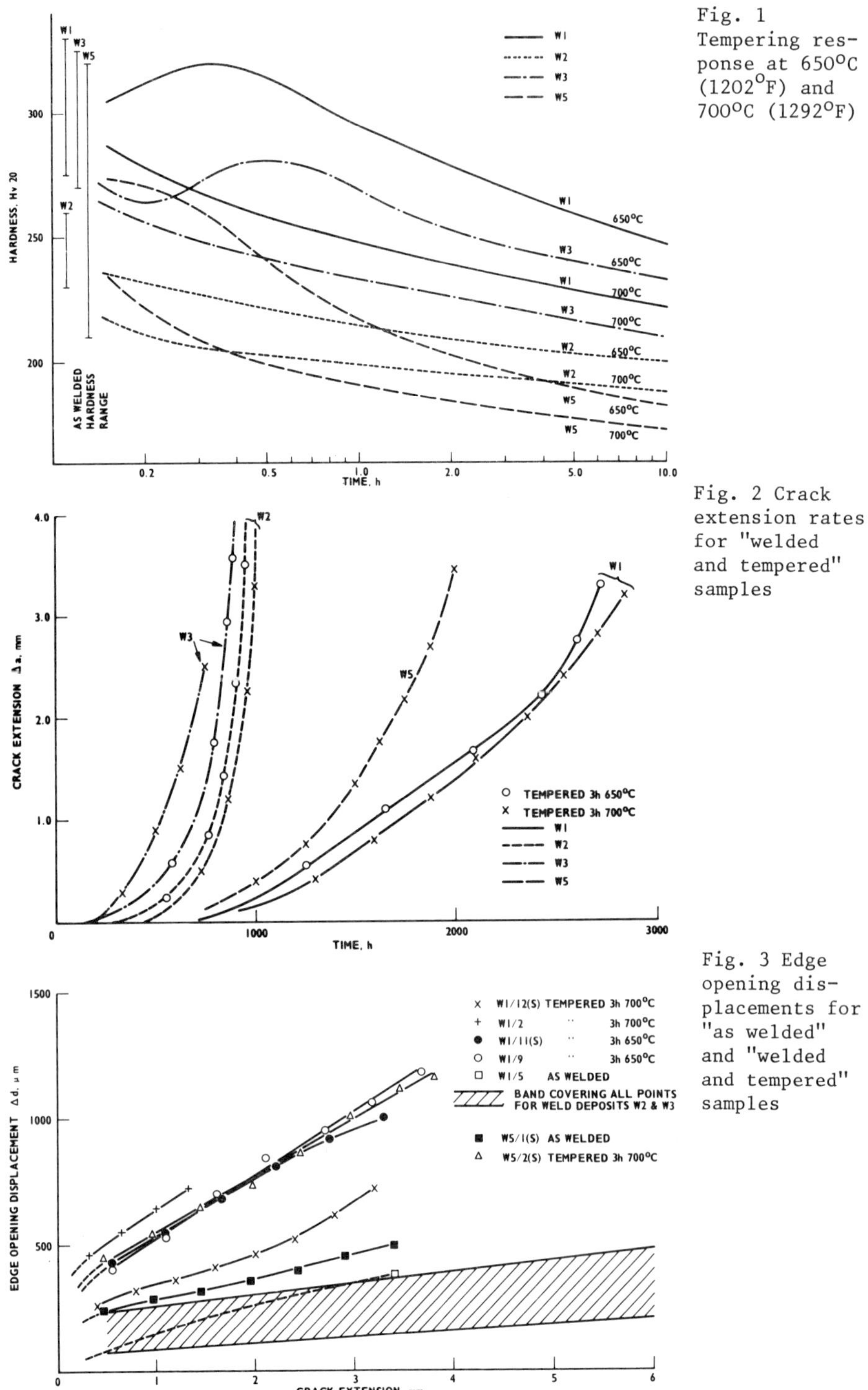

Fig. 1 Tempering response at 650°C (1202°F) and 700°C (1292°F)

Fig. 2 Crack extension rates for "welded and tempered" samples

Fig. 3 Edge opening displacements for "as welded" and "welded and tempered" samples

still shorter than for the welded and tempered condition. The 100μm grain size materials, retransformed at 1300°C (2372°F), however, exhibited failure times significantly longer than for the welded and tempered deposit.

In the case of W5, failure times of the retransformed structures were very similar to the welded and tempered specimen although the displacements in the former cases were somewhat lower.

Crack extension rates for quenched and tempered structures are shown in Fig. 4 and illustrate clearly that for W2 and W3 the 100μm grain size material is substantially more crack resistant than the 40μm. The same trend is apparent for W5 although in this case the difference is marginal. Similarly Fig. 5 shows that the edge opening displacements accompanying crack growth are generally larger for the 100μm than for the 40μm specimens.

The tensile strains induced during stress relaxation for 3 hours from the 0.25% proof stresses at 650°C (1202°F) or 700°C (1292°F) are shown in Table 3. The higher strains accumulated at 650°C (1202°F) are due primarily to the higher proof stresses and consequent greater elastic strain energy stored in the testing machine.

The failure times and EODs during the subsequent crack growth tests are shown in Table 2 and Fig. 6. The lives of W3 after relaxation were not significantly lower than for the tempered specimens, although the aspect ratios were higher by between ∿70% and 90%. This apparent inconsistency is resolved by the exhaustion of the primary stage of creep during the pre-strain, thus resulting in slower displacement rates during the subsequent crack growth test compared with the tempered specimens. In the case of W2 stress relaxation resulted in a reduction of failure times by ∿40% and an increase in aspect ratios of 34% (700°C, 1292°F) and 100% (650°C, 1202°F).

The effect of the 650°C (1202°F) pre-strain on W1 was similar, giving a decrease in failure time of 25% and an increase in aspect ratio of ∿200%; however, the behaviour after the 700°C (1292°F) treatment was very different. Although the life was less than half that of the correspondingly tempered deposit, matrix softening resulting from the pre-strain treatment provided large displacements during crack growth, exceeding those in the tempered material by 2.5 times and those in the as-welded structure by a factor of 13. The rapid softening of W1 during the 700°C (1292°F) pre-strain is reflected in the very low proof stress at 700°C (1292°F) compared with 650°C (1202°F). The other three weld deposits showed a substantially smaller drop (Table 3).

TABLE 3. Summary of Pre-Strain Data and Effect on Crack Growth

	Stress Relaxation			Crack Growth at 565°C (1049°F)	
Ref	T°C(°F)	Proof Stress MNm^{-2} (ksi)	Elongation after 3h	Ratio of Failure Times (relaxed/ tempered)	Ratio of Aspect Ratios (relaxed/ tempered)
W1	650 (1202)	366 (53.1)	1.91%	0.75	2.93
W2	"	320 (46.4)	2.00%	0.67	2.00
W3	"	330 (47.9)	2.30%	0.86	1.89
W5	"	356 (51.7)	2.62%	0.44(E)	3.94(E)
W1	700 (1292)	114 (16.5)	1.01%	0.40	0.40
W2	"	228 (33.1)	1.64%	0.52	1.34
W3	"	216 (31.3)	1.81%	0.97	1.67
W5	"	165 (23.9)	1.04%	1.52	1.22

Note (E) Estimated

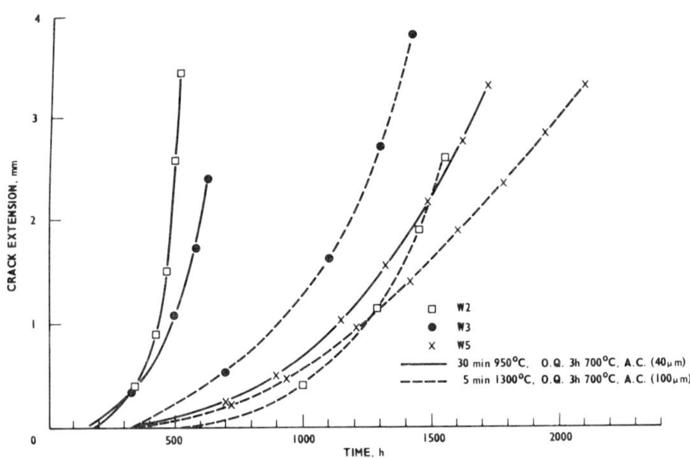

Fig. 4 Crack extension rates in retransformed structures

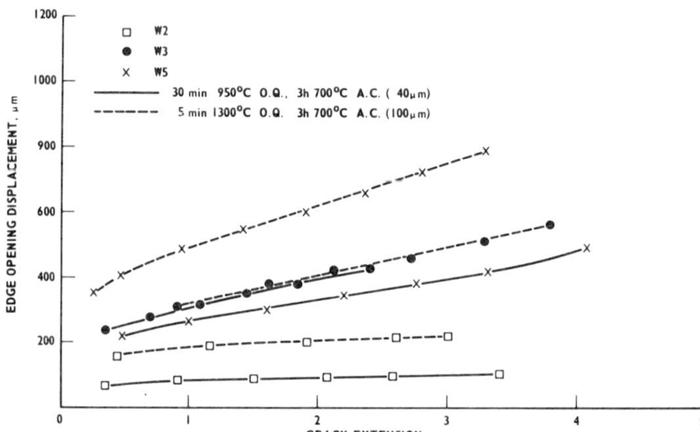

Fig. 5 Edge opening displacements for retransformed structures

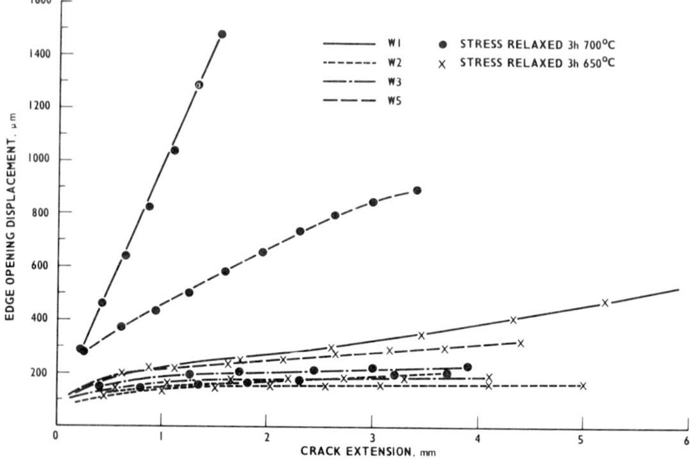

Fig. 6 Edge opening displacements for pre-strained specimens

W5 was severely embrittled by the 650°C (1202°F) pre-strain, the failure time being reduced by ∿50% and the aspect ratio being increased by ∿300%. However, the life of the sample relaxed from 700°C (1292°F) exceeded that of the tempered specimen by ∿50% and there was very little difference between the displacements. As in the case of W3, this may be a consequence of the exhaustion of primary creep during pre-strain.

In spite of the differing effects of pre-strain on the four weld deposits, Fig. 6 shows that the trends in the relationship between EOD and crack extension observed for the welded, welded and tempered, and retransformed structures are maintained. That is to say that the EODs of W1 and W5 during crack growth are always higher than those of W2 and W3. In addition the EODs of the 700°C (1292°F) relaxed samples are always greater than those relaxed at 650°C (1202°F) the differences being most striking in the case of W1 and W5.

The greater strains accumulated at the lower relaxation temperature in the present tests are clearly contributory to the reduced EODs during subsequent crack growth. However, many other factors are also important as is evident from the increased EODs of W1 after relaxation at 700°C (1292°F). Crack growth rates will be increased and EODs decreased by cavitational damage accumulated during pre-strain. The extent of damage accumulation will depend among other things, upon the proof stress, relaxation rate, impurity level and detailed microstructure. However, strain softening will <u>increase</u> EODs and may either increase or decrease overall propagation rates depending on whether crack growth is controlled by overall specimen displacement rates or by localised crack tip events. In the former the ultimate failure is essentially a creep rupture mechanism whereas in the latter case strain softening facilitates strain accommodation at the crack tip thereby reducing grain boundary tensile stresses and inhibiting cavity formation.

In the tempered structures of W1 both displacement and crack propagation rates were reduced by the increased constraint due to the presence of sidegrooves. This is caused by the inhibition of matrix flow by the increased triaxiality of the stress state in sidegrooved specimens. In the comparatively harder, untempered refined structures of W1 no significant effect of sidegrooving was observed. Similarly in the more brittle weld deposits W2 and W3 crack growth rates were not retarded by sidegrooves; indeed there is some evidence (Table 2) that the lower constraint conditions of plain sided specimens may in this case have reduced crack propagation. For example the longest failure time for any sample of W3 was achieved with the plain sided specimen tempered at 650°C (1202°F). The different behaviour of W2 and W3 as opposed to W1 may be compared with standard uniaxial notch rupture creep tests where brittle materials tend to exhibit notch weakening characteristics whereas ductile materials are frequently notch strengthening.

<u>CREEP TESTS</u>

The creep rupture data obtained for tempered samples of all four weld metals is summarised in Table 4 and Fig. 7.

In general the rupture strengths and ductilities of W1 and W5 were similar at all stress levels. The exception to this was at 120 MN m^{-2} (17.4 ksi) when W5 had a rupture life significantly greater than W1. No significant difference was detected between samples of W1 tempered at 650°C (1202°F) or 700°C (1292°F).

The rupture lives of W2 and W3 tempered at 700°C (1292°F) were substantially greater at high stresses but approached those of W1 and W5 at lower levels. In addition the rupture ductilities of W2 and W3 were much lower at all stresses. In contrast to W1, tempering at the lower temperature increased the rupture life by ∿35%.

Specimen No.	Tempering	Stress MNm^{-2}	Elongation %	Reduction in Area %	Minimum Creep Rate h^{-1}	Rupture Life h.
W1/1	3 h 650°C	105	25	65	2.2×10^{-5}	2469
W1/2	3 h 700°C	120	40	79	3.3×10^{-5}	1890
W1/6	3 h 700°C	135	40	86	1.1×10^{-4}	640
W1/11	3 h 650°C	150	45	88	1.6×10^{-4}	472
W1/12	3 h 700°C	165	30	77	2.6×10^{-4}	297
W2/2	3 h 700°C	120	14	47	1.5×10^{-5}	1872
W2/3A	3 h 700°C	135	19	50	2.3×10^{-5}	1322
W2/6	3 h 650°C	150	15	40	1.0×10^{-5}	1480
W2/3B	3 h 700°C	165	20	55	5.0×10^{-5}	627
W3/1	3 h 700°C	120	21	37	1.3×10^{-5}	2875
W3/3A	3 h 700°C	135	22	40	1.6×10^{-5}	2398
W3/6	3 h 650°C	150	12	20	1.1×10^{-5}	2374
W3/3B	3 h 700°C	165	30	50	5.2×10^{-5}	924
W5/A	3 h 700°C	120	35	62	1.7×10^{-5}	3054
W5/B	3 h 700°C	135	45	80	6.7×10^{-5}	820
W5/C	3 h 700°C	150	40	90	1.5×10^{-4}	385
W5/D	3 h 700°C	165	55	85	1.8×10^{-4}	357

Table 4 (above). Uniaxial creep rupture data at 565°C (1049°F)

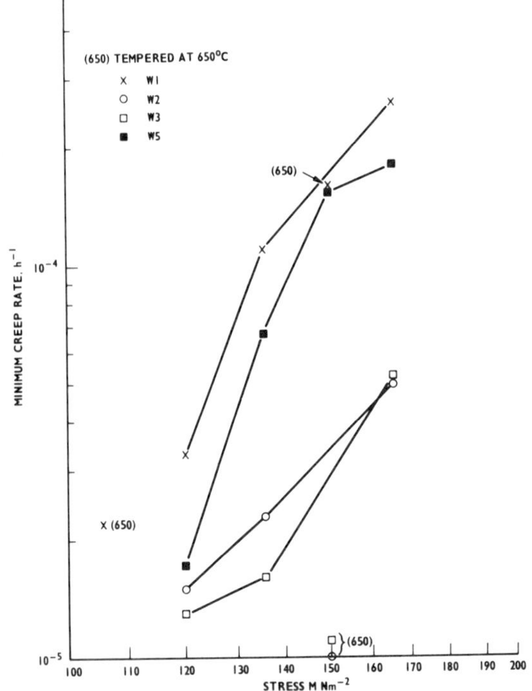

Fig. 7 (left) Minimum creep rates at 565°C (1049°F) for welds tempered at 700°C (1292°F) for 3h

The minimum creep rate data (Fig. 7) shows similar trends to the rupture data, W1 and W5 being substantially weaker at 165 MN m^{-2} (23.9 ksi) but approaching W2 and W3 at lower stresses.

Comparison with the crack growth data shows that the best resistance to crack growth occurs in the weld deposits (W1 and W5) with high secondary creep rates and high rupture ductilities, which are not simply related to carbon, phosphorus or arsenic levels or to room temperature hardness values.

METALLOGRAPHY

In general, cracks propagated in an intercrystalline manner perpendicular to the tensile stress axis. However, in retransformed regions of weld deposits the prior austenite boundaries were often difficult to locate and it is possible that the crack propagated in a transgranular mode. This was particularly true in the case of W1 and W5 where extensive plasticity accompanied crack growth.

Severe columnar boundary cavitation and microcracking was present on columnar boundaries in the as-welded specimen of W1, the damage extending several millimetres on either side of the crack plane (Fig. 8(a)). Cavitational damage was also present in the refined region although to a much lesser extent that in the columnar regions. In the as-welded structures of W2, W3 and W5 the degree of damage was far less severe and did not extend as far to either side of the crack (Fig. 8(b), 8(c), 8(d)).

Prior tempering reduced the susceptibility to widespread cavitational damage especially in the case of W1 and to a lesser degree in W5. For W2 and W3 the effect was minimal and is consistent with the similarity of crack growth behaviour between tempered and as-welded structures in these deposits (Table 2).

Similar trends between the four weld metals were observed in both the fine and coarse retransformed structures to the welded and tempered samples.

In the 650°C (1202°F) pre-strained structure of W1 separation of the columnar boundaries had occurred within a region extending approximately 2mm on either side of the crack path and there was evidence of cavitational damage extending to considerably greater distances. In contrast, pre-strain at 700°C (1292°F) produced little general grain boundary damage and subsequent crack growth was accompanied by extensive ductility (Plate 9(a)).

The appearance of W5 after pre-strain at 700°C (1292°F) and subsequent crack growth was essentially similar to W1 in that there was extensive plastic deformation and crack blunting. The 650°C (1202°F) pre-strained sample was also similar to W1 except that the extent of fine cavitation away from the primary crack was significantly less.

Extensive grain boundary damage, in both columnar and refined regions was observed in all pre-strained samples of W2 and W3 but was still not as severe as for W1 after the 650°C (1202°F) prestrain. In W3 the damage was predominantly in the form of individual cavities or groups of cavities whereas in W2 these frequently had linked to form microcracks on columnar boundaries. This may be a contributory factor to the more severe embrittlement of W2 compared with W3 in terms of failure times (although not in terms of aspect ratio).

DISCUSSION

Very few investigations have previously been performed on the creep ductility of 2CrMo weld metals. The work usually cited is that of Bruscato[2] who looked at the creep rupture properties of two commercial weld metals with high and low levels of P, Sb, As and Sn. He found that the impure weld deposit showed much lower ductilities at 565°C (1049°F) and a similar but less pronounced effect at 482°C (900°F) and attributed this to

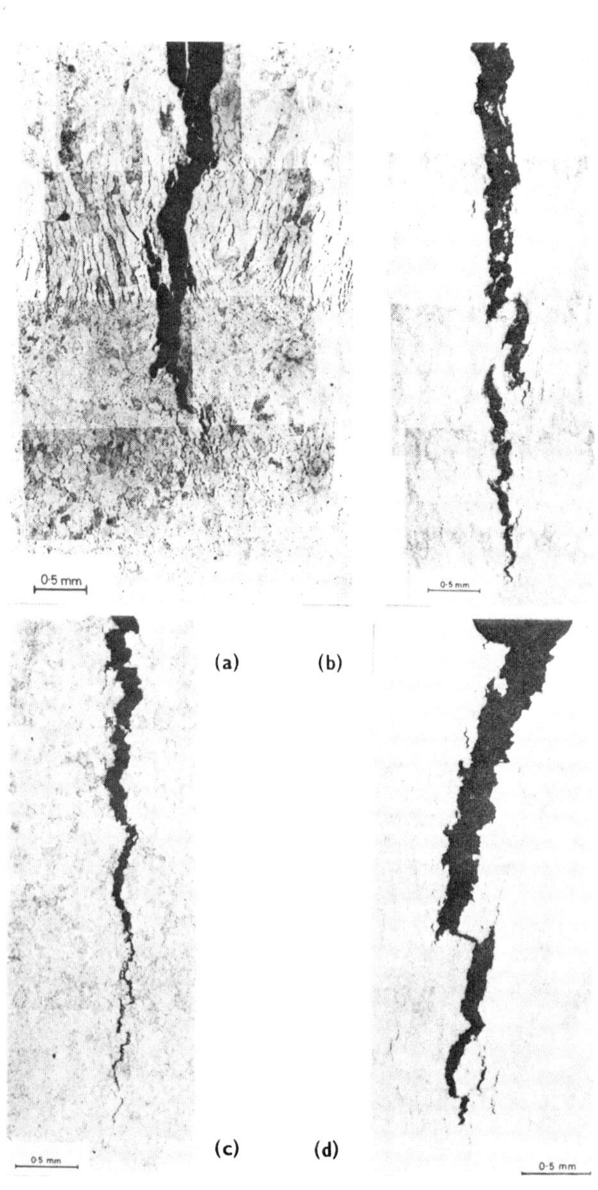

Fig. 8 Crack propagation in "as welded" structures (a)W1/5, (b)W2/5, (c)W3/1, (d) W5/1

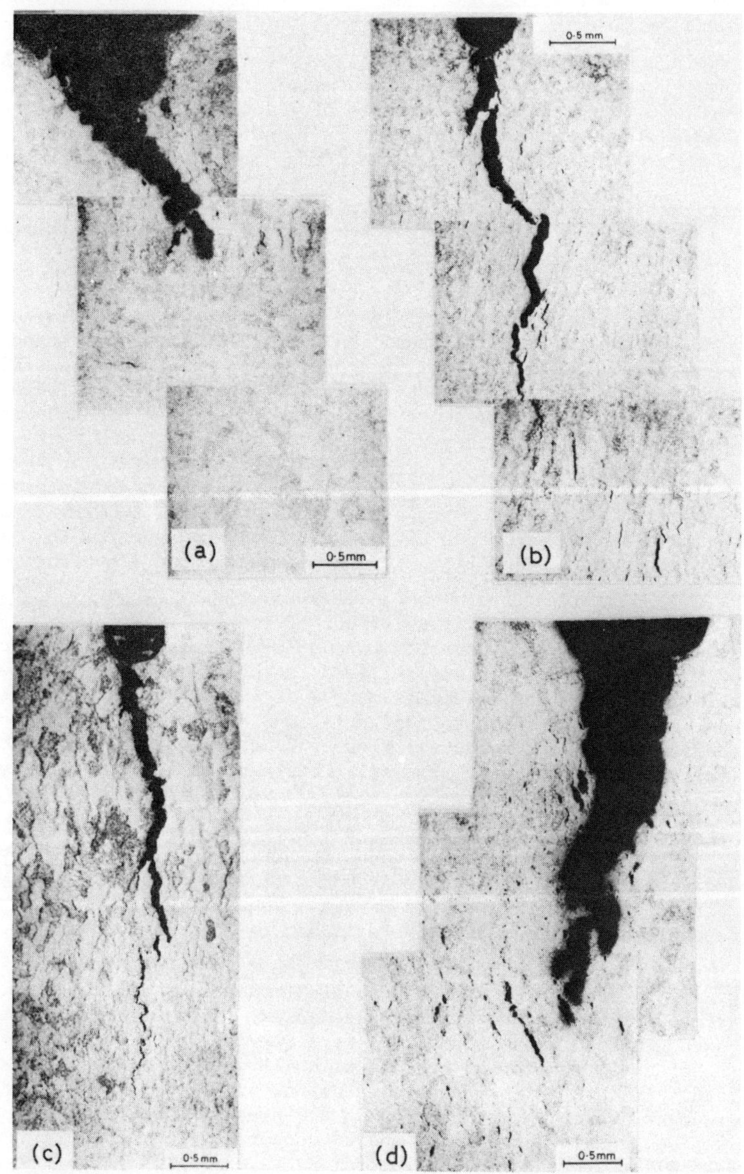

Fig. 9 Crack propagation in structures pre-strained at 700°C (1292°F)

(a)W1/7, (b)W2/11, (c)W3/11, (d)W5/5

the effect of the residual elements. However, the chromium level of the pure deposit was as low as 1.87% compared with 2.35% for the impure and the rupture lives at 482°C (900°F) of the impure deposit were greater by up to a factor of two. This infers that the creep resistance of the pure deposit was lower and this may have been a contributory factor to the higher ductility.

Wolstenholme[7] investigated the creep properties of a high phosphorus deposit, believed to be from the same electrode batch as W1 in the present work and compared them with those of a low phosphorus deposit from the same manufacturer. He found that after 1% elongation there were approximately 60% more cavities in the high phosphorus material. Nevertheless the minimum creep rate and rupture ductility of the high phosphorus deposit were also significantly higher. In later work Wolstenholme[8] also found that there was little effect of phosphorus on reheat cracking susceptibility but that the major effect was that of stress relief temperature.

In the present work it is clear that the dominant factor controlling the creep crack growth is creep resistance, although the factors affecting the creep resistance are less obvious. In particular the creep properties of W1 and W3 are very different but the chemical compositions are similar in all respects other than the phosphorus and arsenic levels, which are substantially higher in the less creep resistant W1. The effects of residual element levels on secondary creep rates other than those due directly to cavitational damage have not been systematically investigated but the work of Viswanathan[9] on experimental melts of a 1¼Cr½Mo steel is of interest. This work showed that the secondary creep rates at 538°C (1000°F) of a cast containing 300 ppm phosphorus and 97 ppm antimony were approximately six times higher than a cast containing 20 ppm phosphorus and 5 ppm antimony but being otherwise similar.

If there is an embrittling effect of residual elements in 2¼CrMo weld metals, as has been demonstrated for Cr-Mo-V steels[3], it is a secondary effect fot the test conditions investigated in the present study. However, it is possible that residual elements may be of importance in more creep resistant weld deposits at longer test durations, especially since it is well established that rupture ductilities of 2¼CrMo steels decrease substantially with increasing test time as a result of cavitational damage at grain boundaries[10].

The observation of widespread cavitational damage away from the main crack path in the high residual deposit (W1) tested in either the as-welded condition or after pre-strain at 650°C (1202°F) supports the hypothesis that the effect of impurity elements on grain boundary cavitation is most pronounced when the matrix is highly resistant to creep deformation.

Miller and Batte[11] found that the important microstructural factor governing reheat cracking susceptibility of 2¼CrMo welds was the proportion of well defined prior austenite grain boundaries, including both columnar and equi-axed grains. These boundaries were very susceptible to crack formation whereas retransformed regions with ill-defined boundaries were very resistant to cavitational damage. This observation is consistent with the faster crack growth rates observed in the present work in solution treated samples of weld deposits W2 and W3 compared with the as-welded structures. All the boundaries in the former were clearly visible whereas as-welded samples contained large proportions of granular bainite in which the boundaries were difficult to locate. A similar effect has previously been observed during crack growth studies on ½Cr½Mo¼V steel[12]. It is believed that irregular prior austenite boundaries provide greater resistance to grain boundary sliding and shear which contribute to cavitation nucleation and growth processes.

It is known that in CrMoV steels a decrease in grain size results in reduced crack propagation rates and higher accompanying displacements[3, 13]. In addition transverse weld metal cracking in service situations generally originates in columnar regions although it may later grow into retransformed regions. Therefore the greater susceptibility to crack propagation of fine grained compared with coarser grained solution treated samples observed in the present work was unexpected. In terms of crack growth rates this can be rationalised by considering the dependence of minimum creep rates on grain size. Cane[14] showed that in quenched and tempered $2\frac{1}{4}$Cr1Mo steel the secondary creep rate passed through a minimum with increasing grain size. If it is assumed that in $2\frac{1}{4}$Cr1Mo steel crack growth is displacement controlled and that, compared with creep rates, the displacement required for propagation is relatively insensitive to grain size in the 40μm to 100μm range, a crack growth rate which increases as grain size decreases can be explained. However, the lower displacements, albeit marginal, observed for the finer grained material cannot be explained in the same way since Cane[14] also found that rupture ductilities consistently decreased as grain sizes increased.

As well as softening the hardened structures formed in the weldment, post weld heat treatment serves to relax residual welding stresses by local creep deformation. Thus, while tempering, per se, may be beneficial to ductility and crack growth, its effect is opposed by the creep damage produced by stress relaxation to a degree determined by the strain tolerance of the structure and the initial residual stress level.

The degree of creep embrittlement which results will depend upon the relative contributions to the relaxation process of matrix deformation and grain boundary cavitation. Maintenance of a high matrix strength for a longer period by stress relaxation at 650°C (1202°F) as opposed to 700°C (1292°F) will favour cavity formation if the effects of high grain boundary stresses outweigh the opposing effect of slower creep processes, whereas the more rapid softening at 700°C (1292°F) will allow more strain to be accommodated by matrix deformation. Thus, although interpretation of the present results is complicated by the higher strains induced during the 650°C (1202°F) pre-strain treatments it is considered that the results, in particular for W1 and W5, demonstrate the importance of an adequate post weld heat-treatment temperature in counteracting stress relief embrittlement.

It has been shown previously[15] that the failure times of SENT specimens of W1 in the welded and tempered condition could be adequately described using engineering reference stress techniques in conjunction with uni-axial stress rupture data[6]. The effect of sidegrooving on welded and tempered specimens of W1 was to increase failure times by a factor of approximately two. It is reasonable to assume that W5 would show similar behaviour and on this basis failure times of 1000 to 1200 hours would be expected for plain sided specimens at a load corresponding to a reference stress of between ∿120 MN m^{-2} (17.4 ksi) (plane strain) and ∿140 MN m^{-2} (20.3 ksi) (plane stress). This compares favourably with the uni-axial rupture lives for W5 of 820 hours at 135 MN m^{-2} (19.6 ksi) and 3054 hours at 120 MN m^{-2} (17.4 ksi) (Table 5). It must be emphasized, however, that the use of engineering reference stress techniques will not necessarily apply to larger specimens or to weldments where the degree of constraint may be much greater resulting in a change of detailed failure mechanism[16]. It is also evident that this treatment is not appropriate for the more brittle deposits W2 and W3 where the rupture lives of notched specimens were shorter and plain specimens longer than those of W1 and W5 (Tables 3 and 5).

CONCLUSIONS

1. Overall crack growth rates in tempered material were primarily dependent upon creep strength, the fastest rates and lowest accompanying displacements being found for the welds with the highest creep strength and lowest creep ductility.

2. In untempered material and that pre-strained at 650°C (1202°F) the high residual material showed the most extensive columnar boundary cavitation. The major effect of high residual element levels is thus to enhance the susceptibility to columnar boundary damage during post weld heat treatments before the matrix has time to soften.

3. Solution treatment producing 40μm (ASTM 6) or 100μm (ASTM 3-4) grain sizes was beneficial to the overall crack growth behaviour of the high residual material in the as-welded condition but in the other samples either had no effect or was detrimental.

4. Stress relaxation at 650°C (1202°F) invariably resulted in embrittlement when compared with tempered material whereas relaxation at 700°C (1292°F) embrittled the high creep strength welds but softened or had no effect on the lower strength deposits.

5. Overall failure times for the low strength welds in the tempered condition could be adequately predicted using reference stress techniques in conjunction with uni-axial creep rupture data.

ACKNOWLEDGEMENT

The work has been carried out at the Central Electricity Research Laboratories and is published by permission of the Central Electricity Generating Board.

REFERENCES

1. Wolstenholme, D.A., C.E.G.B. Report No. R/M/N746.
2. Bruscato, R., 1970, Welding Journal, 49, 4, 148s.
3. King, B.L., 1978, To be published.
4. Myers, J. and Price, A.T., 1977. Metals Technology 4, 406.
5. Freed, C.N. and Krafft, J.M., 1966, J. Materials 1, 770.
6. Haigh, J.R. and Richards, C.E., 1973, C.E.G.B. Report RD/L/N19/73.
7. Wolstenholme, D.A., 1976, Proc. Spring Residential Conf., Institute of Metallurgists, Ser. 3, No. 5.
8. Wolstenholme, D.A., 1977, Unpublished work.
9. Viswanathan, R., 1975, Metals Engineering Quarterly, 15, 50.
10. Swift, R.A. and Rogers, H.C., 1976, Welding Journal, 55, 188s.
11. Miller, R.C. and Batte, A.D., To be published.
12. Gooch, D.J., 1977, Mater. Sci. Eng. 29, 227.
13. Gooch, D.J., 1977, Mater. Sci. Eng. 23, 55.
14. Cane, B.J., 1976, Metal Sci., 10, 29.
15. Gooch, D.J., King, B.L. and Briers, H.D., 1978, Mater. Sci. Eng., 32, 81.
16. Gooch, D.J., Haigh, J.R. and King, B.L., 1977, Metal Sci., 11, 545.

DISCUSSION

W. Borges (Supship Boston): (1) What type filler metal was used? (2) Have you accomplished any studies using 1¼ Cr-½ Mo (type filler metal) for repairs?

Author: (1) The filler metals used were commercially available 2¼Cr1Mo manual metal arc electrodes, details of which are given in the manuscript. (2) We have not performed any studies using 1¼Cr-½Mo filler metal.

J. J. Pepe (General Electric Company): Your data suggest that high strength, low ductility materials have the highest crack growth rate. However, the high strength materials have the lowest minimum creep rate. Since crack growth probably is dependent on a parameter which includes both creep rate and ductility your data seems to suggest that ductility is the more dominant factor. Would you care to comment on this?

Author: In low alloy ferritic steels in which creep failure is by a grain boundary cavity nucleation and growth process, ductility and creep strength are intimately related. The susceptibility to cavitational failure is related to the relative "strengths" of the matrix and grain boundary regions. Increasing the matrix strength relative to that of the grain boundary results in an increase in the strain which is accommodated in grain boundary regions per unit total creep strain. This is reflected in increased cavitation and lower creep ductility. Therefore, crack growth by this mechanism depends both on creep rate and ductility, as you suggest, but inversely on matrix creep rate rather than directly. This assumes of course, that other factors which influence grain boundary cavitation such as segregation effects and precipitate free zone widths remain unaltered. The main point of this paper is that, under the conditions investigated, the effect of low matrix creep strength in increasing ductility outweighs the opposing effects of tramp elements in reducing ductility.

THE STRUCTURE OF WELDMENTS AND ITS RELEVANCE TO HIGH TEMPERATURE FAILURE

M.C. Coleman

Central Electricity Generating Board
Marchwood Engineering Laboratories
Marchwood, Southampton, SO4 4ZB. U.K.

ABSTRACT

The microstructures and associated mechanical properties in weldments can be extremely complex and can result in failure occurring by cracking in the heat affected zone or weld metal. Using a simple model the microstructures arising in low alloy ferritic steel weldments used by the CEGB in high temperature power plant are described for a single weld bead and for the multiple bead situation that exists in full size weldments. Developments of this work enable welding procedures to be defined to produce optimum or specific microstructural distributions in heavy section weldments.

The common modes of failure occurring in such weldments, arising as circumferential cracking in the HAZ and transverse cracking in the weld metal, are then reviewed and related to the weldment microstructure. Certain microstructures present in these weldments are susceptible to crack initiation and growth and research work studying the mechanisms and mechanics of high temperature cracking in heavy section weldments is described. Various aspects of this work are presented including the use of controlled welding procedures to generate predetermined microstructures in large welded components for creep testing to failure.

INTRODUCTION

Weldments can be considered at their simplest to consist of three interacting component parts, the weld metal, its attendant heat affected zone (HAZ) and the adjacent parent material. The detailed microstructure is however extremely complex and is controlled by the interaction of thermal fields, produced by the heat input from the welding process, and the phase transformation and grain growth characteristics of the materials being welded. These microstructures, which generally vary from wrought parent material through transformed HAZ's to cast weld metal, can have greatly different mechanical properties. As a consequence, cracking can arise both in the weld metal and HAZ when the weldment is subjected to the stresses and strains associated with heat treatment and/or service conditions.

The Central Electricity Generating Board (CEGB) operate power generating plant constructed using fusion welding as the major fabrication process. Of these welds a large proportion are produced using the shielded metal arc (SMA) technique. This is particularly true for welds in main steam lines which are usually ½CrMoV ferritic steel pipes welded using 2CrMo electrodes. These weldments, like others, are microstructurally complex but their design life at high temperature is invariably based only on the parent material creep and stress rupture properties. The required life is in excess of 100,000 hours but creep failures, characterised by the formation of macroscopically large cracks though not usually resulting in complete failure of the component, can occur after very much shorter periods of service. Invariably these are associated with the weldment and in particular with certain microstructural regions of the HAZ and weld metal.

This paper reviews how the microstructural distribution observed in low alloy ferritic steel weldments arises by considering first a single weld bead. The model is then extended to show the effect of multiple beads and their associated thermal cycles. The common modes of failure arising in such heavy section weldments result from circumferential cracking around the HAZ and transverse cracking across the weld metal. These are described and related to the weldment microstructure. Finally, research being conducted at the Marchwood Engineering Laboratories (MEL) of the CEGB to study the mechanisms and mechanics of high temperature crack growth in heavy section weldments is described. In particular, this considers the use of controlled welding procedures to generate predetermined microstructure distributions and the location of defects in weldments forming part of a large pressure vessel component intended for creep testing to failure.

WELDMENT MICROSTRUCTURE

Single Bead

For a single bead on plate the microstructure of the "weldment" varies from cast weld metal through transformed HAZ to the parent material as illustrated in Figure 1a. The cast weld metal consists mainly of columnar grains with the range of grain sizes depending mainly on the heat input used and the consequent cooling rate. In almost all practical cases the structure of 2CrMo weld metal deposited using the SMA technique is fully bainitic. The hardness of these deposits is related primarily to the amounts of the silicon and manganese deoxidants present[1].

The HAZ in ½CrMoV material adjacent to such a bead contains a range of microstructures which depends mainly on the peak temperature of the weld thermal cycle experienced. The regions closest to the fusion boundary experience peak temperatures sufficiently high to completely austenitise the adjacent parent material and this transforms on cooling to give a fully bainitic structure. Across this region the prior austenite grain size decreases almost linearly with decreasing peak temperature[2]. Adjacent to the fine grained side of the fully bainitic structure and further away from the fusion boundary is a region where the thermal cycle peak temperatures experienced have produced some but not

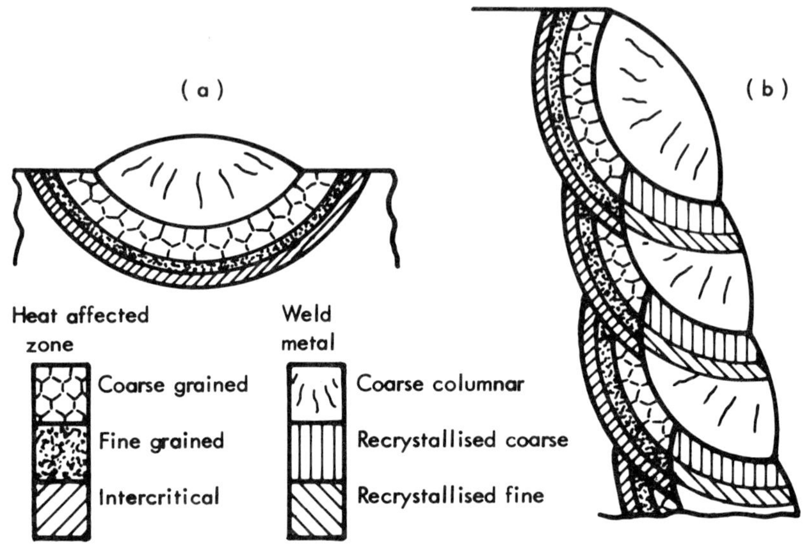

FIGURE 1. SCHEMATIC REPRESENTATION OF THE MICROSTRUCTURES IN a) SINGLE BEAD b) MULTIPLE BEAD WELDMENTS

complete transformation to austenite. The resulting microstructures in this intercritically transformed region are a mixture of ferrite and bainite, containing predominantly bainite at the high temperature end and predominantly ferrite, with only small amounts of bainite decorating the parent material grain boundaries, at the low temperature end. In the region experiencing thermal cycle peak temperatures just below this intercritical range some degradation of the structure occurs, mainly shown by increased vanadium carbide precipitation. Optically however, the microstructure appears very similar to the immediately adjacent parent material which consists of ferrite and about 5% bainite, produced by the original normalising heat treatment.

The exact structural distribution arising in any particular weld depends on the transformation characteristics of the material involved and the heating and cooling rates of the weld thermal cycles which in turn depend on the material composition and the heat input respectively. Knowing these factors, the HAZ thermal cycles can be determined from heat flow equations[3], which can then be used to predict both the prior austenite grain sizes[4] and the final microstructures arising in the HAZ[5]. For example, in a large component weldment using 5 mm diameter (6 swg) electrodes the fully bainitic region would extend from the fusion boundary at about 1520°C to a distance corresponding to a HAZ peak temperature of about 950°C with the prior austenite grain size decreasing from 140 μm to less than 10 μm. The intercritical region in this case then occurs between 950°C and 850°C. For the purpose of this paper however, the general features of the single bead situation described above are shown schematically in Figure 1a taking the weld metal as predominantly coarse columnar bainite and the HAZ as coarse grained bainite (prior austenite grain size > 40 μm), fine grained bainite and intercritical microstructures.

Multiple Beads

The structures arising with muliple beads can be described with reference to Figure 1b. Considering first the weld metal, as each bead is deposited it generates a range of thermal cycles in the adjacent weld beads. The major effect of this is to produce a band of equiaxed bainitic grains varying from coarse to fine and running parallel to each weld bead fusion boundary. This corresponds to the region where the original columnar grains have transformed and recrystallised during thermal cycling. Also, by analogy with the parent material HAZ, some region of partial transformation is also present but since the original microstructure is bainitic and columnar the intercritical regions are often not as apparent as those in the parent material HAZ. The only weld bead remaining fully columnar is the very last bead deposited which in Figure 1b, is the top bead of those illustrated.

In the parent material HAZ, the effect of multiple beads can be identified where the thermal cycles from one bead overlap the HAZ from the previous bead. Consequently each of the microstructural regions of the original HAZ will at some location be subjected to subsequent thermal cycles resulting in grain coarsening, grain refinement or intercritical transformations. This is illustrated by considering the original coarse grained bainite region. In one location it will be subjected to an identical grain coarsening thermal cycle, while in an adjacent location it is refined or, at lower temperatures, partially transformed. Thus, traversing the HAZ parallel and adjacent to the fusion boundary the microstructure which originally was coarse grained bainite repeatedly changes in sequence through fine grained bainite and partially transformed bainite back to coarse grained bainite for each pair of adjacent beads.

The models presented so far only relate to single layers of beads but clearly by controlling the extent of overlap of adjacent beads the proportion of each microstructure can be varied. In addition, by careful control of the size and distribution of subsequent layers of beads the HAZ's from these can be arranged either to penetrate or not penetrate the interface layer of weld beads. This means that the parent material HAZ indicated in Figure 1b can be further transformed into many combinations of the microstructures described. This, in principle, forms a basis for the production of weldments with controlled HAZ microstructures.

To derive the maximum benefit from this approach it is necessary to know the mechanical properties of each microstructural type and how to combine these to give a satisfactory weldment. A problem here is that the scale on which the microstructure changes in a real weldment is such that conventional mechanical testing specimens will not produce results relating to one microstructure. However, thermal simulation techniques[6] enable the constituent parts of the HAZ to be reproduced in sufficient bulk to carry out mechanical tests. For the low alloy ferritic steel weldments considered here, the refined bainite microstructure appears to have the optimum combination of creep strength and rupture ductility for elevated temperature operation. This is currently exploited by the CEGB where weldments associated with repairs or susceptible materials are often produced using small diameter electrodes overlaid with larger diameter electrodes to produce highly refined HAZ microstructures[7]. This is similar, though not identical, to the "half bead" repair technique, stipulated by the ASME XI Boiler and Pressure Vessel Code[8], which will almost certainly produce a highly refined HAZ. In general however, many weldments exist in operating plant that have been produced without controlling the microstructure. It is these weldments and the failure phenomena arising in them that will now be considered.

WELDMENT CRACKING AND FAILURE PHENOMENA

The probability of failure of a component in service would be expected to follow the form shown in Figure 2 as a continuous line[9]. This results from high temperature designs being based on parent material

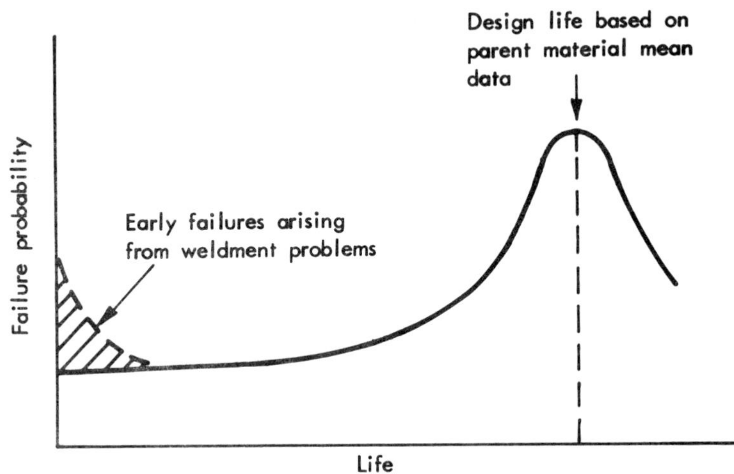

FIGURE 2. SCHEMATIC REPRESENTATION OF THE PROBABILITY OF FAILURE DURING THE LIFE OF A WELDED COMPONENT

creep and rupture properties. However, as shown by the dashed line in Figure 2, for weldments, and particularly for the low alloy ferritic steels considered in this paper, there exists a period early in the life of the component where the probability of macroscopic crack formation, sufficient to constitute failure, is high. This cracking occurs primarily in two modes, circumferentially in the HAZ or transversely in the weld metal. In both cases initiation and growth of the cracks may occur during stress relief heat treatment or early in service, and is primarily the result of creep controlled processes. These failure modes can be related specifically to the weldment microstructure though the significance of the additional factors, composition and residual stress, must also be considered.

Circumferential HAZ Cracking

This mode of cracking occurs in the HAZ immediately adjacent to the fusion boundary and may run fully around the circumference of the component. Macroscopically, in cross section the cracking parallels the perturbations of the fusion boundary, as shown in Figure 3, and can penetrate to the bore in places. On a microscopic scale the cracking initiates as cavitation, nucleating on the prior austenite grain boundaries in the coarse grained region of the HAZ.

The initial stages of cavitation and crack formation are shown in Figure 4. These are usually observed first in the HAZ adjacent to the outer surface weld beads and overall crack growth occurs towards the bore. However, some HAZ cavitation and cracking can occur adjacent to weld beads lower in the weld and consequently ahead of the main crack, with intervening regions of refined microstructure remaining crack free. This demonstrates the value of the refined microstructure which in some cases is considered to contribute to the complete arrest of crack growth[10]. Clearly the exact proportions of coarse and fine microstructures adjacent

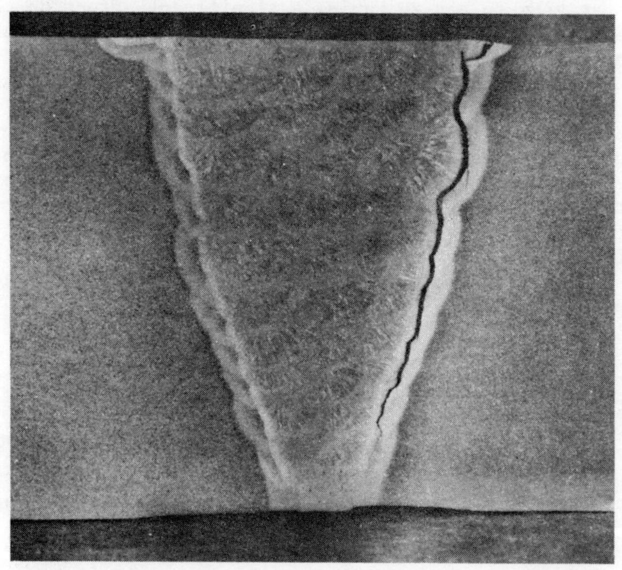

FIGURE 3. CIRCUMFERENTIAL HAZ CRACKING IN A WELDMENT

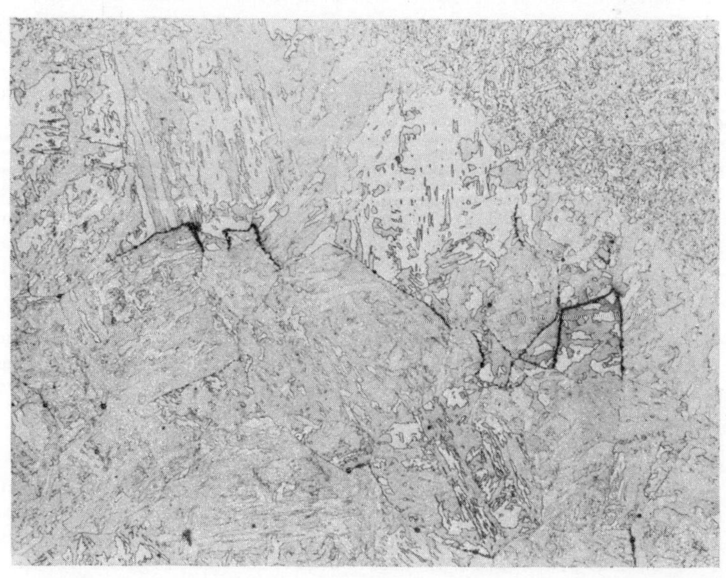

FIGURE 4. CAVITATION AND MICROCRACKING IN A HAZ COARSE GRAINED BAINITE REGION

to the fusion boundary are important since with coarse grained structures the cracks can grow to a size having a significant effect on the component life, while in the other extreme of total refinement, even crack initiation may be prevented.

Examination of failures observed in operating plant shows that the proportions of HAZ coarse grained microstructure are often in excess of 80% and linking of cracks across the short regions of refinement easily occurs[11]. Final failure from this mode of cracking can lead to complete separation of the component along the weldment HAZ before significant steam leakage occurs at any one position. Furthermore, while the major failure mode involves creep crack growth the final stages may involve fatigue, brittle fracture, or plastic collapse since the main steam lines are subjected to load and thermal transients at various stages of plant operation.

Transverse Weld Metal Cracking

This mode of failure occurs as cracking orientated transversely across and radially through the weld metal. The most severe examples traverse the width of the weld completely and in some isolated cases penetrate the parent material HAZ. In longitudinal cross section, Figure 5, a full distribution of crack sizes occurs, with their frequency decreasing as the crack size increases. Thus many small cracks of up to a bead diameter are found, with a few extending over several beads and the occasional crack growing to the full weld width and depth[12].

Microscopically the cracking clearly initiates as creep cavitation, linking to give cracks on the prior austenite grain boundaries as seen in Figure 6. Detailed examination has shown that initiation occurs first in the coarse columnar regions of the side wall beads in the top third of the weld. The growth of these cracks may be arrested by the refined microstructure arising in the weld metal HAZ. Cracks that extend past this stage can then grow to the full weld dimension, further growth usually occurring much more slowly in the more ductile parent material. In such a sequence the significant microstructural feature is the region of refinement which is more ductile and resistant to cracking than the coarse columnar grains. Again, as in the case of HAZ cracking, the proportions of these microstructures are important and in this case are determined primarily by the SMA electrode size welding technique and heat input.

Final failure with this mode of cracking has never been catastrophic although steam leakage has occurred. There does however exist the possibility of an extensive longitudinal split which would involve the crack propagating well into the parent material. However, steam leakage through the large transverse cracks in the weld metal should indicate component failure in a relatively safe manner.

Composition and Residual Stress

Composition and residual stress are two additional factors which will affect the cracking and failure of main steam line weldments. Changes in composition influence the mechanical properties of the constituent parts of the weldment whereas residual stresses result in additional creep strain in the weldment. While it is not the main purpose of this paper to discuss these factors in detail, the salient features must be mentioned.

HAZ and weld metal cracking starts in the coarse grained region of the weldment primarily because these regions although stronger in creep are less ductile than the refined regions. However, the exact differences in the properties of the coarse and fine regions depend on the material composition. Consequently some HAZ's or weld metals may contain high levels of alloying elements [1,13] or impurities[14,15] that make the coarse regions particularly susceptible to cracking. Alternatively, the basic compositon of the parent material and weld metal may be such that the coarse grained regions are not susceptible to cracking and life is not affected.

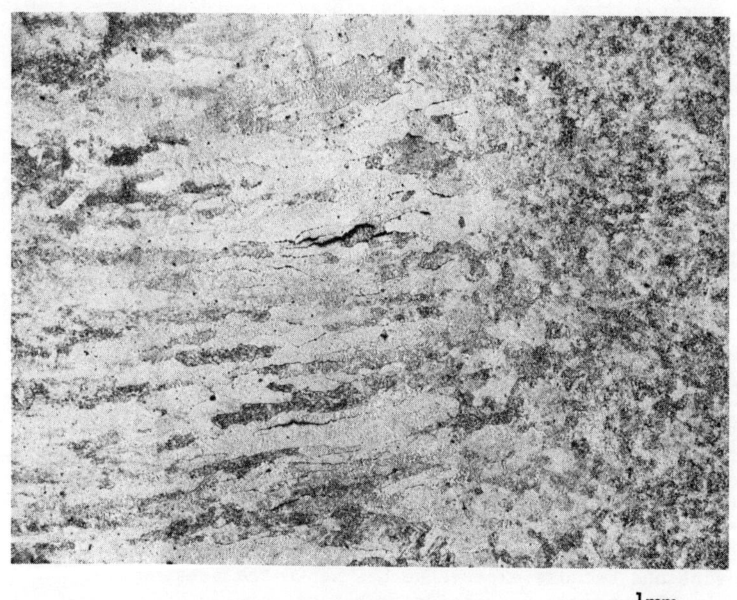

FIGURE 5. TRANSVERSE WELD METAL CRACKING IN A LONGITUDINAL CROSS SECTION OF A WELDMENT

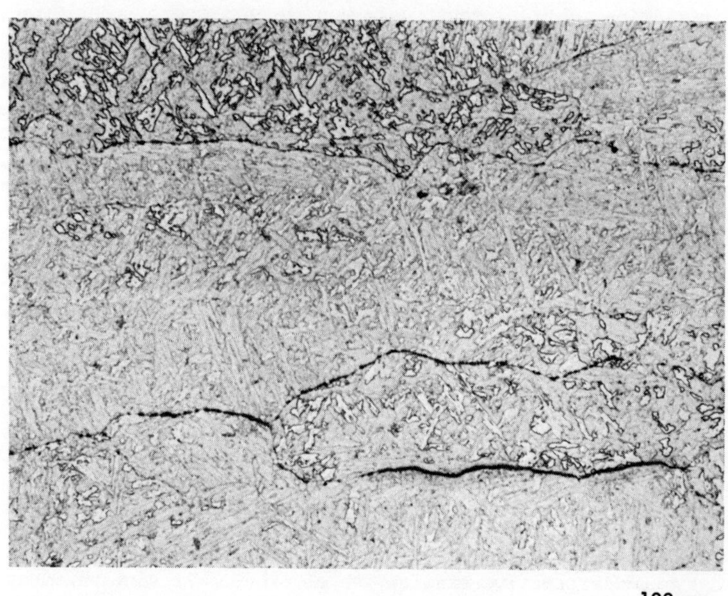

FIGURE 6. CAVITATION AND CRACKING ON THE WELD METAL COLUMNAR GRAIN BOUNDARIES

The presence of a residual stress distribution in a weldment generates a range of elastic strains which at high temperature result in creep strains and therefore damage in addition to that arising directly from the pressure stress and strain. The significance of residual stress can be demonstrated by considering the location of crack initiation. Both modes initiate in the top third of the weldment and residual stress work has shown that this is the region of maximum tensile stress[16]. Consequently, the local strains associated with stress relaxation, whether occurring during stress relief heat treatment or in service, are highest in these regions and when imposed on the coarse grains of microstructures can result in early cracking. In fact, the combination of microstructure, composition and residual stress can be such that severe cracking and failure occur during stress relief heat treatment.

PRESSURE VESSEL RESEARCH

Where crack like defects arise in a weldment it is necessary to consider their significance with respect to the integrity of the welded component. At present the CEGB assessment procedure utilises an approach where the creep crack growth rate is related to the linear elastic stress intensity factor K, and a single relationship is applied to HAZ and weld metal defects[17]. There are, however, objections to this approach, centred particularly on fundamental arguments against describing a time dependent creep process through a linear elastic parameter and on size effects that may result in unknown errors when assessing crack growth in a full size component using data from small scale uniaxial test pieces. For these reasons and because experience has shown the present approach to be conservative, a pressure vessel research programme is underway at MEL to study creep crack growth in full size components.

The initial phases of this work were concerned with crack growth and failure in ½CrMoV pipe[18,19] and the most recent work relates to the circumferential HAZ mode of cracking in weldments. This considers part circumferential cracks growing from the outer surface of the weldment HAZ and represents the situation existing in service after the early period of crack initiation. While growth data can be obtained by continuously monitoring such cracks in pressure vessels[20], interpretation of the data depends critically on the microstructural distribution and residual stresses present, accepting composition as a constant and known parameter. Accordingly, preliminary work has involved defining these parameters for the pressure vessel under examination.

Welding trials, utilising the HAZ model principles, produced a procedure that consistently gave a specific microstructural distribution. The procedure used involved a buttered layer of 5 mm diameter (6 swg) electrodes on which a layer of 2.5 mm diameter (12 swg) electrodes was first deposited followed by a layer of 3.25 mm diameter (10 swg) electrodes. The spacing of the largest electrodes was so selected to produce a HAZ containing about 80% coarse grained bainite, equal to that arising in plant weldment failures[11]. The purpose of the smaller electrode deposits was to avoid "burn through" refinement of the coarse grained bainite. This can be seen in Figure 7 and successfully provided a controlled HAZ for the pressure vessel experiments.

The residual stress distribution in a weldment is more difficult to define. The stresses present in the surface layers can be determined by direct measurement and related to specific microstructures[21]. However, since crack growth occurs away from the outer surface it is necessary to understand how the through thickness residual stress varies. This can be obtained by utilising the through thickness distribution determined by the destructive examination of other full size weldments[16]. Even so, it must be stated that the effects of residual stress and their possible re-distribution during creep crack growth are not yet fully understood. Indeed one of the objectives of the present work is to clarify this position. Consequently, to accentuate any effects of residual stress, the pressure vessel experiment incorporates four weldments, two in the

FIGURE 7. BUTTERED INTERFACE BACKING LAYERS AND RESULTING HAZ AS USED IN THE PRESSURE VESSEL MANUFACTURE

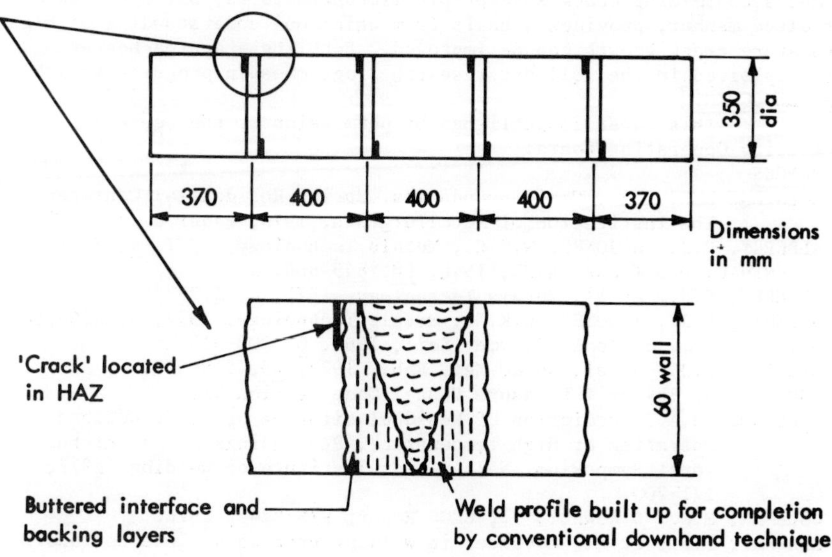

FIGURE 8. GENERAL FEATURES OF THE PRESSURE VESSEL AND SCHEMATIC REPRESENTATION OF THE WELDMENT AND HAZ CRACK

as-welded condition to retain high residual stresses in untempered microstructures, and two in the stress relieved condition, 3 hours at 700°C, to contain low residual stresses in tempered microstructures.

The final pressure vessel design is given in Figure 8 and shows a schematic representation of the weldment, with the HAZ's and "cracks" orientated normal to the pipe surface. There are two part circumferential HAZ cracks in each weldment, centred 180° apart and on opposite sides of each weld. These have been produced using spark machining techniques and the initial crack dimensions cover a range of depths from 6 mm (0.24 in) to 18 mm (0.71 in) and surface lengths from 60 mm (2.4 in) to 180 mm (7.1 in). The vessel, which has now been completed, will be taken to failure at a temperature of 565°C and a pressure appropriate to parent material creep life of 15,000 hours. Throughout its life crack growth and strain data will be continuously monitored so that they may be related to the HAZ microstructure, uniaxial data, stresses and final failure.

CONCLUDING REMARKS

The microstructures and consequent properties arising in weldments are very complex. However, for the case of SMA deposits, by reducing the weldment to its simplest form of a single bead and its attendant HAZ, the structures arising in a multibead weld can be adequately described. Furthermore such an approach enables the microstructural features associated with different failure modes observed in steam generating plant low alloy ferritic steel weldments to be clearly identified.

Having identified such structures and the factors responsible for their formation it is possible to use this knowledge in two quite different ways. In new welds and repair situations, careful control of the welding procedure can ensure that the most desirable features of the microstructure are produced to such an extent that the integrity of the weldment in service is maximised. Alternatively, knowing how to produce weldments containing crack susceptible microstructures, but again in a controlled manner, provides a basis from which our understanding of high temperature crack growth can be improved. Both these approaches are being exploited in the weldment research programmes in progress at MEL.

ACKNOWLEDGEMENT

This paper is published by permission of the Central Electricity Generating Board.

REFERENCES

1. WOLSTENHOLME, D.A., Grain Boundaries, Spring Residential Conference, The Institution of Metallurgists, 1976, Paper C7.
2. ALBERRY, P.J., & JONES, W.K.C., Metals Technology, 1977, 4, 557-566.
3. ROSENTHAL, D., Trans. ASME, 1946, 68, 849-866.
4. ALBERRY, P.J., et al, Metals Technology, 1977, 4, 317-325.
5. ALBERRY, P.J., & JONES, W.K.C., Metals Technology, 1977, 4, 360-364.
6. SAVAGE, W.F., J. Appl. Polymer Sci., 1962, 6, 303-315.
7. ALBERRY, P.J., et al, Weld. Met. Fab., 1977, 45, 549-553.
8. ASME, 1974, Boiler & Pressure Vessel Code Section XI.
9. WILLIAMS, J.A., Prediction of Residual Lifetime of Constructions Operating at High Temperature, Proceedings of an International Symposium, Netherlands Institute of Welding, 1977, 171-200.
10. COLEMAN, M.C., & ROWLEY, T., CEGB Report R/M/R242, 1976.
11. ALBERRY, P.J., et al, Advances in Welding Processes, International Conference, The Welding Institute 1978, 105-116.
12. WOLSTENHOLME, D.A., CEGB Report R/M/R190, 1973.
13. MURRAY, J.D., British Welding Journal, 1967, 14, 447-456.
14. MYERS, J., & PRICE, A.T., Metals Technology, 1977, 4, 406-410.
15. GOOCH, D.J., & KING, B.L., CEGB Report RD/L/R1976, 1978.
16. FIDLER, R., CEGB Report R/M/R261, 1977.
17. NEATE, G.J., & SIVERNS, M.J., Creep and Fatigue in Elevated Temperature

Applications, Proceedings of an International Conference, Institution of Mechanical Engineers, 1973, Paper C234.
18. COLEMAN, M.C., et al, Fracture 1977, Proceedings of the Fourth International Conference, 1977, 2, 649-662.
19. COLEMAN, M.C., Fractography in Failure Analysis, ASTM STP 645, 1978, 297-311.
20. COLEMAN, M.C., et al, Detection and Measurement of Cracks, The Welding Institute, 1976, 40-44.
21. FIDLER, R., & JERRAM, K., Ferritic Steels for Fast Reactor Steam Generators, Proceedings of an International Conference, 1977, 2, 466-469.

DISCUSSION

J. Feldstein (Babcock and Wilcox): Have you examined the application of the (ASME Section XI) half bead repair procedure for the repairs you described?

Author: The ASME Section XI half bead repair prodedure has been considered, but the integrity of repairs produced using this technique is critically dependent on close control of the grinding procedure. It is our opinion that it is not adequate to specify "approximately one half the thickness of the buttering layer shall be removed by grinding before depositing the second layer", since this will not guarantee that the second layer refines or even tempers the buttered layer HAZ. Instead, we consider that more satisfactory results can be achieved by closely controlling the buttered and second layer deposition sequence. Hence, we use a buttered layer of 3.25 mm (1/8") diameter electrodes and a second layer of 4.00 mm (5/32") diameter electrodes to give highly refined HAZ microstructures in our repairs. In practice, our experience of this procedure has been entirely satisfactory.

Microstructural Observations on Time-dependent Failure in Weld-deposited Type 16-8-2 Stainless Steel*

D. T. Raske and D. R. Diercks

Materials Science Division
ARGONNE NATIONAL LABORATORY
Argonne, Illinois 60439

Abstract

Type 16-8-2 stainless steel weld metal is less resistant to rupture than Type 316 stainless steel base metal under monotonic creep conditions, but is substantially more resistant to failure under conditions of fatigue and combined creep and fatigue. Fractographic and metallographic analyses indicate that specimens from the two types of tests differ in failure modes. In monotonic-creep tests, the Type 16-8-2 weld-metal specimens failed by intergranular and interphase separation followed by microvoid growth and coalescence. Under creep-fatigue conditions, failure occurred by a mixture of matrix and interphase cracking, with substantial transgranular cracking present even for tests with a 1.8×10^4-s (300-min) tension hold time per cycle. The present results indicate that the resistance of Type 16-8-2 weld metal to creep-fatigue failure is directly related to its resistance to grain-boundary and interphase cracking.

---

*Work supported by the U.S. Department of Energy.

Microstructural Observations on Time-dependent Failure in Weld-deposited Type 16-8-2 Stainless Steel*

D. T. Raske and D. R. Diercks

Materials Science Division
ARGONNE NATIONAL LABORATORY
Argonne, Illinois 60439

Introduction

For more than 20 years, Type 16-8-2 (16% Cr-8% Ni-2% Mo) stainless steel weld metal has been used for the fabrication of main steam piping in fossil-fueled power-generating plants. The alloy offers good elevated-temperature mechanical properties, immunity to embrittlement on aging, and resistance to solidification cracking.[1] It is presently under consideration for use as a filler metal in joining Liquid Metal Fast Breeder REactor (LMFBR) components fabricated from Type 316 stainless steel. The microstructural, creep-rupture, and monotonic tensile properties of Type 16-8-2 weld metal have been reasonably well characterized,[2-9] and a recent study has investigated the elevated-temperature creep-fatigue and cyclic stress-strain properties of the alloy.[10,11] The present paper is concerned with microstructural aspects of the fracture behavior of Type 16-8-2 stainless steel weld metal under fatigue and creep-fatigue loading conditions.

Background

Before the results of the present study are described, it is appropriate to review results previously reported on the creep and creep-fatigue behavior of Type 16-8-2 stainless steel weld metal. The creep properties of this alloy have been studies by King et al.[5] at temperatures between 482 and 649°C (900 and 1200°F), using manual and automatic V-groove test welds joining 12.7-mm-thick (0.5-in) Type 316 stainless steel base metal plates. The tests employed creep stresses of 110 to 379 MPa (16 to 55 ksi) and times to failure ranged from 97 to >11,000 hr. A majority of the tests had failure times between 100 and 1000 hr. Since Type 16-8-2 weld metal is commonly used in commercial practice to join Type 316 stainless steel base metal, the relative creep properties of the

*Work supported by the U.S. Department of Energy.

weld and base metal are of particular interest. King et al. have made this comparison using a Larson-Miller parameter approach to account for temperature effects on time to rupture. They found that, in general, the Type 16-8-2 weld metal creep-rupture lives fell within the 1.65-standard-deviation scatter band defined by tests on a number of heats of Type 316 stainless steel, but lay fairly consistently to the low side of this scatter band. The observed creep strains at the onset of tertiary creep were generally well below those associated with Type 316 stainless steel base metal. In a later study using 25.4-mm-thick (1-in) plates welded together by the automatic submerged-arc (ASA) process, King et al.[6,7] found that the creep-rupture lives for the Type 16-8-2 weld metal at both 566 and 650°C (1050 and 1200°F) fell below the -1.65-standard-deviation curve for the base metal.

The above observations may be contrasted with the results obtained by Raske[10,11] on the relative resistances of Type 16-8-2 weld metal and Type 316 base metal to failure under cyclic deformation and creep-fatigue conditions. Raske conducted strain-controlled zero- and tension-hold-time fatigue tests at 593°C (1100°F) on ASA-deposited Type 16-8-2 weld metal. His observed cyclic lives are compared with the mean behavior of Type 316 stainless steel base metal (heat number 65808)[12,13] in Fig. 1.

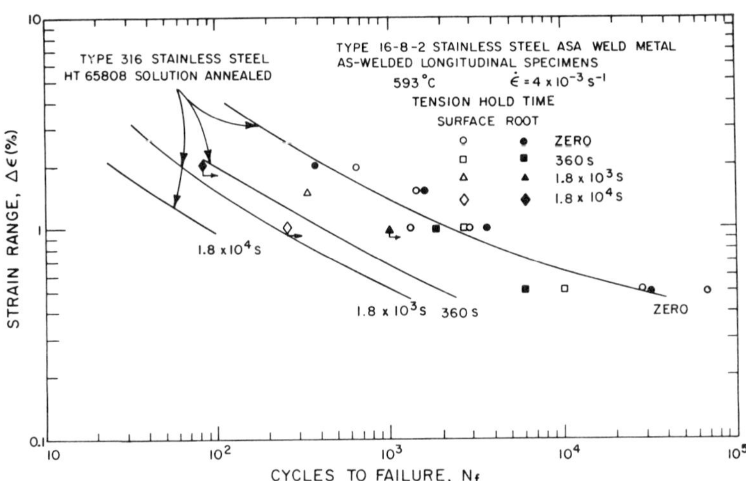

Fig. 1. Comparison of Fatigue and Creep-fatigue Behavior of Type 16-8-2 Stainless Steel Weld Metal (Data Points) with Mean Behavior of Type 316 Stainless Steel Base Metal (Curves). Neg. No. MSD-64049.

For longitudinal test specimens taken from both the surface and root of the weld, the Type 16-8-2 weld metal displays somewhat longer cyclic lives under zero-hold-time conditions than the Type 316 base metal and substantially longer lives under conditions where a hold time at the peak tensile strain is inserted into each cycle. For the longest hold-time tests conducted (1.8×10^4 s or 300 min), the cyclic lives of the weld metal exceeded those of the base metal by a factor of three or more. Under these latter conditions, the "creep" contribution to failure associated with stress relaxation during the hold time may be expected to be substantial.

The above results immediately raise the question of why Type 16-8-2 weld metal is noticeably less resistant to monotonic creep than Type 316 stainless steel base metal, but is substantially more resistant to creep-fatigue. The two behaviors are, in general, thought to be closely related; in fact, the failure criterion (i.e., linear summation of damage) for creep-fatigue loading conditions used in ASME Code elevated-temperature nuclear design rules[14] is predicated upon this assumption. The microstructural features of the failure processes in Type 16-8-2 weld metal during fatigue, creep-fatigue, and monotonic-creep loading conditions are examined in the present paper, and some insight into the observed differences in macroscopic behavior is provided. The specimens examined were those used in the above mechanical-properties studies of Raske[10,11] and King et al.[7]

Experimental Procedure

The fatigue-test specimens examined here and the testing procedures employed have been described previously,[10,11] and only the essential points will be repeated. The test specimens were machined from weldments provided by Babcock and Wilcox. Type 316 stainless steel plates 25.4 mm (1 in.) in thickness were joined with Type 16-8-2 stainless steel filler metal using the ASA process. The weld joint was a single V-groove with a 19-mm (0.75-in.) root opening and a 20° included angle. Approximately 20 passes were required to complete the weld. A chemical analysis of the Type 16-8-2 stainless steel weld metal at both the weld center and the final pass is given in Table I.[6] The microstructure of the weld metal consisted of cellular austenite grains oriented parallel to the direction of heat flow and separated by approximately 0.8 to 3.0% of a second phase nominally identified as delta ferrite distributed in a cellular-dendritic pattern (Fig. 2). The grain boundaries in the microstructure were frequently difficult to distinguish. By way of comparison, the microstructure of the Type 316 stainless steel base metal is shown in Fig. 3.

The cyclic-loading tests were conducted on uniform-gage specimens under axial strain control. The specimens were machined from the surface and root of the weld with the specimen axis parallel to the welding direction (longitudinal orientation), and were tested in the as-welded condition. Total axial strain ranges of 0.5 to 2.0% were employed, and the total strain rate was controlled at $4 \times 10^{-3}\ s^{-1}$. For some tests, hold times of 360 to 1.8×10^4 s (6 to 300 min) were inserted into the strain waveform at the peak tensile strain.

The monotonic creep-rupture specimens of Type 16-8-2 weld metal examined in the present study were obtained from tests conducted by King et al.[7] As noted above, these specimens were machined from ASA V-groove test welds joining 25.4-mm (1.0-in.)-thick Type 316 stainless steel base metal plates. The specimens examined were taken longitudinally from both the surface and root of the weld and were tested at 566 and 650°C (1050 and 1200°F) and at stresses ranging from 138 to 276 MPa (20 to 40 ksi). Times to failure ranged from 209 to 700 hr.

Results and Discussion

As noted above, the microstructure of the Type 16-8-2 weld metal (Fig. 2) appears to consist of an austenitic matrix containing interdendritic regions apparently composed largely of delta ferrite. However, since it was felt that the interdendritic phase(s) present could have a significant influence on the fracture behavior, the nature of this phase was examined in more detail. Figure 4 shows a typical microstructure of

Table I. Chemical Composition (weight percent) of Type 16-8-2 Stainless Steel ASA Weld Metal[6]

	C	Mn	P	S	Si	Cu	Ni	Cr	Mo	Ti	Nb	Ta	Co	N	B	V
Weld Center	0.047	1.48	0.026	0.012	0.89	0.09	10.03	14.85	1.76	0.010	0.030	0.03	0.09	0.028	<0.001	0.04
Final Pass	0.047	1.39	0.029	0.014	0.87	0.04	9.27	14.43	1.80	0.013	0.003	0.03	0.09	0.028	<0.001	0.04

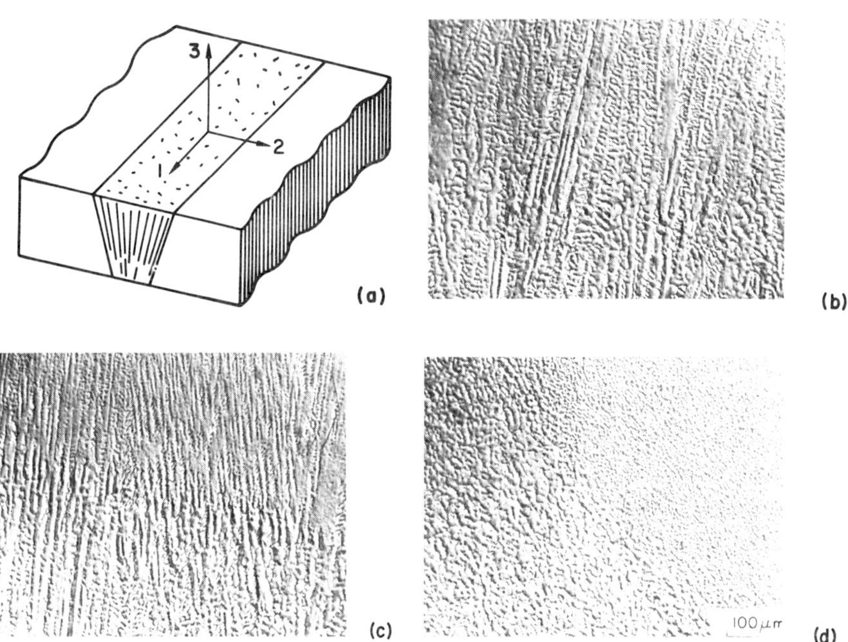

Fig. 2. Microstructure of ASA-deposited Type 16-8-2 Weld Metal (Electrochemically Etched in Oxalic Acid and Water). (a) Coordinate-axes system, (b) section parallel to 2-3 plane, (c) section parallel to 3-1 plane, (d) section parallel to 1-2 plane. ANL Neg. No. 306-77-647.

Fig. 3. Microstructure of Type 316 Stainless Steel Base Metal, Heat No. 65808, in Solution-Annealed Condition (Electrochemically Etched in Oxalic Acid and Water). ANL Neg. No. 306-77-631.

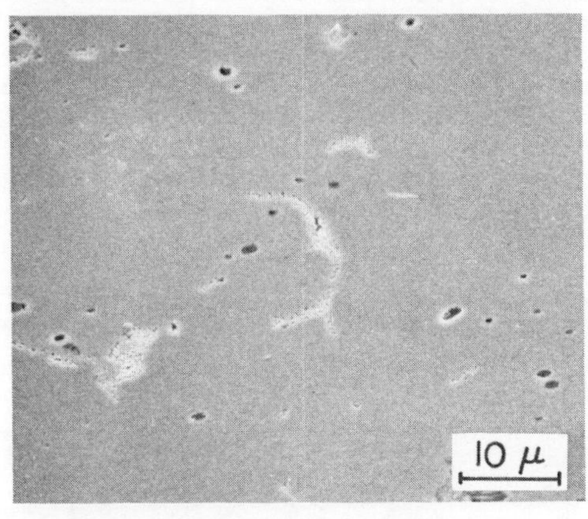

Fig. 4. Scanning Electron Micrograph of Interdendritic Region in Fatigue Specimen Tested at 1.5% Total Strain Range with 1.8 x 10^3-s (30-min) Tension Hold Time (Electrochemically Etched in 50% Solution of HCl in Water).

an interdendritic region as seen at relatively high magnification in the scanning electron microscope. The electrolytic etch with a 50% solution of HCl in water attacks both the ferrite and carbide phases more rapidly than the austenite matrix, and the mottled appearance displayed by the interdendritic region in the resulting microstructure suggests that both ferrite and carbide are present in this region.[15] This suspicion was confirmed when the same specimen was repolished and electroetched in a solution of 10% HCl in methanol. This etching treatment tends to dissolve both the austenite and ferrite phases and leave the $M_{23}C_6$ carbides standing in relief;[16] this effect may be seen in Fig. 5, which depicts essentially the same region shown in Fig. 4. The above phase identifications were further verified by analyzing the characteristic x-ray spectra obtained from the austenite, ferrite, and carbide regions.

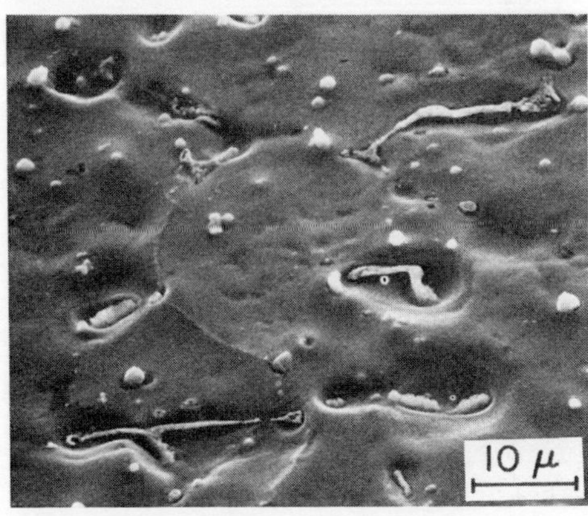

Fig. 5. Scanning Electron Micrograph of Region Similar to that Shown in Fig. 4, Electrochemically Etched in 10% Solution of HCl in Methanol.

In the cyclically loaded specimens, the crack-propagation relative to the grain and interphase boundaries was found to vary with loading conditions. Relatively few secondary cracks were observed to emanate from the sides of the specimens subjected to continuous cyclic loadings (zero hold time). The few secondary cracks present, and the primary crack as well, do not appear to be significantly influenced by the grain and interphase boundaries (Fig. 6). Fine fatigue striations were observed near the crack origin, but the fracture surface away from the origin tended to be relatively featureless, with an occasional area displaying the dimpled appearance characteristic of ductile fracture. For continuous-cycling specimens tested at lower strain ranges, the fractographic features present tended to be similar, except that the striated areas were somewhat more extensive and the striations a little coarser. However, relatively few regions exhibited well-defined striations.

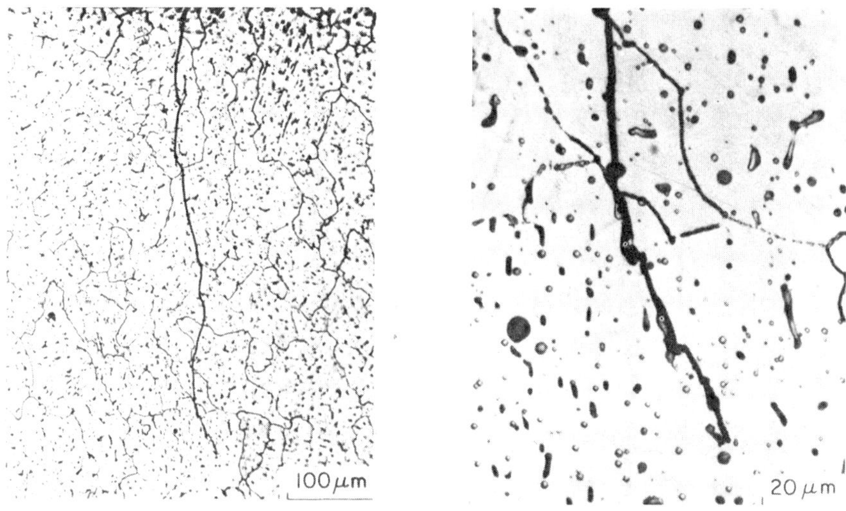

Fig. 6. Photomicrographs of Secondary Crack in Fatigue Specimen Tested at 0.5% Total Strain Range with Zero Hold Time (Electrochemically Etched in Oxalic Acid and Water). ANL Neg. No. 306-77-644.

Specimens from tension-hold-time tests displayed a greater number of surface cracks and a greater amount of internal cracking, or damage, and the extent of this surface and internal cracking increased with the length of the hold time. In addition, for a given hold time, the internal cracking was greater in the specimens tested at the lower strain ranges, where the total test times to failure were greater. The nature of the internal grain-boundary and interphase cracking is shown in Fig. 7 for a specimen tested at a 1.0% strain range with a 360-s (6-min) tension hold time. The more extensive surface and internal cracking present at the longer hold times is illustrated in Fig. 8 for a specimen tested at a 1.5% strain range and a 1.8×10^4-s (300-min) hold time. The surface of this specimen also displayed extensive rumpling parallel to the specimen axis after testing.

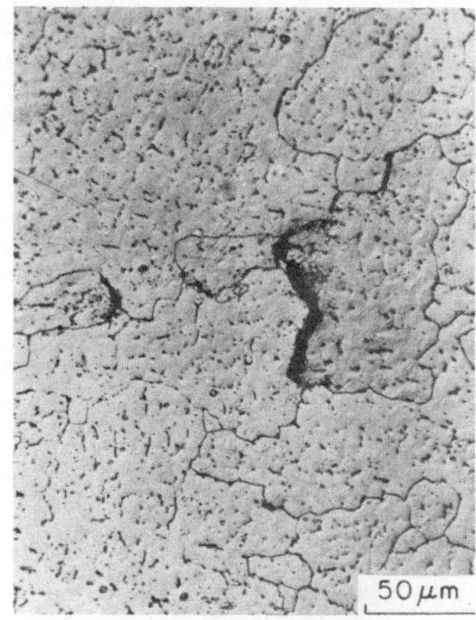

Fig. 7. Photomicrograph of Internal Cracking Present in Fatigue Specimen Tested at 1.0% Total Strain Range with 360-s (6-min) Tension Hold Time (Electrochemically Etched in Oxalic Acid and Water).

Only one of the tension-hold-time specimens, tested at a total strain range of 0.5% with a 360-s hold time, exhibited a significant number of fatigue striations on its fracture surface. However, even the longest hold-time specimens exhibited at least some indications of transgranular fracture between regions of interphase separation. Energy-dispersive x-ray analyses of the fracture surface of one such specimen indicated that areas exhibiting striations had a lower chromium and higher nickel content than areas identified as interphase boundaries. The striated regions were thus almost certainly regions of transgranular fracture through the austenite phase. It is significant that such mixed-mode fracture was seen in the very long hold-time Type 16-8-2 weld-metal fatigue specimens; the fracture mode observed in austenitic stainless steel base-metal specimens at similar temperatures is typically entirely intergranular for tension hold times of the order of 10 min or greater.[17-19]

In contrast to the mixed fracture mode observed in the creep-fatigue specimens, crack propagation in the monotonic-creep specimens was observed to occur almost entirely along interphase and grain boundaries. The features of the fracture surfaces suggest that after sufficient reduction in cross-sectional area had taken place in this manner, final failure in tertiary creep occurred by the link-up of these intergranular and interphase cracks through microvoid formation and coalescence. Figure 9 illustrates the type of internal intergranular damage typically observed near the fracture surface in the creep specimens. Again, microstructural examinations revealed no obvious reasons for the relative absence of transgranular crack growth in these specimens.

It is interesting to note that the observations described above concerning the relative resistances of Type 16-8-2 weld metal and Type 316 base metal to creep and creep-fatigue are borne out in another way. King et al.[5] have made a plot of the type suggested by Monkman and Grant,[20] relating the logarithms of the time to rupture and the minimum creep rate in an approximately linear scatter band having a slope of less than -0.7. (It should be noted that in Fig. 13 of Ref. 5 the coordinates are reversed

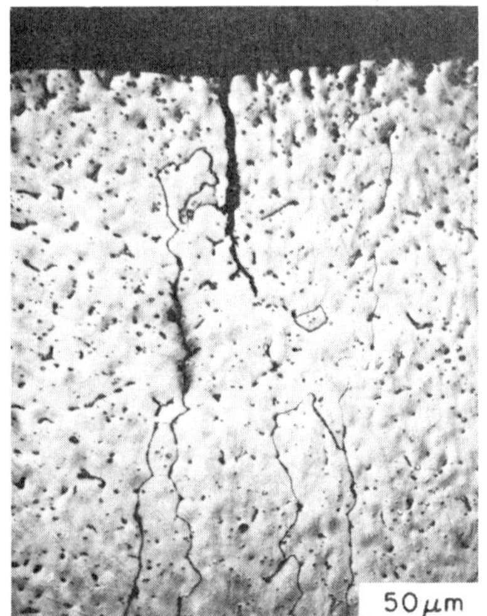

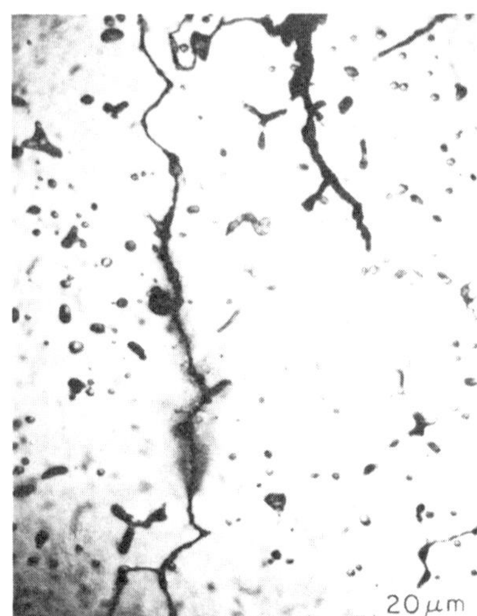

Fig. 8. Photomicrographs of Secondary Crack and Internal Damage in Fatigue Specimen Tested at 1.5% Total Strain Range with 1.8×10^3-s (30-min) Tension Hold Time (Electrochemically Etched in Oxalic Acid and Water). ANL Neg. No. 306-77-628.

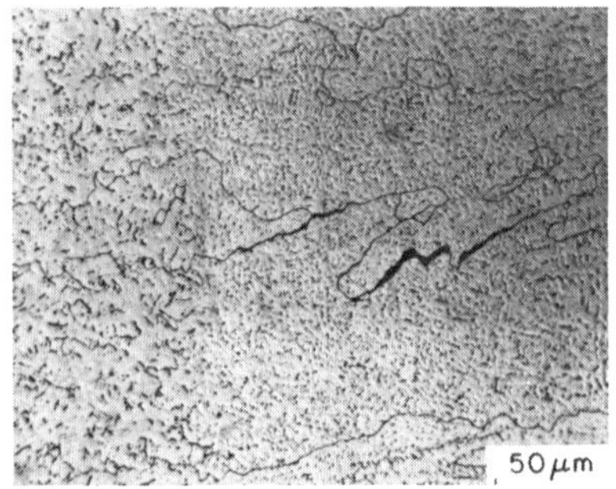

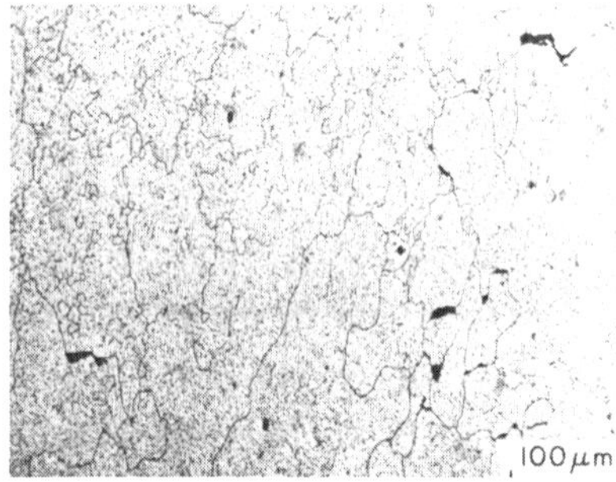

Fig. 9. Photomicrographs of Internal Damage in Monotonic Creep Specimens (Electrochemically Etched in Oxalic Acid and Water). (a) Specimen from weld root tested at 566°C (1050°F) and 276 MPa (40 ksi) with failure time of 209 h, (b) specimen from weld root tested at 650°C (1200°F) and 138 MPa (20 ksi) with failure time of 609 h. ANL Neg. No. 306-77-642.

from the usual way of plotting these data.) This is an unusually shallow slope compared with the value of ~ -0.9 to -1.0 typically observed for austenitic stainless steel base metal.[20,21] This difference in slopes reflects the greater strain-rate sensitivity of the rupture process in the Type 16-8-2 weld metal. Thus, in the very low strain-rate region, corresponding to the monotonic-creep results reported here, rupture occurs more rapidly in the Type 16-8-2 weld metal than in the Type 316 stainless steel base metal. However, at somewhat higher strain rates, corresponding to those present during the stress-relaxation process in the cyclic hold-time tests considered here, the two curves have crossed over and the Type 16-8-2 weld metal is now more resistant to rupture.

The results described in this section strongly suggest that the relatively good resistance of Type 16-8-2 weld metal to failure in creep-fatigue is related to its resistance to intergranular fracture in this loading mode. A similar observation has been made by Maiya and Majumdar[22] in connection with the variation in creep-fatigue properties exhibited by several heats of Type 304 stainless steel. However, the detailed microstructural mechanisms responsible for this resistance to boundary cracking in Type 16-8-2 weld metal are not apparent. One may speculate that the reversed loading present during creep-fatigue reduces the tendency toward the linkup of grain- and phase-boundary cracks by microvoid formation and coalescence. In addition, the higher stresses present during creep (actually relaxation) in the creep-fatigue loadings as compared with the monotonic-creep loadings considered provide a greater driving force for the departure of a growing crack from a purely intergranular path, particularly at points of reentrant angles along the boundaries.

Conclusions

This paper has described the results of a microstructural and fractographic study of Type 16-8-2 stainless steel weld-metal specimens that were tested in monotonic creep and in cyclic creep-fatigue. The study was made in an attempt to better understand the observation that Type 16-8-2 weld metal is less resistant to failure in monotonic creep than Type 316 stainless steel base metal but is significantly more resistant to failure in creep-fatigue. The results obtained here lead to the following conclusions:

1. The failure process during monotonic creep appears to be grain-boundary and interphase separation, followed by microvoid growth and coalescence apparently associated with tertiary creep.

2. Under creep-fatigue loading conditions, failure occurs by a combination of grain- and phase-boundary separation and transgranular crack growth. Significant transgranular cracking is present even in specimens subjected to a 1.8×10^4-s (300-min) tension hold time per cycle.

3. In the Monkman-Grant plots for Type 16-8-2 weld metal and Type 316 base metal, the strain-rate sensitivity of the failure process in the weld metal is seen to be substantially greater than for the base metal. This is consistent with the observations concerning the relative resistances of the two alloys to failure in monotonic creep and in creep-fatigue.

4. The relatively good resistance of Type 16-8-2 weld metal to failure in creep-fatigue appears to be associated with its resistance to intergranular and interphase cracking in this loading mode. The microstructural mechanisms responsible for this resistance require further study.

Acknowledgments

The authors gratefully acknowledge helpful discussion with S. Majumdar, P. S. Maiya, and A. P. L. Turner. W. F. Burke and J. C. Tezak assisted in the experimental aspects of the program. Finally, the authors wish to thank R. W. Weeks for his support and encouragement.

References

1. O. R. Carpenter and R. D. Wylie, Met. Prog. *70*(5), 65-73 (1956).

2. A. L. Ward, J. E. Irvin, L. D. Blackburn, G. M. Slaughter, G. M. Goodwin, and N. C. Cole, *Preparation, Characterization, and Testing of Austenitic Stainless Steel Weldments*, Hanford Engineering Development Laboratory, HEDL-TME-71-118 (Aug. 1971).

3. G. M. Goodwin, N. C. Cole, and G. M. Slaughter, Weld. J., Miami, Fla., Research Supplement *51*(9), 425s-429s (1972).

4. A. L. Ward, *Austenitic Stainless Steel Weld Materials--A Data Compilation and Review*, Hanford Engineering Development Laboratory, HEDL-TME-74-25 (May 1974).

5. R. T. King, N. C. Cole, R. G. Berggren, and G. M. Goodwin, "Properties and Structure of 16-8-2 Stainless Steel Weld Metal," *Second National Congress on Pressure Vessels & Piping*, San Francisco, Jung 23-27, 1975, ASME Publication No. 75-PVP-45.

6. R. T. King and D. P. Edmonds, in *Mechanical Properties Test Data for Structural Materials, Quarterly Progress Report for Period Ending April 30, 1976*, Oak Ridge National Laboratory, ORNL-5150, pp. 135-146.

7. R. T. King, D. P. Edmonds, and E. Bolling, in *Mechanical Properties Test Data for Structural Materials, Quarterly Progress Report for Period Engine October 31, 1976*, Oak Ridge National Laboratory, ORNL-5237, pp. 150-155.

8. A. L. Ward and L. D. Blackburn, J. Eng. Mater. Technol., Trans. ASME, Series H *98*(3), 213-220 (1976).

9. C. R. Brinkman, V. K. Sikka, and R. T. King, in *Mechanical Properties of LMFBR Primary Piping Materials*, Oak Ridge National Laboratory, ORNL-5199 (Nov. 1976).

10. D. T. Raske, "Low-cycle Fatigue and Cyclic Deformation Behavior of Type 16-8-2 Weld Metal at Elevated Temperature," in *Fatigue Testing of Weldments*, ASTM STP 648, American Society for Testing and Materials, Philadelphia, 1978 (in press).

11. D. T. Raske, *Cyclic-deformation Resistance of Weld-deposited Type 16-8-2 Stainless Steel at 593°C*, Argonne National Laboratory, ANL-77-72 (Aug. 1977).

12. C. R. Brinkman, G. E. Korth, and J. M. Beeston, *Comparison of the Fatigue and Creep-Fatigue Properties of Unirradiated and Irradiated Type 304 and 316 Stainless Steel at 593°C (1100°F)*, Aerojet Nuclear Company, ANCR-1078 (Aug. 1972).

13. R. W. Weeks, D. R. Diercks, and C. F. Cheng, *ANL Low-cycle Fatigue Studies--Program, Results, and Analysis*, Argonne National Laboratory, ANL-8009 (Nov. 1973).

14. Cases of the ASME Boiler and Pressure Vessel Code, Case 1592-7, American Society of Mechanical Engineers, New York, 1975.

15. *ASM Metals Handbook*, 8th ed., Vol. 8, American Society for Metals, Metals Park, Ohio, 1973, p. 106.

16. P. S. Maiya and D. E. Busch, Met. Trans. *6A*(2), 409-415 (1975).

17. J. T. Berling and J. B. Conway, "Effect of Hold Time on the Low-cycle Fatigue Resistance of 304 Stainless Steel at 1200°F," *Proc. First International Conf. on Press. Vessel Techn., Part II - Materials and Fabrication*, Delft, Holland, 1970, pp. 1233-1246.

18. C. R. Brinkman, G. E. Korth, and R. R. Hobbins, Nucl. Technol. *16*(8), 297-307 (1972).

19. C. Y. Cheng and D. R. Diercks, Met. Trans. *4*(2), 615-617 (1973).

20. F. C. Monkman and N. J. Grant, Proc. ASTM *56*, 593-605 (1956).

21. F. Garofalo, R. W. Whitmore, W. F. Domis, and F. von Gemmingen, Trans. AIME *221*(4), 310-319 (1961).

22. P. S. Maiya and S. Majumdar, Met. Trans. *8A*(11), 1651-1660 (1977).

DISCUSSION

G. A. Knorovsky (Sandia Labs): A few days ago Tom Devine presented an interesting paper detailing the microstructural evolution of Type 308 weld metal, though he was primarily interested in response to intergranular corrosion.

May I suggest that putting your heads together may be fruitful to come up with an explanation for the reversal of creep and creep fatigue strengths for your 16-8-2 weld-metal and 316.

Author: I unfortunately missed Mr. Devine's paper. However, I will plan to discuss this with him. Thank you for the suggestion.

M. C. Coleman (CEGB): Do you know from the plant operating conditions whether it is the creep, creep-fatigue or fatigue properties of the weldments that will be the factor controlling life?

Author: Liquid Metal Fast Breeder Reactor components operate at temperatures where time-dependent effects play an important role in the design for cyclic loadings. Therefore, both the creep and creep-fatigue properties control life.

ATTENDEES

Fifth Bolton Landing Conference

J.S. Ahearn
Martin Marietta Labs
1450 S. Rolling Road
Baltimore, MD 21227

John B. Ballance
Journal of Metals
P.O. Box 430
Warrendale, PA 15086

H. Curtis Barber
General Electric Co.
Box 1021
106 Erie Blvd.
Schenectady, NY 12301

Doreen Ball
Rensselaer Polytechnic Inst.
Troy, NY 12181

S.R. Bates
University of Florida
Gainesville, FL 32601

Charles L. Bauer
Dept. of Metallurgy & Mat.
Carnegie-Mellon University
Pittsburgh, PA 15213

Malcolm Bibby
Carleton University
Colonel By Drive
Ottawa, Canada

Martin Birnby
Knolls Atomic Power Lab.
Building A-3 Room 200
P.O. Box 1072
Schenectady, NY 12301

Warren Borges
U.S. Navy Supships Boston
666 Summer Street
Boston, MA 02210

Joseph Borrino
Amoco International Oil Co.
P.O. Box 8368
200 East Randolph Drive
Chicago, IL 60680

Thomas J. Bosworth
Boeing Marine Systems
20703 7th Pl. 50
Seattle, WA 98148

Philip E. Bretz
Lehigh University
Dept. of Met. & Mats. Eng.
Whitaker Lab #5
Bethlehem, PA 18015

Robin P. Brobst
Knolls Atomic Power Lab.
P.O. Box 1072
Building C-1 Room 162
Schenectady, NY 12301

A.J. Bryhan
Climax Molybdenum
1600 Huron Parkway
Ann Arbor, MI 48105

Larry H. Burck
Univ. Wisconsin-Milwaukee
Materials Department
Milwaukee, WI 53201

Joseph C. Burge
General Electric Company
Evendale, OH 45215

Philip Callan
Lipton Steel & Metal Prods.
458 South Street
Pittsfield, MA 16447

Julian D. Carey
General Electric 10A-115
1 River Road
Schenectady, NY 12345

Ralph E. Carter
General Electric Company
Corporate Research & Devel.
P.O. Box 8
Schenectady, NY 12301

Dr. Chun Chen
Carnegie Mellon University
Dept. of Metallurgy & Mat.
Pittsburgh, PA 15213

R.M. Chrenko
General Electric Company
Corporate Research & Devel.
P.O. Box 8
Schenectady, NY 12301

R. Christoffel
General Electric Company
Building 55
Schenectady, NY 12345

M. Cocca
General Electric Company
Schenectady, NY 12301

Dr. M.C. Coleman
CEGB
Marchwood Eng. Labs.
Marchwood Southampton
SQ46JE United Kingdom

William V. Coll
General Electric LSTG
Building 273 Room 238
1 North Avenue
Schenectady, NY 12345

Gerald R. Crawmer
General Electric 10A-117
1 River Road
Schenectady, NY 12345

S.A. David
Oak Ridge National Lab.
Oak Ridge, TN 37830

Daniel B. Dawson
Sandia Laboratories
Division 8314
Box 969
Livermore, CA 94550

Peter Dembowski
General Electric, MAO
Box 1021
Schenectady, NY 12301

John E. DeMoss
General Electric Ordi. System
Room 2282 100 Plastics Avenue
Pittsfield, MA 01201

Jaydev Desai
General Electric Company
MIS
120 Erie Boulevard
Schenectady, NY 12305

Thomas Devine
General Electric Company
Corporate Research & Devel.
Schenectady, NY 12301

Roland deWit
National Bureau of Standard
Fracture & Deformation Div.
Washington, DC 20234

Dr. David Dickinson
Republic Steel Corp.
6801 Brecksville Road
Independence, OH 44131

R.J. Diefendorf
Rensselaer Polytechnic Inst.
Troy, NY 12181

Vitomir Djordjevic
Mechanical Faculty
11000 Beograd
27 marta 80
Ivana Milutinovica 76

Thomas W. Eager
M. I. T.
Cambridge, MA 02139

Leon Eriv
Westinghouse Elevator Co.
P.O. Box 190
Randolph, NJ 07801

Richard A. Farrara
Watervliet Arsenal
Watervliet, NY 12189

Joel Feldstein
Babcock and Wilcox
P.O. Box 835
Alliance, OH 44601

Marv Fishman
G.E. Gas Turbine
1365 Wemple Lane
Schenectady, NY 12309

Gary L. Fogle
Argonne National Lab.
9700 S. Cass Avenue
Argonne, IL 60439

Mrs. Frances W. Fraser
Naval Research Lab. Code 6311
Washington, DC 20378

James Fawley
General Electric Company
Building 55 Room 115
Schenectady, NY 12345

R.L. Fullman
General Electric Company
Corporate Research & Devel.
P.O. Box 8
Schenectady, NY 12301

Dr. Peter Furrer
Swiss Aluminium Ltd.
Research and Development
CH-8212 Neuhausen
a. Rhf./ SWITZERLAND

Stanley S. Glickstein
Westinghouse-Bettis Lab.
P.O. Box 79
W. Mifflin, PA 15122

A.G. Glover
Canadian Welding Institute
232 Merton Street
Toronto, Ontario, Canada
M45 1A1

D.J. Gooch
Central Electricity Research Laboratories
Kelvin Ave., Leatherhead
Surrey, UNITED KINGDOM

T.G. Gooch
Welding Institute
Abington Cambridge CB16AL
UNITED KINGDOM

G.M. Goodwin
Oak Ridge National Lab.
Oak Ridge, TN 37830

Gordon E. Grotke
Westinghouse R&D Center
Pittsburgh, PA 15235

Francis B. Gurtner
Consultant Engineers
1100 Beech Drive
Middle River, MD 21220

Meryl M. Hall, Jr.
Argonne National Lab.
9700 S. Cass Avenue
Argonne, IL 60441

Paul Handt
Dow Chemical
Building 824
Midland, MI 48640

David W. Harvey
ARCOS Corporation
1500 South 50th Street
Philadelphia, PA 19143

John S. Haydon
General Electric Co. GTEMD
Building 53 Room 317
1 River Road
Schenectady, NY 12345

Harry J. Heckler
General Electric Company
Aircraft Engine Bus. Gr. E-2
Cincinnati, OH 45215

Richard W. Hertzberg
Lehigh University
Dept. of Met. & Mats. Eng.
Whitaker Lab. #5
Bethlehem, PA 18015

Allan G. Hins
Argonne National Lab. D212
9700 S. Cass Avenue
Argonne, IL 60439

James J. Hunter
Knolls Atomic Power Lab.
Building A-3 Room212
Box 1072
Schenectady, NY 12301

R.J. Jesseman
Research & Technology
Armco Inc.
Middletown, OH 45043

Marshall G. Jones
General Electric Company
Building 37 Room 1020
P.O. Box 43
Schenectady, NY 12301

J.T. Justice
Trans Canada Pipe Lines
P.O. Box 54
Commerce Court West M5L 1C2
Toronto, Ontario, Canada

Andrew S. Kao
Carnegie Mellon University
Dept. of Mech. Eng. HH37
Pittsburgh, PA 15213

R.A. Kelsey
Alcoa Labs
Alcoa Center, PA 15069

H.W. Kerr
University of Waterloo
Dept. of Mechanical Eng.
Waterloo, Ontario CANADA

G. Knorovsky
Sandia Labs
P.O. Box 5800
Albuquerque, NM 87185

Joseph B. Koch
Carpenter Technology Corp.
P.O. Box 662
Reading, PA 19603

Joseph M. Kostoss
General Electric Co. 59W-209
1 River Road
Schenectady, NY 12345

Edgar Landerman
Westinghous PWR
Box 355
Pittsburgh, PA 15230

W. Lawrence
Rockwell Science Center
P.O. Box 1085
Thousand Oaks, CA 91360

A.C. Lingenfleter
Huntington Alloys
Huntington, WV 25720

John C. Lippold
Rensselaer Polytechnic Inst.
Troy, NY 12181

R.V. London
G.E. ASD
Lakeside Avenue
Burlington, VT 05402

Patrick Luttrell
Sandy Hill Corporation
27 Allen Street
Hudson Falls, NY 12839

Dr. Charles Lyman
Rensselaer Polytechnic Inst.
Troy, NY 12181

Joseph C. Maciora
General Electric Company
Ordnance Systems
100 Plastics Avenue
Pittsfield, MA 01201

Kenneth MacKay
Dominion Engineering Works
P.O. Box 220 Montreal
Quebec H3C255 CANADA

John Majka
General Electric Company
P.O. Box 5000
Binghamton, NY 13902

Steve Matthews
Stellite Division
1020 Park Avenue
Kokomo, IN 46901

J.J. McCarthy
Rensselaer Polytechnic Inst.
Troy, NY 12181

J.T. McGrath
Department of Energy Mines
and Resources
568 Booth Street
Ottawa, Ontario, CANADA

John Milligan
Eaton Corporation
463 N. 20th
Battle Creek, MI 49016

K.S. Milliken
Government of Canada, DEMR
568 Booth Street
Ottawa, Canada K1A)G1

G. Scott Mills
Rockwell International
P.O. Box 464
Golden, CO 80401

Joris Naiman
Rensselaer Polytechnic Inst.
Troy, NY 12181

Thomas Natale
Rensselaer Polytechnic Inst.
Troy, NY 12181

Ernest F. Nippes
Rensselaer Polytechnic Inst.
Troy, NY 12181

A.K. Oros
Ford Motor Company
24500 Glendale
Detroit, MI 48234

Chris M. Pacula
United Nuclear Corp.
67 Sandy Desert Road
Uncasville, CT 06382

J.Y. Park
Argonne National Lab.
Building 212
Argonne, IL 60439

Joseph Pepe
General Electric Company
Building 55
Schenectady, NY 12301

Irwin L. Peters
G.E. AESPD MD-263
French Road
Utica, NY 13503

L.M. Petrick
Ebasco Services, Inc.
2 Rector Street
New York, NY 10006

Dr. Martin Prager
Consultant
125 East 87th Street
New York, NY 10028

William D. Quinn
General Electric Company
Gas Turbine Div. Lab. 262-105
Schenectady, NY 12345

D.T. Raske
Argonne National Lab.
9700 S. Cass Avenue
Argonne, IL 60439

Ann Ritter
General Electric Company
Corporate Research & Devel.
P.O. Box 8
Schenectady, NY 12301

Harry C. Rogers
Drexel University
Department of Materials Eng.
Philadelphia, PA 19104

A.B. Rothwell
Foothills Pipelines Yukon Ltd.
205 Fifth Ave. SW
Box 9083 T2P 2W4
Calgary, Alb. Canada

Clayton O. Ruud
Denver Research Institute
University of Denver
Denver, CO 80208

Jay M. Samuel
Rensselaer Polytechnic Inst.
Troy, NY 12181

L. Sandor
Sun Shipbuilding and Dry Dock
Chester, PA 19013

Z.P. Saperstein
Modine Manufacturing Co.
1500 DeKoven Avenue
Racine, WI 53401

Dr. Warren F. Savage
Rensselaer Polytechnic Inst.
Troy, NY 12181

D.W. Schaaf
E.I. duPont de Nemours
Louviers 13E53
Wilmington, DE 19898

George J. Schaefer
General Electric Company
P.O. Box 1021
Schenectady, NY 12301

Richard Schneeberger
Scientific Design Co.
4 Topaz Lane
Iselin, NJ 08830

Ronald Schneeberger
Lafayette College
4 Topaz Lane
Iselin, NJ 08830

Stojan Sedmak
Technolosko-Metalurski Fakul
11000 Beograd Karnegijeva 4
Jogoslavija

S.R. Shatynski
Rensselaer Polytechnic Inst.
Troy, NY 12181

Dr. Charles B. Shaw, Jr.
Rockwell International
Science Center
P.O. Box 1085
Thousand Oaks, CA 91360

Michael Silverman
General Electric Company
Corporate Research & Devel.
P.O. Box 8
Schenectady, NY 12301

Galen R. Smolik
EG&G Idaho, Inc.
Idaho National Eng. Lab.
Idaho Falls, ID 83401

Harvey D. Solomon
General Electric Company
Corporate Research & Devel.
P.O. Box 8
Schenectady, NY 12301

J.R. Spingarn
Sandia Laboratories
Box 969
Livermore, CA 94550

Roger Stanton
ARRADCOM
RD #4 Box 4026
Norton Road
Stroudsburg, PA 18360

G. Strabiel
General Electric Company
Evendale, OH 45215

John Tack
ARMCO Inc.
703 Curtis Street
Middletown, OH 45043

David S. Taylor
General Electric Power Trans
100 Woodlawn Avenue
Building 16-210
Pittsfield, MA 01201

R.G. Trechel
General Electric Company
Bldg. 59 West, Roo, 201
North Avenue
Schenectady, NY 12345

J.L. VanUllen
General Electric Company
Bldg. 55, Room 113
1 River Road
Schenectady, NY 12345

Rose Marie Wadsworth
General Electric Company
Corporate Research & Devel.
Schenectady, NY 12301

James L. Walker
General Electric Company
Corporate Research & Devel.
Schenectady, NY 12301

Stanley Weiss
Univ. of Wisconsin-Milwaukee
3200 N. Cramer Street - CEAS
Milwaukee, WI 53201

R.D. Williamson
General Electric Company
P.O. Box 2913
Bloomington, IL 61701

L.A. Wojcik
General Electric Company
Corporate Research & Devel.
Schenectady, NY 12301

George A. Young
Knolls Atomic Power Lab.
P.O. Box 1072
Building C-1, Room 162
Schenectady, NY 12301